# Advanced Geological Researches

# Advanced Geological Researches

Edited by **Karl Seibert**

R CALLISTO REFERENCE

New York

Published by Callisto Reference,
106 Park Avenue, Suite 200,
New York, NY 10016, USA
www.callistoreference.com

**Advanced Geological Researches**
Edited by Karl Seibert

International Standard Book Number: 978-1-63239-018-9 (Hardback)

Printed in the United States of America.

# Contents

# Preface

The study of geological researches is the field of science that studies about three primary areas of research: study of environmental geosciences, study of evolution of orogenic systems, and researches on global climate change. These are the primary areas, towards which most of the geological studies are based. Some of the key topics and research areas to work on, include: Economic Geology and Fossil Fuels, Geophysics and Seismology, Petroleum Geology, Dynamics of Ocean and Atmosphere, Geochemistry, Oceanography and Paleoclimatology, Satellite Oceanography, Paleontology and Paleoecology, Earth Science Education, Geomorphology, Surface Processes and Sedimentation, Tectonics, Structural Geology, Hydrogeology, Oceanography, Coastal and Estuarine Processes, Hard Rock Petrology, Basin Evolution, Sedimentary Petrology, Stratigraphy and Sequence Stratigraphy.

This book contains studies that cover the basic and some advances made in the field of geology. It encompasses relevant case studies to examine the field. This book can attempts to cover all the aspects of such a vast topic like geological researches, authors who contributed in this book tried to explain the most significant topics through this book and all the individual chapters do justice to this attempt. Multiple authors, with internationally recognized expertise, have generously participated in this book and I am extremely grateful to all for their time and energy to contribute to new and innovative topics on geology. I wish to thank my friends and family for supporting me in this endeavor.

**Editor**

# Mechanical Erosion in a Tropical River Basin in Southeastern Brazil: Chemical Characteristics and Annual Fluvial Transport Mechanisms

**Alexandre Martins Fernandes,[1] Murilo Basso Nolasco,[1] Christophe Hissler,[2] and Jefferson Mortatti[1]**

[1] Centro de Energia Nuclear na Agricultura, Universidade de São Paulo, Avenida Centenário, 303, 13416-970 Piracicaba, SP, Brazil
[2] Département Environnement et Agrobiotechnologies, Centre de Recherche Public Gabriel Lippmann, 41, rue du Brill, Grand-Duchy of Luxembourg, 4422 Belvaux, Luxembourg

Correspondence should be addressed to Alexandre Martins Fernandes, afernandes@cena.usp.br

Academic Editor: Steven L. Forman

This study aims to evaluate the mechanical erosion processes that occur in a tropical river basin, located in the São Paulo state, southeastern Brazil, through the chemical characterization of fine suspended sediments and the transport mechanisms near the river mouth, from March 2009 to September 2010. The chemical characterization indicated the predominance of $SiO_2$, $Al_2O_3$, and $Fe_2O_3$ and showed no significant seasonal influences on the major element concentrations, expressed as oxides. The concentration variations observed were related to the mobility of chemical species. The evaluation of the rock-alteration degree indicated that the physical weathering was intense in the drainage basin. The fine suspended sediments charge was influenced by the variation discharges throughout the study period. The solid charge estimate of the surface runoff discharge was four times higher in the rainy season than the dry season. The transport of fine suspended sediments at the Sorocaba River mouth was $55.70\,t\,km^{-2}\,a^{-1}$, corresponding to a specific physical degradation of $37.88\,m\,Ma^{-1}$, a value associated with the mechanical erosion rate that corresponds to the soil thickness reduction in the drainage basin.

## 1. Introduction

The mechanical erosion processes that occur in a drainage basin tend to reduce the thickness of the soils and the volume of the bedrock, by which dislocated and degraded particles are exported by the rivers to the sedimentation areas (lakes and oceans). The sediment production, transport, and deposition at the drainage basin scale are principally controlled by natural processes that may be intensified due to human activities, mainly agriculture and urbanization [1]. It is now recognized that in areas where urbanization grew drastically during the last century, the contribution of urban-originated sediments to the suspended particulate matter of water bodies is significant [2].

Soil erosion and its impacts on agricultural productivity, water quality, and siltation in rivers and reservoirs have been frequently discussed with regard to sustainable management of the services that these ecosystems deliver to the population [3]. In this context, the knowledge of mechanical erosion processes that occur in a tropical river basin is of outmost importance. It is now essential to reduce the loss of surface soil layers in agricultural areas and to anticipate the accumulation of deposited sediments at the industrial/urbanized areas. The intensity of mechanical erosion in a drainage basin can be assessed by monitoring the fine suspended sediments (FSS) dynamic in a given period [2].

This study aims at evaluating the main aspects of the mechanical erosion processes that occur in the Sorocaba River basin, located in the São Paulo state, southeastern Brazil, through the chemical characterization of fine suspended sediments and mechanisms of river transport close

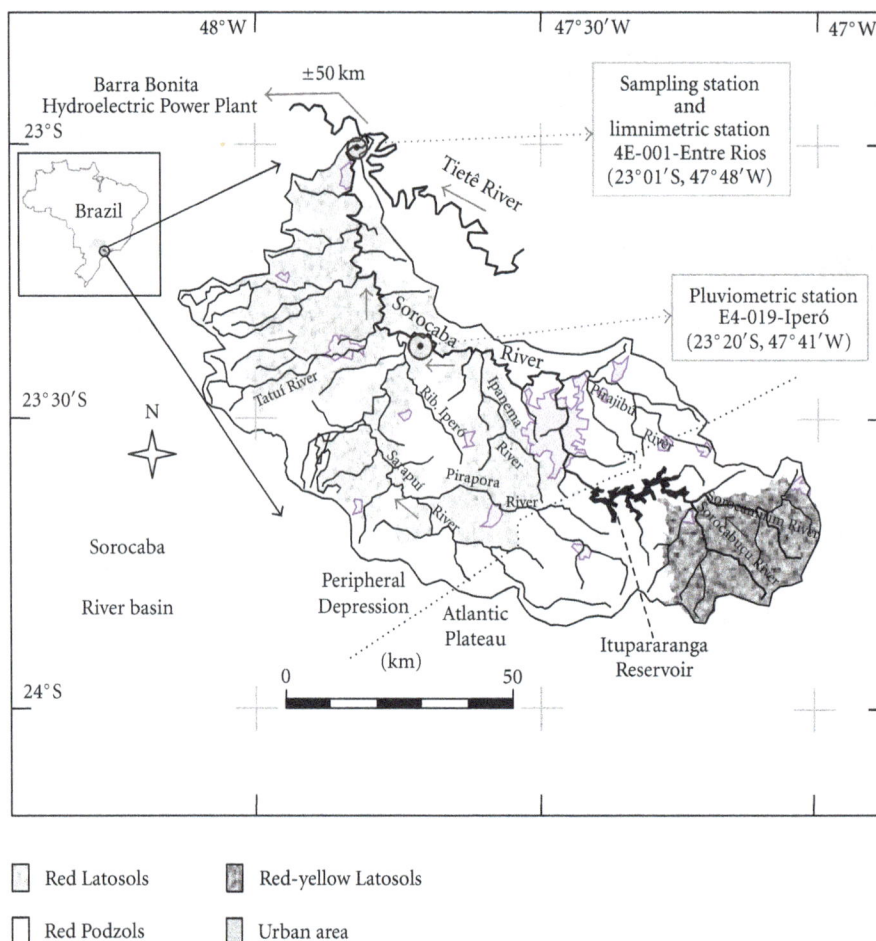

FIGURE 1: Drainage basin of the Sorocaba River with the location of the sampling station, near to the mouth, of the limnimetric station (4E-001—Entre Rios) and the pluviometric station (E4-019—Iperó). The main soils type, urban areas, and the division between Atlantic Plateau and Peripheral Depression (dotted line) are shown.

to its mouth, in the Laranjal Paulista city, from March 2009 to September 2010.

The knowledge of mechanical erosion processes that occur in the Sorocaba River basin is needed because the 50 km downstream of the confluence with the Tietê River is located the Barra Bonita Hydroelectric Power Plant Reservoir, and the change in the drainage system river system provided by the dam, from lotic to lentic, favors the processes of sediment deposition, which in turn affects the water storage capacity and the lifetime of the reservoir.

## 2. Study Area

The Sorocaba River, 227 km long, is the most important left bank tributary of the Middle Tietê River basin (Figure 1). It is located in the southeastern part of São Paulo state, between 23° and 24° S parallels and 47° and 48° W meridians, covering an area of 5,269 km². This basin extends over 18 municipalities, with a total population of 1.2 million inhabitants, approximately [4].

The Sorocaba River basin is divided in two geomorphological units. The upstream part corresponds to the *Planalto*

*Atlântico* (Atlantic Plateau), and this headwaters region is situated in the morphological unit of the Plateau Ibiúna/São Roque, where high hills with acute and convex tops are predominant. The elevations range between 800 and 1000 m and steep slopes above 20%. The lithology is represented by migmatites and granite, and the predominant soils are red-yellow Latosol and red Podzol. The downstream part of the basin is situated in the *Depressão Periférica* (Peripheral Depression), after Itupararanga Reservoir. This drainage basin area is located in the Depression of the Middle Tietê morphological unit, which presents a relief of hills with broad tabular and convex tops. The elevations range between 500 and 650 m and the slopes between 5% and 10%. The lithology is represented by diabases and sandstones, and the soils are red Latosols and red Podzols [5–7].

The drainage system of the Sorocaba River basin is dendritic pattern, and drainage density has a potential fragility from medium to high, an important morphometric parameter of analysis in hydrographic basins related to erosion susceptibility [6]. Most of the original vegetation, characterized by woods, barns, fields, savanna-type grasslands, and wetlands, has been replaced by agriculture and urban areas.

Mechanical Erosion in a Tropical River Basin in Southeastern Brazil: Chemical Characteristics and Annual
Fluvial Transport Mechanisms

3

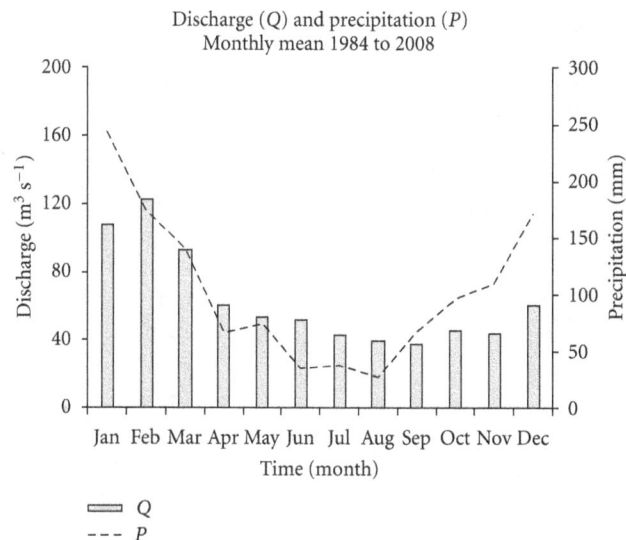

Discharge ($Q$) and precipitation ($P$)
Monthly mean 1984 to 2008

FIGURE 2: Average monthly discharge ($Q$) and precipitation ($P$) in the Sorocaba River basin for the 1984–2008 period.

The region of the headwaters, in Atlantic Plateau, is better preserved (60% with original vegetation) and the Peripheral Depression region concentrated agricultural and urban areas, with 70% and 10% of its total area, respectively [8].

According to Koppen classification [9], the climate is Cwb in the upstream part and Cwa in the Peripheral Depression area, both indicating the predominance of summer rains and dry winters. Figure 2 illustrates the annual variability of the precipitation and the river discharge for the 1984–2008 period. The average calculations were performed from daily data (personal communication DAEE/CTH in 2009) for the limnimetric station 4E-001 Entre Rios (23°01′S, 47°48′W) and the pluviometric station E4-019 Iperó (23°20′S, 47°41′W), located in the central region of the drainage basin. The annual averages calculated for the 25-year period were $63.1\,\mathrm{m}^3\,\mathrm{s}^{-1}$ for the discharge and 1243.1 mm for the precipitation, with a higher monthly average discharge in February ($122.1\,\mathrm{m}^3\,\mathrm{s}^{-1}$) and the lowest discharge in September ($37.5\,\mathrm{m}^3\,\mathrm{s}^{-1}$).

## 3. Material and Methods

The fine suspended sediments (FSS) transported by rivers are usually characterized as particles smaller than $63\,\mu\mathrm{m}$ [10], whose composition is a complex mixture of organic and inorganic particles originating from various sources, such as atmospheric deposition, eroded soil and rock, and domestic and industrial sewages [11]. Eighteen FSS samples were collected at the Sorocaba River basin mouth station, from March 2009 to September 2010. The FSS concentrations were determined from 1000 mL of river water composite samples, collected in the left and right margins of the river and in the main axis of the current river at a 1.5 m depth using a single-stage sampling point [12]. From these composite samples, 300 mL were filtered through previously weighed cellulose membrane filters ($0.45\,\mu\mathrm{m}$), and the FSS obtained plus the

membrane filters were dried at 60°C to constant weight. The FSS concentration was quantified gravimetrically according to (1):

$$C_{\mathrm{FSS}} = \frac{\left(m_f - m_i\right)}{V_L}, \qquad (1)$$

where $C_{\mathrm{FSS}}$ is the FSS concentration ($\mathrm{mg\,L}^{-1}$), $m_f$ the final mass of the filter (mg), $m_i$ the initial mass filter (mg), and $V_L$ the volume filtered (L).

The average FSS concentration normalized by the discharge ($C_{\mathrm{MNQ}}$, in $\mathrm{mg\,L}^{-1}$) was calculated according to the procedure established by Probst [13], as shown in (2):

$$C_{\mathrm{MNQ}} = \frac{\sum(Q_i \times C_i)}{\sum Q_i}, \qquad (2)$$

where $Q_i$ is the discharge measured on the sampling day ($\mathrm{m}^3\,\mathrm{s}^{-1}$), $C_i$ the FSS concentration for the sampling day, and $\sum Q_i$ the sum of all discharges measured on the sampling days ($\mathrm{m}^3\,\mathrm{s}^{-1}$).

Additionally, river water composite samples were collected from the left and right margins of the river and in the main axis of the current river. 30 L of water samples are stored in polypropylene containers to obtain sufficient FSS mass for chemical characterization. After the settlement of the FSS, the recovered sediments were air-dried and sieved using a $63\,\mu\mathrm{m}$ nylon mesh. The sediments retained in the sieve ($>63\,\mu\mathrm{m}$) were separated and with a portion of FSS ($<63\,\mu\mathrm{m}$) reserved for mineralogical analysis. Other portions of the FSS samples were calcinated at 1000°C to remove all the organic matter and were digested using the alkaline fusion method proposed by Samuel et al. [14]. 100 mg of the sample was mixed with lithium tetraborate and lithium metaborate ($2:1$) in a platinum crucible and heated at 1000°C for 30 minutes. After cooling, the melt was dissolved in 20 mL of HCl (1 M) under agitation and heating at 40°C, and the volume completed to 50 mL with MilliQ water. This analysis was performed in triplicate using the inductively coupled plasma optical emission spectrometry (ICP-OES). The limits of detection, in $\mathrm{mg\,L}^{-1}$, were Si (0.020), Al (0.020), Fe (0.020), Mn (0.002), Ca (0.050), Mg (0.050), Na (0.050), and K (0.050). We controlled the purity of the chemical reagents and the blanks with bank ground measurements below the precision limits. To control the quality of the extraction process and the chemical analyses, we used the international reference material Soil-7 of the International Atomic Energy Agency (IAEA). We obtained the following average % of recovery for the extraction and analysis in triplicate: Si (91.7%), Al (90.1%), Fe (107.3%), Mn (101.2%), Ca (92.3%), Mg (89.7%), Na (89.2%), and K (91.0%).

## 4. Results and Discussions

*4.1. Chemical Characterization of the FSS.* The chemical characterization of the FSS showed a predominance of $\mathrm{SiO}_2$ (48.60%), $\mathrm{Al}_2\mathrm{O}_3$ (21.05%), and $\mathrm{Fe}_2\mathrm{O}_3$ (8.23%) for the analyzed elements (Table 1). Analysis of the FSS mineralogical

TABLE 1: Major element concentrations (expressed as oxide %) for the FSS sampled at the Sorocaba River basin mouth during the study period.

| Sampling date | Discharge ($m^3 s^{-1}$) | Major element concentrations (expressed as oxide %) | | | | | | | |
|---|---|---|---|---|---|---|---|---|---|
| | | $SiO_2$ | $Al_2O_3$ | $Fe_2O_3$ | MnO | CaO | MgO | $Na_2O$ | $K_2O$ |
| **Mar-10-09** | **51.50** | **41.22** | **19.69** | **9.82** | **0.54** | **0.83** | **0.86** | **0.44** | **2.06** |
| **Apr-07-09** | **59.30** | **45.74** | **22.45** | **10.47** | **0.49** | **0.74** | **0.89** | **0.40** | **2.15** |
| **May-12-09** | **46.83** | **46.03** | **19.44** | **7.92** | **0.39** | **0.71** | **0.82** | **0.38** | **2.02** |
| **Jun-17-09** | **37.47** | **42.12** | **19.50** | **9.30** | **0.40** | **0.70** | **0.74** | **0.39** | **2.02** |
| **Jul-21-09** | **43.71** | **44.60** | **23.48** | **7.61** | **0.31** | **0.55** | **0.81** | **0.38** | **2.30** |
| Aug-25-09 | 123.23 | 47.80 | 21.43 | 6.83 | 0.23 | 0.56 | 0.81 | 0.32 | 2.22 |
| Sep-22-09 | 99.84 | 43.20 | 20.26 | 6.99 | 0.31 | 0.57 | 0.78 | 0.31 | 2.07 |
| Nov-10-09 | 170.01 | 49.17 | 18.74 | 5.97 | 0.16 | 0.51 | 0.88 | 0.33 | 2.03 |
| Dec-12-09 | 363.35 | 48.95 | 19.65 | 6.33 | 0.22 | 0.52 | 0.82 | 0.34 | 2.14 |
| Jan-15-10 | 279.15 | 51.65 | 21.35 | 7.71 | 0.34 | 0.68 | 1.04 | 0.37 | 2.46 |
| Jan-20-10 | 366.47 | 49.71 | 21.72 | 7.03 | 0.21 | 0.54 | 0.97 | 0.28 | 2.17 |
| Feb-23-10 | 184.04 | 53.23 | 20.01 | 7.60 | 0.50 | 0.64 | 0.85 | 0.37 | 2.45 |
| Mar-30-10 | 151.30 | 56.55 | 21.61 | 6.65 | 0.20 | 0.56 | 0.86 | 0.33 | 2.38 |
| Apr-27-10 | 107.64 | 50.48 | 22.78 | 8.37 | 0.48 | 0.67 | 0.87 | 0.34 | 2.59 |
| **May-27-10** | **70.21** | **52.07** | **22.92** | **10.46** | **0.51** | **0.76** | **0.90** | **0.39** | **2.68** |
| **Jun-24-10** | **42.15** | **46.48** | **23.05** | **12.32** | **0.69** | **0.93** | **0.86** | **0.35** | **2.35** |
| **Jul-27-10** | **54.62** | **47.81** | **24.23** | **9.22** | **0.53** | **0.75** | **0.86** | **0.34** | **2.58** |
| **Sep-01-10** | **26.56** | **57.98** | **16.53** | **7.61** | **0.73** | **1.08** | **0.80** | **0.46** | **2.05** |
| **Average** (%) | **48.04** | **47.12** | **21.26** | **9.42** | **0.51** | **0.78** | **0.84** | **0.39** | **2.25** |
| **SD** | **12.76** | **5.16** | **2.56** | **1.57** | **0.14** | **0.15** | **0.05** | **0.04** | **0.25** |
| Average (%) | 205.00 | 50.08 | 20.84 | 7.05 | 0.29 | 0.58 | 0.88 | 0.33 | 2.28 |
| SD | 105.13 | 3.69 | 1.25 | 0.74 | 0.12 | 0.06 | 0.08 | 0.03 | 0.20 |
| CV (%) | 85.9 | 9.5 | 9.3 | 20.6 | 41.9 | 22.3 | 8.1 | 12.3 | 9.7 |

The average and the standard deviations (SD) are shown for two periods, low water and high water, and the coefficient of variation (CV) for the study period (all samples). The bold font represents the low water period and the light face font represents the high water period.

TABLE 2: Mobility estimation of the elements during the rock-alteration process in the Sorocaba River basin.

| Element | Crust $Ct$ (%) | Sorocaba River Basin | |
|---|---|---|---|
| | | $Ct$ (%) | % loss (−) or gain (+) |
| Al | 6.93 | 11.14 | 0 |
| Fe | 3.59 | 6.40 | 11 |
| Mn | 0.07 | 0.31 | 177 |
| Na | 1.42 | 0.27 | −88 |
| Ca | 4.50 | 0.49 | −93 |
| Mg | 1.64 | 0.52 | −80 |
| K | 2.44 | 1.88 | −52 |
| Si | 27.50 | 22.71 | −49 |

composition, performed by X-ray diffraction, confirmed this predominance, because a significant amount of quartz, kaolinite, and magnetite was found in the fine fraction (<63 $\mu$m), and quartz, potassium feldspar, and plagioclase were found in the larger fraction (>63 $\mu$m). The concentrations of Fe and Al in the FSS are attributed to the presence of oxides that originated from the red lateritic soils that cover the basin. This observation was also verified for the Piracicaba River basin [15], another important tributary of the Tietê River.

The coefficients of variation showed a greater dispersion for Mn (41.9%), Ca (22.3%), and Fe (20.6%) and around 10% for other elements. These results reflect an apparent seasonal influence, which may be linked to the different origins of the FSS [16]. Some studies showed that the chemical composition of river sediments can provide information about the rock alteration processes and particularly the origin of the FSS particles. This chemical information may indicate element mobilities during the weathering process at the drainage basin scale [16–18].

In the Sorocaba River basin, this mobility is estimated considering the ratio between the FSS chemical composition ($Ct$) and the mean composition of surface rocks of the crust, described by Martin and Meybeck [16]. We applied a normalization, by the respective concentrations of Al, assuming its conservative behavior during the rock-alteration processes [19], according to (3):

$$\left\{ \left[ \frac{(Ct/Al)_{sed}}{(Ct/Al)_{crust}} \right] \times 100 \right\} - 100. \qquad (3)$$

The results in Table 2 indicate a conservative behavior for Fe and a significant gain for Mn. Oxides and hydroxides secondary minerals of Fe and Mn may be formed after the rock-alteration processes, especially when predominant land use is agriculture [19]. Na, Ca, Mg, and K levels indicate

Mechanical Erosion in a Tropical River Basin in Southeastern Brazil: Chemical Characteristics and Annual
Fluvial Transport Mechanisms

5

reductions from 49% to 93% and illustrate a high degree of mobility that can be related to silicate-alteration processes [19]. Such processes release these elements in the same range that is already found in the alteration process of carbonates and evaporites, which are more easily weathered [20]. For instance, Si, the main element in the alteration of silicate rocks, showed a significant reduction of 49%. This may be related to the solubilization in the rock-alteration process where the mass loss in rocks is associated with Si removal [21].

During the rock-alteration process in the drainage basin, the evolution of primary to secondary minerals is associated with a relative enrichment in Al and Fe and depletion in Si. The increase in $Al_2O_3$ and $Fe_2O_3$ concentrations to the detriment of $SiO_2$ provides an indication of the rock-alteration degree, expressed by the $R$ index, according to (4) [22]. The $R$ index allows comparison of the rock alteration intensity between drainage basins a lower $R$ value indicates a higher rock-alteration process in the drainage basin [23]:

$$R = \frac{SiO_{2(\%)}}{(Al_2O_3 + Fe_2O_3)_{(\%)}}. \qquad (4)$$

The $R$ value calculated for the Sorocaba River basin was 1.7, indicating an intense rock-alteration process when compared with other drainage basins, such as the Patagonia rivers (2.9 to 4.3) in Argentina, the Ill (3.1) and Garonne (5.1) rivers in France, and the Sebou river (2.8 to 4.8) in Africa [23]. However, this result was similar to that observed for the Piracicaba River basin (1.8) [24], another important tributary of the Middle Tietê basin.

This rock-alteration process was also assessed by the chemical maturity index (ChM), established by Konta [25], which relates the Al with the cationic bases (5). The longer a rock is altered, it becomes more mature and the higher the ChM index:

$$ChM = \frac{Al_2O_{3(\%)}}{(Na_2O + MgO + CaO)_{(\%)}}. \qquad (5)$$

The ChM index, calculated for the Sorocaba River basin, was significant (11.1) and similar to the one reported by Mortatti and Probst [15] for the Piracicaba River basin (10.4) and higher than that observed for the Patagonia rivers (2.0 to 4.8), the Ill (4.6), the Garonne (5.1), and the Sebou (3.4 to 5.3) rivers [23], confirming the high rock-alteration degree determined by $R$ index.

*4.2. Estimate of the Suspended Load of Quick Surface Runoff.* The quick surface runoff is directly associated with mechanical erosion processes that occur in a drainage basin. It is responsible for the transport of particulate matter into the river channel throughout the drainage basin. The particulate matter load that reaches the river channel by quick surface runoff is diluted by river waters from the subsurface and hypodermic and underground compartments, whose particulate matter load can be considered negligible [22]. The relationship responsible for the mechanical erosion between the surface and subsurface reservoirs can be described according to (6) [22, 26]. Using the relationship between total discharge

TABLE 3: Estimation of the particulate matter load of quick surface runoff ($Cr$, $mg\,L^{-1}$) depending on the total concentration of FSS ($Ct$, $mg\,L^{-1}$) and the respective total discharge ($Qt$, $m^3\,s^{-1}$) observed at the Sorocaba River mouth, in the study period, with respective averages and standard deviations.

| Sampling date | $Qt$ ($m^3\,s^{-1}$) | $Ct$ ($mg\,L^{-1}$) | $Cr$ ($mg\,L^{-1}$) |
|---|---|---|---|
| Mar-10-09 | 51.50 | 16.33 | 43.11 |
| Apr-07-09 | 59.30 | 20.67 | 54.24 |
| May-12-09 | 46.83 | 25.33 | 67.16 |
| Jun-17-09 | 37.47 | 14.33 | 38.46 |
| Jul-21-09 | 43.71 | 46.50 | 123.69 |
| Aug-25-09 | 123.23 | 54.33 | 139.90 |
| Sep-22-09 | 99.84 | 45.33 | 117.21 |
| Nov-10-09 | 170.01 | 145.00 | 371.56 |
| Dec-12-09 | 363.35 | 87.17 | 221.86 |
| Jan-15-10 | 279.15 | 47.67 | 121.54 |
| Jan-20-10 | 366.47 | 101.50 | 258.33 |
| Feb-23-10 | 184.04 | 37.83 | 96.85 |
| Mar-30-10 | 151.30 | 178.00 | 456.84 |
| Apr-27-10 | 107.64 | 51.00 | 131.66 |
| May-27-10 | 70.21 | 22.67 | 59.25 |
| Jun-24-10 | 42.15 | 14.50 | 38.64 |
| Jul-27-10 | 54.62 | 28.83 | 75.92 |
| Sep-01-10 | 26.56 | 17.67 | 48.60 |
| Average (%) | 126.52 | 73.55 | 188.41 |
| SD | 108.62 | 44.96 | 114.67 |

and quick surface runoff presented by Fernandes et al. [27] for the 1984–2008 period (7), it has been possible to estimate the particulate matter load transported by the Sorocaba River associated with the quick surface runoff (Table 3):

$$Ct \times Qt = Cr \times Qr, \qquad (6)$$

where $Ct$ is the total concentration of FSS transported by the total discharge ($Qt$) and $Cr$ is the particulate matter load concentration transported by quick surface runoff ($Qr$):

$$Qr = 0.3952\,Qt - 0.8423. \qquad (7)$$

During dry periods, both preceding the rise of the water level (March–July 2009) and after the recession (May–September 2010), the flow-weighted average concentrations calculated for $Cr$ were similar, with values of 64.61 and 57.97 $mg\,L^{-1}$, respectively. In the flood period, from December 2009 to January 2010, the flow-weighted average concentration was approximately four times higher (231.03 $mg\,L^{-1}$). It should be emphasized that the highest $Cr$ concentration occurred in March 2010 (456.84 $mg\,L^{-1}$), during the receding period of Sorocaba River, being associated with erosion processes of the river banks, which is difficult to document.

*4.3. Dynamics of Fine Sediments in Suspension.* The FSS concentrations seem to be influenced by the variation of the Sorocaba River discharge during the study as illustrated in

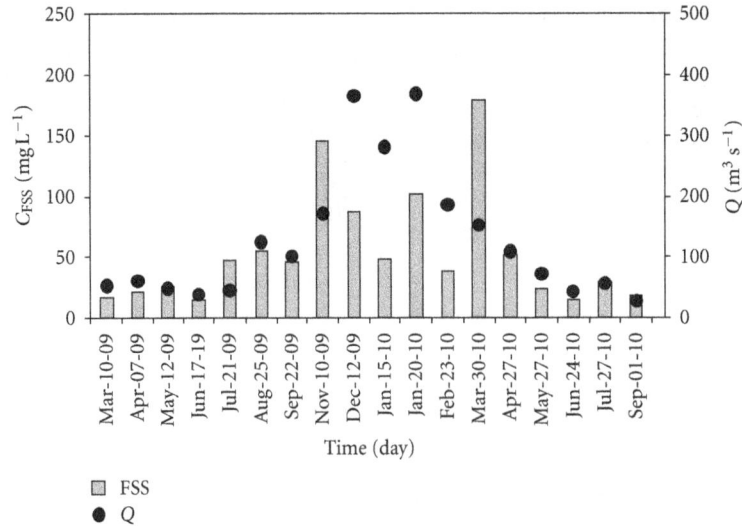

FIGURE 3: Temporal variability of the FSS concentration (mg L$^{-1}$) and respective discharges ($Q$, m$^3$ s$^{-1}$) at the mouth of Sorocaba River basin.

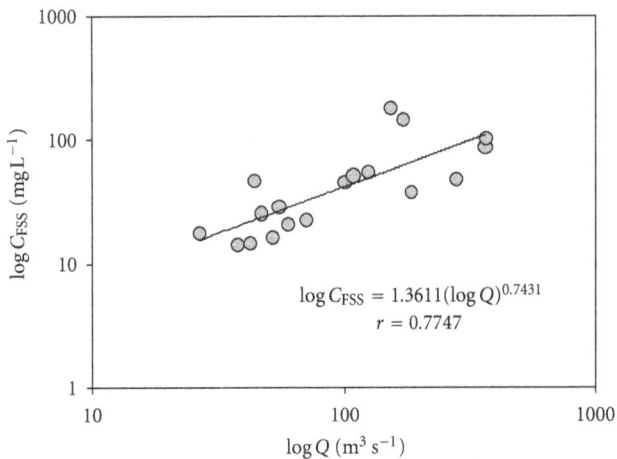

FIGURE 4: Logarithmic correlation between the FSS concentration (mg L$^{-1}$) and the respective discharges ($Q$, m$^3$ s$^{-1}$) at the Sorocaba River basin mouth.

Figure 3. We emphasize that during the rise of the water level (November 2009) and low river levels (March 2010), the highest FSS peaks are observed in the Sorocaba River.

The relationship between the suspended sediment concentration ($C$) and stream discharge ($Q$) can be evaluated with the use of the logarithmic model ($C = aQ^b$) [28], where the $a$ and $b$ values vary between drainage basins, reflecting the different physical, climatic, and hydrological characteristics of each basin contribution area. For most rivers of the world, the $b$ exponent is positive, indicating the increase in suspended sediment concentration caused by an increased discharge [29]. The logarithmic correlation model established for the Sorocaba River, using our sampling strategy, was significant at 1% (Figure 4).

The $b$ exponent values obtained for the Sorocaba River (0.7431) were lower than those observed in the literature, which range between 1 and 2, according to the drainage basin characteristics [22]. This difference is confirmed when compared to the results obtained for the Tietê (1.7788) and the Piracicaba (1.4954) Rivers, two other rivers in the Middle Tietê basin region [30].

Although the annual hydrological regime, with the seasonal distribution of precipitation and discharge, as well as the land use (predominantly agricultural) of the Sorocaba River is similar to Tietê and Piracicaba Rivers [31], the comparison of the $b$ exponent indicates a different distribution pattern of the suspended sediment concentration according to the discharge variation of these three rivers, which can be partly explained by distinct physical characteristics of the drainage basins, mainly associated with the relief and the arrangement and contribution of the tributaries. The dispersion of points observed in Figure 4 indicates the possible influence of the dynamic processes on the remobilization and sedimentation, characteristic to small and medium sized rivers [2, 22, 29, 32].

*4.4. Transport of FSS.* The sediment load carried by the river is evaluated in terms of the FSS, assuming that suspended load represents approximately 90% of the total sediment river flow [33]. Thus, the total transport of river sediments ($T$, in t a$^{-1}$) was estimated considering the average FSS concentration normalized by the discharge ($C_{MNQ}$, in mg L$^{-1}$) and the average discharge of the study period ($Q_m$, in m$^3$ s$^{-1}$), using the stochastic method [34], according to (8):

$$T = C_{MNQ} \times Q_m \times f, \tag{8}$$

where the $f$ factor corresponds to the mass and time correction in the calculation of total transport for results in tons per year and is equal to 31.536.

Mechanical Erosion in a Tropical River Basin in Southeastern Brazil: Chemical Characteristics and Annual
Fluvial Transport Mechanisms

7

The total transport of FSS calculated for the Sorocaba River mouth was $293 \times 10^3$ t a$^{-1}$, which corresponds to a specific transport of 55.70 t km$^{-2}$ a$^{-1}$. This value is in accordance with that reported for the Tietê and the Piracicaba Rivers, 59.60 and 55.50 t km$^{-2}$ a$^{-1}$, respectively [30]. Compared to international drainage basins of the same magnitude, the specific transport of FSS observed for the Sorocaba River is similar to the Chena River, USA (5125 km$^2$ and 54,75 t km$^{-2}$ a$^{-1}$) and the Chaudiere River, France (5805 km$^2$ and 56,58 t km$^{-2}$ a$^{-1}$) [35].

*4.5. Specific Physical Degradation.* The physical degradation degree of a drainage basin, understood as the mechanical erosion rate in meters per million years, can be evaluated as a function of specific transport of FSS ($F_{FSS}$, in t km$^{-2}$ a$^{-1}$) and mean density of soils ($d$, in t m$^{-3}$) of the basin studied [36], according to (9):

$$EM = \frac{F_{FSS}}{d}. \tag{9}$$

The mean density of the Sorocaba River basin soils was determined by the classical method reported by Blake and Hartge [37], the core method, considering three distinct portions of the drainage basin established in accordance with the distribution of predominant soil types. In the headwater region, where the red-yellow Latosol prevails, the mean density soil was estimated at $1.24 \pm 0.02$ t m$^{-3}$; in the middle part, with the predominance of red Podzols the mean density for this soil was $1.65 \pm 0.01$ t m$^{-3}$; and at the downstream part of the basin, with a predominance of red Latosols, the mean density soil was $1.53 \pm 0.01$ t m$^{-3}$. The mean density calculated from the Sorocaba River basin soil was $1.47 \pm 0.01$ t m$^{-3}$.

The mechanical erosion rate calculated for the Sorocaba River basin is 37.99 m Ma$^{-1}$. Compared to other drainage basins with similar areas and land use predominantly agricultural, the physical degradation degree of the Sorocaba River basin was similar to that observed by Bortoletto Junior [30] for the Tietê (42.6 m Ma$^{-1}$) and the Piracicaba (37.0 m Ma$^{-1}$) Rivers, five times higher than the degradation of the Congo River, with 7.4 m Ma$^{-1}$ [38], and thirteen times higher than the degradation of the Niger River, with values ranging from 2.4 to 3.4 m Ma$^{-1}$ [39]. These differences could be related to land use processes in the respective drainage basins, particularly the agricultural practices of each region, such as soil management and types of crops, and other human activities developed there.

## 5. Conclusions

A focus of this study is to derive a better understanding of the mechanical erosion processes that occur in a tropical river basin of southeastern Brazil, where the predominant land use is agricultural, with increased urbanization in the past decade. The FSS load is influenced by the variation in discharge river during the study period, with the same seasonal effect. This situation is emphasized by suspended matter load associated with the quick surface runoff, which was four times higher in the high water period than in dry periods.

However, the FSS chemical characterization for major oxides does not indicate that seasonal influence and the variability of concentrations observed during the study period were related to the mobility of these elements during rock weathering process. The mechanical erosion in the Sorocaba River basin was 55.70 t km$^{-2}$ a$^{-1}$, representing a physical degradation rate of 37.88 m Ma$^{-1}$, in the same magnitude of the other two major rivers that comprise the Middle Tiete basin.

## Acknowledgments

The authors thank the Fundação de Apoio à Pesquisa do Estado de São Paulo (Process 08/57104-4 and 08/09369-9) and the Conselho Nacional de Desenvolvimento Científico e Tecnológico (Process 134169/2009-3) for the financial support granted, as well as the Centro Tecnológico de Hidráulica e Recursos Hídricos of the Departamento de Águas e Energia Elétrica, for providing the time series of daily discharge and rainfall data for the Sorocaba River basin.

## References

[1] M. R. L. Siviero and E. M. Coiado, "A produção de sedimentos a montante de uma seção do rio Atibaia associada à descarga sólida transportada," in *Proceedings of the 13th Simpósio Brasileiro de Recursos Hídricos*, Belo Horizonte, 1999.

[2] F. F. B. Ferraz and J. Mortatti, "Avaliação do processo erosivo mecânico em bacia subtropical desenvolvida pela análise de sedimentos finos em suspensão," *Geociências*, vol. 21, no. 1-2, pp. 113–120, 2002.

[3] F. F. Pruski, L. N. Rodrigues, and D. D. Silva, "Modelo hidrológico para estimativa do escoamento superficial em áreas agrícolas," *Revista Brasileira de Engenharia Agrícola e Ambiental*, vol. 5, no. 2, pp. 301–307, 2001.

[4] Instituto Brasileiro de Geografia e Estatística-IBGE, "Dados do Censo 2010 publicados no Diário Oficial da União do dia 04/11/2010," 2010, http://www.ibge.gov.br/censo2010/dados_divulgados/index.php/uf=35.

[5] J. L. S. Ross and I. C. Moroz, *Mapa Geomorfológico do Estado de São Paulo*, Laboratório de Geomorfologia/Departamento de Geografia/FFLCH-USP, Laboratório de Cartografia Geotécnica-Geologia Aplicada-IPT, FAPESP, Mapas e Relatórios, São Paulo, Brazil, 1997.

[6] J. B. Oliveira, M. N. Camargo, M. Rossi, and B. Calderano Filho, *MaPa Pedológico do Estado de São Paulo: Legenda Expandida*, Instituto Agronômico, Campinas, São Paulo, and Rio de Janeiro: Embrapa-Solos, Rio de Janeiro, Brazil, 1999.

[7] M. M. Perrota, E. D. Salvador, R. C. Lopes et al., *Mapa Geológico do Estado de São Paulo, Scale 1:750,000*, Programa Geologia Brasil-PGB, CPRM, São Paulo, Brasil, 2005.

[8] Instituto de Pesquisas Tecnológicas-IPT, *Atualização do Relatório de Situação dos Recursos Hídricos 1995 da Bacia do Sorocaba e Médio Tietê (Relatório Zero) Como Subsídio à Elaboração do Plano de Bacia*, vol. 1, (Relatório Técnico no. 80 401-205), Agrupamento de Geologia Aplicada ao Meio Ambiente-AGAMA/Divisão de Geologia-DiGeo/Instituto de Pesquisas Tecnológicas-IPT, Paulo, Brazil.

[9] W. P. Köppen, *Climatologia: Com um Estúdio de los Climas de la Tierra*, Fondo de Cultura Economica, Español. México, 1st edition, 1948.

[10] D. E. Walling and P. W. Moorehead, "The particle size characteristics of fluvial suspended sediment: an overview," *Hydrobiologia*, vol. 176-177, no. 1, pp. 125–149, 1989.

[11] E. T. Degens, S. Kempe, and J. E. Richey, *Biogeochemistry of Major World Rivers*, John Wiley & Sons, Chichester, UK, 1990.

[12] J. Mortatti, *Erosão na Amazônia: processos, modelos e balanço [Sci thesis]*, Escola Superior de Agricultura "Luiz de Queiroz", Universidade de São Paulo, Piracicaba, São Paulo, Brazil, 1995.

[13] J. L. Probst, "Géochimie et hydrologie de l'érosion continentale. Mécanisms, bilan global actuel et fluctuations au cours des 500 derniers millions d'annés," *Sciences Géologiques*, no. 94, p. 161, 1992.

[14] J. Samuel, R. Rouault, and Y. Besnus, "Analyse multiémentaire standardisée des materiaux géologiques en spectrométrie d'émission par plasma a couplage inductif," *Analusis*, vol. 13, no. 7, pp. 312–317, 1985.

[15] J. Mortatti and J. L. Probst, "Characteristics of heavy metals and their evaluation in suspended sediments from Piracicaba river basin (São Paulo, Brazil)," *Revista Brasileira De Geociências*, vol. 40, no. 3, pp. 375–379, 2010.

[16] J. M. Martin and M. Meybeck, "Elemental mass-balance of material carried by major world rivers," *Marine Chemistry*, vol. 7, no. 3, pp. 173–206, 1979.

[17] R. F. Stallard, "Weathering and erosion in the humid tropics," in *Physical and Chemical Weathering in Geochemical Cycles*, A. Lerman and M. Meybeck, Eds., pp. 225–246, Kluwer Academic, 1988.

[18] J. I. Drever, *The Geochemistry of Natural Watersed*, Prentice-Hall, New Jersey, NJ, USA, 2nd edition, 1988.

[19] D. E. Canfield, "The geochemistry of river particulates from the continental USA: major elements," *Geochimica et Cosmochimica Acta*, vol. 61, no. 16, pp. 3349–3365, 1997.

[20] M. Meybeck, "Global chemical weathering of surficial rocks estimated from river dissolved loads," *American Journal of Science*, vol. 287, no. 5, pp. 401–428, 1987.

[21] J. M. Edmond, M. R. Palmer, C. I. Measures, B. Grant, and R. F. Stallard, "The fluvial geochemistry and denudation rate of the Guayana Shield in Venezuela, Colombia, and Brazil," *Geochimica et Cosmochimica Acta*, vol. 59, no. 16, pp. 3301–3325, 1995.

[22] J. L. Probst and A. Bazerbachi, "Transports en solution et en suspension par la Garonne supérieure," *Scence Géologique Bulletin*, vol. 39, pp. 79–98, 1986.

[23] L. Leleyter, *Spéciation chimique des éléments majeurs, traces et des terres rares dans les matières en suspension et dans lês sédiments de found dês cours d'eau: application aux fleuves de Patagonia (Argentine), à La Piracicaba (Brésil), à Sebou (Maroc) et à l'Ill (France) [Ph.D. thesis]*, Université Louis Pasteur, Strasbourg, France, 1998.

[24] J. Mortatti, M. C. Bernardes, J. L. Probst, and L. Leleyter, "Composição química dos sedimentos fluviais em suspensão na bacia do rio Piracicaba: extração seletiva de elementos traço," *Geochimica Brasiliensis*, vol. 16, no. 2, pp. 123–141, 2002.

[25] J. Konta "Mineralogy and chemical maturity of suspended matter in major world rivers sampled under the Scope/Unep project," in *Transport of Carbon and Minerals in Major World Rivers, Part 3*, E. T. Degens, Ed., vol. 58, pp. 569–592, Mitt. Geology Paläontology, Ins. Universität Hamburg, Scope/Unep Sonderband, 1985.

[26] J. L. Probst and N. Sigha, "Estimation de l'écoulement superficiel et de as charge en suspension sur quelques grands bassins fluviaux du monde," *Comptes Rendus de L'Acadadémie des Sciences*, vol. 309, no. 4, pp. 357–363, 1989.

[27] A. M. Fernandes, M. B. Nolasco, and J. Mortatti, "Estimativa do escoamento superficial rápido e sua carga em suspensão com a aplicação de do modelo de separação de hidrogramas com filtros numéricos: bacia do rio Sorocaba-SP," *Geociências*, vol. 29, no. 1, pp. 49–57, 2010.

[28] M. Meybeck, *Les fleuves et le cycle géochimique des elements [Sci. thesis]*, Université de Paris VI, Paris, France, 1984.

[29] J. L. Probst, "Dissolved and suspended matter transported by the Girou River (France): mechanical and chemical erosion rates in a calcareous molasse basin," *Hydrological Sciences Journal/Journal des Sciences Hydrologiques*, vol. 31, no. 1, pp. 61–79, 1986.

[30] J. M. Bortoletto Junior, *Características hidrogeoquímicas e processos erosivos mecânicos e químicos nas bacias de drenagem dos rios Tietê e Piracicaba [Ph.D. thesis]*, Centro de Energia Nuclear na Agricultura, Universidade de São Paulo, São Paulo, Brasil, 2004.

[31] J. Mortatti, M. J. Bortoletto Junior, L. C. E. Milde, and J. L. Probst, "Hidrologia dos rios Tietê e Piracicaba: séries temporais de vazão e hidrogramas de cheia," *Revista Ciência e Tecnologia*, vol. 12, no. 23, pp. 55–97, 2004.

[32] J. Mortatti, J. L. Probst, and M. J. Bortoletto Junior, "Piracicaba river basin: mechanical and chemical erosions," *Geociências*, vol. 22, no. 1, pp. 75–81, 2003.

[33] D. E. Walling and D. Fang, "Recent trends in the suspended sediment loads of the world's rivers," *Global and Planetary Change*, vol. 39, no. 1-2, pp. 111–126, 2003.

[34] J. L. Probst, P. Amiotte-Suchet, and Y. Tardy, "Global continental erosion and fluctuations of atmospheric $CO_2$ consumed during the last 100 years," in *Proceedings of the 7th International Symposium on Water-Rock Interaction*, pp. 483–486, 1992.

[35] M. Meybeck, L. Laroche, H. H. Dürr, and J. P. M. Syvitski, "Global variability of daily total suspended solids and their fluxes in rivers," *Global and Planetary Change*, vol. 39, no. 1-2, pp. 65–93, 2003.

[36] J. Mortatti, R. L. Victória, and Y. Tardy, "Balanço de alteração e erosão química na bacia amazônica," *Geochimica Brasiliensis*, vol. 11, no. 1, pp. 2–13, 1997.

[37] G. R. Blake and K. H. Hartge, "Bulk densityin," in *Method of Soil Analysis. Part 1. Physical and Mineralogical Methods*, A. Klute, Ed., pp. 363–375, American Society of Agronomy, Soil Science Society of America, Madison, Wis, USA, 2nd edition, 1986.

[38] R. R. Nkounkou and J. L. Probst, "Hydrology and geochemistry of the Congo river system," *Mitt. Geology Paläontology, Ins. Universität Hamburg, Scope/Unep Sonderband*, vol. 64, pp. 483–508, 1987.

[39] J. L. Boeglin and J. L. Probst, "Physical and chemical weathering rates and $CO_2$ consumption in a tropical lateritic environment: the upper Niger Basin," *USDA Forest Service—General Technical Report RMRS-GTR*, vol. 148, no. 3-4, pp. 137–156, 1998.

# 2

# Recycling of Badger/Fox Burrows in Late Pleistocene Loess by Hyenas at the Den Site Bad Wildungen-Biedensteg (NW, Germany): Woolly Rhinoceros Killers and Scavengers in a Mammoth Steppe Environment of Europe

**Cajus Diedrich**

*PaleoLogic Private Research Institute, Petra Bezruce 96, 26751 Zdice, Czech Republic*

Correspondence should be addressed to Cajus Diedrich; cdiedri@gmx.net

Academic Editor: Atle Nesje

The Late Pleistocene (MIS 5c-d) Ice Age spotted hyena open air den and bone accumulation site Bad Wildungen-Biedensteg (Hesse, NW, Germany) represents the first open air loess fox/badger den site in Europe, which must have been recycled by *Crocuta crocuta spelaea* (Goldfuss, 1823) as a birthing den. Badger and fox remains, plus remains of their prey (mainly hare), have been found within the loess. Hyena remains from that site include parts of cub skeletons which represent 10% of the megafauna bones. Also a commuting den area existed, which was well marked by hyena faecal pellets. Most of the hyena prey bones expose crack, bite, and nibbling marks, especially the most common bones, the woolly rhinoceros *Coelodonta antiquitatis* (NISP = 32%). The large amount of woolly rhinoceros bones indicate hunting/scavenging specializing on this large prey by hyenas. Other important mammoth steppe hyena prey remains are from *Mammuthus primigenius, Equus caballus przewalskii, Bison/Bos, Megaloceros giganteus, Cervus elaphus,* and *Rangifer tarandus.* The few damaged bone remains of a scavenged cave bear *Ursus spelaeus* subsp. are unique for an open air situation. Abundant micromammal, frog, and some fish remains were concentrated in "pellets" that contain mainly mammoth steppe micromammals and also frog and fish remains that seem to originate from the nearby river/lake.

## 1. Introduction

Late Pleistocene European bone assemblages have been produced mainly by late Ice Age spotted hyena *Crocuta crocuta spelaea* [1] and were first recognized by Buckland [2] in the "Kuhloch Cave" (König-Ludwigs Cave, Bavaria, Germany) and the Kirkdale Cave (Kent, England). More recent studies provide information on the hyena prey bone assemblages (e.g., [3–10]) as well as on the new subdivided fossil hyena den types (e.g., [11]). These identifications of three classified Ice Age den forms are particularly important also to distinguish bone accumulations made by hyenas from those accumulated by Middle Palaeolithic humans (e.g., [9, 12–15]).

Few contemporary used hyena and Neanderthal sites have been described from hyena dens in mammoth steppe lowlands and adjacent cave-rich region environments of north-central Europe, in England and Germany [9, 16]. The degree of prey bone damage and presence/absence of "nibbling sticks" and faecal pellets or hyena population structure and their individual amount allow the reconstruction, much better, of the ethology of the last hyenas of Europe. The discussions for nonarchaeological sites no longer focus only on the human/carnivore origin discussion. Although hyena cave-den sites predominate in the European fossil record (e.g., Germany in [17]), open air sites may have been much more common throughout the mammoth steppe lowlands of Europe, but have been overlooked or not identified as such (cf. Westeregeln or Bottrop sites in [10, 18]).

Open air hyena den sites in loess deposits without human impact are not analyzed in Germany, as yet, whereas other bone accumulation sites on river terraces have been analyzed along the Emscher River near Bottrop in the Westphalian mammoth steppe lowland [10]. Recently many open air hyena den sites (loess, gypsum karst, river terraces: Saalfeld, Bottrop, Westeregeln, Sewecken-Berge, Thiede, and others) from Germany have been described [17, 19–21], whose density overlaps with the Middle Palaeolithic Neanderthal occupation and open air and cave sites in Germany, even in the famous Neanderthal valley [22, 23]. Additionally, the review of lion localities in northern Germany [24] demonstrates not only quite hard competition conditions about megafauna prey between those two top predators killing and consuming each other, but also competition with human Neanderthals during the Late Pleistocene. In Germany, additionally, mostly hyena den cave sites have been described and newly identified, also partly overlapping with human camp sites, for example, Balve Cave [17, 22, 23, 25–27]. The herein reviewed hyena den site Bad Wildungen-Biedensteg is not far from a Middle Palaeolithic site Buhlen (Micoquien to Late Moustérien: [28]), but has no evidence of human impact.

*History of the Bad Wildungen Hyena Den Site.* First Ice Age fauna remains in the clay pit site "Ziegeleigrube Biedensteg" in Bad Wildungen-Biedensteg of northern Hesse (Central Germany, Figure 1, GPS coordinates: long. 9°8′24.32″E, lat. 51°7′16.44″N) were discovered in 1932 by the hobby paleontologist/archaeologist Pusch, who excavated and rescued many macromammal bones. In 1952 Jacobshagen and Lorenz found a micromammal-rich "pellet horizon" and two hyena skulls [29]. Jacobshagen described in 1963, briefly, this fauna, but wrote mainly about the micromammals. Huckriede and Jacobshagen [30] published the first section, which was studied with an addition of new sedimentological results by Semmel [31] and Kulick [32]. The last micropalaeontological research was performed by Storch [33] on pellet material. First thoughts about hyena gnawing and bone deposits were mentioned by Jacobshagen [34] with new research being published about the hyenas, woolly rhinoceros, and cave bears [35]. Here, the complete megafauna and hyena den site analyses are presented in more broad comparisons to many other new analyzed Late Pleistocene hyena dens studied these past years in Germany and Czech Republic (Figure 1).

## 2. Material and Methods

The main collection (including coll. Pusch, coll. Lorenz) is owned by the Rudolf-Lorenz-Stiftung (coll. no. Bi-52/1-237) and was partly presented in the "Stadtmuseum of Bad Wildungen." Additionally, a few macromammal bones from the collection in the "University of Marburg" were integrated in this study, which was also mentioned in the article of Jacobshagen [34]. This collection was partly rediscovered by Dr. Fichter, who kindly helped by donating the important micromammal collection to the "Kurmuseum Bad Wildungen." Only Kulick [32] made a small systematic excavation at the site, which produced mainly micromammals from pellets.

Comparative bone material was used in many different collections. The most important is the woolly rhinoceros skeleton from Petershagen (NW-Germany) in the Museum Natur und Mensch Bielefeld (MNMB). Another mounted skeleton cast in the Museum für Ur- und Ortsgeschichte Eiszeithalle Quadrat Bottrop (EMOB) was used for the skeleton redrawing and comparison of the bone positions in the skeleton of the Bad Wildungen-Biedensteg material. Skeletons of the extinct Przewalski horse (*Equus caballus*

*przewalskii*) were studied in the Julius-Kühn Museum Halle/Saale (JKMH; see also [36]), reindeer (*R. tarandus*) and arctic fox (*V. lagopus*) skeletons in the collection of the University of Alberta Department of Biological Sciences (UADBS); mammoth (*M. primigenius*) remains and cave bear (*U. spelaeus*) and red fox (*V. vulpes*) bones were compared to skeletal material in the Geologisch-Paläontologische Museum der Westfälischen Wilhelms-Universität Münster (GPIM). Finally, recent badger (*M. meles*) or common hare (*L. europaeus*) and the Pleistocene hyena materials from the Srbsko-Chlum were used in the collection of the National Museum Prague (NMP) and from the Perick Caves of the Staatliche Naturhistorische Sammlungen Dresden (SNSD). The open air gypsum karst site Westeregeln material was studied in the Martin-Luther-University Halle/Saale (MLU.IFG) and the Natural History Museum of the Humboldt-University Berlin (MB).

## 3. Sedimentary Geology, Paleoenvironment, and Dating

The geological situation at the hyena site "Lehmgrube Biedensteg" was published by Huckriede and Jacobshagen [30], Semmel [31], and Kulick [32]. The overview of the redrawn sketch of the outcrop section, with a combination of all published results and new interpretations about the hyena deposits, is presented in Figure 2.

The Wilde River gravels at the base of the section are of the Eemian Interglacial period. They consist of Red Bunter sandstone and claystone, lydite, quartz, or diabase pebbles. These deposits are overlain by a palaeosoil resulting from solifluction. In this "Eemian Soil" the river pebbles are resedimented with reddish-brown loess. The "Lower Loess" is from the early to middle Lower Weichselian (MIS 5c-d), and after Semmel [31], a product of the first part of the glaciation (early Late Pleistocene, Figure 2), where, in this mountainous region, loess was deposited in a mammoth steppe environment. Some snails were found in the Lower Loess by Jacobshagen [34], the mentioned loess soil snail *Pupilla muscorum* (Müller) fitting to the cold period climatic and environmental mammoth steppe interpretation.

In the middle and at the end of the Late Pleistocene a climatic stagnation resulted in a palaeosoil along the Wilde River gravels which were, at that time, on the shore of a small lake. This lake was caused by subsurface salt dissolution and positioned in a large-scaled sinkhole structure. The lake was filled up by the Wilde River, indicated by the presence of many aquatic vertebrate species, such as frogs (*Rana agiloides* Brunner), but mainly by salmonid fish (cf. [34]) that lived in fluent water.

The muddy area at the Wilde River or lake shore was used by the Ice Age spotted hyenas as prey deposit sites [35]. Bones from animals of the mammoth steppe macrofauna were deposited here, whereas "bone nests" were mentioned in the publication of Jacobshagen [34]. The sedimentary depression structures in the bone-rich loess horizon described by Kulick [32] as "cryoturbation and channels" also could be partially of bioturbation origin and were possibly caused by the hyenas who deposited animal prey remains in the soft soil, only in

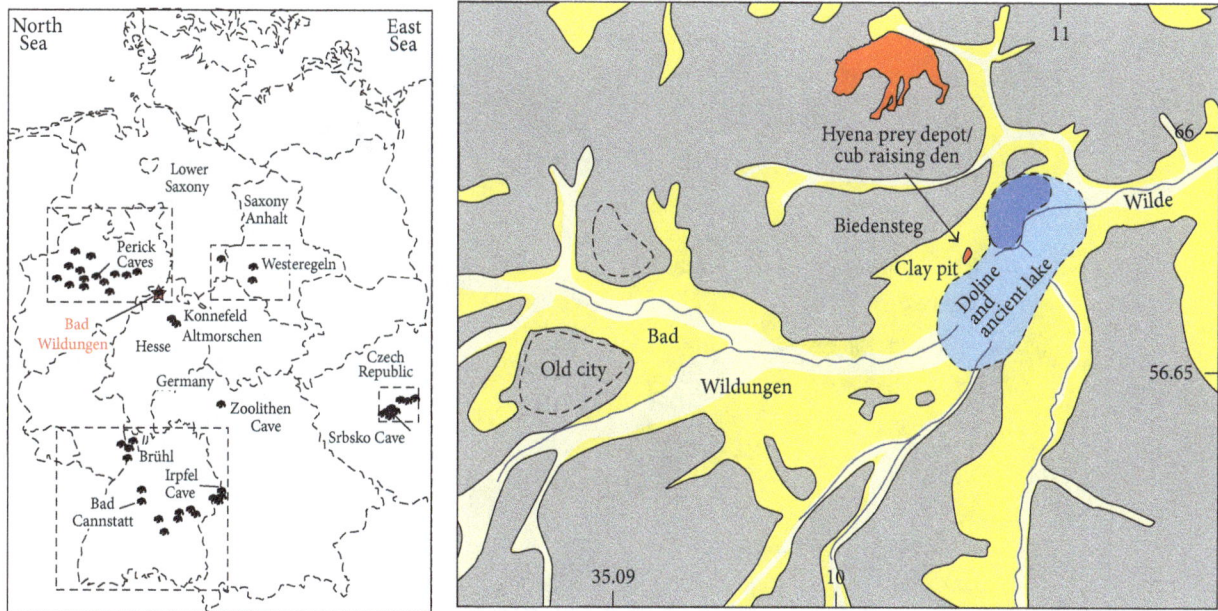

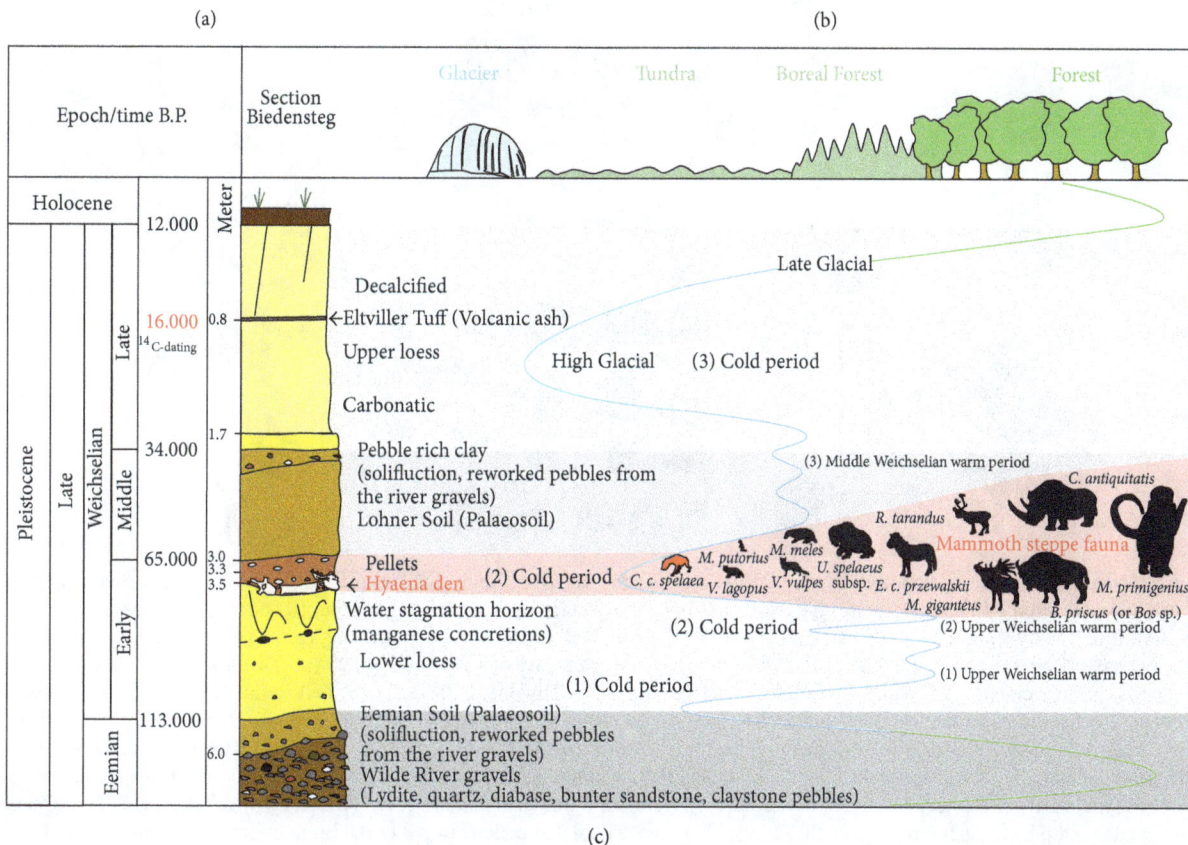

FIGURE 1: (a) Topographic position of the Ice Age spotted hyena *Crocuta crocuta spelaea* birth and commuting den site Bad Wildungen-Biedensteg (Hesse, NW-Germany). (b) The prey was deposited at the margin of an ancient small lake and muddy area of the Pre-Wilde River that filled up a doline during the Late Pleistocene. (c) Generalized section at the Ice Age spotted hyena *C. c. spelaea* prey deposit site Biedensteg (Bad Wildungen, Hesse, NW-Germany).

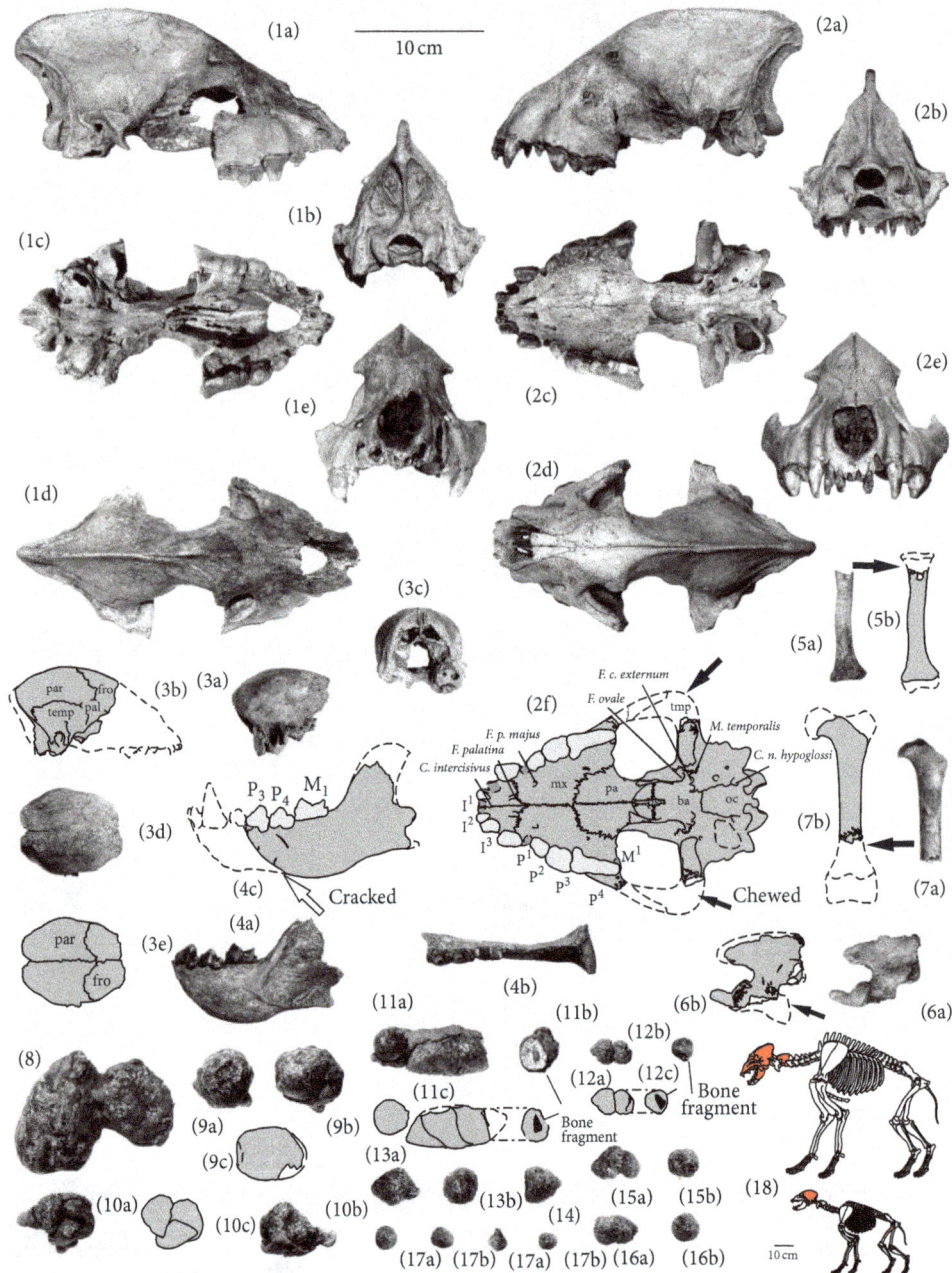

FIGURE 2: Remains of the Ice Age spotted hyena *Crocuta crocuta spelaea* from the hyena open air site Bad Wildungen-Biedensteg (Hesse, NW-Germany). (1) High adult female deformed skull (no. Bi-10at), (a) lateral, occipital, (c) ventral, (d) dorsal, (e) frontal. (2) Early adult female skull (no. Bi-52/45), (a) lateral, (b) occipital, (c) ventral, (d) dorsal, (e) frontal, (f) redrawing (pmx: premaxillary, mx: maxillary, pa: palatine, ba: basis occipital, oc: occipital, j: jugal, tmp: temporal). (3) Brain case of a very young cub (no. Bi-10ev), (a)-(b) lateral, (c) caudal, (d)-(e) dorsal. (4) Left cracked mandible of an adult female, (b) dorsal, (c) labial (no. Bi-52/51). (5) Left radius of a young cub (no. Bi-10ew), craniolateral. (6) Axes of an adult animal (no. Bi-52/234), lateral. (7) Left femora of a young cub cranial (no. Bi-10em). (8)–(17) Coprolites from the hyena open air site Bad Wildungen-Biedensteg (Hesse, NW-Germany). (8) Two large pellets, partly encrusted by caliche (no. Bi-52/221). (9) Large oval pellet, partly encrusted by caliche at which originally another pellet was attached (no. Bi-52/213). (10) Three articulated pellets of different shape, partly encrusted by caliche (no. Bi-52/214). (11) Four partly articulated pellets, encrusted by caliche. Pellet D is broken; the end exposes a small prey bone fragment (no. Bi-52/210). (12) Two articulated pellets. In pellet B a bone fragment is present (no. Bi-52/219). (13) Sigmoid drop shaped and pointing single pellet (no. Bi-52/209). (14) Cone shaped and basal flat single pellet, that was originally attached to another pellet (no. Bi-52/220). (15) Irregular u-shaped pellet that was originally attached to other pellets (no. Bi-52/207). (16) Irregular shaped pellet that was originally attached to other pellets (no. Bi-52/212). (17) Small flat drop shaped pellet (no. Bi-52/218). (11) Small drop shaped single pellet (no. Bi-52/211). (18) Bone remains (red are represent) from an adult female, an early juvenile cub of few weeks of age and coprolites from the hyena freeland prey deposit site Bad Wildungen-Biedensteg near Hesse (NW-Germany).

Recycling of Badger/Fox Burrows in Late Pleistocene Loess by Hyenas at the Den Site Bad Wildungen-Biedensteg (NW, Germany): Woolly Rhinoceros Killers and Scavengers in a Mammoth Steppe Environment of Europe

13

summer times, when the permafrost soil was soft in the upper parts.

The bioturbation interpretation would fit into the "hyena commuting/prey storage site," but can no longer be studied because of the nonopen loess pit Biedensteg. In this section (Figure 2) such depressions are figured as hyena prey depots. Possibly, a later cryoturbation, a result of permafrost soils fitting into the environment and climatic situation of that time, was responsible for secondary overprint of the primary sediment structures. Bioturbation by mammoths on lake shores, which left depressions of their footprints, must be taken into account, as is discussed for other sites (cf. [37]).

The "pellet horizon" is figured differently in the publications (of Jacobshagen et al., 1963, [32]). The section of Kulick [32] indicates that the pellets and the macromammal bones are mixed in a single horizon. Proof for that might be caliche concretions around hyena coprolites in which micromammal bones and teeth are also cemented in. The "hyena prey depot site" and the "pellet horizon" are from the same period and are dated relatively (no absolute data) into the late Middle Late Pleistocene or Weichselian (65.000–90.000 BP, MIS 5c-d, Figure 2).

The bone-rich horizon is overlain by another palaeosoil, the "Lohner Soil," which can be found in the region at different sections [31, 32]. After their interpretations a solifluction of Loess and Wilde river gravel material took place in the middle Late Pleistocene warm period (Figure 2). *V. vulpes* and *M. meles* were the dominating faunal elements, besides *L. europaeus*. This fauna fits to *Meles/Vulpes* den burrow sites in loess soils, in front of which they often left some prey bones.

Finally the upper loess was deposited within the LGM, and after, the upper part was decalcified during the Holocene period. The "Eltviller Tuff" is a one to two centimeter thin layer in the upper loess and the only absolute dated horizon with an age of around 16.000 BP ([31], Figure 1(c)).

## 4. Small Carnivore Fox Den and Mustelid Bone Assemblage

*Meles meles* (Linné 1758) (Figure 4(13)–(32)) (Table 5) is known by one skull of an adult male (Figure 4(13)) and a second brain case of a juvenile. Several postcranial bones consist of the forelimb (Figure 4(14)–(20)) and hind limb bones (Figure 4(21)–(29)), although vertebrae are missing (cf. Table 1).

*Vulpes vulpes* (Linné 1758) (Figure 4(1)–(9)) remains consist of 13 common fox bones (Table 3) including a skull. This skull is incomplete, as most of the anterior part with its dentition is missing. The last three teeth are in the left maxillary (Figure 4(1)). From a right forelimb the scapula, humerus, and radius were found, which seem to belong to one individual (Figure 4(2)–(4)). From a hind limb, not only the left femur shaft and incomplete tibia but also a right calcaneus and a metatarsus III are represented (Figure 4(5)–(8)). A fragment of a metapodial is missing its proximal joint. Finally a lumbar vertebra and one rib are preserved. The pelvis is missing its left part (Figure 4(9)). A second pelvis fragment is again incomplete. Material from two individuals is present, indicated by the pelvis remains. Possibly most of

the bones belong to only one individual. All postcranial bones show a complete fuse of the symphyses and are from either a single animal or several adult animals.

*Vulpes lagopus* (Linné 1758) (Figure 4(10)–(12)) (Table 4) was found with a nearly complete skull, without the jugal arches, but with the right mandible (Figure 4(10)-(11)). The skull sutures are not fully fused and teeth are barely used; therefore it was a young adult individual, as only a single individual can be estimated from the bone material. The postcranial material is present with a femur shaft and pelvic fragment (Figure 4(12)).

*Mustela putorius* Linnaeus 1758 (Figure 4(33)) (Table 6) is present with a single half skull (Figure 4(33)) of which the anterior part with most of the dentition is preserved.

*Lepus europaeus/timidus* Linné 1758 (Figure 14(1)–(9)) (Table 14) is represented by 28 bones which are cranial fragments, two are mandibles and the rest are postcranial bones (Table 13). There is an articulated pedal skeleton (Figure 14(9)) and an articulated pelvis with lumbar vertebral column (Figure 14(5)). The figured material (Figure 14) seems to be from one individual, which is indicated by the bone preservation and articulations. Another argument is the individual adult's age and the fresh fractures of the humerus, radius, the right femur and left tibia, or some processes of the vertebrae, which were caused during the excavations. Bones from other individuals of young and adult age are also preserved and have been completely disarticulated. 25% of the remains are from young animals; 75% are from adult hares. Three animals can be estimated by the tibia as minimum individual number.

## 5. The Hyena Population and Coprolite Remains

The Ice Age spotted hyena *Crocuta crocuta spelaea* [1] (Figure 2) skeletal remains consist of four skulls, three mandibles, one radius, and a femur (Table 1). Additionally, there are 16 coprolites which were rescued.

From the first skull (Figure 2(1)) deformations do not allow exact metric data. The second skull (Figure 2(2)) is 290 mm in total length and measures 265 mm between the incisive and condyle. The largest height is behind the frontal processes (114 mm). The distances between the canines and $P^4$ are about 68 mm. The width of the frontals (zygomatic processes) measures 90 mm. Finally the outer distance between the canines is 58 mm. The largest diameter of the canines in the middle of the tooth is 18 mm. The brain case symphyse of the third animal (Figure 2(3)) is slightly fused and articulated. The parietal, frontal, palatine, and temporal are incomplete. The maximum width measured, between the temporal, 73 mm, whereas it is preserved in 76 mm in length.

One left mandible (Figure 2(4)) is of an adult animal and might belong to one of both individual adult skulls, which show a similar tooth use stage. The jaw was cracked by hyenas between the $P_2$ and $P_3$; the $P_{3-4}$ and $M_1$ are present. The ramus was damaged during excavations.

A few postcranial bones are represented with one axis of an adult animal exposing bite damage marks (Figure 2(6)).

TABLE 1: Bones of *Crocuta crocuta spelaea* (Goldfuss 1823) from the open air site Bad Wildungen-Biedensteg (Hesse, NW-Germany).

| No. | Coll.-No. | Bone type | Commentary | Left | Right | Age | Bite marks | Collection |
|---|---|---|---|---|---|---|---|---|
| 1 | 52/45 | Skull | Nearly complete, female | | | Early adult | x | Rudolf-Lorenz-Stiftung |
| 2 | 10at | Skull | Nearly complete, female | | | High adult | x | Rudolf-Lorenz-Stiftung |
| 3 | / | Skull | — | | | ? | | (Mentioned in [34], missing) |
| 4 | 10ev | Skull | Brain case | | | Cub | x | Rudolf-Lorenz-Stiftung |
| 5 | 52/51 | Mandibula | — | x | | Early adult | x | Rudolf-Lorenz-Stiftung |
| 6 | ? | Mandibula | — | | | ? | | (Mentioned in [34], missing) |
| 7 | ? | Mandibula | — | | | ? | | (Mentioned in [34], missing) |
| 8 | 52/234 | Cervical vertebra | Axes | | | Adult | x | Rudolf-Lorenz-Stiftung |
| 9 | 10ew | Radius | Without joints | x | | Cub | x | Rudolf-Lorenz-Stiftung |
| 10 | 52/249 | Thoracic vertebra | Disc | | | Cub | | Rudolf-Lorenz-Stiftung |
| 11 | 10em | Femur | Without joints | x | | Cub | x | Rudolf-Lorenz-Stiftung |
| 12 | 52/209 | Coprolite | Single pellet | | | | | Rudolf-Lorenz-Stiftung |
| 13 | 52/220 | Coprolite | Single pellet | | | | | Rudolf-Lorenz-Stiftung |
| 14 | 52/207 | Coprolite | Single pellet | | | | | Rudolf-Lorenz-Stiftung |
| 15 | 52/210 | Coprolite | Single pellet | | | | | Rudolf-Lorenz-Stiftung |
| 16 | 52/212 | Coprolite | Single pellet | | | | | Rudolf-Lorenz-Stiftung |
| 17 | 52/206 | Coprolite | Single pellet, with prey bone fragment | | | | | Rudolf-Lorenz-Stiftung |
| 18 | 52/218 | Coprolite | Single pellet | | | | | Rudolf-Lorenz-Stiftung |
| 19 | 52/226 | Coprolite | Single pellet, with prey bone fragment | | | | | Rudolf-Lorenz-Stiftung |
| 20 | 52/208 | Coprolite | Single pellet, with prey bone fragment | | | | | Rudolf-Lorenz-Stiftung |
| 21 | 52/225 | Coprolite | Single pellet, with prey bone fragment | | | | | Rudolf-Lorenz-Stiftung |
| 22 | 52/211 | Coprolite | Single pellet | | | | | Rudolf-Lorenz-Stiftung |
| 23 | 52/214 | Coprolite | Three articulated pellets | | | | | Rudolf-Lorenz-Stiftung |
| 24 | 52/213 | Coprolite | Single large pellet | | | | | Rudolf-Lorenz-Stiftung |
| 25 | 52/219 | Coprolite | Two articulated pellets, with prey bone fragment | | | | | Rudolf-Lorenz-Stiftung |
| 26 | 52/221 | Coprolite | Two large articulated pellets | | | | | Rudolf-Lorenz-Stiftung |
| 27 | 52/237 | Coprolite | Single pellet | | | | | Rudolf-Lorenz-Stiftung |

A left radius and a left femur (Figure 2(5) and (7)) are from one very young cub, both being incomplete as a result of scavenging activities by large carnivores.

*Coprolite Material.* The hyena coprolites are generally white inside and the pores are filled with iron and manganese minerals. The coprolites show a moderate variability and even bone contents (Figure 2(8)–(17)). The largest one (Figure 2(8)) is a double pellet being connected by caliche incrustations. It seems to represent a fossilized, originally softer and humid, faecal pellet. The other pellets have repeating shapes and have attached 3–5 smaller pellets (Figure 2(9)–(12)), representing possibly more dry dung. Single pellets have often defined shapes. The most represented one is the "drop shaped pellet" (Figure 2(13)–(15)). They can point to both sides or can end round to flat on one side as a result of attachment to another pellet. Other pellets are "unshaped" and irregular. These were often found in the non-spindle-like pellet aggregations (Figure 2(10)). In the material from Biedensteg each coprolite contains several bone fragments, which are often visible on the surfaces (Figure 2(11)-(12)). These are small pieces, well rounded by stomach acid, and are mainly from the bone compacta, but also are isolated pieces of bone spongiosa. This spongiosa is very thin walled and should have been completely dissolute. These spongiosa pieces are most comparable to the bone spongiosa of the woolly rhinoceros, but might also refer to other megamammals.

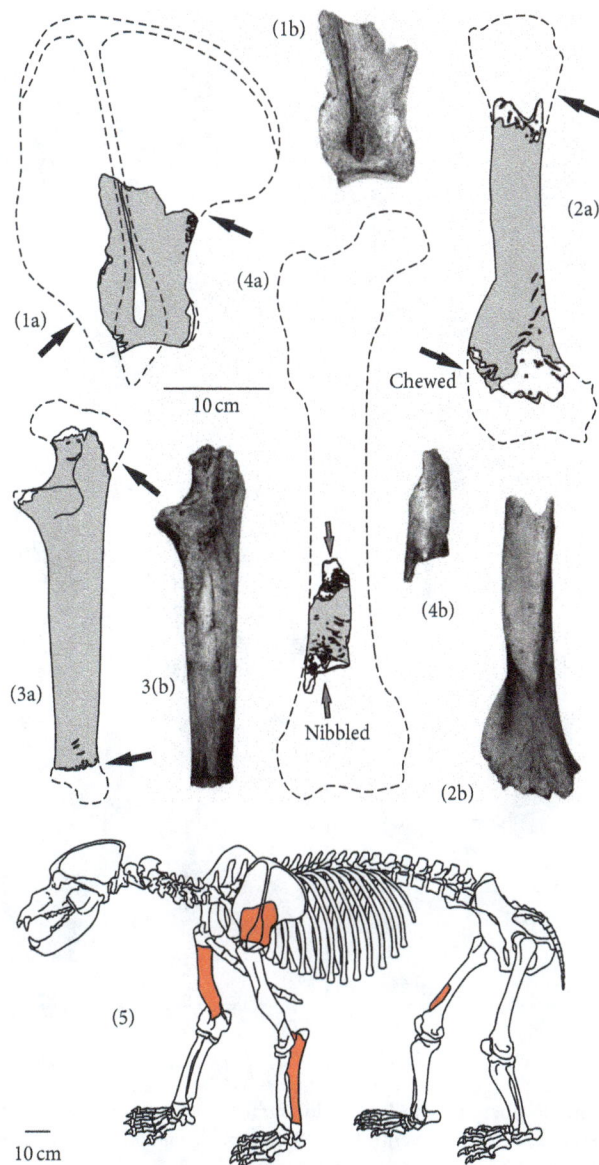

FIGURE 3: *Ursus spelaeus* subsp. bones from the hyena open air site Bad Wildungen-Biedensteg (Hesse, NW-Germany). The bones seem to belong to one adult female individual. (1) Left scapula fragment (no. Bi-52/227), lateral. (2) Right humerus shaft (no. Bi-52/2), cranial. (3) Left ulna shaft (no. Bi-52/241), lateral. (4) Right femora shaft fragment and "nibbling stick" (no. Bi-52/242), cranial. (5) Present bones (red) of an adult female *Ursus* cf. *spelaeus* Rosenmüller, 1794, from the hyena open air site Bad Wildungen-Biedensteg (Hesse, NW-Germany).

## 6. Hyena Megafauna Prey Remains

*Ursus spelaeus* Rosenmüller 1794 subsp. (Figure 3) is represented by four cave bear bones and fragments. The left scapula (Tables 2 and 3(1)), which lacks all distal parts seems to be destroyed by hyenas. Large carnivore gnawing and bite marks are visible at the glenoid. A right humerus shaft (Figure 3(2)) is missing the joints as a result of heavy carnivore chewing. At the shaft ends and in the lower middle, bite marks are present. The diameter of the bone shaft is small, being only 49 mm. From one left incomplete ulna (Figure 3(3)) the distal joints were chewed and also some bite marks are visible. The 50 mm maximum width ulna has, again, small proportions.

Finally, a fragment of a femur shaft (Figure 3(4)) with heavy chewing damage indicate the cracking and further use of the bone fragment as a typical hyena "nibbling stick" (for teething purposes of hyena cubs).

*Mammuthus primigenius* (Blumenbach 1799) (Figure 12 (1)–(3)) is represented by three remains consisting of a tooth lamella fragment from a juvenile animal, a thoracic vertebra neural arch and centrum fragment, and a long bone fragment used as a nibbling stick (Table 7). The material is from adolescent elephants.

*Coelodonta antiquitatis* (Blumenbach 1799) (Figures 5–11) is the most abundant, listed in Table 8. The cranial elements consist of a middle part of a skull from a young calf (Figure 9).

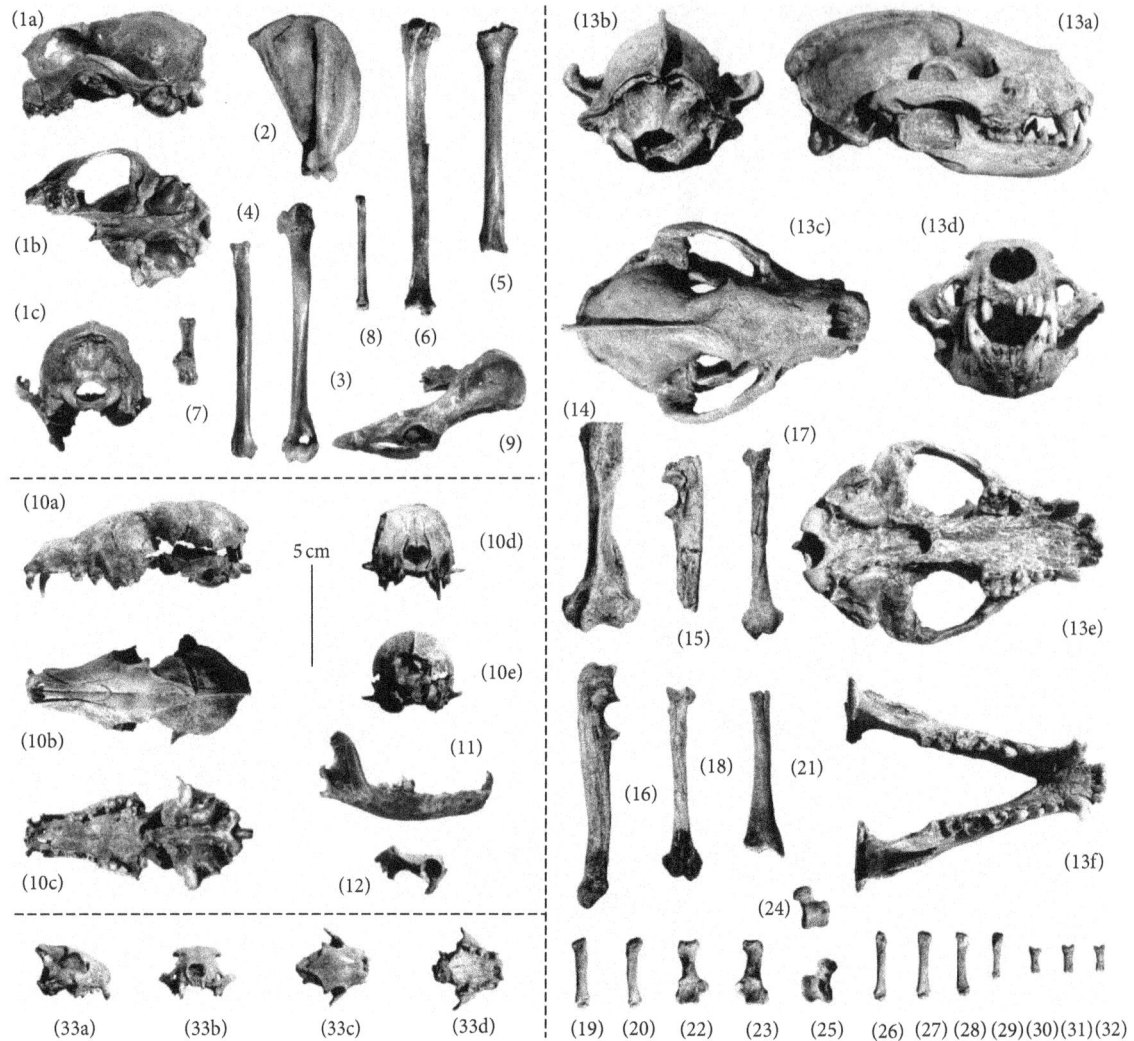

FIGURE 4: Small carnivores from the hyena open air site Bad Wildungen-Biedensteg (Hesse, NW-Germany). (1)–(9) *Vulpes vulpes*: (1) incomplete skull (no. Bi-52/39), (a) lateral, (b) caudal, (c) ventral. (2) Right scapula (no. Bi-52/235), lateral. (3) Right humerus (no. Bi-52/10), cranial. (4) Right radius (no. Bi-52/24), cranial. (5) Left femur without joints (no. Bi-52/104), cranial. (6) Left tibia (no. Bi-52/105a), cranial. (7) Left calcaneus (no. Bi-52/238), cranial. (8) Right metatarsus III (no. Bi-52/239), cranial. (9) Incomplete pelvis (no. Bi-52/127), lateral. (10)–(12) *Vulpes lagopus*: (10) nearly complete skull with right lower jaw (no. Bi-10bh), (a) lateral, (b) dorsal, (c) ventral, (d) frontal, (e) occipital. (11) Right mandible (no. Bi-52/243), lateral. (12) Pelvis, acetabulum (no. Bi-10bn), lateral. (13)–(32) *Meles meles*: (1) skull with lower jaw (no. Bi-10ah), (a) lateral, (b) occipital, (c) dorsal, (d) frontal, (e) ventral, (f) lower jaw dorsal. (14) Left Humerus (no. Bi-10ap), cranial. (15) Left ulna shaft (no. Bi-10ao), lateral. (16) Right ulna (no. Bi-10av), lateral. (17) Left radius (no. Bi-10aw), lateral. (18) Right radius (no. Bi-10ao), lateral. (19) Left mt III (no. Bi-10bf), dorsal. (20) Left mt V (no. Bi-10bb), dorsal. (21) Left tibia (no. Bi-52/85), cranial. (22) Right calcaneus (no. Bi-10an), cranial. (23) Left calcaneus (no. Bi-10at), cranial. (24) Rigth astragal (no. Bi- BadW-2), dorsal. (25) Left astragal (no. Bi-10ay), dorsal. (26) Rigth mt IV (no. Bi- BadW-5), dorsal. (27) Rigth mt I (no. Bi- BadW-8), dorsal. (30) Phalanx II (no. BadW-2), dorsal. (31) Phalanx II (no. BadW-3), dorsal. (32) Phalanx II(no. BadW-4), dorsal. (33) *Mustela putorius,* anterior part of a skull (no. Bi-10bs), (a) lateral, (b) frontal, (c) dorsal, (d) ventral.

TABLE 2: Bones of *Ursus spelaeus* subsp. Rosenmüller 1794 from the open air site Bad Wildungen-Biedensteg (Hesse, NW-Germany).

| No. | Coll.-No. | Bone type | Commentary | Left | Right | Age | Bite marks | Collection |
|-----|-----------|-----------|------------|------|-------|-----|------------|------------|
| 1 | 52/227 | Scapula | Without distal part | x | | Adult | x | Rudolf-Lorenz-Stiftung |
| 2 | 52/2 | Humerus | Shaft | | x | Adult | x | Rudolf-Lorenz-Stiftung |
| 3 | 52/241 | Ulna | Incomplete | x | | Adult | x | Rudolf-Lorenz-Stiftung |
| 4 | 52/242 | Femur | Fragment | | ? | Adult | x | Rudolf-Lorenz-Stiftung |

Recycling of Badger/Fox Burrows in Late Pleistocene Loess by Hyenas at the Den Site Bad Wildungen-Biedensteg (NW, Germany): Woolly Rhinoceros Killers and Scavengers in a Mammoth Steppe Environment of Europe

17

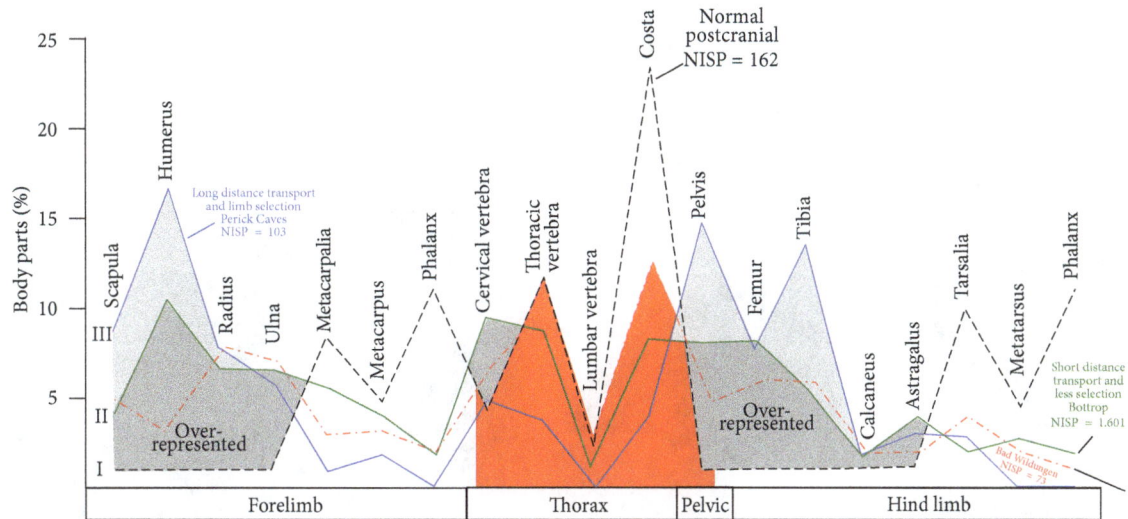

(a)

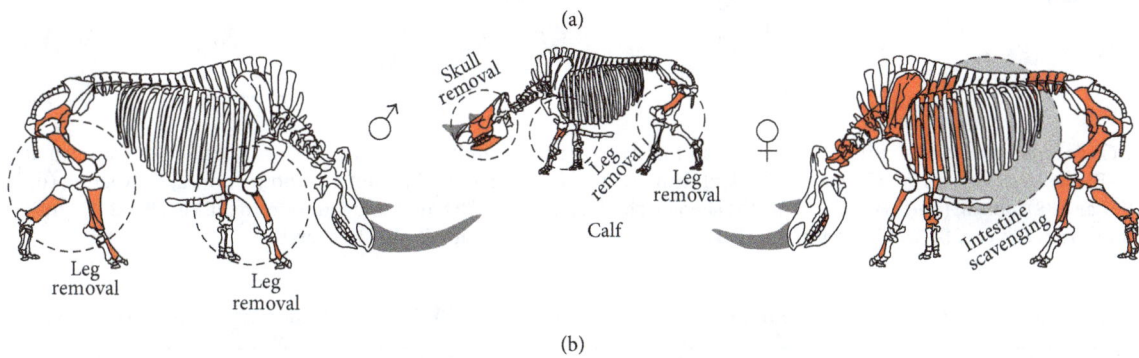

(b)

(c)

FIGURE 5: Present bone material of one (or several) male, a female skeleton remain, and calf skeleton remain of the woolly rhinoceros *Coelodonta antiquitatis* from Bad Wildungen-Biedensteg (Hesse, NW-Germany) open air hyena den.

The connection in-between the maxillas were restored in former times. Originally, the maxillary part between the teeth was damaged by hyenas. All three $dm^{1-3}$ milk teeth on both sides are present (Figure 6(1a)–(1d)). Both $m^1$'s are breaking through, whereas the $m^2$'s were still in the maxillary.

These are not present, but the alveolar grooves are preserved. This skull was badly damaged by the hyenas, especially at the anterior part and the brain case. The latter shows a very interesting large carnivore brain case opening. There are some bite marks, but thin parallel long scratch

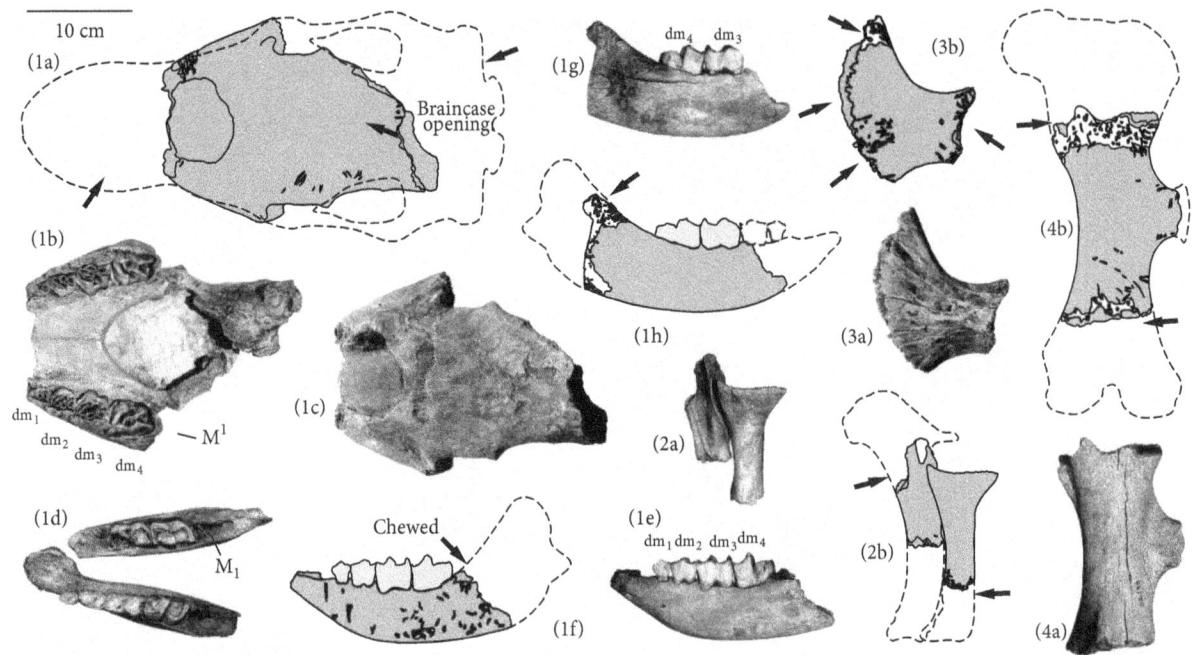

FIGURE 6: *Coelodonta antiquitatis* remains of a less than half-year-old calf with hyena chewing marks from the hyena open air site Bad Wildungen-Biedensteg (Hesse, NW-Germany). (1) Skull with brain case opening (no. Bi-10ac), (a) and (c) dorsal, (b) ventral, (d) lower jaw (no. Bi-52/37 and 38), dorsal, (e)-(f) lateral left, (g)-(h) lateral right. (2) Articulated left ulna and radius from calf (no. Bi-52/47 and 42), lateral. (4) Left femur from calf (no. Bi-52/43), cranial. (3) (1) Left ileum remain of a calf (no. Bi-52/13), lateral.

TABLE 3: Bones of *Vulpes vulpes* 1758 from the open air prey deposit site Bad Wildungen-Biedensteg (Hesse, NW-Germany).

| No. | Coll.-No. | Bone type | Commentary | Left | Right | Age | Bite marks | Collection |
|-----|-----------|-----------|------------|------|-------|-----|------------|------------|
| 1 | 52/39 | Cranium | Incomplete | | | Adult | | Rudolf-Lorenz-Stiftung |
| 2 | 52/35 | Scapula | Nearly complete | | x | Adult | | Rudolf-Lorenz-Stiftung |
| 3 | 52/10 | Humerus | Complete | | x | Adult | | Rudolf-Lorenz-Stiftung |
| 4 | 52/24 | Radius | Complete | | x | Adult | | Rudolf-Lorenz-Stiftung |
| 5 | 52/104 | Femur | Nearly complete | x | | Adult | | Rudolf-Lorenz-Stiftung |
| 6 | 52/105a | Tibia | Complete | x | | Adult | | Rudolf-Lorenz-Stiftung |
| 7 | 52/238 | Calcaneus | Complete | x | | Adult | | Rudolf-Lorenz-Stiftung |
| 8 | 52/127 | Pelvis | Nearly complete | | | Adult | | Rudolf-Lorenz-Stiftung |
| 9 | 52/128 | Pelvis | Fragment | | x | Adult | | Rudolf-Lorenz-Stiftung |
| 10 | 52/239 | Metatarsus III | Complete | | x | Adult | | University of Marburg |
| 11 | 52/240 | Metatarsus | Without proximal joint | | | Adult | | University of Marburg |
| 12 | 52/21 | Lumbar vertebra | Nearly complete | | x | Adult | | Rudolf-Lorenz-Stiftung |
| 13 | 52/105b | Costa | Nearly complete | | x | Adult | | Rudolf-Lorenz-Stiftung |

marks on the right maxillary in the high of the $dm^{2\text{-}3}$ could have resulted from other smaller carnivores or hyena cubs. Both mandibles of the lower jaw (Figure 6(1e)–(1h)) fit to the skull by the identical milk dentition of the $dm_{1-3}$ and the tooth rising of the $m_1$. Both jaws were cracked in the symphyses area and have old fractures. Additionally, they are lacking the rami and have large carnivore chewing and gnawing marks (Figure 6(1e)–(1h)). The left jaw possesses the $dm_{1-3}$ and the $m_1$. The right mandible was damaged by the excavations and because of this is lacking the anterior part, including the $dm_{1-2}$. Other cranial material was described and partly refigured by Jacobshagen [34]. He refigured some

lower jaw teeth of one individual (right $P_{3-4}$, $M_1$, and left $M_{2-3}$). The little use of the $M_3$ indicates an origin of an early adult animal. It is suggested here that these belonged most probably to the skeleton of an early adult female individual (Figure 5(b)). Scapulae are preserved with one nearly complete left shoulder blade (Figure 7(1)). Some parts from the left side and joint area, destroyed by the excavations, were restored. Bite marks were found only distally. Here, hyenas left typical chewing marks in the very soft scapula. The margin is therefore typically irregular, resulting from cracked bone material. The scapula seemed to belong to the female skeleton. A second fragment of a scapula is

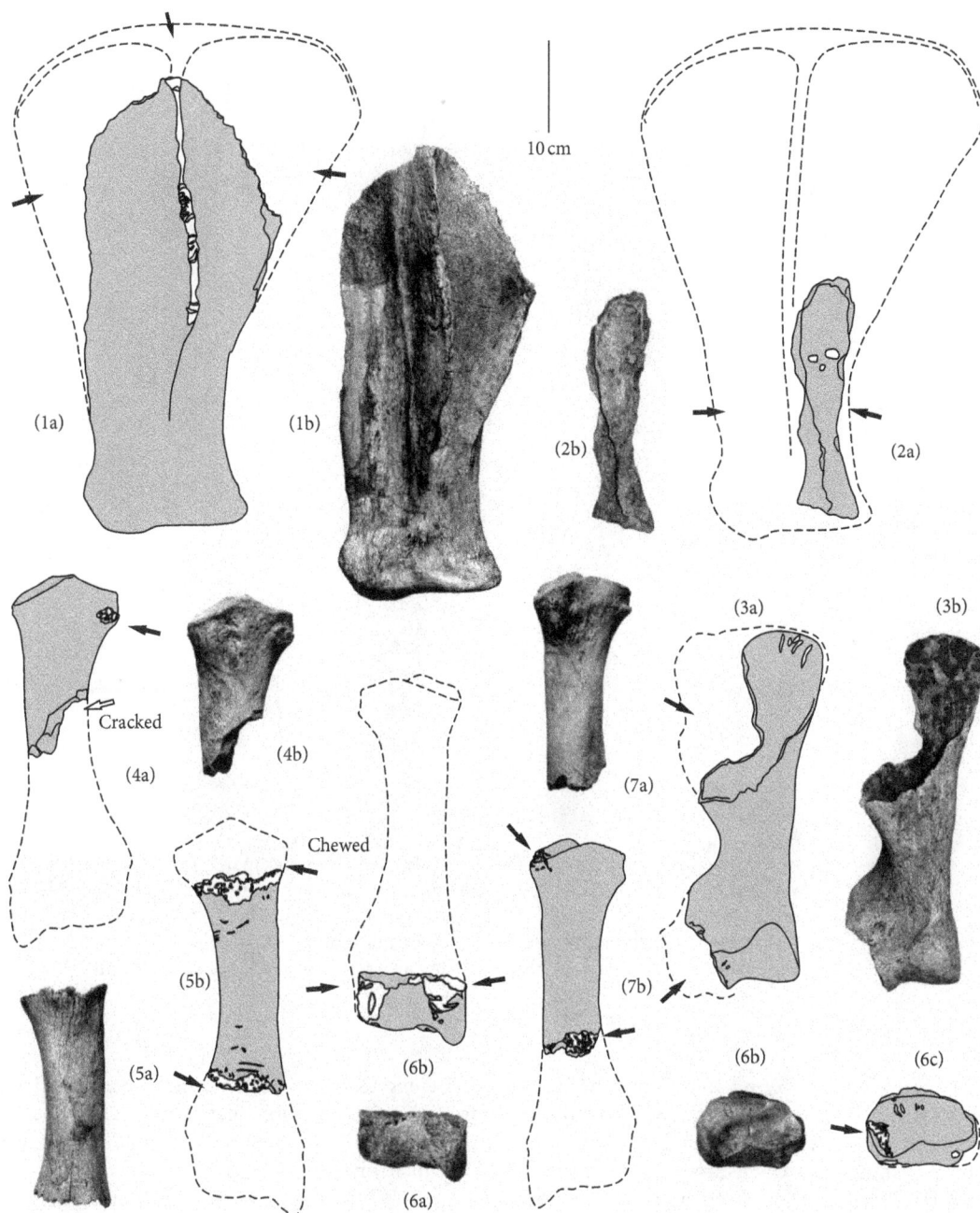

FIGURE 7: *Coelodonta antiquitatis* fore leg remains of adolescent and grown up animals with hyena chewing marks from the hyena open air site Bad Wildungen-Biedensteg (Hesse, NW-Germany). (1) Left scapula from an adult individual (no. Bi-52/20), lateral. (2) Scapula fragment (no. Bi-52/88), lateral. (3) Rigth humerus (no. Bi-180c), caudal. (4) Left radius from an adult male individual (no. Bi-52/30), cranial. (5) Right radius from an adult male individual (no. Bi-52/44), cranial. (6) Right distal radius joint from an early adult female individual (no. Bi-52/224), (a) cranial, (b)-(c) ventral. (7) Left radius from an adult female individual (no. Bi-52/49), cranial.

TABLE 4: Bones of *Vulpes lagopus* 1758 from the open site Bad Wildungen-Biedensteg (Hesse, NW-Germany).

| No. | Coll.-No. | Bone type | Commentary | left | right | Age | Bite marks | Collection |
|---|---|---|---|---|---|---|---|---|
| 1 | 10bh | Cranium | Nearly complete with right lower jaw | | | Senile | | Stadtmuseum Bad Wildungen |
| 2 | 52/243 | Mandibula | Fragment with P4 | x | | Adult | | University of Marburg |
| 3 | 10eh | Femur | Shaft | x | | | | Stadtmuseum Bad Wildungen |
| 4 | 10bn | Pelvis | Fragment, acetabulum | x | | Adult | | Stadtmuseum Bad Wildungen |

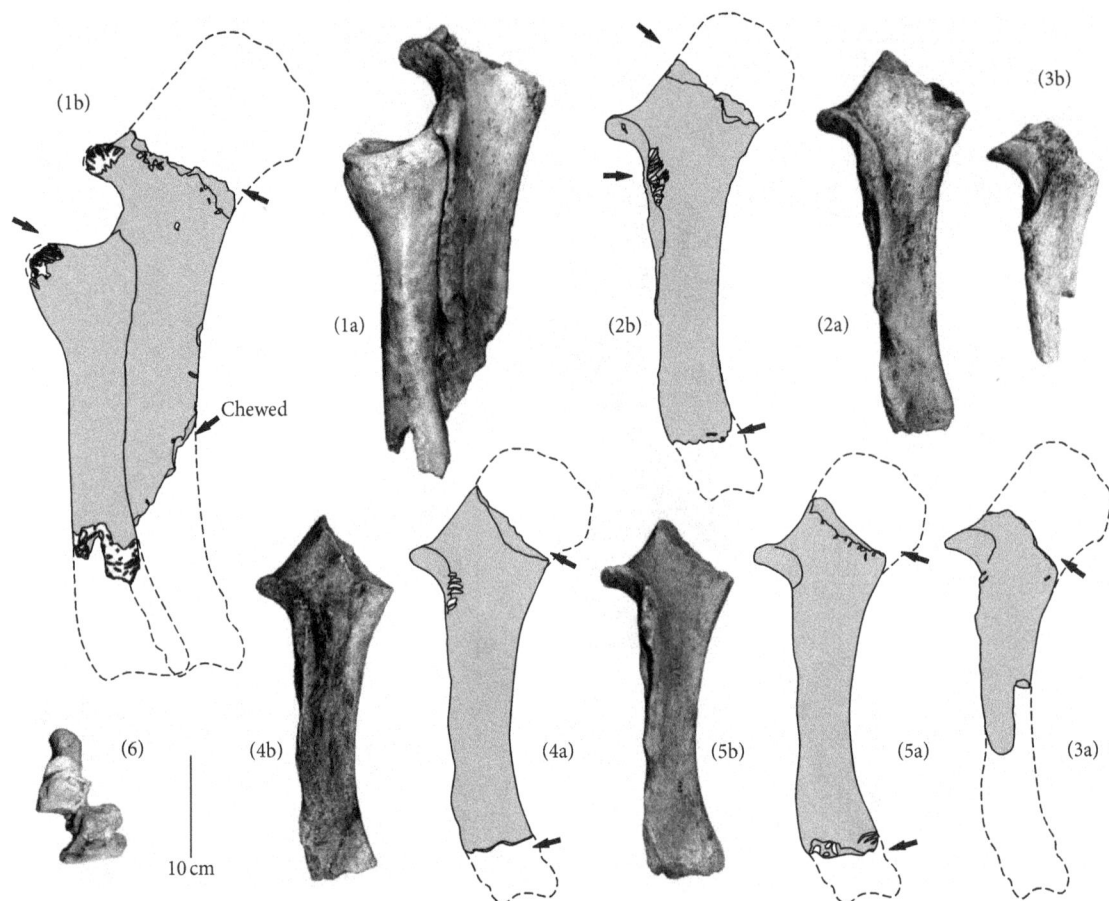

FIGURE 8: *Coelodonta antiquitatis* fore leg remains of grown up animals with hyena chewing marks from the hyena open air site Bad Wildungen-Biedensteg (Hesse, NW-Germany). (1) Articulated right ulna and radius from an early adult female individual (no. Bi-52/111 and 116), lateral. (2) Right ulna from an adult female individual (no. Bi-52/53), lateral. (3) Right ulna from an early adult female individual (no. Bi-10a), lateral. (4) Right ulna from an adolescent/adult individual (no. 10p), lateral. (5) Right ulna from an adolescent/adult female individual (no. 52/143), lateral. (6) Articulated intermedium and carpale 3 from an adult individual (no. Bi-52/34 and 235), cranial.

in preservation and could be found in a lower horizon. One humerus is described by Jacobshagen [34], which can no longer be located. It was a right humerus that was chewed on the proximal joint. Ulnae are present with five bones (Figure 8(1)–(4)) from different old animals. The most juvenile, a neonate to young, animal's left ulna must have been articulated to one radius (Figure 6(2)). This result is from the comparison to an articulated right ulna/radius from a young adult to adult animal whose joints are chewed away in the same way (Figure 8(1)). The latter might belong to the young adult female rhinoceros (Figure 5(b)), of which also other bones were found partly articulated. At least seven radii (Figure 7(4)–(6), MNI = 7) were found, of which four are from young adult to adult animals and the last from the neonate to very young individual. The four pelvis remains are typical rests of hyena feeding activities (Figure 10(1)–(3)). The acetabular and surrounding two acetabular fragments are from different animals. The one figured (Figure 10(1)) has not only hyena, but also arctic fox, wolf or hyena cub, and even small rodent nibbling marks. The fourth pelvis remain is only a part of the ileum (Figure 6(3)) and seems

to belong to the juvenile animal, because it is also chewed from the acetabular region. It is also heavily chewed at the soft distal part with irregular margin. Four femora are preserved, of which one is a fragment, a second is from a juvenile animal (Figure 6(4)), and a third and fourth are from an adult *C. antiquitatis* (Figure 10(4)-(5)). Another fragment is of an adolescent, with strong chewing marks (Figure 10(6)). As described by Jacobshagen [34], there was a right femur (Figure 10(4)) found in articulation with a tibia (Figure 11(2)). Only one nearly complete left patella (Figure 11(9)) was excavated and might belong also to the female skeleton's hind leg (Figure 5(b)). The tibia has very typical hyena caused damages and is in an early stage (stage 1) of destruction. Also this fits well with the partly articulated female skeleton carcass. Three tibiae are very massive and have a strong width in the shaft (Figure 11(3)–(5)). All tibiae compared indicate a sexual dimorphism with males being stronger and more massive in their bones. Mostly the proximal joint was chewed away first, although at the distal part in a middle stage (stage 2 of three) of bone feeding, two groves were left, which is documented at all three tibiae (Figure 11(3)–(5)). Two fibula

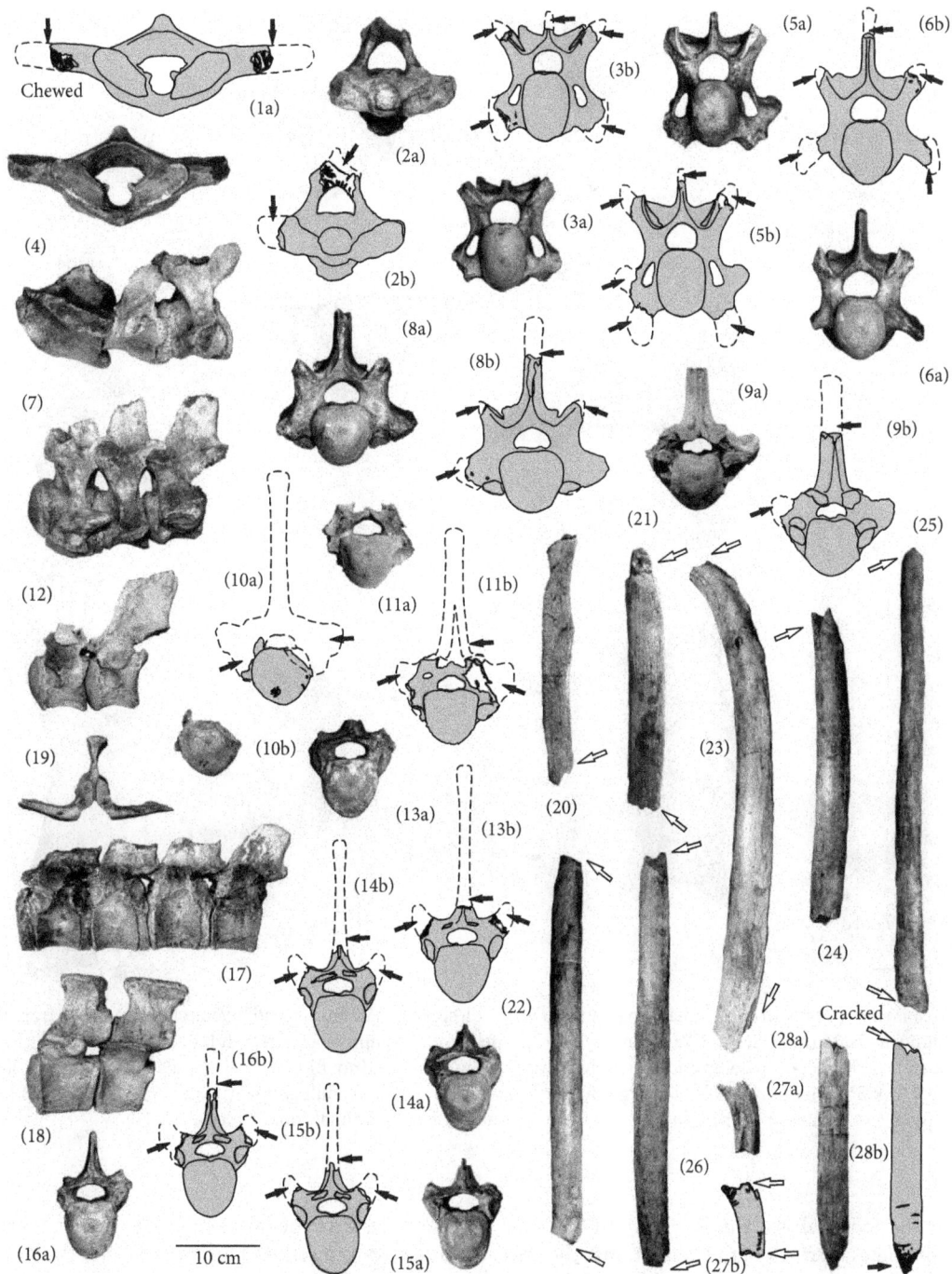

FIGURE 9: *Coelodonta antiquitatis* thoracic remains, all of which are most probably from one adolescent female animal with hyena chewing marks from the hyena open air site Bad Wildungen-Biedensteg (Hesse, NW-Germany). (1) First cervical vertebra (atlas) (no. Bi-52/9), caudal. (2) Second cervical vertebra (axes) (no. Bi-52/1), cranial. (3) Third cervical vertebra (no. Bi-52/11), cranial. (4) First three cervical vertebrae (no. Bi-52/107-1 to 3), lateral. (5) Sixth cervical vertebra (no. Bi-52/107-1), cranial. (6) Seventh cervical vertebra (no. Bi-52/107-2), cranial. (7) Articulated last cervical to second thoracic vertebrae (no. Bi-52/107-1 to 3), lateral. (8) First thoracic vertebra (no. 108-3), cranial. (9) Third thoracic vertebra (no. Bi-52/10m), cranial. (10) Fourth thoracic vertebra (no. Bi-52/152), cranial. (11) Second thoracic vertebra (no. Bi-52/10j), cranial. (12) Articulated second and third thoracic vertebrae (no. Bi-52/10j to m), lateral. (13) Sixth thoracic vertebra (no. Bi-52/107-1), cranial. (14) Seventh thoracic vertebra (no. Bi-52/107-2), cranial. (15) Eighth thoracic vertebra (no. Bi-52/107-3), cranial. (16) Ninth thoracic vertebra (no. Bi-52/107-4), cranial. (17) All four articulated sixth to ninth thoracic vertebrae (1)–(4) (no. Bi-52/107-1 to 4), lateral, (18) Articulated last thoracic and first lumbar vertebra (no. Bi-52/10l and 10h), lateral. (19) Lumbar vertebra neural arch (no. Bi without no.), cranial. (20) Posterior right costa fragment (no. Bi-10ad). (21) Anterior costa fragment (no. Bi-10v). (22) Middle right costa fragment (no. Bi-52/52). (23) Anterior left costa fragment (no. Bi-10q). (24) Anterior right costa fragment (no. Bi-52/100). (25) Middle left costa fragment (no. Bi-52/15). (26) Middle left costa fragment (no. Bi-52/156). (27) Upper costa fragment with chewing marks (no. Bi-52/3). (28) Anterior right costa fragment, distally chewed (no. Bi-52/3a), cranial.

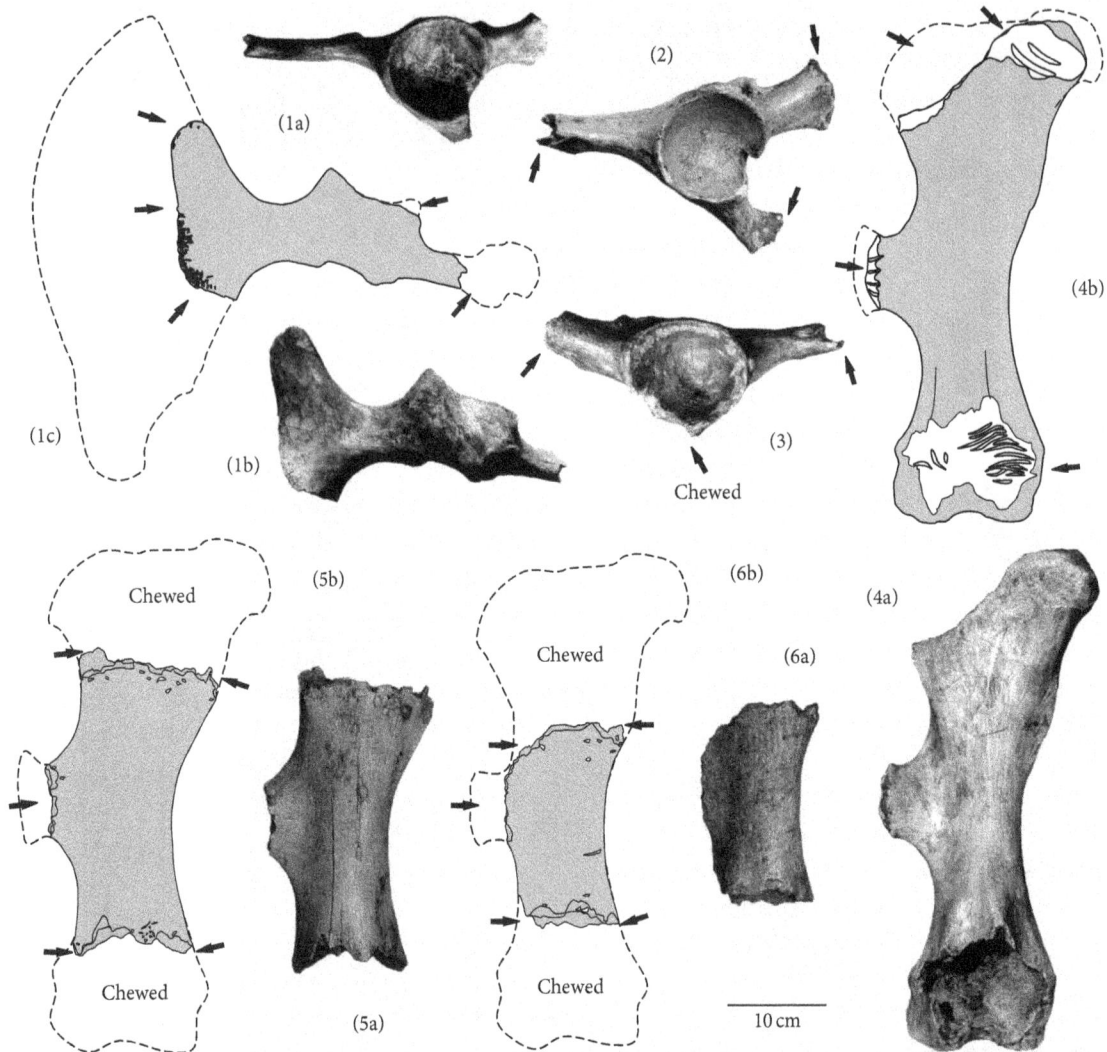

FIGURE 10: *Coelodonta antiquitatis* thoracic remains partly from one adolescent female animal with hyena chewing marks from the hyena open air site Bad Wildungen-Biedensteg (Hesse, NW-Germany). (1) Left pelvic acetabulum of an early adult to adult individual (no. Bi-52/48), (b) acetabular, (c) lateral. (2) Right pelvic ileum and acetabulum of an adult individual (no. Bi-52/82), acetabular. (3) Left pelvic ileum remain of an adult individual (no. Bi-52/10e), acetabular. (4) Right femur from an early adult to adult animal (no. Bi-10ab), cranial. (5) Left femur shaft of a grown-up animal (no. Bi-10aya), cranial. (6) Left femur shaft of an adolescent animal (no. Bi-10ea), cranial.

remains are in the material, with one (Figure 11(7)) being proximally incomplete as a result of the excavations. That one was articulated to one tibia in the stage of hyena chewing and seems to belong to the female carcass (Figure 5(b)). The distal part shows long bite scratches. The second fibula was cracked away from a tibia and was left with the middle shaft with bite marks at both ends (Figure 11(6)). Only one astragalus and calcaneus are in the material (Figure 11(8)) also most probably belonging to the hind leg of the female skeleton (Figure 5(b)). They fit perfectly together, indicated additionally by overlapping bite scratch marks which are crossing both bones. After the descriptions by Jacobshagen [34] there were three complete metatarsals (2–4) that also fit for the female skeleton (Figure 5(b)), although it is unclear whether they are from the right or left side. All vertebrae show the typical hyena chewing by the lack of nearly all processes.

They seem to be all from one nearly adult individual, indicated by a series of articulation and the similar degree of nonfusing of the caudal vertebra centrum disc. The cranial disc, in contrast, is already fused completely at all vertebrae. From the vertebral column, the first three cervical vertebrae were found connected (Figure 9(4)). Atlas (Figure 9(1)), axes (Figure 9(2)), and the third cervical vertebra (Figure 9(3)) have bite marks on the damaged processes. The next articulated vertebral column part is the vertebra from the sixth cervical to the first thoracic (Figure 9(7)). Articulated cervical vertebrae no. 6 (Figure 9(5)) and no. 7 (Figure 9(6)) and thoracic vertebra no. 1 (Figure 9(8)) are also lacking most of their processes, especially the dorsal ones. Two more articulated vertebrae are the second (Figure 9(9)) and third (Figure 9(11)) thoracic vertebrae which are heavily chewed (Figure 9(12)). The fourth thoracic vertebra (Figure 9(10))

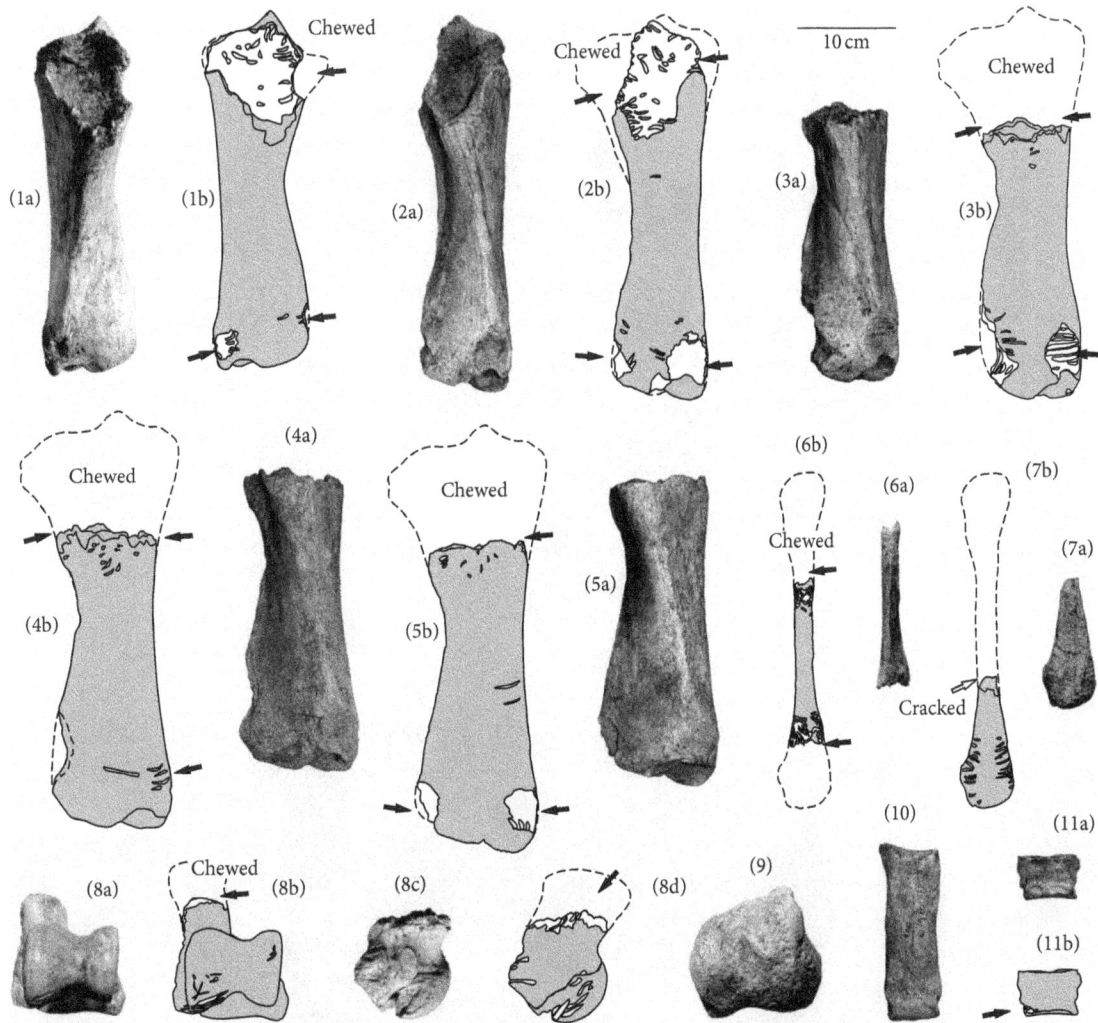

FIGURE 11: *Coelodonta antiquitatis* hind limb remains partly from one adolescent female animal with hyena chewing marks from the hyena open air site Bad Wildungen-Biedensteg (Hesse, NW-Germany). (1) Left tibia of a female individual (no. Bi-10c), cranial. (2) Right tibia from a female individual (no. Bi-52/7), cranial. (3) Right tibia from a female individual (no. Bi-10e), cranial. (4) Right tibia from a male individual (no. 52/9), cranial. (5) Right tibia from a male individual (no. Bi-52/201), cranial. (6) Left fibula shaft (no. Bi-52/4), lateral. (7) Left fibula fragment (no. Bi-52/16), lateral. (8) Articulated right calcaneus and astragal (no. Bi-10f, g), (a)-(b) dorsal, (c)-(d) lateral. (9) Right patella from an early adult to adult animal (no. Bi-52/228), cranial. (10) Right metatarsus III (no. Bi-140b), cranial. (11) First phalanx of an adult animal (no. Bi-52/101), cranial.

was only a centrum that was found in nonarticulation with other vertebrae. The complete neural arch was eaten. Parts of the left side were cut by excavation activities. The longest articulated vertebral column part exists from the sixth to ninth thoracic vertebrae (Figure 9(17)). Typical for the hyena scavenging activities are the chewed dorsal spines. Finally, the articulated last thoracic and first lumbar vertebra were found connected (Figure 9(18)). Also, the first lumbar vertebra is lacking parts of the proc. transversus. The ribs generally have no hyena bite marks, but obviously they were removed from the carcass (Figure 9(20)–(28)). All costae have cracking fractures at both ends; all joints are lacking. Only one small rib fragment (Figure 9(28)) has distally small bite marks. Nibbling by a small carnivore, such as a young hyena, wolf, or arctic fox, has caused a pointed distal end. A small fragment was used for nibbling by young hyenas ("nibbling stick" no. 3,

Figure 9(27)). The present rib fragments are from the anterior part around the forelimb, and a few are from the last thoracic vertebrae.

*Bison/Bos* (Figure 12(4)–(9)) remains consist of 13 bones (Table 9), two of which are teeth, the others being postcranial bones, which are all incomplete as a result of large carnivore activities. Most bones are limb bones, especially from the hind limbs. The teeth are two M1's, one from the upper and the other from the lower jaw. The strong tooth use indicates an individual of adult to older adult age. From the forelimb a metacarpal fragment (Figure 12(4)) was found. The metacarpal shows a typical hyena cracking preservation; the distal part has sharp edges. Most bones are from the hind limbs. Both femora were cracked in the middle of the shaft but also the distal joints were heavily eaten and nibbled (Figure 12(5)-(6)). One middle shaft of a cracked tibia

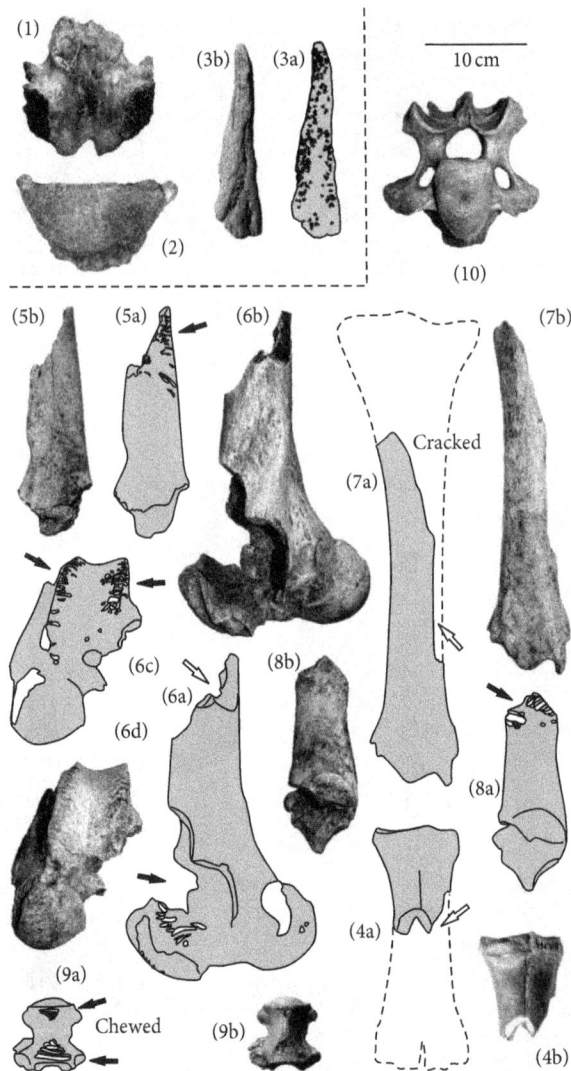

FIGURE 12: Elephant and bison/bos remains of adult animals from the hyena open air prey deposit site Bad Wildungen-Biedensteg near Hesse (NW-Germany). (a) Redrawing, (b) photo. (1)–(3) *Mammuthus primigenius*. (1) Dorsal vertebra neural arch (no. Bi-52/116), ventral. (2) Thoracic vertebra centrum (no. Bi-52/149), cranial. (3) Long bone fragment nibbling stick (no. Bi-52/222). (4)–(10) *Bos/Bison*. (4) Left metacarpal (no. Bi-10af), cranial. (5) Right femur shaft (no. Bi-10o), lateral. (6) Left distal femur (no. Bi-52/205), lateral, ventral. (7) Right tibia (no. Bi-52/236), cranial. (8) Right calcaneus (no. Bi-52/12), lateral. (9) Thoracic vertebra centrum (no. Bi-52/17), ventral. (10) Middle cervical vertebra (no. Bi without no.).

and one proximally chewed calcaneus (Figure 12(8)) and two femur fragments seem to originate of the rigth hind limb of one animal. Finally, there is one thoracic vertebra centrum (Figure 12(9)) and one cervical vertebra (Figure 12(10)). The processes were chewed, and also some deep scratch bite marks can be found ventrally. All bones belonged to one, or possibly a few adult individuals.

*Equus caballus przewalskii* Poljakoff 1881 (Figure 13(4)–(15)) consists of 19 bones, of which two are mandible fragments, one cranial fragment and a single tooth, although mainly leg remains are represented (Table 10). The one metacarpus is 236 mm in length and distally 50 mm in width (Figure 13(8)) and falls within the small Przewalskii horse metapodial osteometry (cf. [9–11, 18, 37–44]). The same is for one complete metatarsus (Figure 13(15)) with its 257 mm

length and 53 mm distal width. Also, there is the nearly complete lower jaw of a male horse (Figure 13(4)), as well as other small-sized bones from the smaller Przewalskii horse. There are bones from young horses (21%), with all others being from adult individuals (79%).

*Megaloceros giganteus* (Blumenbach 1799) (Figure 13(1)) was found with only seven bones, including one mandible fragment and three teeth, all from adult animals (Table 11). The material described and figured from Jacobshagen [34] is lost.

*Cervus elaphus* Linné 1758 (Figure 13(2)-(3)) is present with only two remains (Table 13). From the cranium, a right maxillary fragment with two $M^{1-2}$ shows the $M^2$ not in a developed state, although, the $M^3$ alveolar is opened and the tooth is in change. Another remain is a metatarsus

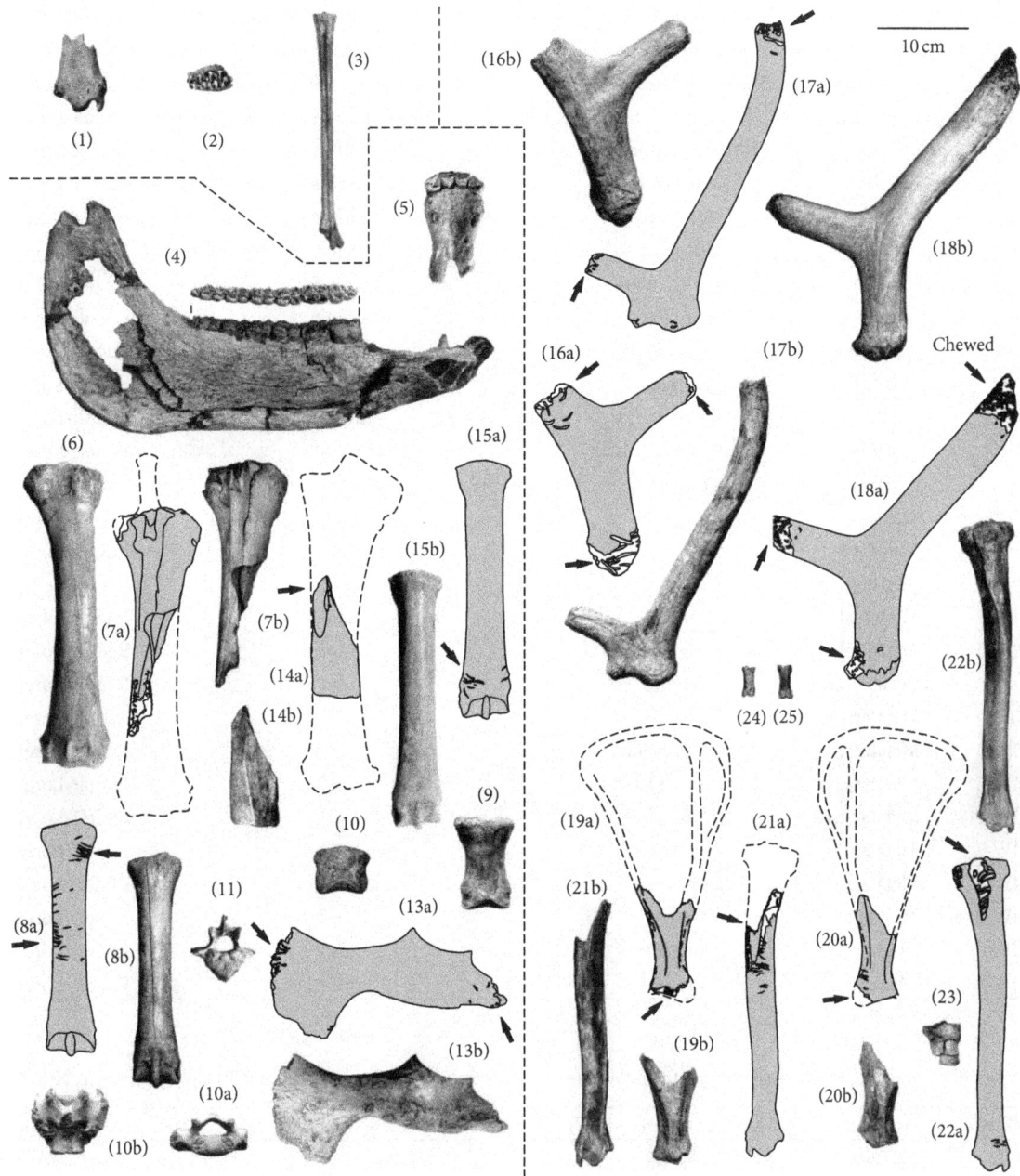

FIGURE 13: (1)-(2) Cervid and Equiid remains from the hyena open air den site Bad Wildungen-Biedensteg (Hesse, NW-Germany). (a) Redrawing, (b) photo. (1) *Megaloceros giganteus* tibia fragment (no. Bi-52/32), cranial. (2)-(3) *Cervus elaphus*. (2) Maxillary of a young animal (no. Bi-10ep), ventral. (3) Metatarsus of a young animal (no. Bi-52-113-1), cranial. (4)–(15) *Equus caballus przewalskii*. (4) Nearly complete lower jaw with both mandibles from an adult male. This jaw was broken into some pieces as a result of sediment pressure and not of hyena cracking activities (no. Bi-52/204), (a) lateral right mandibula, (b) dentition dorsal. (5) Anterior symphyseal part of a lower jaw from a juvenile less than one-year-old male (no. Bi-52-27), dorsal. (6) Radius/ulna of an adult animal (no. Bi-10aa), cranial. (7) Radius/ulna of an adult animal (no. Bi-52/50), caudal. (8) Metacarpus of an adult animal (no. Bi-52/112), cranial. (9) Phalanx 1 of an adult animal (no. Bi-52/14), cranial. (10) Phalanx 2 of an adult animal (no. Bi-52/78), cranial. (11) Lumbar vertebra no. 4 of a juvenile animal (no. Bi-10eq), cranial. (12) Anterior part of the sacrum of a juvenile animal, belonging to the vertebra of Figure 9 (no. Bi-10ad), dorsal. (13) Right pelvis remain (ileum, ischium) of an adult animal (no. Bi-10i), lateral. (14) Tibia fragment (no. Bi-52/51), caudal. (15) Metatarsus of an adult animal (no. Bi-10lt), cranial. (16)–(25) *Rangifer tarandus*. (16) Antler base of an adult animal (no. Bi-52/41). (17) Antler base of an adult animal (no. Bi-52/40). (18) Antler base of an adult animal (no. Bi-52/33), all craniolateral. (19) Right scapula (no. Bi-52/132), lateral. (20) Left scapula (no. Bi-52/126), lateral. (21) Left tibia (no. BI-52/10), cranial. (22) Right tibia (no. BI-52/151), cranial. (23) Articulated metatarsalia bones (no. Bi-52/115-4, 5, 7, 8), cranial. (24) Forelimb phalanx II of a juvenile (no. BI-52/246), dorsal. (25) Hind limb phalanx II (BI-52/246), dorsal.

TABLE 5: Bones of *Meles meles* Linné 1758 from the open air site Bad Wildungen-Biedensteg (Hesse, NW-Germany).

| No. | Coll.-No. | Bone type | Commentary | Left | Right | Age | Bite marks | Collection |
|---|---|---|---|---|---|---|---|---|
| 1 | 10ah | Cranium | Skull with lower jaws | | | Senile | | Rudolf-Lorenz-Stiftung |
| 2 | BadW-1 | Cranium | Skull with lower jaws | | | Juvenile | | Rudolf-Lorenz-Stiftung |
| 3 | 64/1 | Humerus | Without joints | | | Juvenile | | Rudolf-Lorenz-Stiftung |
| 4 | 10ap | Humerus | Without proximal joint | x | | Adult | | Rudolf-Lorenz-Stiftung |
| 5 | 10ao | Ulna | Incomplete | x | | Adult | | Rudolf-Lorenz-Stiftung |
| 6 | 10av | Ulna | complete | | x | Adult | | Rudolf-Lorenz-Stiftung |
| 7 | 10aw | Radius | Complete | x | | Adult | | Rudolf-Lorenz-Stiftung |
| 8 | 52/84 | Radius | Without joints | | | Juvenile | | Rudolf-Lorenz-Stiftung |
| 9 | 10ao | Radius | complete | | x | Adult | | Rudolf-Lorenz-Stiftung |
| 10 | 52/87 | Radius | Without joints | | | Juvenile | | Rudolf-Lorenz-Stiftung |
| 11 | 10bd | Pisiform | Complete | | | Adult | | Rudolf-Lorenz-Stiftung |
| 12 | 52/86 | Femur | Without joints | | x | Juvenile | | Rudolf-Lorenz-Stiftung |
| 13 | 52/85 | Tibia | Without joints, half | x | | Juvenile | | Rudolf-Lorenz-Stiftung |
| 14 | 10aq | Tibia | Fragment | | | Adult | | Rudolf-Lorenz-Stiftung |
| 15 | 10at | Calcaneus | Complete | x | | Adult | | Rudolf-Lorenz-Stiftung |
| 16 | 10an | Calcaneus | Complete | | x | Adult | | Rudolf-Lorenz-Stiftung |
| 17 | 10ay | Astragal | Complete | x | | Adult | | Rudolf-Lorenz-Stiftung |
| 18 | BadW-2 | Astragal | Complete | | x | Adult | | Rudolf-Lorenz-Stiftung |
| 19 | 10qr | Astragal | Complete | | x | Adult | | Rudolf-Lorenz-Stiftung |
| 20 | 10lm | Intermedium | Complete | | | Adult | | Rudolf-Lorenz-Stiftung |
| 21 | 10bf | Metatarsus | III, complete | x | | Adult | | Rudolf-Lorenz-Stiftung |
| 22 | 10bb | Metatarsus | V, complete | x | | Adult | | Rudolf-Lorenz-Stiftung |
| 23 | BadW-5 | Metatarsus | IV, complete | | x | Adult | | Rudolf-Lorenz-Stiftung |
| 24 | BadW-6 | Metatarsus | III, complete | | x | Adult | | Rudolf-Lorenz-Stiftung |
| 25 | BadW-7 | Metatarsus | II, complete | | x | Adult | | Rudolf-Lorenz-Stiftung |
| 26 | BadW-8 | Metatarsus | I, complete | | x | Adult | | Rudolf-Lorenz-Stiftung |
| 27 | BadW-2 | Phalanx II | Complete | | | Adult | | Rudolf-Lorenz-Stiftung |
| 28 | BadW-3 | Phalanx II | Complete | | | Adult | | Rudolf-Lorenz-Stiftung |
| 29 | BadW-4 | Phalanx II | Complete | | | Adult | | Rudolf-Lorenz-Stiftung |

TABLE 6: Bones of *Mustela putorius* Linnaeus 1758 from the open air prey deposit site Bad Wildungen-Biedensteg (Hesse, NW-Germany).

| No. | Coll.-No. | Bone type | Commentary | Left | Right | Age | Bite marks | Collection |
|---|---|---|---|---|---|---|---|---|
| 1 | 10bs | Cranium | Nearly complete | | | Senile | | Stadtmuseum Bad Wildungen |
| 2 | 52/247 | Pelvis | Fragment | | | Adult | | University of Marburg |

(Figure 13(3)). All remains are from possibly a single calf, approximately 1.5 years old.

*Rangifer tarandus* Linné 1758 (Figure 13(16)–(25), Table 12) is more common, with 24 remains. The rest of the bone material, such as a right metatarsus, a phalanx 1 and phalanx 2 proximal joint disc, and a right radius distal joint fit in the nonfusing of the joints to one young animal. The dropped antlers are from males and are all from sheds, which must have been collected by hyenas. Similar damages are present on the distal ends where large triangular-oval bite impact marks and elongated scratches indicate large carnivore damage (Figure 13(15)–(17)).

## 7. Discussion

*7.1. The Badger/Fox Types and Den Micromammals and Pellet Accumulators.* At open air badger den sites, typically, most skulls and massive long bones were found, although such long-term used badger loess den systems are described [45]. In those, bone accumulations are dominated by skull remains, being figured, for example, for the Schneehalle Cave (South Germany, [46]). Commonly, badgers die in their dens [46–48], explaining their bone accumulations in burrows and caves. The amount of bones, mainly of senile and very young badgers of Bad Wildungen, fit into such a scheme. Bite marks

TABLE 7: Bones of *Mammuthus primigenius* (Blumenbach 1799) from the open air site Bad Wildungen-Biedensteg (Hesse, NW-Germany).

| No. | Coll.-No. | Bone type | Commentary | Left | Right | Age | Bite marks | Collection |
|-----|-----------|-----------|------------|------|-------|-----|-----------|------------|
| 1 | 10ex | Dens | Fragment of lamella | | | Early juvenile | | Rudolf-Lorenz-Stiftung |
| 2 | 52/116 | Thoracic vertebra | Neural arch | | | ? Adult | x | Rudolf-Lorenz-Stiftung |
| 3 | 52/149 | Thoracic vertebra | Centrum | | | ? Adult | x | Rudolf-Lorenz-Stiftung |
| 4 | 52/222 | Long bone | Fragment, "nibbling stick" | x | | Adult | x | Rudolf-Lorenz-Stiftung |

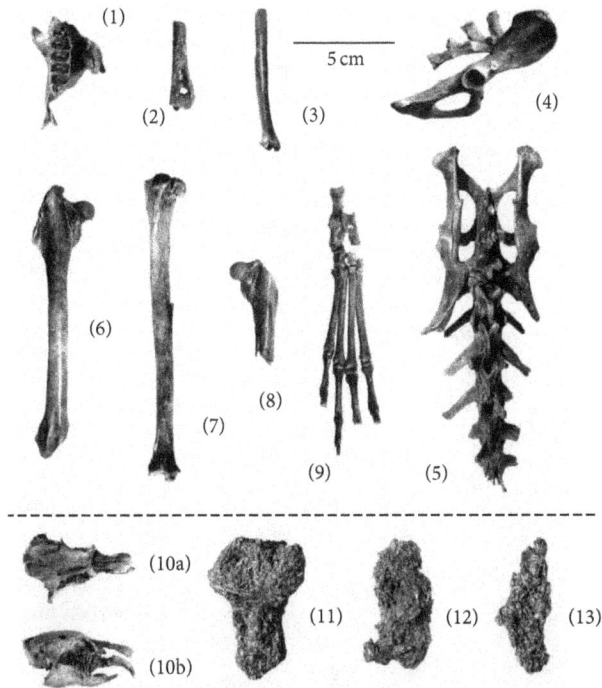

FIGURE 14: (1)–(9). *Lepus europaeus/timidus* remains from the hyena open air prey deposit site Bad Wildungen-Biedensteg (Hesse, NW-Germany) possibly belonging to one individual. (1) Left maxillary with dentition (no. Bi-63h), ventral. (2) Right humerus fragment (no. Bi-63g), cranial. (3) Right radius fragment (no. Bi-63f), cranial. (4) Pelvis (no. Bi-63a), lateral. (5) Pelvis of Figure 4 with five articulated lumbar vertebrae (no. Bi-63a), dorsal. (6) Right femur (no. Bi-63b), cranial. (7) Right tibia (no. Bi-63d), cranial. (8) Left femur fragment (no. Bi-63c), cranial. (9) Right incomplete pedal skeleton (no. Bi-63e), cranial. (10) *Spermophilus rufescens*, skull with lower jaw (52/257), (a) dorsal, (b) lateral. (11)–(13) Pellets with frog and micromammal remains (Bi-52/243, 52/244, 52/245).

and missing joints in a humerus and tibia might be the result of badger cannibalism [47] or even hyena activities. The skull and postcranial material can be referred to the Asian species *Meles meles* cf. *leucurus* (cf. [49, 50]), and the skull seems to be of male origin (cranial sexual dimorphism; see [51]). This is so far important, because this subspecies seem to have immigrated to Europe from Asia during the Late Pleistocene, where it is nowaday's extinct [50]. The badger, with its diet (cf. [52]), was not responsible for the bone accumulations of medium-sized mammals and anures, or reptiles, but of micromammals (cf. [53]), also at the Bad Wildungen-Biedensteg open sir site.

Foxes (*V. lagopus* and *V. vulpes*) might have reused the badger burrows [48]. Fox bones and skulls are typically found at those fox den sites and would explain, additionally, the presence of smaller mammal fox prey remains, especially hares and the micromammal pellets generally found at modern fox dens (cf. [48]).

Quaternary small mustelids in central Europe are rare in the fossil record outside caves (cf. [54, 55]). Their pellets can contain anure or fish bones. Frog or fish remains from Bad Wildungen seem to be partly of prey deposits of *Mustela putorius*. The small marten type is storing along small rivers or lakes, fishes, frogs, and other animals [48].

A especially high amount of frog bones must have resulted, additionally, from other large water birds and/or other predators which also left pellets and bone remains at the river and along the lake.

*7.2. Hyena Population and Cannibalism.* The hyena skulls from Bad Wildungen-Biedensteg are from female hyenas which are similar to many other skulls of central Europe (cf. [17]) and are anatomically interesting in their dentition (partly absence of $M^1$), but fall into the variability of *C. c. spelaea*. A brain case, two incomplete limb bone shafts, a left radius, and a left femur are fitting for a single cub, which are very small in their proportions. They also have bite marks and must have been chewed, as compared to other cannibalistic damaged hyena long bone finds from Europeans caves (cf. [11, 22, 23, 25, 27, 56]). Their proportions fit best for a very young cub, maybe only of a few days or weeks in age, compared to the cub material from the Srbsko-Chlum-Komin Cave, Czech Republic [11]. The young hyena was possibly eaten cannibalistically, possibly by another cub, due to competition (cf. modern African hyenas in [57–59]). All bones of the Bad Wildungen hyena population and even the skulls have nibbling, chewing, and cracking marks of hyenas. The lack of the jugals and temporal parts of the skulls is the result of cracking the lower jaws from their joints, which is demonstrated for many skull finds in Europe (cf. [17]). The scavenging of their own species leaves dominantly cranial remains at not only den sites, such as the skulls, lower jaws, and teeth, but also the long bones (e.g., modern spotted hyenas, [60, 61]). Scavenging of their own is best documented in the Srbsko-Chlum-Komin Cave [11]. The dominance of cranial material at Bad Wildungen hyena den site is comparable not only to the German Perick Caves and Rösenbeck Cave and other Sauerland Karst hyena dens, but also to other caves, such as the Czech Sloup Cave, Výpustek, in the Bohemian and Moravian Karst regions [7, 17, 56]. Vertebrae and rib bones are underrepresented at most hyena den sites (especially at birthing dens and prey storage den types), the exceptions being where complete articulated skeletons are found at prey storage sites, such as were found

TABLE 8: Bones of *Coelodonta antiquitatis* (Blumenbach) from the open air site Bad Wildungen-Biedensteg (Hesse, NW-Germany).

| No. | Coll.-No. | Bone type | Commentary | Left | Right | Age | Bite marks | Collection |
|---|---|---|---|---|---|---|---|---|
| 1 | 10ac | Cranium | Middle part with $dm^{1-3}$, $M^1$ dentition | | | Early juvenile | x | Rudolf-Lorenz-Stiftung |
| 2 | 52/37 | Mandible | Milk dentition, with $dm_3$, $M_1$ | | x | Early juvenile | x | Rudolf-Lorenz-Stiftung |
| 3 | 52/38 | Mandible | Milk dentition, with $dm_{1-3}$, $M_1$ | x | | Early juvenile | x | Rudolf-Lorenz-Stiftung |
| 4 | Ma 1 | Dens | Milk tooth, upper jaw | | | Early juvenile | | University of Marburg |
| 5 | Ma 2 | Dens | Milk tooth, upper jaw | | | Early juvenile | | University of Marburg |
| 6 | Ma 3 | Dens | Milk tooth, upper jaw | | | Early juvenile | | University of Marburg |
| 7 | Ma 4 | Dens | P3 | | x | Early adult | | University of Marburg |
| 8 | Ma 5 | Dens | P4 | | x | Early adult | | University of Marburg |
| 9 | Ma 6 | Dens | M1 | | x | Early adult | | University of Marburg |
| 10 | Ma 7 | Dens | M2 | x | | Early adult | | University of Marburg |
| 11 | Ma 8 | Dens | M3 | x | | Early adult | | University of Marburg |
| 12 | 10l | Scapula | Fragment | x | | ? Adult | x | Rudolf-Lorenz-Stiftung |
| 13 | 52/20 | Scapula | Without distal joint | x | | Adult | x | Rudolf-Lorenz-Stiftung |
| 14 | 52/200 | Scapula | Incomplete | | x | ? Adult | | Rudolf-Lorenz-Stiftung |
| 15 | 52/88 | Scapula | Fragment | | | ? Adult | | Rudolf-Lorenz-Stiftung |
| 16 | 180c | Humerus | Incomplete | | x | Adult | x | (Mentioned in [34], missing) |
| 17 | 10v | Humerus | Incomplete | x | | Adult | x | Rudolf-Lorenz-Stiftung |
| 18 | 52/47, 42 | Ulna/radius | Shafts, articulated | x | | Juvenile | x | Rudolf-Lorenz-Stiftung |
| 19 | 52/116, 111 | Ulna/radius | Shafts, articulated | | x | Early adult | x | Rudolf-Lorenz-Stiftung |
| 20 | 52/143 | Ulna | Shaft | | x | ? Adult | x | Rudolf-Lorenz-Stiftung |
| 21 | 10p | Ulna | Shaft | | x | ? Adult | x | Rudolf-Lorenz-Stiftung |
| 22 | 52/53 | Ulna | Shaft | | x | Adult | x | Rudolf-Lorenz-Stiftung |
| 23 | 10a | Ulna | Shaft | | x | Adult | x | Rudolf-Lorenz-Stiftung |
| 24 | 52/49 | Radius | Without distal joint | x | | Adult | x | Rudolf-Lorenz-Stiftung |
| 25 | 52/44 | Radius | Shaft | | x | Adult | x | Rudolf-Lorenz-Stiftung |
| 26 | 52/30 | Radius | Proximal joint | x | | Adult | x | Rudolf-Lorenz-Stiftung |
| 27 | 52/224 | Radius | Distal joint | | x | Early adult | x | Rudolf-Lorenz-Stiftung |
| 28 | 10a | Radius | Proximal joint | | x | Adult | x | Rudolf-Lorenz-Stiftung |
| 29 | 52/235 | Intermedium | Nearly complete | | x | Adult | | Rudolf-Lorenz-Stiftung |
| 30 | 52/34 | Carpale 3 | Nearly complete | | x | Adult | | Rudolf-Lorenz-Stiftung |
| 31 | Ma 11 | Metacarpale 3 | Nearly complete | | | Adult | | University of Marburg |
| 32 | Ma 12 | Metacarpale 3 | Nearly complete | | | Adult | | University of Marburg |
| 33 | 52/101 | Phalanx | Complete | | | Adult | | Rudolf-Lorenz-Stiftung |
| 34 | 52/43 | Femur | Shaft | x | | Juvenile | x | Rudolf-Lorenz-Stiftung |
| 35 | 52/153 | Femur | Shaft, fragment | x | | | | Rudolf-Lorenz-Stiftung |
| 36 | 10ab | Femur | Incomplete | | x | Adult | x | Rudolf-Lorenz-Stiftung |
| 37 | 10ea | Femur | Shaft | x | | Early adult | x | Rudolf-Lorenz-Stiftung |
| 38 | 10aya | Femur | Shaft | x | | Early adult | x | Rudolf-Lorenz-Stiftung |
| 39 | 52/228 | Patella | Complete | | x | Adult | | Rudolf-Lorenz-Stiftung |
| 40 | 52/7 | Tibia | Incomplete | | x | Adult | x | Rudolf-Lorenz-Stiftung |
| 41 | 52/201 | Tibia | Without proximal joint | | x | Adult | x | Rudolf-Lorenz-Stiftung |
| 42 | 10c | Tibia | Incomplete | x | | Adult | x | Rudolf-Lorenz-Stiftung |
| 43 | 52/9 | Tibia | Without proximal joint | | x | Adult | x | Rudolf-Lorenz-Stiftung |
| 44 | 10t | Tibia | Without proximal joint | | x | Adult | x | Rudolf-Lorenz-Stiftung |
| 45 | 52/4 | Fibula | Distal joint | x | | Adult | x | Rudolf-Lorenz-Stiftung |
| 46 | 52/16 | Fibula | Shaft | x | | Adult | x | Rudolf-Lorenz-Stiftung |

Recycling of Badger/Fox Burrows in Late Pleistocene Loess by Hyenas at the Den Site Bad Wildungen-Biedensteg
(NW, Germany): Woolly Rhinoceros Killers and Scavengers in a Mammoth Steppe Environment of Europe

29

TABLE 8: Continued.

| No. | Coll.-No. | Bone type | Commentary | Left | Right | Age | Bite marks | Collection |
|---|---|---|---|---|---|---|---|---|
| 47 | 10f | Calcaneus | Incomplete | | x | Adult | x | Rudolf-Lorenz-Stiftung |
| 48 | 10g | Astragalus | Incomplete | | x | Adult | x | Rudolf-Lorenz-Stiftung |
| 49 | 52/140b | Metatarsus | III, complete | | x | Adult | | Rudolf-Lorenz-Stiftung |
| 50 | Ma 13 | Metatarsale 2 | Proximal joint | | | Adult | | University of Marburg |
| 51 | Ma 14 | Metatarsale 3 | Nearly complete | | | Adult | | University of Marburg |
| 52 | Ma 15 | Metatarsale 4 | Nearly complete | | | Adult | | University of Marburg |
| 53 | 52/48 | Pelvis | Incomplete | x | | Adult | x | Rudolf-Lorenz-Stiftung |
| 54 | 52/82 | Pelvis | Incomplete | | x | Adult | x | Rudolf-Lorenz-Stiftung |
| 55 | 52/13 | Pelvis | Ilium, fragment | x | | Adult | x | Rudolf-Lorenz-Stiftung |
| 56 | 10e | Pelvis | Incomplete | x | | Adult | x | Rudolf-Lorenz-Stiftung |
| 57 | 52/9 | Cervical vertebra | Atlas | | | Early adult | x | Rudolf-Lorenz-Stiftung |
| 58 | 52/1 | Cervical vertebra | Axes | | | Early adult | x | Rudolf-Lorenz-Stiftung |
| 59 | 52/11 | Cervical vertebra | No. 3 | | | Early adult | x | Rudolf-Lorenz-Stiftung |
| 60 | 52/18 | Cervical vertebra | No. 5 | | | Early adult | | Rudolf-Lorenz-Stiftung |
| 61 | 52/107-1 | Cervical vertebra | No. 6 | | | Early adult | x | Rudolf-Lorenz-Stiftung |
| 62 | 52/107-2 | Cervical vertebra | No. 7 | | | Early adult | x | Rudolf-Lorenz-Stiftung |
| 63 | 52/107-3 | Thoracic vertebra | No. 1 | | | Early adult | x | Rudolf-Lorenz-Stiftung |
| 64 | 10m | Thoracic vertebra | No. 2 | | | Early adult | x | Rudolf-Lorenz-Stiftung |
| 65 | 10j | Thoracic vertebra | No. 3 | | | Early adult | x | Rudolf-Lorenz-Stiftung |
| 66 | 52/152 | Thoracic vertebra | Centrum, No. 4 | | | Early adult | x | Rudolf-Lorenz-Stiftung |
| 67 | 52/108-1 | Thoracic vertebra | No. 6 | | | Early adult | x | Rudolf-Lorenz-Stiftung |
| 68 | 52/1808-2 | Thoracic vertebra | No. 7 | | | Early adult | x | Rudolf-Lorenz-Stiftung |
| 69 | 52/108-3 | Thoracic vertebra | No. 8 | | | Early adult | x | Rudolf-Lorenz-Stiftung |
| 70 | 52/108-4 | Thoracic vertebra | No. 9 | | | Early adult | x | Rudolf-Lorenz-Stiftung |
| 71 | 10l | Thoracic vertebra | No. 18 | | | Early adult | x | Rudolf-Lorenz-Stiftung |
| 72 | 10h | Lumbar vertebra | No. 1 | | | Early adult | x | Rudolf-Lorenz-Stiftung |
| 73 | 10r | Lumbar vertebra | Neural arch | | | Early adult | x | Rudolf-Lorenz-Stiftung |
| 74 | 52/3 | Costa | Fragment | | | ? | x | Rudolf-Lorenz-Stiftung |
| 75 | 52/5 | Costa | Fragment | | | ? | | Rudolf-Lorenz-Stiftung |
| 76 | 52/156 | Costa | Fragment | | | Early adult | | Rudolf-Lorenz-Stiftung |
| 77 | 52/58 | Costa | Anterior, 2, distally incomplete | x | | Early adult | | Rudolf-Lorenz-Stiftung |
| 78 | 52/57 | Costa | Middle, approx. 6 to 8 | x | | Early adult | x | Rudolf-Lorenz-Stiftung |
| 79 | 52/52 | Costa | Middle, approx. 4–6 | | x | Early adult | x | Rudolf-Lorenz-Stiftung |
| 80 | 52/15 | Costa | Middle, approx. 7–9 | x | | Early adult | x | Rudolf-Lorenz-Stiftung |
| 81 | 52/100 | Costa | Anterior, approx. 2–3 | | x | Early adult | x | Rudolf-Lorenz-Stiftung |
| 82 | 52/3a | Costa | Anterior, approx. 3-4 | | x | Early adult | x | Rudolf-Lorenz-Stiftung |
| 83 | 10q | Costa | Anterior, approx. 4–6 | x | | Early adult | x | Rudolf-Lorenz-Stiftung |
| 84 | 10v | Costa | Anterior, approx. 3-4 | x | | Early adult | x | Rudolf-Lorenz-Stiftung |
| 85 | 10ad | Costa | Posterior | | x | Early adult | x | Rudolf-Lorenz-Stiftung |

at the Czech Výpustek Cave, Koněprusy Cave and Srbsko-Chlum-Komin Cave [9, 40].

*7.3. Hyena Den Type and Recycling of Badger/Fox Dens.*
Hyena dens are identified starting in the Pliocene to Middle Pleistocene (e.g., [12, 62, 63]). In the Late Pleistocene the hyena den site record is much higher (e.g., [3–6, 8, 17, 64–66]) and more details about the "den type" can be studied.

The large bone enrichment at Bad Wildungen was already identified as a product of the activities of *C. c. spelaea* [35]. The comparison of different Late Pleistocene *C. c. spelaea* hyena cave and open air den sites in Europe allows a classification of the den type, by separating three main age classes: (1) cubs, (2) adolescents, and (3) adult-senile individuals (Figure 15). The high presence of cubs indicates, similarly as in modern spotted hyenas [57, 67–69], birthing dens. Other

TABLE 9: Bones of *Bison priscus* (Bojanus 1827) from the open air site Bad Wildungen-Biedensteg (Hesse, NW-Germany).

| No. | Coll.-No. | Bone type | Commentary | Left | Right | Age | Bite marks | Collection |
|---|---|---|---|---|---|---|---|---|
| 1 | / | Dens | M1, upper jaw | | | | | (Mentioned in [34], missing) |
| 2 | / | Dens | M1, lower jaw | | | | | (Mentioned in [34], missing) |
| 3 | / | Scapula | | | | | | (Mentioned in [34], missing) |
| 4 | BadW-9 | Scapula | Proximal half | | | Adult | | Rudolf-Lorenz-Stiftung |
| 5 | 10af | Metacarpus | Proximal joint | x | | Adult | x | Rudolf-Lorenz-Stiftung |
| 6 | / | Carpale 3 + 4 | | | | | | (Mentioned in [34], missing) |
| 7 | 52/205 | Femur | Distal joint and shaft fragment | x | | Adult | x | Rudolf-Lorenz-Stiftung |
| 8 | 10o | Femur | Shaft | | x | Adult | x | Rudolf-Lorenz-Stiftung |
| 9 | 10k | Femur | Distal joint, fragment | | x | Adult | x | Stadtmuseum Bad Wildungen |
| 10 | 52/236 | Tibia | Without proximal joint | | x | Adult | x | Museum Korbach, (Stadtmuseum Bad Wildungen) |
| 11 | 52/12 | Calcaneus | Nearly complete | | x | Adult | x | Rudolf-Lorenz-Stiftung |
| 12 | 52/17 | Thoracic vertebra | Centrum | | | Adult | x | Rudolf-Lorenz-Stiftung |

TABLE 10: Bone material list of *Equus caballus przewalskii* Poljakoff 1881 from the open air prey deposit site Bad Wildungen-Biedensteg (Hesse, NW-Germany).

| No. | Coll.-No. | Bone type | Commentary | Left | Right | Age | Bite marks | Collection |
|---|---|---|---|---|---|---|---|---|
| 1 | 52/221 | Mandibula | Nearly complete | x | | Adult | | Rudolf-Lorenz-Stiftung |
| 2 | 52/27 | Mandibula | Anterior part, male | | | Juvenile | | Rudolf-Lorenz-Stiftung |
| 3 | 52/203 | Cranium | Occipital, fragment | | | | | Rudolf-Lorenz-Stiftung |
| 4 | 52/147 | Dens | C, male | | | Adult | | Rudolf-Lorenz-Stiftung |
| 5 | 52/50 | Ulna/radius | Incomplete | | x | Adult | x | Rudolf-Lorenz-Stiftung |
| 6 | 10aa | Ulna/radius | Nearly complete | | x | Adult | x | Rudolf-Lorenz-Stiftung |
| 7 | 52/112 | Metacarpus, length = 236 mm, distal width = 50 mm | Nearly complete | x | | Adult | x | Rudolf-Lorenz-Stiftung |
| 8 | 52/155 | Metacarpus | Distal joint | x | | Adult | | Rudolf-Lorenz-Stiftung |
| 9 | 52/14 | Phalanx 1 | Complete | | | | | Rudolf-Lorenz-Stiftung |
| 10 | 52/78 | Phalanx 2 | Complete | | | | | Rudolf-Lorenz-Stiftung |
| 11 | 10lt | Metatarsus, length = 257 mm, distal width = 53 mm | Complete | | x | Adult | x | Rudolf-Lorenz-Stiftung |
| 12 | 52/51 | Tibia | Fragment | | | | x | Rudolf-Lorenz-Stiftung |
| 13 | 52/28 | Pelvis | Fragment, ilium | x | | Adult | x | Rudolf-Lorenz-Stiftung |
| 14 | 10i | Pelvis | Fragment, ilium | | x | Adult | x | Rudolf-Lorenz-Stiftung |
| 15 | 52/131 | Cervical vertebra | Fragment, neural arch | | | Adult | | Rudolf-Lorenz-Stiftung |
| 16 | 52/202 | Cervical vertebra | Fragment, neural arch | | | Adult | x | Rudolf-Lorenz-Stiftung |
| 17 | 10eq | Lumbar vertebra | No. 4, without processi | | | Juvenile | x | Rudolf-Lorenz-Stiftung |
| 18 | 10ad | Pelvis | Sacrum, incomplete | | | Juvenile | x | Rudolf-Lorenz-Stiftung |
| 19 | 52/157 | Costa | Fragment | | | | x | Rudolf-Lorenz-Stiftung |

indicators for such birthing dens are "nibbling sticks." At Bad Wildungen there are three such chewed bone fragments: one of a mammoth, whose bone fragments are found at birthing dens [70] for teething purposes of hyena cubs [7]; the other nibbling sticks are from *Coelodonta* and *Ursus* bone fragments. These birthing dens are generally recycled from medium-sized carnivore, such as porcupines, or by hyenas own excavated burrows, which can be situated nearby commuting dens (cf. modern in [71]). Bad Wildungen must have also been this type of den, where higher amounts of prey remains were accumulated, or even stored (prey storage den type). Similar large bone accumulations at commuting den sites have been reported in Africa from *C. c. crocuta* (cf. [61, 68, 71–81]).

TABLE 11: Bones of *Megaloceros giganteus* (Blumenbach 1799) from the open air site Bad Wildungen-Biedensteg (Hesse, NW-Germany).

| No. | Coll.-No. | Bone type | Commentary | Left | Right | Age | Bite marks | Collection |
|---|---|---|---|---|---|---|---|---|
| 1 | / | Mandibula | Fragment with M1–3 | | x | | | (Mentioned in [34], missing) |
| 2 | / | Dens | P1, upper jaw | | | | | (Mentioned in [34], missing) |
| 3 | / | Dens | M2, upper jaw | x | | | | (Mentioned in [34], missing) |
| 4 | / | Dens | M3, upper jaw | x | | | | (Mentioned in [34], missing) |
| 5 | / | Cervical vertebra | Atlas | | | | | (Mentioned in [34], missing) |
| 6 | / | Cervical vertebra | Axes | | | | | (Mentioned in [34], missing) |
| 7 | 52/32 | Tibia | Distal joint | | x | Adult | x | Rudolf-Lorenz-Stiftung |

TABLE 12: Bones of *Rangifer tarandus* Linné 1758 from the open air site Bad Wildungen-Biedensteg (Hesse, NW-Germany).

| No. | Coll.-No. | Bone type | Commentary | Left | Right | Age | Bite marks | Collection |
|---|---|---|---|---|---|---|---|---|
| 1 | / | Dens | — | | | | | (Mentioned in [34], missing) |
| 2 | / | Dens | — | | | | | (Mentioned in [34], missing) |
| 3 | 52/40 | Antler | Dropped antler with base, fragment | x | | Adult | x | Rudolf-Lorenz-Stiftung |
| 4 | 52/41 | Antler | Dropped antler with base, fragment | | x | Adult | x | Rudolf-Lorenz-Stiftung |
| 5 | 52/33 | Antler | Dropped antler with base, fragment | x | | Adult | x | Rudolf-Lorenz-Stiftung |
| 6 | 52/132 | Scapula | Incomplete | | x | Adult | x | Rudolf-Lorenz-Stiftung |
| 7 | 52/126 | Scapula | Incomplete | x | | Adult | x | Rudolf-Lorenz-Stiftung |
| 8 | 52/115-1 | Ulna | Proximal joint | | | Juvenile | | Rudolf-Lorenz-Stiftung |
| 9 | 52/115-3 | Radius | Distal joint | | x | Juvenile | | Rudolf-Lorenz-Stiftung |
| 10 | 52/115-4 | Radiale | Complete | | x | Juvenile | | Rudolf-Lorenz-Stiftung |
| 11 | 52/115-5 | Intermedium | Complete | | x | Juvenile | | Rudolf-Lorenz-Stiftung |
| 12 | 52/115-8 | Carpale | Complete | | x | Juvenile | | Rudolf-Lorenz-Stiftung |
| 13 | 52/115-7 | Carpale 4 | Complete | | x | Juvenile | | Rudolf-Lorenz-Stiftung |
| 14 | 52/117 | Metacarpus | Distal joint | | | Juvenile | | Rudolf-Lorenz-Stiftung |
| 15 | 52/52 | Pelvis | Acetabulum, fragment | | | Adult | | Rudolf-Lorenz-Stiftung |
| 16 | 52/57 | Pelvis | Acetabulum, fragment | | | Adult | | Rudolf-Lorenz-Stiftung |
| 17 | 52/115-2 | Phalanx 1 | Without proximal joint, forelimb | | | Juvenile | | Rudolf-Lorenz-Stiftung |
| 18 | 52/115-6 | Phalanx 2 | Proximal joint, forelimb | | | Juvenile | | Rudolf-Lorenz-Stiftung |
| 19 | 52/74 | Tibia | Fragment, distal | | | Juvenile | | Rudolf-Lorenz-Stiftung |
| 20 | 52/151 | Tibia | Nearly complete | | x | Adult | x | Rudolf-Lorenz-Stiftung |
| 21 | 52/10 | Tibia | Without proximal joint | x | | Adult | x | Rudolf-Lorenz-Stiftung |
| 22 | 10lz | Phalanx 1 | Without proximal joint, hind limb | | | Juvenile | | Rudolf-Lorenz-Stiftung |
| 23 | 52/246 | Phalanx 1 | Without proximal joint, hind limb | | | Juvenile | | Rudolf-Lorenz-Stiftung |
| 24 | 4.4/54 | Phalanx 1 | Complete | | | Adult | | Rudolf-Lorenz-Stiftung |

*7.4. Hyena Den Marking.* In most cases, pellets of the Late Pleistocene spotted hyenas have repeating shapes, which were found recently at several reported den sites [3, 5–7, 11, 22, 41, 82]. Exact documented excrement markings on a gypsum karst open air den were recently published at the site Westeregeln, Central Germany [9]. A first terminology was published for the pellet shape types [44]. The hyena pellets from Bad Wildungen fall within the hyena pellet shape types. Several smaller pellets are attached to each other, forming spindle-like, or irregular accumulated aggregations, similar to modern African spotted hyena excrements [9]. Modern spotted hyenas are using faecal pellets to mark their territory, especially their den sites [83]. The Ice Age spotted hyenas must have done the same. Well

documented examples are found in Germany at two open air sites: Bad Wildungen-Biedensteg [35] and the gypsum karst site Morschen-Konnefeld [84]. Similar abundant pellets are found in caves of France [6] and Czech Republic [5].

*7.5. Bone Assemblage and Fauna Statistics.* The high amount (10%) of hyena bone remains is typical for Late Pleistocene hyena dens (e.g., [8, 11, 65, 66]).

A high percentage of hyena prey bone remains at the site Bad Wildungen-Biedensteg (Figure 16) do not represent the real percentages of the prey. It is more demonstrated, for example, at other hyena open air sites, as a result of taphonomy and selection [9]. The bones of the woolly rhinoceros are extremely massive, and, in contrast to nearly all other large

TABLE 13: Bones of *Cervus elaphus* Linné 1758 from the open air site Bad Wildungen-Biedensteg (Hesse, NW-Germany).

| No. | Coll.-No. | Bone type | Commentary | Left | Right | Age | Bite marks | Collection |
|---|---|---|---|---|---|---|---|---|
| 1 | 10ep | Cranium | Maxillar, with M1-2 | | x | Juvenile | | Rudolf-Lorenz-Stiftung |
| 2 | 52-113-1 | Metatarsus | Without distal joint | | x | Juvenile | | Rudolf-Lorenz-Stiftung |

TABLE 14: Bones of *Lepus* sp. from the open air site Bad Wildungen-Biedensteg (Hesse, NW-Germany).

| No. | Coll.-No. | Bone type | Commentary | Left | Right | Age | Bite marks | Collection |
|---|---|---|---|---|---|---|---|---|
| 1 | 98 | Cranium | Brain case, frontals, parietals, incomplete | | | Adult | | University Marburg |
| 2 | 14 | Cranium | Maxillar | x | | Adult | | University Marburg |
| 3 | 63h | Cranium | Maxillar | | x | Adult | | Rudolf-Lorenz-Stiftung |
| 4 | 12 | Mandibula | Incomplete | x | | Adult | | University Marburg |
| 5 | 11 | Mandibula | Incomplete | | x | Juvenile | | University Marburg |
| 6 | 63g | Humerus | Half, from skeleton | | x | Adult | | Rudolf-Lorenz-Stiftung |
| 7 | 63f | Radius | Half, from skeleton | | x | Adult | | Rudolf-Lorenz-Stiftung |
| 8 | 15 | Radius/ulna | Without joints | | x | Adult | x | University Marburg |
| 9 | 63b | Femur | Distal joint incomplete, from skeleton | | x | Adult | | Rudolf-Lorenz-Stiftung |
| 10 | 63c | Femur | Half without distal joint, from skeleton | x | | Adult | x | Rudolf-Lorenz-Stiftung |
| 11 | 10 | Femur | Without joints | x | | Juvenile | | University Marburg |
| 12 | 63d | Tibia | Proximal joint incomplete, from skeleton | | x | Adult | | Rudolf-Lorenz-Stiftung |
| 13 | 9 | Tibia | Without proximal joint | | x | Juvenile | | University Marburg |
| 14 | 13 | Tibia | Without middle shaft | | x | Adult | x | University Marburg |
| 15 | 63e | Pes | Nearly complete articulated, from skeleton | | x | Adult | | Rudolf-Lorenz-Stiftung |
| 16 | 52-105c | Calcaneus | Complete | | x | Adult | | Rudolf-Lorenz-Stiftung |
| 17 | 3 | Pelvis | Fragment, acetabulum | x | | Adult | | University Marburg |
| 18 | 5 | Pelvis | Fragment, acetabulum | x | | Adult | | University Marburg |
| 19 | 52/10 | Femur | Incomplete | x | | Juvenile | | University Marburg |
| 20 | 52/248 | Lumbar vertebra | Incomplete | | | Juvenile | | University Marburg |
| 21 | 63a | Pelvis and lumbalvertebra | Articulated from skeleton | | | Adult | | Rudolf-Lorenz-Stiftung |
| 22 | 52/249 | Calcaneus | Incomplete | | x | Juvenile | | Rudolf-Lorenz-Stiftung |
| 23 | 52/252 | Pes | Incomplete, articulated | x | | Adult | | Rudolf-Lorenz-Stiftung |
| 24 | 52/253 | Metatarsus IV | Complete | x | | Adult | | Rudolf-Lorenz-Stiftung |
| 25 | 52/254 | Astragalus | Complete | | x | Adult | | Rudolf-Lorenz-Stiftung |
| 26 | 52/251 | Metacarpus | 3 incomplete | | | Juvenile | | Rudolf-Lorenz-Stiftung |
| 27 | 52/256 | Ulnar | 2 complete | x | x | Adult | | Rudolf-Lorenz-Stiftung |
| 28 | 52/255 | Tarsalia | 2 complete | | x | Adult | | Rudolf-Lorenz-Stiftung |

mammal bones, completely filled with the spongiosa. The long bones were difficult or impossible to crack and hyenas always left, in a last stage (stage 3), the bone shaft of long bones or massive bones which are classified in three damage stages [10].

The open air site Bad Wildungen-Biedensteg has delivered only a very few mammoth bones (2% of the prey bones) which are typical at middle high mountainous hyena dens of Europe, where mammoths seem to have been absent or rare [7]. Hyenas specialized there on cave bear scavenging ([42], Figure 16). The amount of Przewalski horse remains (8%) is as usual high. In most open air sites and middle mountainous elevated European caves the small Przewalski horse is the

main or second dominant prey (up to 50%; [7, 9–11, 18, 37, 40–44]). If all the small carnivores are excluded from the statistics, then the horse remains represent the second largest prey (cf. [85]). Bones of those horses are recorded with small proportioned forms (see metapod discussion) attributed to *E. c. przewalskii* in Germany or Czech Republic at other hyena den sites of early to middle Late Pleistocene age [7, 85]. Late Palaeolithic archaeological sites have the youngest records from the Late Magdalénian [86] or Epipalaeolithic/Early Mesolithic [87]. Finally, trackways have been described from the German Volcanic ashes of the Laacher Volcano to be of Przewalski horse origin [37, 88]. Additionally, archaeologists have discussed intensive horse figurations in cave and mobile

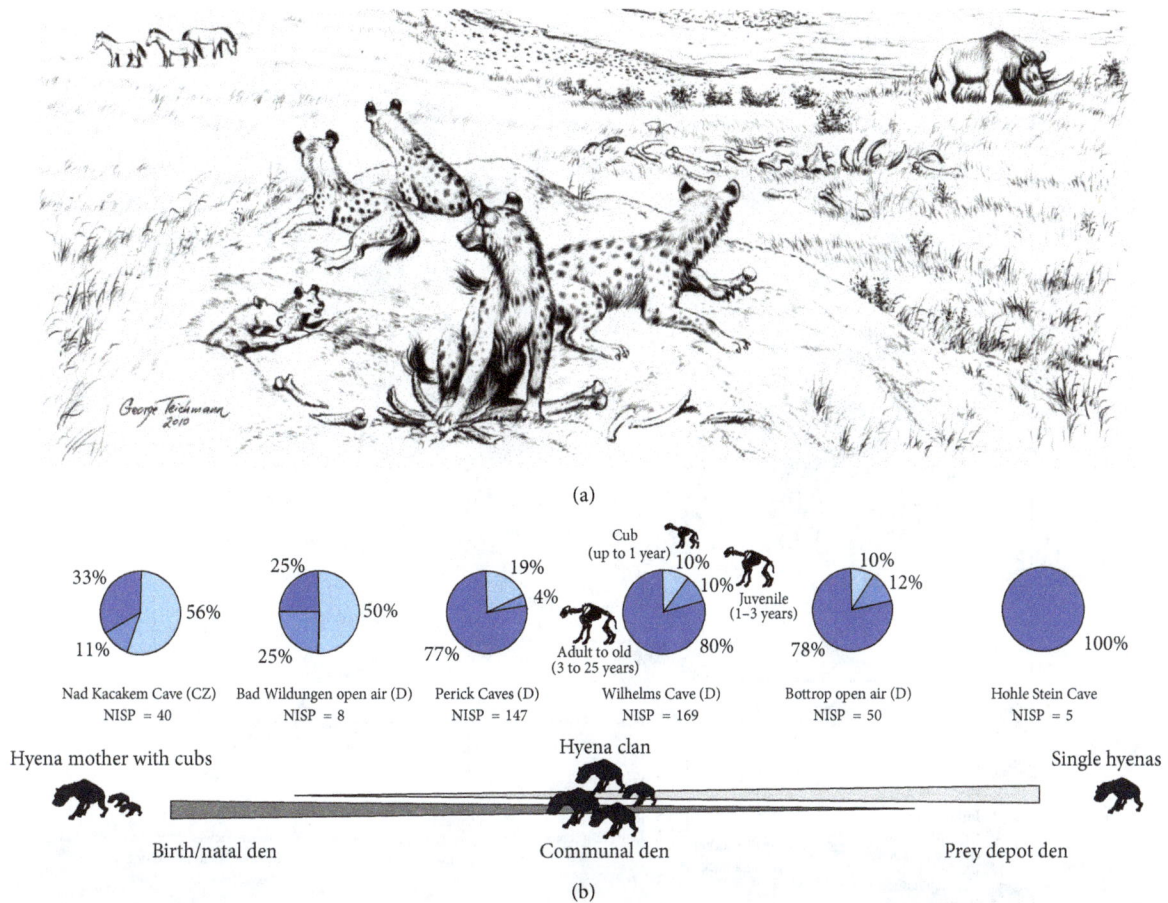

FIGURE 15: (a) Late Pleistocene spotted hyena clan at the birth/natal den at the open air loess site Bad Wildungen-Biedensteg (Hesse, NW-Germany). (b) Population structure comparisons of hyena den sites of central Europe in which Bad Wildungen falls intermediately within the birth/natal den and communal sites.

art and identified also the horses by the unique "M-sign" (resulting from fur colour and fur change) and "uplifted mane" (only in those horses, not in modern present horses) to represent obviously Przewalski horses within the Late Palaeolithic times (cf. e.g., [86, 89]) and especially within the cold periods of the Late Pleistocene.

*7.6. Woolly Rhinoceros as Main Prey for Hyenas.* Most remains are from the woolly rhinoceros (32%), which corresponds well to several other northern Germany open air hyena den sites, such as Bottrop, Westeregeln, or cave sites on the mountain slope regions, such as Hohle Stein Cave or Teufelskammer Cave ([9, 10, 22, 82], Figure 16). All bones have medium to massive nibbling, chewing, and gnawing marks, mainly produced by the Ice Age spotted hyenas, as compared to other den sites [10, 91] and modern spotted hyenas [92, 93]. Scratches deep into the spongiosa of the joints are very typical of hyena origin and can be found at many other European open air and cave sites (e.g., [11, 21–23, 25, 29, 40, 41, 82, 94, 95]). The material from Bad Wildungen consists of a few cranial and mainly postcranial bones of at least five woolly rhinoceros individuals. Remains of a young, less than one-year-old calf, a young adult female, and a few remains of a male adult skeleton can be

distinguished (Figure 5(b)). Besides those, mainly forelimb bones from some other rhinoceros individuals were found. A comparison to a normal bone proportion relation analyses [10] to the material from Bottrop open air site (Figure 5(a)) shows differences mainly in the thoracic (vertebrae, costae) presence. In Bad Wildungen, those thoracic elements are more abundant, similar to those found on nonscavenged skeletons like the Petershagen skeleton [90], which indicates the scavenging of a carcass very nearby the den.

The presence of a carcass is also demonstrated by the articulated vertebral column (Figure 5(b)). To this, most probably, other elements belong. An originally articulated right hind limb (femur and tibia, astragalus, and calcaneus) or forelimb bones, such an ulna and radius, support the original presence of one animal carcass which was decomposed in parts. Such decompositions could have taken days, such as what is known for Late Pleistocene elephant carcasses [43]. The carcass of the most probable female *C. antiquitatis* must have laid on the right side of her body during main carcass feeding activities, because more bones from that side are preserved. The skull is lacking, but it seems as if all isolated teeth found from the lower jaw indicate the complete destruction of the mandibles by the hyenas. Isolated teeth of woolly rhinoceros are typically at hyena den sites (e.g., [10]).

(a)

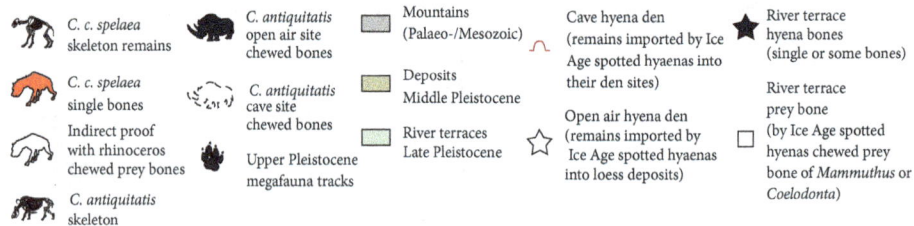

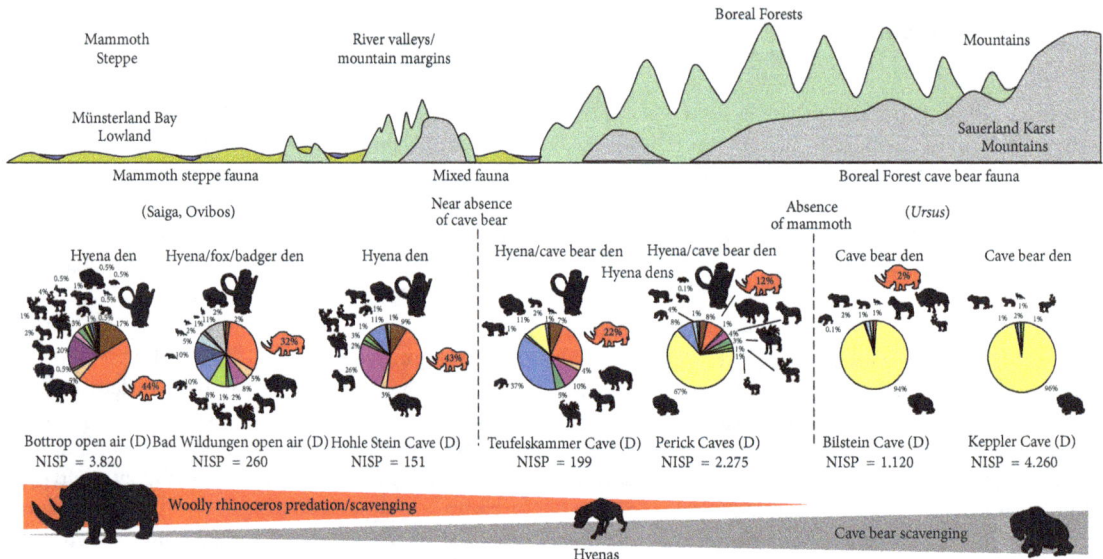

(b)

FIGURE 16: (a) Late Pleistocene spotted hyena sites and dens and woolly rhinoceros remains in NW-Germany. (b) "Cross-section" through the mountain boreal forest cave bear dominated bone assemblages to the mammoth steppe lowland faunal assemblages (composed after [9, 10, 22, 82, 90] and new results).

Maybe the skull was cut off by the hyenas or at least destroyed. A few ribs were only cracked, and nearly all are lacking their joints. The long bone joints were not chewed off completely, because of their articulation. This indicates a fresh carcass that was not completely used by the hyenas and was left in an intermediate stage of carcass destruction (cf. Figure 5(a)). After the bone destruction stages, those are in stage 2 sensu Diedrich [10]. The spongiosa remains of woolly rhinoceros were quite often found in the hyena coprolites at the Bad Wildungen-Biedensteg site [35]. The brain case opening of a calf is similarly figured as an adolescent rhinoceros skull from Selm-Ternsche [10], as figured from rhinoceros skull damages from other sites [96].

The finds of juveniles, such as the few-weeks-old rhinoceros (Figures 5(b) and 6), hyena, or the neonate cave bear, fit for the hunting and main activity time of the hyenas at Biedensteg in the late spring and early summer. Other remains of at least four more rhinoceros individuals and other prey remains were imported, possibly from the Ice Age spotted hyenas.

*7.7. Hyenas as Cave Bear Scavengers.* The cave bear bones might belong to one skeleton of a mature female cave bear [35]. The small diameter, 75 mm, of the scapula glenoid fits for cave bears of the smaller subspecies *U. spelaeus* subsp. of the early/middle Late Pleistocene, compared, for example, to the cave bear population of the Perick Caves in the Sauerland Karst (Figure 1; [97]) or the newer studied cave bear populations and subspecies of the Rübeland Caves [98]. Also, the other bones and femur fragments were compared to some hundred bones from the Perick and Rübeland Caves, all having again smaller proportions, excluding a *U. ingressus* cave bear type of the latest Late Pleistocene. Finally, similarly as figured with the "nibbling stick" in the Perick Caves, some cave bear femora and other bone fragment nibbling sticks are present [70], which only hyenas must have produced by teething cubs (cf. [7]). A scavenging of a cave bear carcass outside a cave is the only clear report of such a scenario [97], but is not exceptional, if compared to the hunting/feeding strategies of the Late Pleistocene spotted hyenas. It is now well known that they scavenged cave bear carcasses in the mountain regions of Europe, such as the Sauerland Caves, the Perick Caves, and Rübeland Caves, and additionally several other cave bear dens all over Europe [42, 70, 98, 99].

*7.8. Fauna Biodiversity and Climatic Mammoth Steppe Indicators.* The faunal statistics demonstrate (Figure 16(b)) that most megafauna bones from Bad Wildungen are related to be of hyena prey origin. Those represent a mammoth steppe megafauna with *Coelodonta antiquitatis* (cf. [29]), *Mammuthus primigenius*, *Bison/Bos*, *Megaloceros giganteus*, *Cervus elaphus*, *Rangifer tarandus*, *Equus caballus przewalskii*, and boreal mountain forest fauna of *Ursus spelaeus* subsp. (cf. [35]). Additionally, the pellets include many mammoth steppe environment rodents such as *Lemmus lemmus*, *Dicrostonyx henseli*, *Microtus gregalis*, or *Allactaga saliens* (cf. [33, 34, 100]). Represented are in higher amounts furthermore birds such as *Lagopus lagopus* and other species (cf. [34]).

## 8. Conclusion

The open air hyena den site Bad Wildungen-Biedensteg (NW-Germany) must have been located at the margin of an ancient small lake and the Wilde River in a mammoth steppe landscape on the eastern slopes of the Sauerland Mountains during the early to middle glaciation (early late Pleistocene or Weichselian, about "65.000–90.000 BP," MIS 5c-d). This shallow lake margin, or at least muddy area, was in the center of a large sinkhole structure, which was caused by subsurface dissolution of Zechstein salt in the underground. The sinkhole received freshwater influence by the early Wilde River, indicated by especially freshwater fish remains, but also some other water related animals such as frogs, which were found accumulated in many pellets. Those are excrements of red/arctic foxes, steppe iltis and large carnivore water birds, or owls. Nearby, a badger/fox den burrow area in loess deposits must have been present, where their bone remains and those of their prey (mainly hare, and micromammals) were accumulated, also in pellets. With Biedensteg, an open air hyena birthing and overlapping communal den with prey deposit can be presented with probably reused badger/red fox burrows for the natal den function. 10% of the NISP are *Crocuta crocuta spelaea* remains, including three grown-up animal skulls, and cranial and postcranial remains of a young cub. Abundant are hyena coprolites (mainly encrusted by caliche), which contain fragments of bones, and most probably quite abundant bone spongiosa fragments from woolly rhinoceros bones. This corresponds to the main hyena prey *Coelodonta antiquitatis* (NISP = 32%). Another main prey is the horse *Equus caballus przewalskii* (8%). This dominance of woolly rhinoceros/horses in the Late Pleistocene bone assemblages in northern Europe was caused solely by those large carnivores and is typical of many hyena open air and cave den bone accumulation sites in northern Germany and Czech Republic (central Europe).

## Acknowledgments

The work is dedicated in the memory of L. Lorenz and her husband R. Lorenz (in former times running the Rudolf-Lorenz-Stiftung), who both supported much rescuing material from the locality, research financing, and care keeping of the collection, which is housed in the Heimatmuseum of the City Bad Wildungen. For the kind hosting during the studies the author thanks Mrs. L. Lorenz, and for the cooperation Dr. V. Brendow, the museums head of the Kurmuseum Bad Wildungen. Dr. I. Wrazlido (Leader Museum Natur und Mensch Bielefeld) kindly allowed the comparison of the woolly rhinoceros skeleton from Petershagen. The author thanks Dr. J. Fichter very much for the help in the donation of the collection that was housed partly in the University of Marburg (coll. E. Jacobshagen). The research and project "Hyena open air prey deposit site Bad Wildungen-Biedensteg" was financially supported by the Rudolf-Lorenz-Stiftung, the Sparkassenstiftung of Bad Wildungen, and the PaleoLogic. Sadly, the responsible Ministry of Culture of Hesse and the office for palaeontological monument survey Wiesbaden did not take care of the site, which was damaged further by a newer house construction, where many finds have been destroyed in the late 90s. Finally the author thanks the reviewers, especially Prof. Dr. Müller-Beck, for their supporting critics of the first paper draft. And last, the author would like to thank S. Stevens for the spell check.

# References

[1] G. A. Goldfuss, "Osteologische Beiträge zur Kenntnis verschiedener Säugethiere der Vorwelt. VI. Ueber die Hölen-Hyäne (*Hyäna spelaea*)," *Nova Acta Physico-Medica Academiea Caesarae Leopoldino-Carolinae Naturae Curiosorum*, vol. 3, no. 2, pp. 456–490, 1823.

[2] W. Buckland, *Reliquiae Diluvianae, or Observations on the Organic Remains Contained in Caves, Fissures, and Diluvial Gravel, and Other Geological Phenomena, Attesting the Action of an Universal Deluge*, J. Murray, London, UK, 1823.

[3] K. T. Liebe, "Die Lindentaler Hyänenhöhle und andere diluviale Knochenfunde in Ostthüringen," *Archiv des Anthropologischen Organs der Deutschen Gesellschaft für Anthropologie, Ethnographie und Urgeschichte*, vol. 9, pp. 1–55, 1876.

[4] K. Ehrenberg, O. Sickenberg, and A. Stifft-Gottlieb, "Die Fuchs- oder Teufelslucken bei Eggenburg, Niederdonau. 1 Teil," *Abhandlungen der Zoologisch-Botanischen Gesellschaft*, vol. 17, no. 1, pp. 1–130, 1938.

[5] R. Musil, "Die Höhle "Sveduv stul", ein typischer Höhlenhyänenhorst," *Anthropos NS*, vol. 5, no. 13, pp. 97–260, 1962.

[6] J. F. Tournepiche and C. Couture, "The hyena den of Rochelot Cave (Charente, France)," *Monographien des Römisch-Germanischen Zentralmuseums*, vol. 42, pp. 89–101, 1999.

[7] C. Diedrich and K. Žák, "Prey deposits and den sites of the Upper Pleistocene hyena *Crocuta crocuta spelaea* (Goldfuss, 1823) in horizontal and vertical caves of the Bohemian Karst (Czech Republic)," *Bulletin of Geosciences*, vol. 81, no. 4, pp. 237–276, 2006.

[8] G. Mangano, "An exclusively hyena-collected bone assemblage in the Late Pleistocene of Sicily: taphonomy and stratigraphic context of the large mammal remains from San Teodoro Cave (North-Eastern Sicily, Italy)," *Journal of Archaeological Science*, vol. 38, no. 12, pp. 3584–3595, 2011.

[9] C. Diedrich, "Late Pleistocene *Crocuta crocuta spelaea* (Goldfuss 1823) clans as prezewalski horse hunters and woolly rhinoceros scavengers at the open air commuting den and contemporary Neanderthal camp site Westeregeln (central Germany)," *Journal of Archaeological Science*, vol. 39, no. 6, pp. 1749–1767, 0212.

[10] C. Diedrich, "The Late Pleistocene *Crocuta crocuta spelaea* (Goldfuss 1823) population from the Emscher River terrace hyena open air den Bottrop and other sites in NW-Germany— woolly rhinoceros scavengers and their bone accumulations along rivers in lowland mammoth steppe environments," *Quaternary International*, vol. 276-277, pp. 93–119, 2012.

[11] C. Diedrich, "The largest Late Pleistocene hyena population from the Srbsko Chlum-Komín Cave (Czech Republic) and its prey in a commuting and prey depot cave den of Central Europe," *Historical Biology*, vol. 24, no. 2, pp. 161–185, 2012.

[12] J. P. Brugal, P. Fosse, and J. P. Guadeli, "Comparative study of bone assemblages made by recent and Plio-Pleistocene Hyaenids (*Hyena, Crocuta*)," in *Proceedings of the 1983 Bone Modification conference, Hot Springs*, A. Hannus, L. Rossum, and R. P. Winham, Eds., pp. 158–187, Arachaeological Laboratory, Augustana Collegue, Sioux Falls, SD, USA, 1997.

[13] T. R. Pickering, "Reconsideration of criteria for differentiating faunal assemblages accumulated by hyenas and hominids," *International Journal of Osteoarchaeology*, vol. 12, no. 2, pp. 127–141, 2002.

[14] M. C. Stiner, "Comparative ecology and taphonomy of spotted hyenas, humans, and wolves in Pleistocene Italy," *Revue de Paleobiologie*, vol. 23, no. 2, pp. 771–785, 2004.

[15] B. F. Kuhn, L. R. Berger, and J. D. Skinner, "Examining criteria for identifying and differentiating fossil faunal assemblages accumulated by hyenas and hominins using extant hyenid accumulations," *International Journal of Osteoarchaeology*, vol. 20, no. 1, pp. 15–35, 2010.

[16] S. Aldhouse-Green, Scott, K. Schwarcz H et al., "Coygan Cave, Laugharne, 1995. South Wales, a Mousterian site and hyaena den: a report on the University of Cambridge excavations," *Proceedings of the Prehistoric Society London*, vol. 61, pp. 37–80, 1995.

[17] C. Diedrich, "A clan of Late Pleistocene hyenas, *Crocuta crocuta spelaea* (Goldfuss 1823), from the Rösenbeck Cave (Germany) and a contribution to cranial shape variability," *Biological Journal of the Linnean Society*, vol. 103, no. 1, pp. 191–220, 2011.

[18] C. Diedrich, "Late Pleistocene *Crocuta crocuta spelaea* (Goldfuss 18) clans as prezewalski horse hunters and woolly rhinoceros scavengers at the open air commuting den and contemporary Neanderthal camp site Westeregeln (central Germany)," *Journal of Archaeological Science*, vol. 39, no. 6, pp. 1749–1767, 2012.

[19] C. Diedrich, "Late Pleistocene hyenas *Crocuta crocuta spelaea* (Goldfuss 1823), from Upper Rhine valley open air sites and the contribution to skull shape variability," *Cranium*, vol. 25, no. 2, pp. 31–42, 2008.

[20] C. Diedrich, "Late Pleistocene *Hystrix (Acanthion) brachyura* Linnaeus, 1758 from the Fuchsluken cave at the Rote Berg near Saalfeld (Thuringia, Germany)—a porcupine and hyena den and contribution to their palaeobiogeography," *The Open Palaeontological Journal*, no. 1, pp. 33–41, 2008.

[21] C. Diedrich, "Seltene Freilandfunde der späteiszeitlichen Fleckenhyäne *Crocuta crocuta spelaea* (Goldfuss 1823) in Sachsen-Anhalt und Beitrag zu Horsttypen der letzten Hyänen im Jung-Pleistozän von Mitteldeutschland," *Zeitschrift für Mitteldeutsche Vorgeschichte*, vol. 94, pp. 1–19, 2013.

[22] C. Diedrich, "*Crocuta crocuta spelaea* (Goldfuss 1823) remains from the Upper Pleistocene hyaena Teufelskammer Cave den site near Hochdahl in the Neander valley (NRW, NW Germany)," *Cranium*, vol. 24, no. 2, pp. 39–44, 2007.

[23] C. Diedrich, "Periodical use of the Balve Cave (NW Germany) as a Late Pleistocene *Crocuta crocuta spelaea* (Goldfuss 1823) den: hyena occupations and bone accumulations vs. human Middle Palaeolithic activity," *Quaternary International*, vol. 233, no. 2, pp. 171–184, 2011.

[24] C. Diedrich, "Late Pleistocene steppe lion *Panthera leo spelaea* (Goldfuss, 1810) footprints and bone records from open air sites in northern Germany—evidence of hyena-lion antagonism and scavenging in Europe," *Quaternary Science Reviews*, vol. 30, no. 15-16, pp. 1883–1906, 2011.

[25] C. Diedrich, "*Crocuta crocuta spelaea* (Goldfuss 1823) remains from the Upper Pleistocene hyaena Teufelskammer Cave den site near Hochdahl in the Neander valley (NRW, NW Germany)," *Cranium*, vol. 24, no. 2, pp. 39–44, 2007.

[26] C. Diedrich, "The holotypes of the upper Pleistocene *Crocuta crocuta spelaea* (Goldfuss, 1823: Hyaenidae) and *Panthera leo spelaea* (Goldfuss, 1810: Felidae) of the Zoolithen Cave hyena

den (South Germany) and their palaeo-ecological interpretation," *Zoological Journal of the Linnean Society*, vol. 154, no. 4, pp. 822–831, 2008.

[27] C. Diedrich, "The *Crocuta crocuta spelaea* (Goldfuss 1823) population and its prey from the Late Pleistocene Teufelskammer Cave hyena den besides the famous Paleolithic Neandertal Cave (NRW, NW Germany)," *Historical Biology*, vol. 23, no. 2-3, pp. 237–270, 2011.

[28] D. Walther, "Eine Siedlungsstelle des Neandertalers bei Buhlen, Kr. Waldeck-Frankenberg," *Geschichtsblätter für Waldeck*, vol. 93, pp. 6–25, 2005.

[29] C. Diedrich, "By ice age spotted hyenas protracted, cracked, nibbled and chewed skeleton remains of *Coelodonta antiquitatis* (Blumenbach, 1807) from the Lower Weichselian (Upper Pleistocene) open air prey deposit site Bad Wildungen-Biedensteg (Hesse, NW Germany)," *Journal of Taphonomy*, vol. 4, no. 4, pp. 173–205, 2006.

[30] R. Huckriede and V. Jacobshagen, "Die Fundschichten," in *Eine Faunenfolge aus dem jungpleistozänen Löß bei Bad Wildungen*, E. Jacobshagen, R. Huckriede, and V. Jacobshagen, Eds., vol. 44 of *Abhandlungen des Hessischen Landesamtes für Bodenforschung*, pp. 93–105, 1963.

[31] A. Semmel, "Studien über den Verlauf jungpleistozäner Formung in Hessen," *Frankfurter Geographische Hefte*, vol. 45, pp. 1–133, 1968.

[32] J. Kulick and Quartär, in *Erläuterungen zur Geologische Karte von Hessen*, M. Horn, J. Kulick, and D. Meischner, Eds., pp. 184–228, Wiesbaden, Germany, 1973.

[33] G. Storch, "Über Kleinsäuger der Tundra und Steppe in jungpleistozänen Eulengewöllen aus dem nordhessischen Löß," *Natur und Museum*, vol. 99, pp. 541–551, 1969.

[34] E. Jacobshagen, "Die Faunen und ihre Bindung an Klima und Umwelt," in *Eine Faunenfolge aus dem jungpleistozänen Löß bei Bad Wildungen*, E. Jacobshagen, R. Huckriede, and V. Jacobshagen, Eds., vol. 44 of *Abhandlungen des Hessischen Landesamtes für Bodenforschung*, pp. 5–92, 1963.

[35] C. Diedrich, "Ice age spotted hyenas? Hunting or only scavenging on a cave bear *Ursus spelaeus* Rosenmüller at the Ice Age spotted hyena freeland den and prey deposit site Bad Wildungen-Biedensteg (Hesse, Germany)," *Scientific Annals, School of Geology Aristotle University of Thessaloniki (AUTH)*, vol. 98, pp. 193–199, 2006.

[36] W. Spöttel, "*Equus przewalskii* Poljakov 1881 mit besonderer Berücksichtigung der im Tierzuchtinstitut der Universität Halle gehaltenen Tiere," *Kühn Archiv*, vol. 11, pp. 89–137, 1926.

[37] C. Diedrich, "Tracking Late Pleistocene steppe lion predation in a mammoth steppe near a spotted hyena open air den of Northwestern Germany—comparisons to modern African lion trackways—new picture of top predator prey specializations, and global Pleistocene megafauna track site review," *Ichnos*, 2013.

[38] A. Forsten, "The small caballoid horse of the Upper Pleistocene and Holocene," *Journal of Animal Breeding Genetics*, vol. 105, pp. 161–176, 1987.

[39] B. Cramer, *Morphometrische untersuchungen an quartären Pferden in Mitteleuropa [Ph.D. dissertation]*, Universität Tübingen, Tübingen, Germany, 2002.

[40] C. Diedrich, "Europe's first Upper Pleistocene *Crocuta crocuta spelaea* (Goldfuss 1823) skeleton from the Koněprusy Caves—a hyena cave prey depot site in the Bohemian Karst (Czech Republic)—Late Pleistocene woolly rhinoceros scavengers," *Historical Biology*, vol. 24, no. 1, pp. 63–89, 2012.

[41] C. Diedrich, "The Ice Age spotted *Crocuta crocuta spelaea* (Goldfuss 1823) population, their excrements and prey from the Late Pleistocene hyena den Sloup Cave in the Moravian Karst; Czech Republic," *Historical Biology*, vol. 24, no. 2, pp. 161–185, 2012.

[42] C. Diedrich, "Cave bear killers and scavengers from the last ice age of central Europe: feeding specializations in response to the absence of mammoth steppe fauna from mountainous regions," *Quaternary International*, vol. 255, pp. 59–78, 2012.

[43] C. Diedrich, "Late Pleistocene Eemian Ice Age spotted hyena feeding strategies and steppe lions on their largest prey—*Palaeoloxodon antiquus* Falconer and Cautley 1845 at the straight-tusked elephant graveyard and Neandertalian site Neumark-Nord Lake 1, Central Germany," *Archaeological and Anthropological Sciences*. In press.

[44] C. Diedrich, "Typology of Ice Age spotted hyena *Crocuta crocuta spelaea* (Goldfuss 1823) coprolite aggregate pellets from the European Late Pleistocene and their significance at dens and scavenging sites," *New Mexico Museum of Natural History and Science, Bulletin*, vol. 57, pp. 369–377, 2012.

[45] W.-D. Heinrich, G. Peters, K.-D. Jäger, and E. G. Böhme, "Erdbaue von Säugetieren—zusammenfassende Kennzeichnung eines neuen Fundstättentyps im baltischen Vereisungsgebiet," *Wissenschaftliche Zeitschrift der Humboldt-Universität zu Berlin, Mathemathisch-Naturwissenschaftliche Reihe*, vol. 6, pp. 777–781, 1983.

[46] T. Rathgeber, "Pleistozäne und holozäne Tierreste aus Höhlen im Kartenblatt 7521 Reutlingen (Schwäbische Alb)," *Laichinger Höhlenfreund*, vol. 43, pp. 27–34, 2008.

[47] P. Lüps and T. J. Roper, "Cannibalism in a female badger (*Meles meles*): infanticide orpredation?" *Journal of Zoology London*, vol. 221, pp. 314–315, 1990.

[48] G. Claußen, *Der Jäger und Sein Wild*, Paul Parey, Berlin, Germany, 1986.

[49] M. Stubbe, "Biometrie und Morphologie des mitteleuropäischen Dachses *Meles meles* (L., 1758)," *Säugetierkundliche Informationen*, vol. 1, pp. 3–26, 1980.

[50] A. V. Abramov and A. Y. Puzachenko, "Sexual dimorphism of craniological characters in Eurasian badgers, *Meles* spp. (Carnivora, Mustelidae)," *Zoologischer Anzeiger*, vol. 244, no. 1, pp. 11–29, 2005.

[51] P. Lüps, "Geschlechtsdimorphismus beim Dachs *Meles meles* L. Ausprägungsgrad und Versuch einer Interpretation. Populationsökologie marderartiger Säugetiere 2," *Wissenschaftliche Beiträge der Universität Halle*, vol. 37, no. P39, pp. 531–542, 1989.

[52] R. Boesi and C. Biancardi, "Diet of the Eurasian badger *Meles meles* (Linnaeus, 1758) in the natural reserve of Lago di Piano, Northern Italy," *Mammalian Biology*, vol. 67, no. 2, pp. 120–125, 2002.

[53] G. Peters, W.-D. Heinrich, P. Beurton, and K.-D. Jäger, "Fossile und rezente Dachsbauten mit Massenanreicherungen von Wirbeltierknochen," *Mitteilungen des Zoologischen Museums Berlin*, vol. 48, pp. 415–435, 1972.

[54] F. Winterfeld, "Über quartäre Mustelidenreste Deutschlands. Zeitschrift der deutschen geologischen Gesellschaft," *Geologischen Gesellschaft*, vol. 37, pp. 826–864, 1885.

[55] A. Nehring, "Übersicht über 24 mitteleuropäische Quartär-Faunen," *Zeitschrift der Deutschen Geologischen Gesellschaft*, vol. 32, pp. 468–509, 1890.

[56] C. Diedrich, "Eine oberpleistozäne Population von *Crocuta crocuta spelaea* (Goldfuss 1823) aus dem eiszeitlichen Flecken-enhyänenhorst Perick-Höhlen von Hemer (Sauerland, NW

Deutschland) und ihr Kannibalismus," *Philippia*, vol. 12, no. 2, pp. 93–115, 2005.

[57] L. G. Frank, "When hyenas kill their own," *New Scientist*, vol. 141, pp. 38–41, 1994.

[58] P. A. White, "Maternal rank is not correlated with cub survival in the spotted hyena, *Crocuta crocuta*," *Behavioral Ecology*, vol. 16, no. 3, pp. 606–613, 2005.

[59] P. P. White, "Maternal response to neonatal sibling conflict in the spotted hyena, *Crocuta crocuta*," *Behavioral Ecology and Sociobiology*, vol. 62, no. 3, pp. 353–361, 2008.

[60] L. Scott and R. G. Klein, "A hyena accumulated bone assemblage from late holocene deposits at Deelpan, Orange Free State, South Africa," *Annals of the South African Museum*, vol. 86, pp. 217–227, 1981.

[61] Y. M. Lam, "Variability in the behaviour of spotted hyaenas as taphonomic agents," *Journal of Archaeological Science*, vol. 19, no. 4, pp. 389–406, 1992.

[62] P. Fosse, "La grotte no. 1 de Lunel-Viel (Hérault, France): repaire d'hyènes du pleistocène moyen," *Paléo*, vol. 8, pp. 47–81, 1996.

[63] A. Arribas and P. Palmqvist, "Taphonomy and palaeoecology of an assemblage of large mammals: hyaenid activity in the lower Pleistocene site at Venta Micena (Orce, Guadix-Baza basin, Granada, Spain)," *Geobios*, vol. 31, supplement 3, pp. 3–47, 1998.

[64] J. L. Guadelli, "Étude taphonomique du repairesd'hyènes de Camiac (Gironde, France), élements de comparisation entrée un site naturel et un gisement préhistorique," *Bulletin de l'Association Française pour l'Étude du Quaternaire*, vol. 2, pp. 91–100, 1989.

[65] P. Fosse, J. P. Brugal, J. L. Guadelli, P. Michel, and J. F. Tournepiche, "Les repaires d' hyenes des cavernes en Europe occidentale: presentation et comparisons de quelques assemblages osseux," in *Economie Prehistorique: Les Comportements de Substance au Paleolithique. Rencontres Internationales d'Archéologie et d'Histoire d'Antibes*, pp. 44–61, APDCA, Sophia Antipolis, France, 1998.

[66] P. Villa, J. C. Castel, C. Beauval, V. Bourdillat, and P. Goldberg, "Human and carnivore sites in the European Middle and Upper Paleolithic: similarities and differences in bone modification and fragmentation," *Revue de Paleobiologie*, vol. 23, no. 2, pp. 705–730, 2004.

[67] M. G. L. Mills and M. Mills, "An analysis of bones collected at hyaena breeding dens in the Gemsbok National Parks," *Annales of the Transvaal Museum*, vol. 30, pp. 145–155, 1977.

[68] S. M. Cooper, "The hunting behaviour of spotted hyaenas (*Crocuta crocuta*) in a region containing both sedentary and migratory populations of herbivores," *African Journal of Ecology*, vol. 28, no. 2, pp. 131–141, 1990.

[69] M. L. East, H. Hofer, and A. Türk, "Functions of birth dens in spotted hyaenas (*Crocuta crocuta*)," *Journal of Zoology London*, vol. 219, pp. 690–697, 1989.

[70] C. Diedrich, "Von eiszeitlichen Fleckenhyänen benagte *Mammuthus primigenius* (Blumenbach 1799)-Knochen und -Knabbersticks aus dem oberpleistozänen Perick-Höhlenhorst (Sauerland) und Beitrag zur Taphonomie von Mammutkadavern," *Philippia*, vol. 12, no. 1, pp. 63–84, 2005.

[71] H. Hofer and M. East, "Population dynamics, population size, and the commuting system of Serengeti spotted hyenas," in *Serengeti II: Dynamics, Management, and Conservation of an Ecosystem*, A. R. E. Sinclair and P. Arcese, Eds., pp. 332–363, University of Chicago Press, Chicago, Ill, USA, 1995.

[72] H. Kruuk, *The Spotted Hyena. A Story of Predation and Social Behavior*, University of Chicago Press, Chicago, Ill, USA, 1972.

[73] J. R. Henschel, R. Tilson, and F. von Blottnitz, "Implications of a spotted hyaena bone assemblage in the Namib Desert," *South African Archaeological Bulletin*, vol. 34, pp. 127–131, 1979.

[74] A. Hill, "A modern hyena den in Amboseli National Park, Kenya, Nairobi," in *Proceedings of the 8th Pan-African Congress on Prehistory and Quaternary Studies*, pp. 137–138, 1980.

[75] A. K. Behrensmeyer and D. Boaz, "The recent bones of Amboseli Park Kenya in relation to East African paleoecology," in *Fossils in the Making: Vertebrate Taphonomy and Paleoecology*, A. K. Behrensmeyer and A. P. Hill, Eds., pp. 72–92, University of Chicago Press, Chicago, Ill, USA, 1980.

[76] C. K. Brain, "Some criteria for the recognition of bone collecting agencies in African caves," in *Fossils in the Making: Vertebrate Taphonomy and Paleoecology*, A. K. Behrensmeyer and A. P. Hill, Eds., pp. 107–130, University of Chicago Press, Chicago, Ill, USA, 1980.

[77] G. Avery, D. M. Avery, S. Braine, and R. Loutit, "Bone accumulation by hyenas and jackals a taphonomic study," *South African Journal of Science*, vol. 80, pp. 186–187, 1984.

[78] J. R. Henschel, J. D. Skinner, and A. S. van Jaarsveld, "The diet of the spotted hyaenas *Crocuta crocuta* in Kruger National Park," *South African Journal of Zoology*, vol. 21, pp. 303–308, 1986.

[79] E. E. Boydston, K. M. Kapheim, and K. E. Holekamp, "Patterns of den occupation by the spotted hyaena (*Crocuta crocuta*)," *African Journal of Ecology*, vol. 44, no. 1, pp. 77–86, 2006.

[80] S. W. Lansing, S. M. Cooper, E. E. Boydston, and K. E. Holekamp, "Taphonomic and zooarchaeological implications of spotted hyena (*Crocuta crocuta*) bone accumulations in kenya: a modern behavioral ecological approach," *Paleobiology*, vol. 35, no. 2, pp. 289–309, 2009.

[81] J. T. Pokines and J. C. Kerbis Peterhans, "Spotted hyena (*Crocuta crocuta*) den use and taphonomy in the Masai Mara National Reserve, Kenya," *Journal of Archaeological Science*, vol. 34, no. 11, pp. 1914–1931, 2007.

[82] C. Diedrich, "Die späteiszeitlichen Fleckenhyänen und deren Exkremente aus Neumark-Nord," *Archäologie in Sachsen-Anhalt Sonderband*, vol. 62, pp. 440–448, 2010.

[83] S. K. Bearder and R. M. Randall, "The use of fecal marking sites by spotted hyaenas and civets," *Carnivore*, vol. 1, pp. 32–48, 1978.

[84] T. Keller and G. Försterling, "Unterbrochene Mahlzeit? Eiszeitliche Knochenfunde aus dem Gipskarst von Morschen-Konnefeld," in *Archäologische und Paläontologische Bodendenkmalpflege des Landesamtes für Denkmalpflege Hessen*, vol. 202, pp. 18–21, 2002.

[85] C. Diedrich, "Specialized horse killers in Europe: foetal horse remains in the Late Pleistocene Srbsko Chlum-Komín Cave hyena den in the Bohemian Karst (Czech Republic) and actualistic comparisons to modern African spotted hyenas as zebra hunters," *Quaternary International*, vol. 220, no. 1-2, pp. 174–187, 2010.

[86] G. Bosinski, "Tierdarstellungen von Gönnersdorf. Nachträge zu Mammut und Pferd sowie die übrigen Tierdarstellungen," *Monographien des Römisch-Germanischen Zentralmuseums*, vol. 72, pp. 1–175, 2008.

[87] R. Springhorn, "A wild horse (*Equus przewalskii* Poljakov 1881) of Mesolithic age from Kempen (Germany, Northrhine-Westphalia, Lippe County)," *Eiszeitalter und Gegenwart*, vol. 52, pp. 40–46, 2003.

[88] M. Baales, "Auf der Fährte spätglazialer Pferde bei Mertloch (Neuwieder Becken)," *Bonner Zoologische Beiträge*, vol. 50, no. 3, pp. 109–133, 2002.

[89]  A. Leroi-Gourhan, "Préhistoire de l'art occidental," *Ars Antiqua*, vol. 1, pp. 1–601, 1971.

[90]  C. Diedrich, "A skeleton of an injured *Coelodonta antiquitatis* (Blumenbach 1807) from the Upper Pleistocene of north-western Germany," *Cranium*, vol. 25, no. 1, pp. 1–16, 2008.

[91]  P. Wernert, "Beutestücke der Höhlenhyänen im anatomischen Verband aus Achenheimer Lössen," *Quartär*, vol. 19, pp. 55–64, 1968.

[92]  A. Hill, "Bone modification by modern spotted hyenas," in *Bone Modification*, R. Bonnichsen and M. H. Sorg, Eds., pp. 169–178, Center for the Study of the First Americans, Orono, Me, USA, 1989.

[93]  J. T. Faith, "Sources of variation in carnivore tooth-mark frequencies in a modern spotted hyena (*Crocuta crocuta*) den assemblage, Amboseli Park, Kenya," *Journal of Archaeological Science*, vol. 34, no. 10, pp. 1601–1609, 2007.

[94]  C. Diedrich, "Ein Schädelfund und von Hyänen angenagter Oberschenkelknochen des Wollnashorns *Coelodonta antiquitatis* Blumenbach 1799 aus den pleistozänen Weserkiesen bei Minden (Norddeutschland)," *Philippia*, vol. 11, no. 3, pp. 211–217, 2004.

[95]  C. Diedrich, "Eingeschleppte und benagte Knochenreste von *Coelodonta antiquitatis* (Blumenbach, 1807) aus dem oberpleistozänen Fleckenhyänenhorst Perick-Höhlen im Nordsauerland (NW Deutschland) und Beitrag zur Taphonomie von Wollnashornknochen in Westfalen," *Mitteilungen der Höhlen und Karstforscher*, vol. 4, pp. 100–117, 2008.

[96]  F. Heller, "Hyänenfraß-Reststücke von Schädeln des Wollhaarigen Nashorns *Coelodonta antiquitatis* blumenbach," *Quartär*, vol. 14, pp. 89–93, 1962.

[97]  C. Diedrich, "Die oberpleistozäne Population von *Ursus spelaeus* Rosenmüller 1794 aus dem eiszeitlichen Fleckenhyänenhorst Perick-Höhlen von Hemer (Sauerland, NW Deutschland)," *Philippia*, vol. 12, no. 4, pp. 275–346, 2006.

[98]  C. Diedrich, "Evolution, Horste, Taphonomie und Prädatoren der Rübeländer Höhlenbären, Harz (Norddeutschland)," *Mitteilungen des Verbandes der deutschen Höhlen- und Karstforscher*, vol. 59, no. 1, pp. 4–29, 2013.

[99]  C. Diedrich, "Cave bear killers, scavengers between the Scandinavian and Alpine Ice shields—the last hyenas and cave bears in antagonism—and the reason why cave bears hibernated deeply in caves," *Stalactite*, vol. 58, no. 2, pp. 53–63, 2009.

[100]  W. von Koenigswald, *Lebendige Eiszeit*, Thorbecke, Darmstadt, Germany, 2002.

# 3

# Geochronology and Tectonic Evolution of the Lincang Batholith in Southwestern Yunnan, China

**Hongyuan Zhang,**[1,2] **Junlai Liu,**[1,2] **and Wenbin Wu**[2]

[1] State Key Laboratory of Geological Processes and Mineral Resources, China University of Geosciences, Beijing 100083, China
[2] Faculties of Earth Sciences and Resources, China University of Geosciences, Beijing 100083, China

Correspondence should be addressed to Hongyuan Zhang, zhang-hong-yuan@263.net

Academic Editor: Quan-Lin Hou

Geochronological research of the Lincang Batholith is one key scientific problem to discuss the tectonic evolution of the Tethys. Two granitic specimens were selected from the Mengku-Douge area in the Lincang Batholith to perform the LA-ICPMS Zircon U-Pb dating based on thorough review of petrological, geochemical, and geochronological data by the former scientists. Rock-forming age data of biotite granite specimen from Kunsai is about 220 Ma, the Norian age. However, the west sample from Mengku shows 230 Ma, the Carnian age. The later intrusion in Kunsai area located east to the Mengku area shows directly their uneven phenomena in both space and time and may indirectly reflect the space difference of the contraction-extension transformation period of the deep seated faults. Considering the former $^{40}Ar/^{39}Ar$ data and the outcrop in Mengku, the Lincang Batholith should have experienced one tectonic exhumation and regional subsidence cycle. The regional subsidence cycle has close relations to the expansion of the Meso Tethys.

## 1. Introduction

The Sanjiang-Indochina region is one key area to study the evolution of the Tethys [1–10] (Figure 1). The Changning-Menglian zone was considered as the major faunal break between Gondwanan assemblage to the west and Cathaysian to the east [11, 12]. The subduction-related magmatism occurred along the western edge of the Lanping-Simao-Indochina terrane throughout the Triassic and the closure of Palaeo-Tethys [12, 13]. The Lincang Batholith extends ~370 km from north to south, covering the Chiangrai-Chiang Mai region of Thailand, with an area of more than 10000 km$^2$ (Figure 2). It has been shown by many authors that there is great mineral potential [14].

The Lingcang Batholith was divided into three lithologic intervals by previous authors, the Xiaojie, Lincang, and Menghai intervals (Figure 2) [15–17]. The major intrusion is porphyritic monzonitic granite, which mainly contains quartz, zonal plagioclase, K-feldspar, and biotite. In addition, geochemical studies have shown that the granite mainly has source features of the crust mixing origin [16, 18].

The Lingcang Batholith was proposed to form in a passive continental margin according to geochemical research [19]. The intrusive timing spans from late Permian to late Triassic based on whole-rock Rb-Sr, mineral $^{40}Ar/^{39}Ar$, whole-rock $^{87}Sr/^{86}Sr$, Rb-Sr, Sm-Nd, and some other methods (Table 1).

But there are also some understanding differences on the crystallization period and evolution time of the Lincang Batholith. Many predecessors discussed mineral diagenesis and tectonic evolution through traditional data [16, 20], although, not from the viewpoint of the plate tectonics. On scientific problems of the Lincang Batholith, some solutions were put forward such as single period of mixed granite [16], quasi-situ metasomatic granite [20], or Neoproterozoic type I granite complex batholith with multiperiod transforming events [19].

Therefore, the formation and evolution of the Lincang Batholith is still the key to discuss the regional tectonic evolution of the Baoshan and Lanping-Simao-Indochina block. In this study, we conducted petrographic analysis of the Lincang Batholith and provide new LA-ICP-MS U-Pb data in order to constrain the tectonic evolution of the

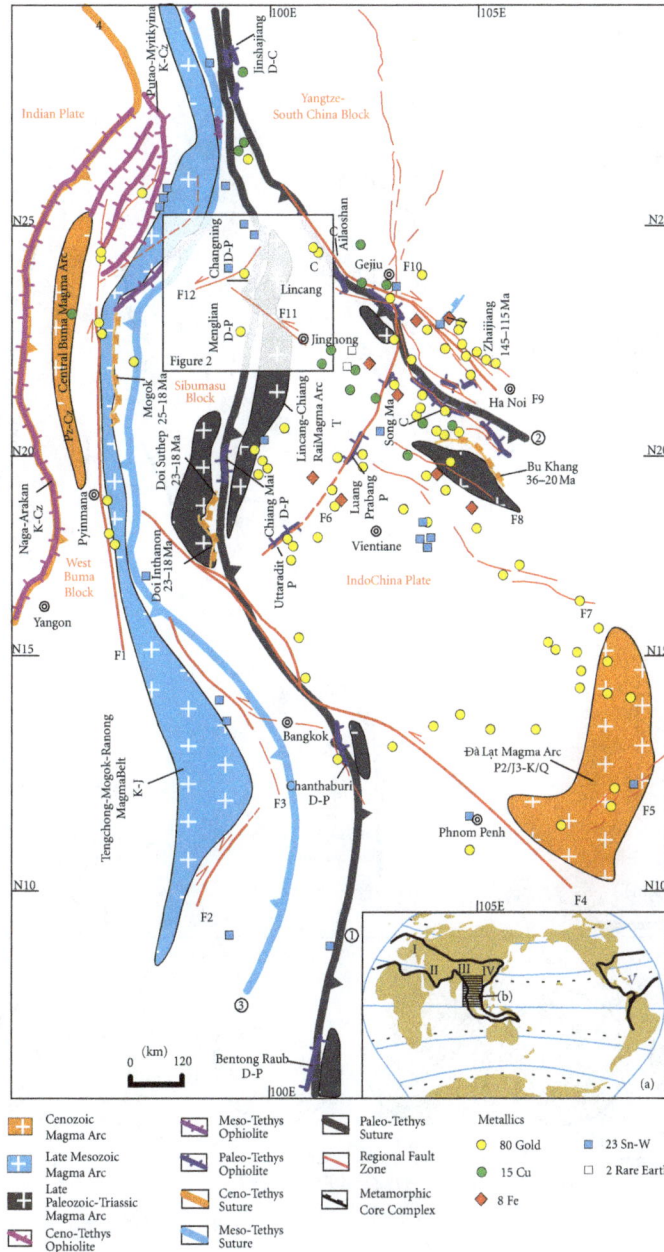

FIGURE 1: Map of plate tectonics and ore-forming metals in the southern Sanjiang and Indochina peninsula. (a) Scope of Tethys and zone divisions; (b) research area and distribution of major metals: (1) Changning-Menglian-Chiangmai PaleoTethys Suture (time and space after [1, 21, 22]); (2) Ailaoshan-Song Ma Paleo-Tethys Suture (time and space after [2, 6, 8]); (3) Bangong Co-Nujiang Meso-Tethys Suture (time and space after [5]); (4) Yanlung Tsangpo-Naga-Arakan New-Tethys Suture (time and space after [22, 23]); (F1) Nujiang-Saging Fault Zone; (F2) Ranong Fault Zone; (F3) Three Tower Fault Zone; (F4) Tonle Sap Fault Zone; (F5) Đà Lạt-Bianhe Fault Zone; (F6) Điên Biên Phú Fault Zone; (F7) Changshan-Da Nang Fault Zone; (F8) Lanjiang Fault Zone; (F9) Ailaoshan-Red River Fault Zone; (F10) Xianshuihe-Xiaojiang Fault Zone; (F11) Lancang-Jinghong Fault Zone.

Batholith. We discuss these results of dating in the context of regional tectonic evolution.

## 2. Dating Methods, Specimens, and Results

*2.1. Dating Methods.* The U-Pb zircon age dating of monzonitic granite samples selected from the Mengku-Douge research area in Lincang segment (Figures 2 and 3) was designed and completed in the Institute of Geology and Geophysics, Chinese Academy of Sciences. Jobs of single zircon grain micro-area U-Pb geochronological analysis were taken through Laser Ablation Inductively Coupled Plasma Mass Spectrometry (LA-ICPMS) of 7500a-type producted by Agilient Co. Ltd., with which the laser beam diameter is $60 \mu m$, the erosion frequency is 8 Hz, the energy density is $15\sim20 J\cdot cm^{-2}$, the erosion time is about 60 s. For detailed

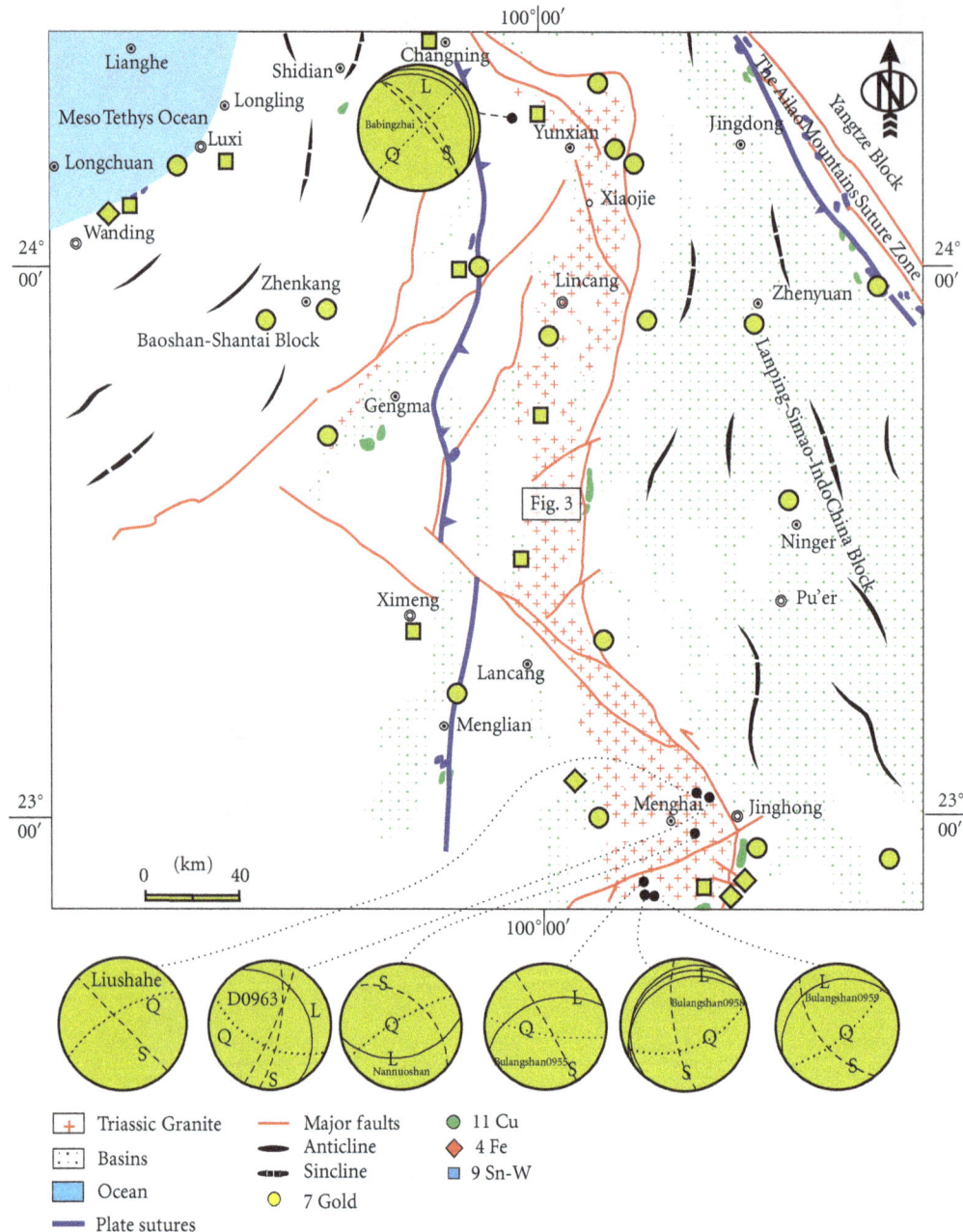

FIGURE 2: Regional map of Indian Period and research locality in western Yunnan. The distribution of metal deposits also indicates that there is an important boundary along the Lincang Granite Batholith. Stereographic maps of lower hemisphere projection indicating the initial occurrence of Lincang granite Batholith are also plotted in the regional map. L, S, Q refer to joints parallel to the initial foliation, parallel to the flow direction but upright to the initial foliation, upright to both the flow direction and initial foliation. Orientation of shear fractures can be generated by local stress system of the Batholith with emplacements [24].

analysis procedures, please see Xu et al. [32]. The Zircon U-Pb isotope and the U, Th data processing are finished by software Glitter4. 0 [33], and computing of U-Pb concordia diagrams, weighted average ages and graphics were completed by software Isoplot3.0 [34].

### 2.2. Specimens

*2.2.1. Specimen from Kunsai Quarries.* The biotite monzonitic granite samples were selected from the Kunsai

quarries to the east of Quannei-Douge migmatite rock belt (Figures 3 and 4(a)). The north part of the rock unit is plunged into the Jurassic basin (Figure 3). Field study shows that both the rock unit and the Jurassic basin have experienced a period of ductile transformation events. The microstructure study shows brittle deformed feldspar (Figures 4(b) and 4(c)), dynamically recrystallized quartz (Figure 4(c)), microdeflected biotite (Figure 4(d)), zircon grains with relatively intact crystal, long column, some

TABLE 1: Isotopic age data for the Lincang Batholite in the earlier stage.

| Lithologic section | Lithology | Measured objects and methods | Dating results, Ma | Sample positions | Testers and the time |
|---|---|---|---|---|---|
| Xiaojie | Granite porphyry | Whole-rock (Rb-Sr) | 169 ± 5 | Laomaocun | Zhang et al., 2006 [25] |
| Xiaojie | Monzonitic granite | Whole-rock ($^{87}Sr/^{86}Sr$) | 279 | Yun County | Chen, 1991 [26] |
| Lincang | Biotite granite | Biotite ($^{40}Ar/^{39}Ar$) | 201.1 ± 2.7 | Near the Lancangjiang River | Dai et al., 1986 [27] |
| Lincang | Monzonitic granite | Whole-rock (Rb-Sr) | 279 | Lincang | Chen, 1991 [26] |
| Lincang | Medium-grained equigranular granodiorite | Whole-rock (Rb-Sr) | 263.8 | Milestone along Lincang-Mengku road 327–387 km | Zhang et al., 1990 [28] |
| Lincang | Unequal-sized biotite granite | Whole-rock (Rb-Sr) | 193 | Shangyun-Xiaotang | Zhang et al., 1990 [28] |
| Lincang | Monzonitic granite | Whole-rock (Rb-Sr) | 275 ± 59 | Lincang | Liu et al., 1989 [18] |
| Menghai | Monzonitic granite | Whole-rock ($^{87}Sr/^{86}Sr$) | 279 | Menghai | Chen, 1991 [26] |
| Menghai | Monzonitic granite | Biotite (Rb-Sr) | 256 | Menghai | Wang, 1984 [29] |

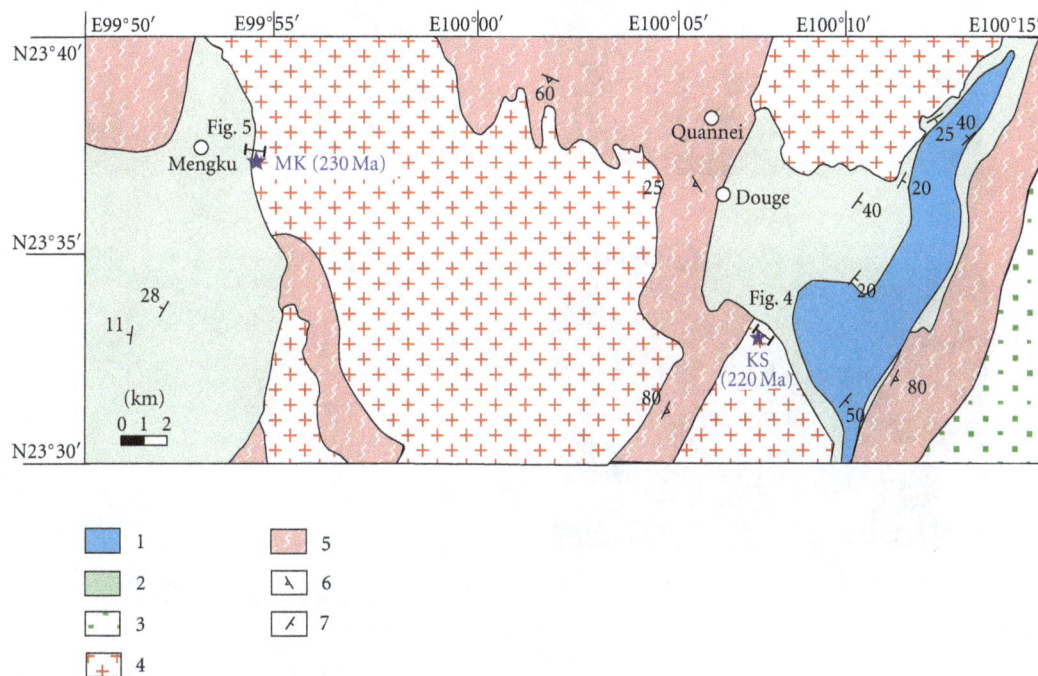

FIGURE 3: Geological sketch map in Mengku-Douge area (revised after [15–17, 30, 31]) (1) Area of ingression sea of the Meso-Tethys; (2) Jurassic terrigenous sediments; (3) Early Jurassic volcanic sedimentary rocks; (4) Lincang Triassic granite rock; (5) deep metamorphic wedge with local granodiorite quality migmatization Late Paleozoic; (6) schistosity occurrence; (7) bedding occurrence.

rounded output and a higher degree of porosity which is symbiotic to microcline grains (Figure 4(b)). The cathode luminescence (CL) images show that the typical magmatic zircon characteristics are colorless and transparent here with oscillatory zoning, the length of about 120 μm–420 μm and the aspect ratio of 1.2 to 4.1 (Figure 4(e)). Totally, for 22 points of 22 zircon grains of sample biotite granodiorite KS-2 from the Kunsai quarries we performed U-Pb isotope analysis in order to statistically gain a feasible dating result (Figure 4(f)). All points were selected on the edge of long- column euhedral grains where the magmatic zircon oscillatory zoning is clear.

*2.2.2. Specimens from Mengku Quarries.* The monzonitic granite sample MK-4 is selected from the Mengku East Quarry at the eastern side of the Mengku Town in the Shuangjiang County, which separates Jurassic red clastic rocks with one bedding bottom granitic conglomerate by the paleo-weathering crust in between (Figures 5(a), 5(b), and 5(c)). Figure 5(a) shows the position photo of the biotite monzonitic granite sample in the Mengku East Quarry. From the field outcrop, both granitic intrusion (lower left) and its weathering top can be found to be cleavaged which had been covered by the upper Jurassic red detrital sediments with terrigenous origin, namely, the Huakaizuo Formation.

(a)

(b)

(c)

(d)

(e)

(f)

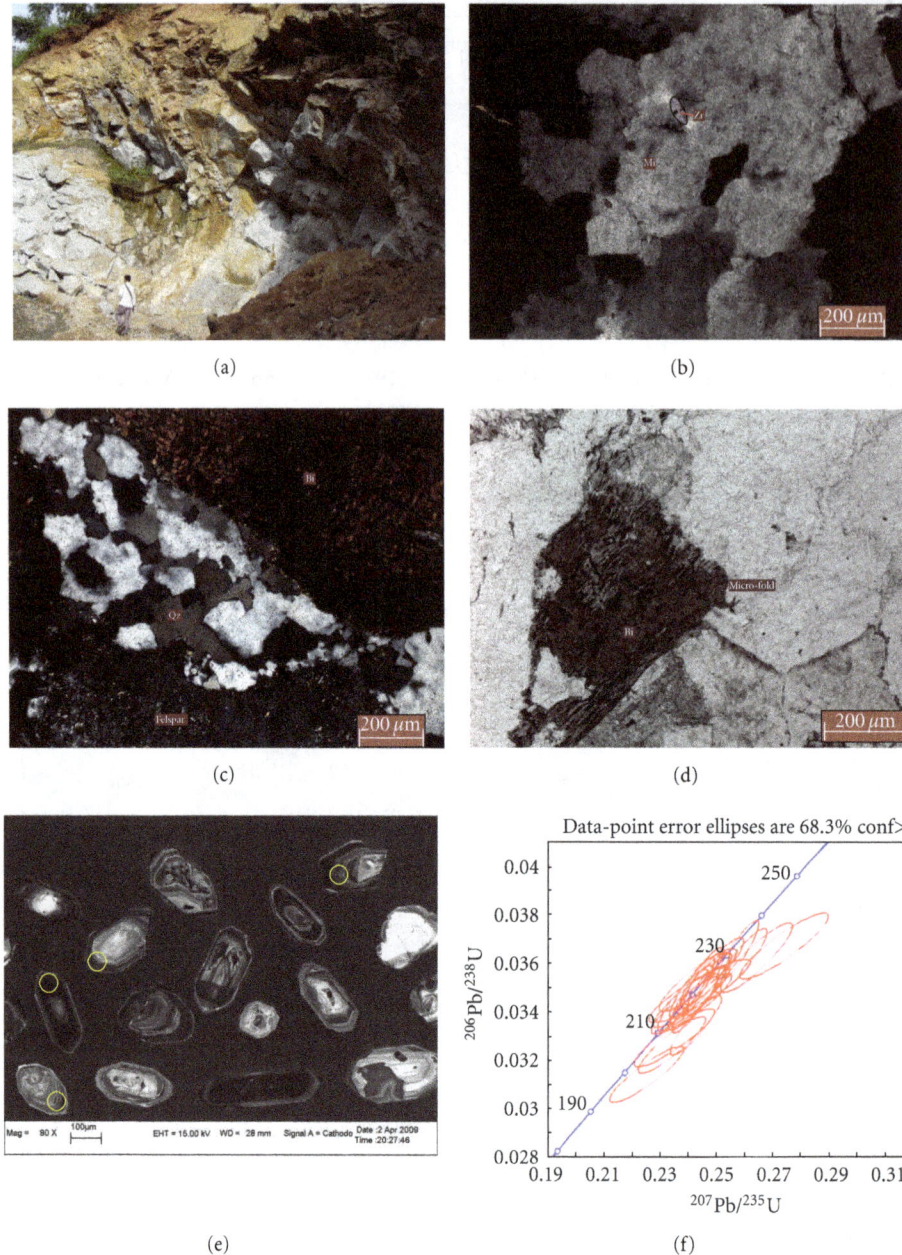

FIGURE 4: Geochronological research of granite sample KS-2. (a) Position photo of the biotite monzonitic granite sample in the field of Kunsai Quarry. (b) The zircon grows together with the microcline CPL. (c) The microstructure of complete recrystallized quartz indicates the rock once experienced high-greenschist facies of deformation, CPL. (d) One biotite crystal is folded, PPL. (e) Magmatic zircon is shown by the SEM cathodoluminescence (CL) image. The circled area is deduced to form during the crystallization of granite and is just the measured area by laser-ICPMS (f) Zircon La-ICPMs U-Pb dating figure, and the average age is about 220 Ma.

The MK-4 zircon grains (Figure 5(d)) change greatly and have different shapes and sizes, such as long column, fan, and irregular granular. However, size of zircon grains is uniform with length between 90 $\mu$m and 150 $\mu$m and length-width ratio in 1.1 : 1–1.8 : 1. Zircon grains develop with quartz, plagioclase in rocks. The rock sample has been affected by weathering, showing retrograde metamorphism (Figure 5(e)). Characteristic oscillatory zoning of typical magmatic zircon is rendered on the cathode CL image. The

idiomorphic degree of zircon is high, but some fragmented. On the sample of MK-4, 18 zircon grains and 18 points were analyzed by ICP-MS. Edge of the long-column idiomorphic or the chipped-hypidiomorphic zircon grains, and the clear zone of magmatic zircon oscillatory were always selected to perform the micro isotope analysis (Figure 5(f)).

*2.2.3. Dating Results.* As shown in Figure 4(f) and Table 2, the overall harmony values of KS-2 zircon age data range

(a)

(b)

(c)

(d)

(e)

(f)

(g)

FIGURE 5: Geochronological research of the granite sample MK-4. (a) Position photo of the biotite monzonitic granite sample in a field Quarry of east Mengku. (b) Granitic intrusion (lower left) and its fossil weathering were cleavaged. (c) The red detrital material of terrigenous origin, Huakaizuo Formation in Jurassic, was normal gradational. (d) Granitic intrusions are selected as sample. (e) The micro characteristics of MK-4 and the zircon setting are shown. The zircon grains are associated with quartz and plagioclase and have been degraded by some weathering process, CPL, (f) Magmatic zircon is shown by the SEM cathodoluminescence (CL) image. The circled area is deduced to form during the crystallization of granite and is just the measured area by laser-ICPMS. (g) Zircon La-ICPMs U-Pb dating figure, and the average age is about 245 Ma.

TABLE 2: U-Pb isotopic dating for the single grain Zircon from granite of Kunsai.

| No. | $^{207}$Pb/$^{206}$Pb Ratio | Error (1σ) | $^{207}$Pb/$^{235}$U Ratio | Error (1σ) | $^{206}$Pb/$^{238}$U Ratio | Error (1σ) | $^{208}$Pb/$^{232}$Th Ratio | Error (1σ) | $^{207}$Pb/$^{206}$Pb Ratio | Error (1σ) | $^{207}$Pb/$^{235}$U Ratio | Error (1σ) | $^{206}$Pb/$^{238}$U Ratio | Error (1σ) | $^{208}$Pb/$^{232}$Th Ratio | Error (1σ) |
|---|---|---|---|---|---|---|---|---|---|---|---|---|---|---|---|---|
| A77 | 0.05789 | 0.00231 | 0.25961 | 0.00949 | 0.03251 | 0.00089 | 0.01383 | 0.00076 | 526 | 38 | 234 | 8 | 206 | 6 | 278 | 15 |
| A78 | 0.05188 | 0.00244 | 0.23562 | 0.00953 | 0.03294 | 0.00079 | 0.01035 | 0.00025 | 280 | 110 | 215 | 8 | 209 | 5 | 208 | 5 |
| A79 | 0.05273 | 0.00141 | 0.24329 | 0.00604 | 0.03345 | 0.00081 | 0.01108 | 0.00055 | 317 | 26 | 221 | 5 | 212 | 5 | 223 | 11 |
| A80 | 0.05213 | 0.00216 | 0.23471 | 0.00801 | 0.03265 | 0.00077 | 0.01025 | 0.00024 | 291 | 97 | 214 | 7 | 207 | 5 | 206 | 5 |
| A81 | 0.05209 | 0.00131 | 0.25938 | 0.00608 | 0.0361 | 0.00086 | 0.01177 | 0.00034 | 289 | 25 | 234 | 5 | 229 | 5 | 237 | 7 |
| A82 | 0.05014 | 0.00118 | 0.24192 | 0.00531 | 0.03498 | 0.00082 | 0.01113 | 0.00044 | 201 | 24 | 220 | 4 | 222 | 5 | 224 | 9 |
| A83 | 0.05126 | 0.00122 | 0.24287 | 0.00541 | 0.03435 | 0.00081 | 0.01155 | 0.0006 | 253 | 24 | 221 | 4 | 218 | 5 | 232 | 12 |
| A84 | 0.05027 | 0.00114 | 0.24534 | 0.00525 | 0.03539 | 0.00083 | 0.01176 | 0.00041 | 207 | 24 | 223 | 4 | 224 | 5 | 236 | 8 |
| A89 | 0.0521 | 0.00237 | 0.22575 | 0.00875 | 0.03143 | 0.00075 | 0.00987 | 0.00023 | 290 | 107 | 207 | 7 | 199 | 5 | 199 | 5 |
| A90 | 0.05305 | 0.00182 | 0.2649 | 0.0084 | 0.03621 | 0.00094 | 0.01376 | 0.00075 | 331 | 33 | 239 | 7 | 229 | 6 | 276 | 15 |
| A91 | 0.05739 | 0.00113 | 0.23972 | 0.00441 | 0.03029 | 0.0007 | 0.00946 | 0.00032 | 507 | 23 | 218 | 4 | 192 | 4 | 190 | 6 |
| A92 | 0.05076 | 0.00135 | 0.24604 | 0.00612 | 0.03515 | 0.00085 | 0.01173 | 0.00045 | 230 | 26 | 223 | 5 | 223 | 5 | 236 | 9 |
| A93 | 0.05089 | 0.00279 | 0.24455 | 0.01197 | 0.03485 | 0.00086 | 0.01098 | 0.00025 | 236 | 127 | 222 | 10 | 221 | 5 | 221 | 5 |
| A94 | 0.05114 | 0.00155 | 0.24377 | 0.00683 | 0.03457 | 0.00087 | 0.01167 | 0.00055 | 247 | 29 | 222 | 6 | 219 | 5 | 235 | 11 |
| A95 | 0.05055 | 0.00109 | 0.24175 | 0.00488 | 0.03469 | 0.00081 | 0.01181 | 0.00045 | 220 | 24 | 220 | 4 | 220 | 5 | 237 | 9 |
| A96 | 0.05018 | 0.00134 | 0.23857 | 0.00594 | 0.03448 | 0.00084 | 0.01149 | 0.00049 | 203 | 26 | 217 | 5 | 219 | 5 | 231 | 10 |
| B02 | 0.05115 | 0.00119 | 0.24774 | 0.00541 | 0.03514 | 0.00084 | 0.01297 | 0.00054 | 248 | 25 | 225 | 4 | 223 | 5 | 260 | 11 |
| B03 | 0.05067 | 0.00224 | 0.24165 | 0.00903 | 0.03459 | 0.00082 | 0.0109 | 0.00025 | 226 | 104 | 220 | 7 | 219 | 5 | 219 | 5 |
| B04 | 0.05104 | 0.00235 | 0.24163 | 0.00953 | 0.03433 | 0.00082 | 0.01081 | 0.00025 | 243 | 109 | 220 | 8 | 218 | 5 | 217 | 5 |
| B05 | 0.05079 | 0.00153 | 0.24968 | 0.00699 | 0.03567 | 0.0009 | 0.01274 | 0.00067 | 231 | 29 | 226 | 6 | 226 | 6 | 256 | 13 |
| B06 | 0.05039 | 0.00095 | 0.24525 | 0.00437 | 0.03531 | 0.00082 | 0.01121 | 0.00038 | 213 | 25 | 223 | 4 | 224 | 5 | 225 | 8 |
| B07 | 0.05429 | 0.00243 | 0.27463 | 0.01035 | 0.03669 | 0.00089 | 0.01147 | 0.00027 | 383 | 103 | 246 | 8 | 232 | 6 | 230 | 5 |
| B08 | 0.05014 | 0.00223 | 0.24399 | 0.00913 | 0.0353 | 0.00085 | 0.01114 | 0.00026 | 201 | 104 | 222 | 7 | 224 | 5 | 224 | 5 |
| B09 | 0.05141 | 0.00105 | 0.25807 | 0.00495 | 0.03643 | 0.00086 | 0.01414 | 0.00053 | 259 | 25 | 233 | 4 | 231 | 5 | 284 | 11 |

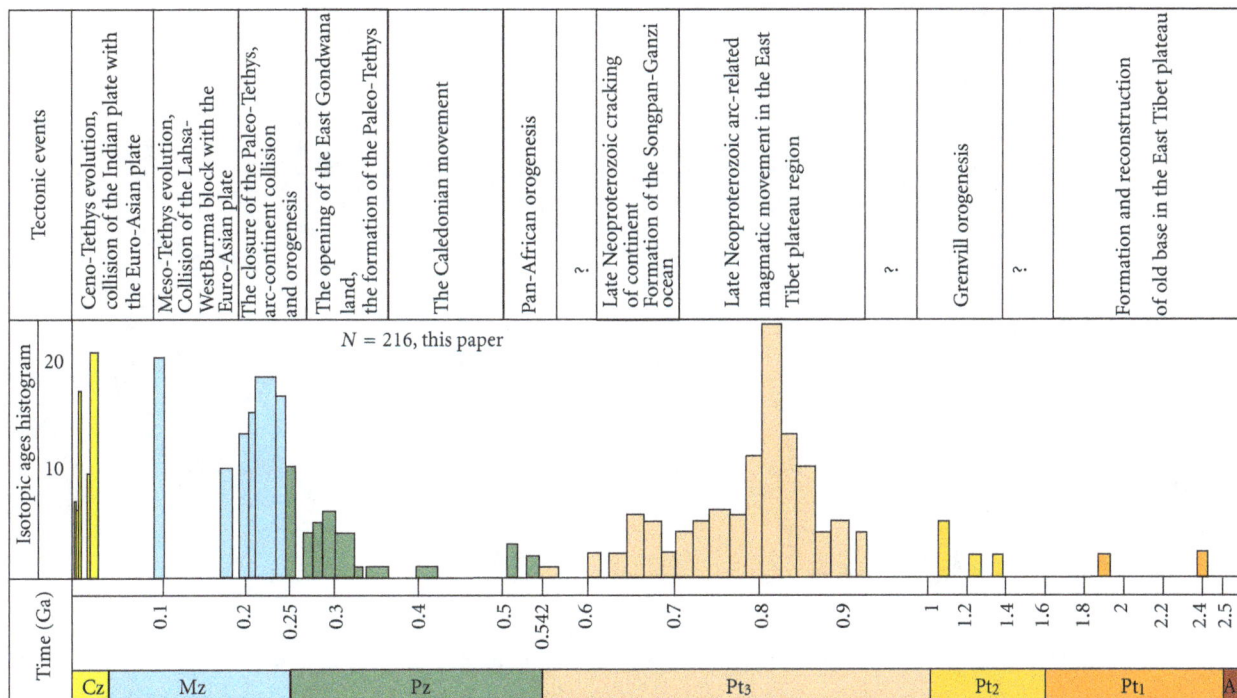

FIGURE 6: Isotopic age histogram showing the major tectonic event in the area east to the Tibet Peateau (after [8]).

from 195 Ma to 245 Ma, but some of them are below the U-Pb harmonic line and deviated from the harmony line. This is probably because the micro region selected is too close to the boundary of the zircon grain. Some zircon grains (Figure 5(f)) also have some cracks in and this may also affect the dating result. From the harmonic visible figure, zircon dating values are focused to a range from 210 Ma to 230 Ma, and the average age is about 220 Ma. This means that the crystallization age of the Kunsai quarries biotite granite is about in the late Triassic Norian period.

As shown in Figure 5(g) and Table 3, the overall harmony values of MK-4 zircon age data range between 190 Ma and 315 Ma. The values are somewhat scattered mainly because the granite of the Mengku quarries has been heavily weathered (Figure 5(d)). During the weathering process, chemical bonds of $ZrO_2$ in zircon grains of our MK-4 sample would be broken. Therefore, some zircon isotope escaped. The average age of 245 Ma might indicate that the late Permian to early-middle Triassic magmatic event happened in the Mengku Quarries area of the Lincang Batholith.

## 3. Tectonic Implications

Previous petrological and geochemical studies [18, 26, 28, 35] indicate that the Lincang granite mainly shows two types of granite with an initial strontium isotope ratio ranging from 0.71 to 0.78 and therefore it belongs to S-type granite and once may have been formed in an environment of tectonic collision. Combined with previous chronological data, we propose a three-stage model on the tectonic evolution between late Permian and Jurassic (Figure 6).

*(1) Regional Collision and Contraction before the Formation of the Lincang Batholith.* Before the formation of the Lincang Batholith, the Paleo-Tethys Ocean and many intraoceanic islands distributed between the Baoshan and the Lanping-Simao-Indosinian blocks [36]. In a regional contractional environment of subduction and collision, it is extremely common to find folding, thrusting, and uplifting phenomena (Figure 7(a)).

*(2) Time Heterogeneity of In Situ Hybrid of the Lincang Batholith.* The Lincang Batholith was controlled by one deep-seated fault along the Lancang Jiang River and proposed in situ hybrid origin of the Batholith in the early to middle Triassic collision between the Baoshan-Shantai and Lanping-Simao-Indochina Blocks [16]. Our study shows an ununiform law result in the aspect of mixing time of the Batholith (Figure 7(b)). The intrusion time in the east Qunsai area was later than in the Mengku area, indicating that the regional contraction in Mengku area was weakened or stopped, and triggered magma poured at ~230 Ma, however, the contraction pattern may still remain in the Qunsai area to the east.

*(3) Tectonic Denudation and Regional Exhumation Happened between Late Triassic and Middle Jurassic, Which Is Supported by ~201 Ma Biotite $^{40}Ar/^{39}Ar$ Age from the Lincang Batholith [27].* This paper research shows that the upper part of the Lincang Batholith experienced long-term denudation, resulting in the development of the weathering crust (Figure 7(a)). The upper crust is the Jurassic Huakaizuo Formation, characterized by carbonate rocks of shallow sea facies, together with some volcano clastic rocks (Figure 5(b)).

TABLE 3: U-Pb isotopic dating for the single grain Zircon from granite of Mengku.

| Analysis no. | $^{207}Pb/^{206}Pb$ Ratio | Error (1σ) | $^{207}Pb/^{235}U$ Ratio | Error (1σ) | $^{206}Pb/^{238}U$ Ratio | Error (1σ) | $^{208}Pb/^{232}Th$ Ratio | Error (1σ) | $^{207}Pb/^{206}Pb$ Ratio | Error (1σ) | $^{207}Pb/^{235}U$ Ratio | Error (1σ) | $^{206}Pb/^{238}U$ Ratio | Error (1σ) | $^{208}Pb/^{232}Th$ Ratio | Error (1σ) |
|---|---|---|---|---|---|---|---|---|---|---|---|---|---|---|---|---|
| 1 | 0.05174 | 0.00244 | 0.32303 | 0.0139 | 0.04528 | 0.00108 | 0.01152 | 0.00053 | 274 | 56 | 284 | 11 | 285 | 7 | 232 | 11 |
| 2 | 0.09115 | 0.00288 | 2.96352 | 0.08448 | 0.23585 | 0.0053 | 0.08061 | 0.00287 | 1450 | 25 | 1398 | 22 | 1365 | 28 | 1567 | 54 |
| 3 | 0.11509 | 0.00332 | 0.50952 | 0.01225 | 0.03212 | 0.00066 | 0.02102 | 0.00068 | 1881 | 19 | 418 | 8 | 204 | 4 | 420 | 13 |
| 4 | 0.04837 | 0.00265 | 0.22077 | 0.01108 | 0.03312 | 0.00085 | 0.01153 | 0.00063 | 117 | 69 | 203 | 9 | 210 | 5 | 232 | 13 |
| 5 | 0.04837 | 0.0022 | 0.27295 | 0.01145 | 0.04095 | 0.00092 | 0.01223 | 0.0006 | 117 | 57 | 245 | 9 | 259 | 6 | 246 | 12 |
| 6 | 0.05351 | 0.00279 | 0.29775 | 0.01415 | 0.04038 | 0.00104 | 0.01371 | 0.00078 | 350 | 62 | 265 | 11 | 255 | 6 | 275 | 16 |
| 7 | 0.0806 | 0.00218 | 2.27431 | 0.05585 | 0.2048 | 0.00394 | 0.05968 | 0.00155 | 1212 | 22 | 1204 | 17 | 1201 | 21 | 1172 | 30 |
| 8 | 0.0802 | 0.00341 | 2.36577 | 0.09085 | 0.2141 | 0.00577 | 0.05352 | 0.00277 | 1202 | 37 | 1232 | 27 | 1251 | 31 | 1054 | 53 |
| 9 | 0.05928 | 0.00248 | 0.29959 | 0.01132 | 0.03668 | 0.00082 | 0.00966 | 0.00046 | 577 | 45 | 266 | 9 | 232 | 5 | 194 | 9 |
| 10 | 0.04989 | 0.00375 | 0.23793 | 0.0163 | 0.03461 | 0.00119 | 0.00907 | 0.00072 | 190 | 96 | 217 | 13 | 219 | 5 | 182 | 14 |
| 11 | 0.07866 | 0.00333 | 0.34683 | 0.01279 | 0.032 | 0.00079 | 0.01233 | 0.00063 | 1164 | 37 | 302 | 10 | 203 | 5 | 248 | 13 |
| 12 | 0.0406 | 0.00217 | 0.22437 | 0.01119 | 0.04011 | 0.00095 | 0.01052 | 0.00055 | -255 | 72 | 206 | 9 | 254 | 6 | 212 | 11 |
| 13 | 0.0639 | 0.00299 | 0.31538 | 0.01317 | 0.03582 | 0.00089 | 0.01126 | 0.00067 | 738 | 48 | 278 | 10 | 227 | 6 | 226 | 13 |
| 14 | 0.07481 | 0.00441 | 0.3517 | 0.01817 | 0.03412 | 0.00108 | 0.01178 | 0.00078 | 1063 | 56 | 306 | 14 | 216 | 7 | 237 | 16 |
| 15 | 0.06821 | 0.00263 | 0.97745 | 0.03376 | 0.10399 | 0.00236 | 0.03233 | 0.00106 | 875 | 37 | 692 | 17 | 638 | 14 | 643 | 21 |
| 16 | 0.05116 | 0.00255 | 0.26673 | 0.01227 | 0.03784 | 0.00088 | 0.01057 | 0.00052 | 248 | 64 | 240 | 10 | 239 | 5 | 213 | 10 |
| 17 | 0.05025 | 0.00304 | 0.25181 | 0.01395 | 0.03636 | 0.00102 | 0.01037 | 0.00064 | 207 | 77 | 228 | 11 | 230 | 6 | 209 | 13 |
| 18 | 0.05379 | 0.003 | 0.27018 | 0.01369 | 0.03645 | 0.00098 | 0.00965 | 0.0006 | 362 | 67 | 243 | 11 | 231 | 6 | 194 | 12 |
| 19 | 0.05083 | 0.00373 | 0.26778 | 0.01802 | 0.03823 | 0.00125 | 0.00987 | 0.00055 | 233 | 96 | 241 | 14 | 242 | 8 | 199 | 11 |
| 20 | 0.06022 | 0.00487 | 0.2998 | 0.0217 | 0.03612 | 0.00142 | 0.01137 | 0.00113 | 611 | 90 | 266 | 17 | 229 | 9 | 229 | 23 |
| 21 | 0.0473 | 0.00214 | 0.20932 | 0.0087 | 0.0321 | 0.0007 | 0.00892 | 0.00038 | 64 | 55 | 193 | 7 | 204 | 4 | 179 | 8 |
| 22 | 0.05405 | 0.00334 | 0.28442 | 0.01602 | 0.03817 | 0.00111 | 0.00858 | 0.0006 | 373 | 75 | 254 | 13 | 241 | 7 | 173 | 12 |
| 23 | 0.05907 | 0.0024 | 0.66212 | 0.02439 | 0.08128 | 0.00181 | 0.03438 | 0.00121 | 570 | 43 | 516 | 15 | 504 | 11 | 683 | 24 |
| 24 | 0.05064 | 0.00245 | 0.21459 | 0.00944 | 0.03072 | 0.00072 | 0.00797 | 0.00046 | 224 | 59 | 197 | 8 | 195 | 5 | 160 | 9 |
| 25 | 0.05594 | 0.00261 | 0.26352 | 0.01113 | 0.03415 | 0.0008 | 0.00964 | 0.00045 | 450 | 53 | 237 | 9 | 216 | 5 | 194 | 9 |
| 26 | 0.06769 | 0.00524 | 0.32217 | 0.02212 | 0.0345 | 0.00134 | 0.01049 | 0.0009 | 859 | 80 | 284 | 17 | 219 | 8 | 211 | 18 |
| 27 | 0.06354 | 0.00206 | 0.75553 | 0.02205 | 0.08618 | 0.00166 | 0.02056 | 0.00084 | 726 | 32 | 571 | 13 | 533 | 10 | 411 | 17 |
| 28 | 0.05014 | 0.00296 | 0.33191 | 0.01801 | 0.04797 | 0.00129 | 0.02068 | 0.0011 | 201 | 77 | 291 | 14 | 302 | 8 | 414 | 22 |
| 29 | 0.23782 | 0.00377 | 6.55914 | 0.08439 | 0.19984 | 0.00325 | 0.01824 | 0.00075 | 3105 | 12 | 2054 | 11 | 1174 | 17 | 365 | 15 |

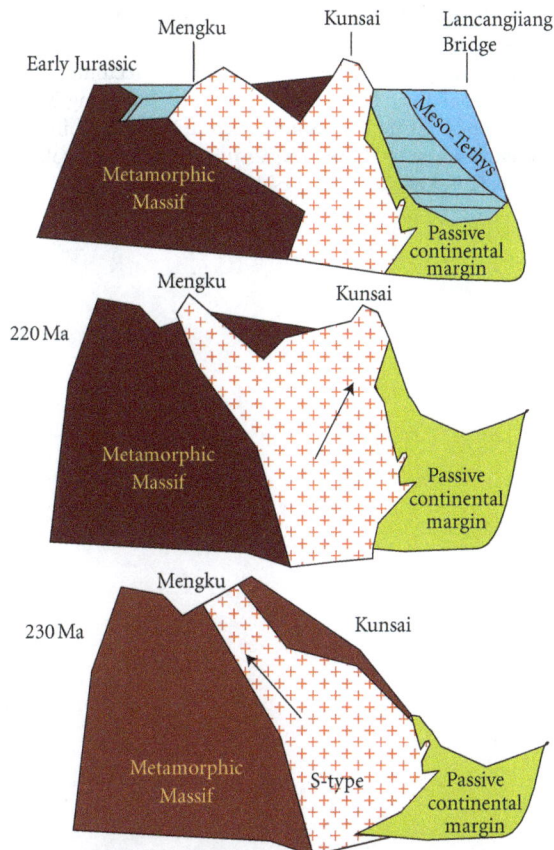

FIGURE 7: Formation and tectonic evolution model for the Lincang Batholith.

Unconformity relationship is shown between the Batholith and the Huakaizuo Formation (Figures 3, 5(a), 5(b), 5(c), and 7(c)). Apparently, the Meso-Tethys expansion also happened in this era and left some effect in the Lincang Area.

## Acknowledgments

The authors appreciate sincerely both the field research and inner discussion together with Professor Yong-Qing Chen from China University of Geosciences (Beijing), Research Fellow Ying-Xiang Lu from Yunnan Bureau of Geology and Mineral Resources, and Research Fellow Fang-Cheng Lin from Chengdu Institute of Geology and Mineral Resources. The authors thank Dr. Pengfei Li from University of Queensland for his perfect academic revision of the paper. The research was supported by the China Geology Survey (no. 200811008, no. 1212011121188), the Ministry of Science and Technology (no. 2006BAB01A03-3), the Chinese National Natural Foundation (no. 90814006) of the People's Republic, and the China University of Geosciences in Beijing (no. 2-9-2001-280).

## References

[1] J. Q. Huang and B. W. Chen, *The Evolution of the Tethys in China and Adjacent Regions*, Geological Publishing House, Beijing, China, 1987.

[2] P. Jian, D. Liu, A. Kröner et al., "Devonian to Permian plate tectonic cycle of the Paleo-Tethys Orogen in southwest China (II): insights from zircon ages of ophiolites, arc/back-arc assemblages and within-plate igneous rocks and generation of the Emeishan CFB province," *Lithos*, vol. 113, no. 3-4, pp. 767–784, 2009.

[3] H. D. Klemme and G. F. Ulmishe, "Effective petroleum source rocks of the world: stratigraphic distribution and controlling depositional factors," *AAPG Bulletin*, vol. 75, no. 12, pp. 1809–1851, 1991.

[4] G. T. Pan, Z. L. Chen, and X. Z. Li, *Geological Structure Formation and Evolution of East Tethys*, Geological Publishing House, Beijing, China, 1997.

[5] G. T. Pan, H. X. Zheng, Y. R. Xu et al., "A preliminary study on Bangong Co-Nujiang Suture," in *Editing Committee of Geological Memoirs of Qinghai-Xizang Plateau(12)-Geological Tectonics of "Sanjiang"*, pp. 229–242, Geological Publishing House, Beijing, China, 1983.

[6] X. Wang, I. Metcalfe, P. Jian, L. He, and C. Wang, "The Jinshajiang-Ailaoshan Suture Zone, China: tectonostratigraphy, age and evolution," *Journal of Asian Earth Sciences*, vol. 18, no. 6, pp. 675–690, 2000.

[7] A. M. C. Sengor and U. J. Botum, "The Tethys orogenic system: an introduction," in *Tectonic Evolution of the Tethyan Region*, A. M. C. Sengor, Ed., pp. 1–22, Kluwer Academic, Boston, Mass, USA, 1989.

[8] Q. Yan, Z. Wang, S. Liu et al., "Opening of the Tethys in southwest China and its significance to the breakup of East Gondwanaland in late Paleozoic: evidence from SHRIMP U-Pb zircon analyses for the Garze ophiolite block," *Chinese Science Bulletin*, vol. 50, no. 3, pp. 256–264, 2005.

[9] P. A. Ziegler, *Evolution of Laurussia: A Study in Late Palaeozoic Plate Tectonics*, Kluwer Academic, Dordrecht, The Netherlands, 1989.

[10] B. P. Liu, Q. L. Feng, C. Chonglakmani et al., "Framework of paleotethyan archipelago ocean of western Yunnan and its elongation towards north and south," *Earth Science Frontiers*, vol. 9, no. 3, pp. 161–171, 2002.

[11] C. S. Hutchison, *Geological Evolution of Southeast Asia*, vol. 13 of *Oxford Monographs on Geology and Geophysics*, Clarendon Press, Oxford, UK, 1989.

[12] H. R. Wu, C. A. Boulter, B. Ke et al., "The Changning-Menglian suture zone: a segment of the major Cathaysian-Gondwana divide in Southeast Asia," *Tectonophysics*, vol. 242, no. 3, pp. 267–280, 1995.

[13] H. Y. Zhang and J. L. Liu, "Plate tectonics and mineralization of the Tethyan ophiolite in the southern Sanjiang and Indo-China peninsular area," *Earth Science, Journal of China University of Geosciences*, vol. 36, no. 2, pp. 262–276, 2011 (Chinese).

[14] P. F. Fan, "Accreted terranes and mineral deposits of Indochina," *Journal of Asian Earth Sciences*, vol. 18, no. 3, pp. 343–350, 2000.

[15] M. J. Lin, "The volcanic rock sequence and its geological age along the bank of Lancang Jiang, West Yunnan" (Chinese), See: Geology Papers of Qinghai-Tibet Plateau (13)-Rocks in Sanjiang Area, pp. 151–158, 1983.

[16] J. Y. Xiong, Y. M. Lin, and S. R. Qin, "A preliminary study on the genesis and the basic characteristics of the Lincang magmatic complex" (Chinese), See: Geology Papers of Qinghai-Tibet Plateau (13)-Rocks in Sanjiang Area, pp. 1–19, 1983.

[17] S. Y. Yu, K. Q. Li, Y. P. Shi, and H. H. Zhang, "A study on the granodiorite in the middle part of Lincang granite batholiths," *Yunnan Geology*, vol. 22, no. 4, pp. 426–442, 2003.

[18] C. S. Liu, J. C. Zhu, X. S. Xu, X. J. Chu, D. K. Cai, and P. Yang, "Study on the characteristics of Lincang Composite Granite Batholith in West Yunnan," *Yunnan Geology*, vol. 8, no. 3-4, pp. 189–212, 1989 (Chinese).

[19] X. L. Li, "Basic characteristics and formation structural environment of Lincang Composite granite batholiths," *Yunnan Geology*, vol. 15, no. 1, pp. 1–18, 1996 (Chinese).

[20] J. C. Chen, "Discussion on the age division and the selects of isotopic age determination for granitic rock in Western Yunnan," *Yunnan Geology*, vol. 6, no. 2, pp. 101–113, 1987 (Chinese).

[21] X. X. Mo, S. Y. Shen, Q. W. Zhu et al., *Volcanic-Ophiolite and Mineralization of Middle-Southern Part in Sanjiang Area of Southwestern China*, Geological Publishing House, Beijing, China, 1998.

[22] D. L. Zhong, *Paleo-Tethys Orogenic Zone in Western Yunnan and Western Sichuan*, Science Press, Beijing, China, 1998.

[23] A. Socquet and M. Pubellier, "Cenozoic deformation in western Yunnan (China-Myanmar border)," *Journal of Asian Earth Sciences*, vol. 24, no. 4, pp. 495–515, 2005.

[24] R. G. Park, *Foundations of Structural Geology*, Chapman & Hall, New York, NY, USA, 2nd edition, 1989.

[25] C. H. Zhang, J. S. Liu, D. L. Liu, and S. Yang, "Geological, geochemical characteristics, age and tectonic setting of Laomaocun small rockbody in South Lancangjiang Zone," *ACTA Mineralogica Sinica*, vol. 26, no. 3, pp. 317–324, 2006.

[26] J. C. Chen, "Characteristics of Pb, Sr isotopic compositions in west Yunnan granites: discussion on the age and nature of the basement in west Yunnan," *Scientia Geologica Sinica*, vol. 2, pp. 174–183, 1991.

[27] T. M. Dai, B. Q. Zhu, Y. Q. Zhang, Z. P. Pu, Q. F. Zhang, and A. S. Hong, "Collision and thermial history of Indian-Sandaland-Eurasian Plates as implicated by $^{40}Ar/^{39}Ar$ age spectra of granodiorites," *Geochimica*, vol. 15, no. 2, pp. 97–107, 1986.

[28] Y. Q. Zhang, Y. W. Xie, and J. W. Wang, "Rb and Sr isotopic studies of granitoids in Tri-river region," *Geochimica*, vol. 19, no. 4, pp. 318–326, 1990 (Chinese).

[29] W. Y. Wang, "Rb-Sr isotopic Data," *Yunnan Geology*, vol. 3, no. 1, pp. 79–96, 1984 (Chinese).

[30] Yunnan Bureau of Geology and Mineral Resources (YBGM), *Geoogical Map of the Jinggu Sheet at a Scale of 1:200, 000*, Ministry of Geology and Mineral Resources, Beijing, China, 1983.

[31] Yunnan Bureau of Geology and Mineral Resources (YBGM), *Geological Map of the Gengma Sheet at a Scale of 1:200, 000*, Ministry of Geology and Mineral Resources, Beijing, China, 1985.

[32] P. Xu, F. Wu, L. Xie, and Y. Yang, "Hf isotopic compositions of the standard zircons for U-Pb dating," *Chinese Science Bulletin*, vol. 49, no. 15, pp. 1642–1648, 2004.

[33] S. E. Jackson, N. J. Pearson, W. L. Griffin, and E. A. Belousova, "The application of laser ablation-inductively coupled plasma-mass spectrometry to in situ U-Pb zircon geochronology," *Chemical Geology*, vol. 211, no. 1-2, pp. 47–69, 2004.

[34] K. Ludwig, *Isoplot/Ex version 2.0: A Geochronological Tool Kit for Microsoft Excel*, Geochronology Center Berkeley, Special Publication, 1999.

[35] B. W. Chappell and A. J. R. White, "I- and S-type granites in the lachlan fold belt, Southeast Australia," in *Geology of Granites and their Metallogenetic Relations*, X. Keqin and T. Guangchi, Eds., pp. 72–78, Science Press, Bejing, China, 1982.

[36] B. P. Liu, Q. L. Feng, N. Q. Fang et al., "Tectonic evolution of Paleo-Tethys Poly-island-ocean in Changning-Menglian and Lancangjiang Belts, south-western Yunnan, China," *Earth Science, Journal of China University of Geosciences*, vol. 18, no. 5, pp. 529–539, 1993 (Chinese).

# 4

# The Mesozoic Tectonic Dynamics and Chronology in the Eastern North China Block

**Quanlin Hou,[1] Qing Liu,[1] Hongyuan Zhang,[2] Xiaohui Zhang,[3] and Jun Li[3]**

[1] *Graduate University of Chinese Academy of Sciences, Beijing 100049, China*
[2] *School of Earth Sciences and Resources, China University of Geosciences, Beijing 100083, China*
[3] *Institute of Geology and Geophysics, Chinese Academy of Sciences, Beijing 100029, China*

Correspondence should be addressed to Quanlin Hou, quhou@gucas.ac.cn

Academic Editor: Yu-Dong Wu

Mesozoic tectonic events in different areas of the eastern North China Block (NCB) show consistency in tectonic time and genesis. The Triassic collision between NCB and Yangtze results in the nearly S-N strong compression in the Dabie, Jiaodong, and west Shandong areas in Middle Triassic-Middle Jurassic. Compression in the Yanshan area in the north part of NCB was mainly affected by the collision between Mongolia Block and NCB, as well as Siberia Block and North China-Mongolia Block in Late Triassic-Late Jurassic. However, in the eastern NCB, compressive tectonic system in Early Mesozoic was inversed into extensional tectonic system in Late Mesozoic. The extension in Late Mesozoic at upper crust mainly exhibits as extensional detachment faults and metamorphic core complex (MCC). The deformation age of extensional detachment faults is peaking at 120–110 Ma in Yanshan area and at 130–110 Ma in the Dabie area. In the Jiaodong area eastern to the Tan-Lu faults, the compression thrust had been continuing to Late Mesozoic at least in upper crust related to the sinistral strike slipping of the Tan-Lu fault zone. The extensional detachments in the eastern NCB would be caused by strong crust-mantle action with upwelling mantle in Late Mesozoic.

## 1. Introduction

In recent years, Mesozoic tectonic regime inversion is becoming one of the focuses in the studies of tectonic evolution and geodynamics. The Mesozoic tectonic dynamics process in the eastern North China is one of the most typical examples. EW-trending structural frame was transformed to the NE-NNE-trending structural framework and a compressive tectonic system to extensional tectonic system during Mesozoic. This intracontinental geodynamic process has attracted sight of geologists from China and around the world. Several hypotheses, such as long-distance effect of Pacific Plate [1], comprehensive constraints of the adjacent blocks (including the subducted and extinct Kula Plate, [2]), adjustment of the stress after the deep subduction of the Dabie area [3], large-scale sinistral strike slip, mantle plume [4, 5], delamination of lithosphere, and/or continental root-plume tectonics, had been put forward during the last decade. These ideas, even though being of great benefit to the study on Mesozoic tectonic regime inversion, have neither given rise a complete

theoretical framework nor concluded a uniform tectonic dynamics process. It should be noticed that China continent stands on the junction of three plates including the Tethyan oceans, the paleo-Asian ocean, and the paleo-Pacific ocean. The relative movement among the three plates, especially the Pacific plate which subducted NW warding beneath China continent, strongly affects China continent in the Mesozoic era. Many other important tectonic events, such as the continental deep subduction along the Dabie mountains, the lithosphere thinning of the eastern China, and the rapid uplifting of crust in western China, took place during a relative short period, although slightly different. With respect to the study of inversion process of tectonic regime in the eastern North China Block (NCB), the correlation and effect of the adjacent tectonic blocks should be considered, especially long-distance effect of the paleo-Pacific in the east and paleo-Tethyan in the west, and mantle upwelling would be considered as a reasonable tectonic model [6, 7]. It is recognized that the Mesozoic tectonic dynamics and chronology is one key problem to understand the Mesozoic

TABLE 1: Mineral $^{40}$Ar-$^{39}$Ar data in the main shear zone of the eastern Dabie Orogenic Belt [11].

| Shear zone | Mineral | Weighted mean plateau age (Ma) |
| --- | --- | --- |
| Xiaotian-Mozitan shear zone | Biotite | 124.17 ± 0.25 |
| | Biotite | 126.91 ± 0.30 |
| Shuihou-Wuhe shear zone | Biotite | 190.59 ± 0.42 |
| Taihu-Mamiao shear zone | Hornblende | 197.41 ± 0.46 |
| | Biotite | 189.42 ± 0.29 |
| | Biotite | 124.87 ± 0.21 |
| Susong-Qingshuihe shear zone | White mica | 194.01 ± 0.36 |
| | Biotite | 127.96 ± 0.30 |

tectonic regime inversion in the eastern NCB. Three areas, the Dabie area and the Jiaodong area at the south margin and the Yanshan tectonic belt at the north part of the eastern NCB, would be discussed here to recognize the Mesozoic tectonic dynamics process in the eastern NCB.

## 2. Mesozoic Tectonic Dynamics at the Dabie Area

The Dabie area, located between NCB and Yangtze block in Early Mesozoic, is at the south margin of the eastern NCB. It is a key region to understand the Mesozoic tectonic regime inversion from compression to extension in the eastern NCB.

*2.1. Collisional Time Constraints during the Early Mesozoic.* The time range could be indicated by three aspects as follows. (1) The ultrahigh-pressure metamorphic assemblages started inversion at about 220–240 Ma and cooled down to 300°C at 180 Ma (time of the $^{39}$Ar-$^{40}$Ar biotite clock start-up, [8, 9]). (2) Geomagnetism studies indicate that visual polar migration curve of Hehuai basin, located at the south margin of NCB, had not been consistent with that of the Yangtze block until 159 Ma [10], which indicated that the two blocks had jointed and the collision orogeny had completed by the time (159 Ma). (3) The sedimentation of the foreland molasse basin in the Badong-Echeng area in Hubei province began at Middle Triassic and finished in the Late Jurassic, which implies that the collisional orogeny developed from the Middle Triassic to the Late Jurassic. Therefore, the collisional orogeny of the Dabie area started at late stage of Early Triassic (~240 Ma) and ended at Middle Jurassic (~180 Ma).

$^{40}$Ar/$^{39}$Ar dating of biotite and hornblende from the Shuihou-Wuhe shear zone and the Taihu-Mamiao shear zone, at the north and south sides of ultrahigh-pressure metamorphic zone in the Dabie area, respectively, was carried out (Table 1 after [11]). The results indicated that the first stage deformation occurred at about 190 Ma, probably reflecting the inversion and cooling of the ultrahigh-pressure metamorphic rocks in the postorogeny. It was suggested that, with the weakening of the orogenic activity and increasing of potential energy of gravity of the orogenic belt in the postorogeny, the principal compressional stress direction changed from horizontal to vertical, leading to the orogenic

laxity or collapse, and the rapid inversion of the ultrahigh-pressure metamorphic rocks; then the nearly completion of collision in the Dabie area is about 190 Ma. Perhaps this 190 Ma event extended to 160 Ma in the late Jurassic because of the subduction of the paleo-Pacific plate [12].

*2.2. The Late Mesozoic Extensional Detachment.* The Northern Dabie metamorphic complex was identified as the crystalline metamorphic core of the Late Mesozoic extensional tectonics, which divided the Dabie area into two extensional tectonic systems, the north extensional tectonic system and the south extensional tectonic system (Figures 1 and 2).

*2.2.1. The North Extensional Tectonic System.* The identification of extensional crenulation cleavage (C′) in the north detachment zones indicated that it is a complete extensional tectonic system, extending NE or NNE along the Xiaotian-Mozitan faults where the principal extensional detachment zone was developed in the early Cretaceous (Figure 1). The shear strain ($\gamma$) in the central part of the north detachment zone is up to about 2.6 and gradually decreased outward. The measurements of rock finite strains in the detachment zone indicated that the shear displacement is at least 56 km.

*2.2.2. South Extensional Tectonic System.* The south extensional tectonic system is composed of, from north to south, the Shuihou-Wuhe shear zone, Taihu-Mamiao shear zone, and Susong-Qingshuihe shear zone, which could represent lower, middle, and upper detachment system, respectively, (Figure 1). Analysis of fabric and dislocation density of deformed quartz indicates that deformation temperature (>700°C, 650–700°C) and differential stress (92 Mpa, 70–84 Mpa) descended from north to south in the south extensional tectonic system. The shear displacement of the upper detachment zone is at least 12 km by measurements of deformed rock finite strains. The directions of the extensional detachment are SSW, SSE, and/or S. The Flinn parameters ($K$) of strain ellipsoids of the lower, middle and upper detachment systems are 0.01–0.1, 1, and >7 respectively, representing accordingly flattened strain ($K$ = 0.01–0.1), plane strain ($K \approx 1$), and extensional strain ($K$ > 7) from north to south, which implies that the extensional detachment could be caused mainly by magma intrusions, consisting with the timing indicates.

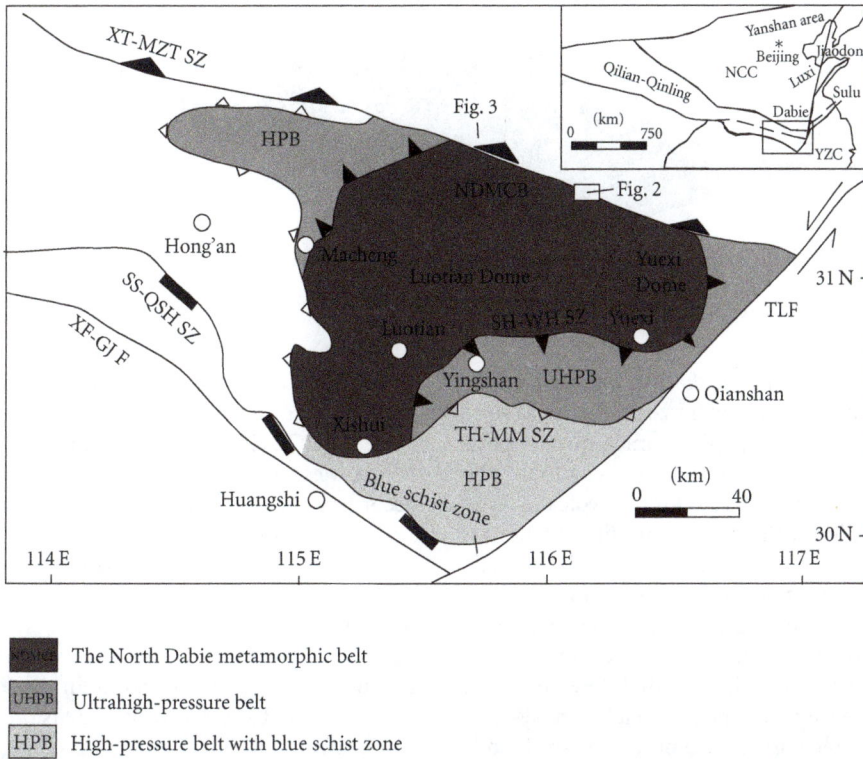

FIGURE 1: Sketch map of Late Mesozoic main shear zones in eastern Dabie Orogenic Belt, Central China (after [13]). NCP: Northern China Block; YZB: Yangzi Block; NHEB: North Huaiyang metamorphic belt; NDMCD: Northern Dabie metamorphic complex belt; UHPB: ultrahigh pressure metamorphic belt; HPB: high-pressure metamorphic belt; SH-WH SZ: Shuihou-Wuhe shear zone; TH-MM SZ: Taihu-Mamiao shear zone; SS-QSH SZ: Susong-Qingshuihe shear zone; XT-MZT SZ: Xiaotian-Mozitan shear zone; TLF: Tancheng-Lujiang fault (Tanlu Fault); XF-GJF: Xiangfan-Guangji fault.

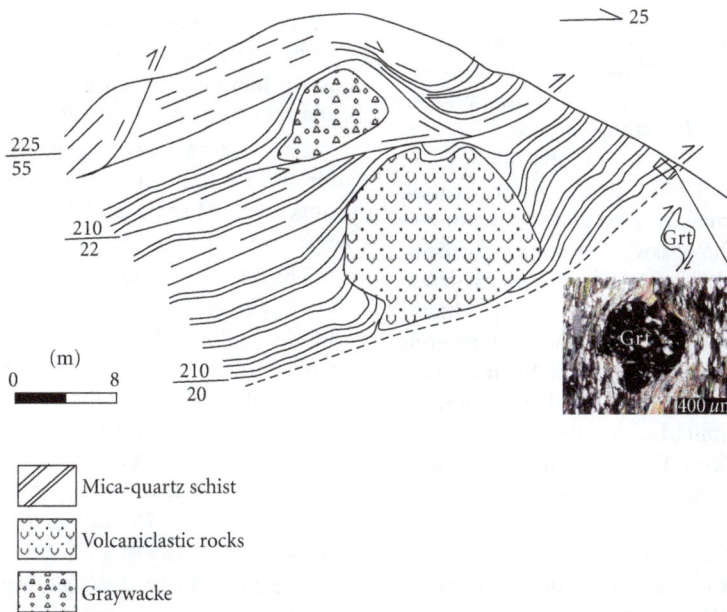

FIGURE 2: Late Jurassic volcaniclastic tectonite blocks were contained in mica-quartz schist (Pt$_2$n), north extensional detachment zone, implying the extension shearing after Late Jurassic. The microstructure picture of mica-quartz schist is shown in the right below the profile, which possibly indicates a dextral rotation of kinematics in the profile. The phenomenon is located near the Zhutang Village of Youdian Town in Jinzhai County of the Anhui Province.

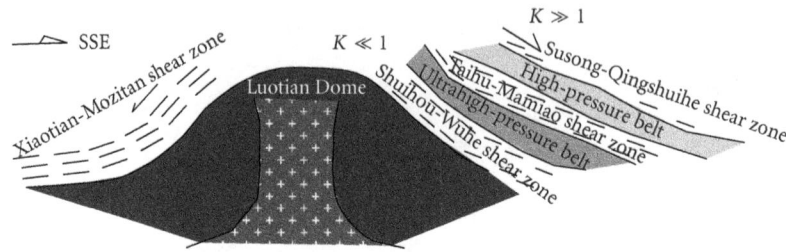

FIGURE 3: Extensional detachment model of Late Mesozoic, the Dabie Orogenic Belt.

*2.3. The Chronology Constraints on the Extensional Detachment.* In north extensional detachment zone, the shear zone developing in the Xinyang Group (Pt$_2$n) mica-quartz schist contains some allochthonous blocks from less than one meter to several meters in size of Late Jurassic volcaniclastic rocks and tuff. In addition, there are some field-scale harmonic recumbent folds in the Late Jurassic volcaniclastic rocks and the Late Proterozoic mica schist. Both Hong'an Group (Pt) and the Yanshanian granite (J$_3$-K$_1$) experienced extensional shear deformation together in the south extensional detachment system, which implies that the extensional detachment of both the south and north detachment systems took place after the Late Jurassic. The biotite and hornblende $^{40}$Ar/$^{39}$Ar data from the north extensional tectonic system and in the Susong-Qingshuihe shear zone in the south detachment system reveal that the age of latter deformation is about 125 Ma (Table 1, after [11] (Figure 1)), which should represent the time of extensional detachment after orogeny. As a whole, the extensional deformation in the Dabie area mainly took place about 120 Ma ago.

*2.4. Genesis of Extensional Detachment.* At about 130–120 Ma, a little earlier than the detachment deformation age, numerous extensive granite and ultrabasic-basic plutons intruded into the Northern Dabie complex core. The $^{40}$Ar-$^{39}$Ar and Rb-Sr isotopic analyses of rocks in the North Dabie also recorded the time at about 110–130 Ma. These chronology data imply that the magma intrusion played a major rule during the extensional detachment, which were supported by the finite strain measure analysis of rocks in the south extensional tectonic system (Figure 3). In addition, the ductile deformation of the plagioclase and pyroxene of pyroxenite and gabbros (115–130 Ma) in the Northern Dabie metamorphic complex (mainly refer to the Luotian Dome) implies that they once reached granulite facies during deformation and then uplift to the surface. Therefore, it is concluded that voluminous magmatic emplacement and delamination of lithosphere in Early Cretaceous induced the rapid uplift of the Northern Dabie as the center area of UHPM rocks and extensional detachment on both sides of the Northern Dabie complex core. The intense denudation results in the voluminous sediments in the Hefei basin. The high-pressure and ultrahigh-pressure eclogites probably emplaced during this process on the basis of our data (Figure 3).

Additionally, the distributions of the platinum group elements (PGEs) in the ultramafic and/or mafic rocks

(~120 Ma) in the Northern Dabie complex core show that their source regional upper mantle in Later Mesozoic enriched PGE (Pd ≈ 6.2 ppb in mantle). The original mantle before Mesozoic, however, depleted the PGE (Pd ≈ 2.3 ppb), according to the PGE distribution in Bixiling ultramafic rocks (older than 240 Ma) in the Northern Dabie complex core [14, 15]. The Later Mesozoic upper mantle with PGE enrichment in the Dabie area is supposed to be contaminated by about 8‰ Earth core materials, as the PGE contents in Earth core are much higher than in upper mantle [14, 15]. If thus, the enrichment mantle in the PGE must have been related to the Later Mesozoic extensional detachment in the Dabie area.

# 3. Mesozoic Tectonic Dynamics in the Jiaodong Area

The Jiaodong area, especially the eastern Jiaodong peninsula is located in the eastern terminal of the Dabie-Sulu ultrahigh pressure metamorphic belt, the eastern part of the south margin of the eastern NCB. As part of the collision orogenic belt between NCB and Yangtze block in Early Mesozoic, the eastern Jiaodong peninsula is significant in regional tectonic dynamics study in the eastern NCB.

The eastern Jiaodong peninsula can be divided into four lithotectonic units by four shear zones, the Shidao, Rongcheng, Mishan, and Mouping faults from SE to NW, respectively (Figure 4).

The Shidao shear zone is a broad ductile shear zone (>15 km in width) thrusting toward north and sinistral strike slipping in ENE-WSW direction. The total shear displacement is more than 90 km, in which the strike displacement is at least 19 km, the thrusting displacement >88 km, and the horizontal shrinkage >85 km, horizontal shrinkage ration about 68% [18]. If Rongcheng, Mishan, and Mouping nappes were taken as an entirety, it turned about 25° clockwise (Figure 4), similar to the Korea peninsula which turned 30.5° clockwise [19].

The Rongcheng shear zone is a sinistral strike slipping and thrusting ductile shear zone dipping S or SSE in early stage (Figure 4). Different rocks, such as porphyritic granite ($\gamma_5^3$, ~120 Ma), granitic gneiss, and hornblende schists (Pt$_1$) of the Weideshan rock bodies, overrode one another as thrusting sheets (Figure 5), indicating a strong thrusting and sinistral strike slipping during late Yanshanian epoch.

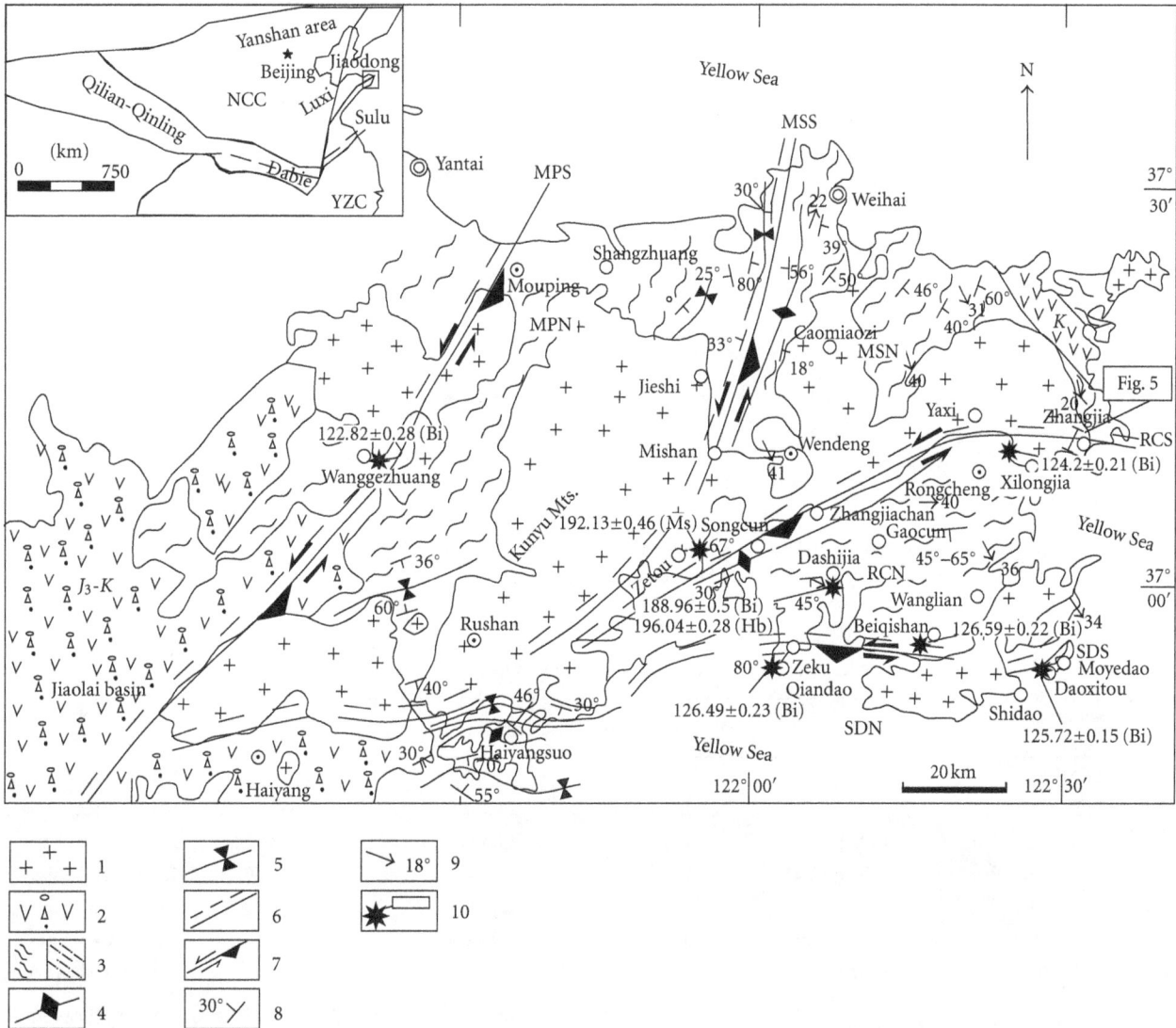

FIGURE 4: Geological map of the eastern Jiaodong peninsula (modified from [16, 17]). (Granite); (volcanic sedimentary rocks from Late Jurassic to Cretaceous); (gneiss rocks and mylonite zones); (syncline); (unticline); (brittle-ductile shear zones); (left lateral and reverse brittle shear sense); (layer occurrence); (lineation occurrence); (sample position and geochronological results). $F_1$, $F_2$, $F_3$, $F_4$: faults, namely, Shidao, Rongcheng, Mishan, and Mouping; I, II, II IV: nappes, namely, Shidao, Rongcheng, Mishan, and Mouping.

However, it could not be excluded that Rongcheng shear zone would experience extensional strike-slip nature at late stage.

The Mishan shear zone is an almost upright sinistral strike slipping fault zone striking NNE, in which a tectonoclastic rock belt of wide was formed. Rocks have a notable change across the fault. A few meters width of mylonite bands was observed in the fault zone, indicating that it could be a ductile shear zone in the early stages, and then was reconstructed later by brittle fractures. Therefore, it is difficult to identify the attitude and nature of the zone. However, it holds somewhat that the later stage of the shear zone was generally in agreement with the brittle shear zones in the Eastern Jiaodong peninsula.

The Mouping shear zone is a compression sinistral strike-slip fault striking NE-SW in steep dip. The shear zone expressed as a brittle-ductile tectonite belt (20–30 km in width) and had extensional characteristics in late stage.

Mouping shear zone could be influenced by Tan-Lu faults, because similar tectonic behavior was proved in both the Mouping shear zone and the Tan-Lu fault zone, such as sinistral slip and long time of fault development.

The paleodifferential stress values of the Shidao, Rongcheng and Mouping shear zones are of 105 Ma, 99 Ma, and 85 MPa, respectively, showing a decrease tendency from south to north. The strain measurement and the quartz c-fabric analysis indicate that the deformation degree of those three shear zones decrease in order from south to north. The Flinn values ($K$) of the strain ellipsoid range between 0.6 and 0.75, implying flattening strain. Therefore, the thrusting in the Eastern Jiaodong peninsula should be from south to north in general.

$^{40}$Ar/$^{39}$Ar chronological analyses on the biotite and hornblende from the Shidao, Rongcheng, and Mouping shear zones (Table 2, [20]) indicated two groups, ~190 Ma and

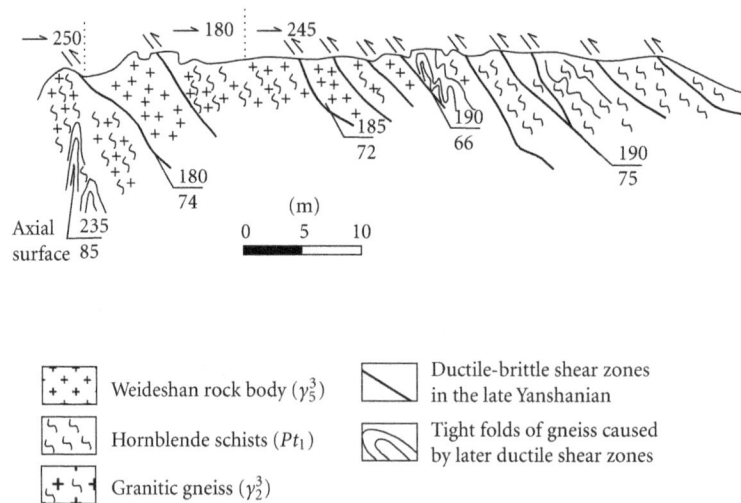

FIGURE 5: Structural outcrop sketch at Zhangjia, Rongcheng. At outcrop scale the porphyritic granite ($\gamma_5^3$), granitic gneiss, and hornblende schists (Pt$_1$) of the Weideshan rock body were found to be overriding together as the style of thrusting sheets.

TABLE 2: $^{40}$Ar/$^{39}$Ar precise dating results of individual minerals in the eastern Jiaodong Peninsula [20].

| Sample position and sample number | Lithology | Minerals selected | Serial number | $^{40}$Ar/$^{39}$Ar plateau age, Ma | Tectonic setting of samples |
|---|---|---|---|---|---|
| Qiandao01120905 | Granite gneiss | Biotite | R03020 | 126.49 ± 0.23 | Shidao shear zone |
| Beiqishan01121002 | Granite gneiss | Biotite | R03002 | 126.59 ± 0.22 | Shidao shear zone |
| Daoxitou01121005 | Granite gneiss | Biotite | R03026 | 125.72 ± 0.15 | Shidao shear zone |
| Dashijia01120908B | Granite gneiss | Biotite | R03005 | 188.96 ± 0.50 | Rongcheng nappe |
| Dashijia01120908H | Granite gneiss | Hornblende | R03019 | 196.04 ± 0.28 | Rongcheng nappe |
| Zetou01120807 | Granite mylonite | Muscovite | R03009 | 192.13 ± 0.46 | Mishan nappe |
| Xilongjia01121101 | Granite gneiss | Biotite | R03008 | 124.20 ± 0.21 | Rongcheng shear zone |
| Wanggezhuang01121504 | Quartzofeldspathic mylonite | Biotite | R03025 | 122.82 ± 0.28 | Mouping shear zone |

~120 Ma, amazingly similar to those of the Dabie area. The two groups might represent the ages of the reversion of ultra-high pressure metamorphic rocks during the postorogenic structural relaxation and the slip compressional deformation after the orogeny, respectively.

As well known, in the Early Cretaceous, almost the whole North China area was tectonically in an extensional or extruding setting. However, the large-scale sinistral strike-slipping of the Tan-Lu fault zone resulted in the sinistral strike-slipping compressional settings at least at the shallow part of the crust in the east side front of the Tan-Lu faults, to form the sinistral strike-slipping thrust nappes and the clockwise rotation of the blocks bounded by the shear zones. The subduction of the west Pacific plate toward NW and/or NNW resulted in the strike slipping and the northward movement at the east side of the Tan-Lu fault zone. Meanwhile, the deep part of the crust in the Jiaodong peninsula might be under the extensional tectonic background, according to the relative data of basins and volcanic rocks.

## 4. Mesozoic Tectonic Dynamics in the Yanshan Area

The Yanshan area, one famous intraplate orogen considered by some authors (e.g., [21, 22]), is located at the north part of the eastern NCB. The Yanshan area is the cradle of Chinese geologists. Since Wong [23] put forward "Yanshan Movement," lots of attentions have been paid by geologists in the aspect of the Mesozoic strong tectonic deformation and severe magma activities. In recent years, the Mesozoic complex intraplate deformation has become one major focus of continental dynamics research in China.

The eastern segment of the Yanshan area is characterized with two ductile shear zones trending EW and NNE respectively, which controlled the tectonic framework of the northern North China. Field observations and structural analyses reveal that the EW trending ductile shear zones are mainly contributed to dextral compressional deformation that resulted from top-to-the-southeast oblique thrust

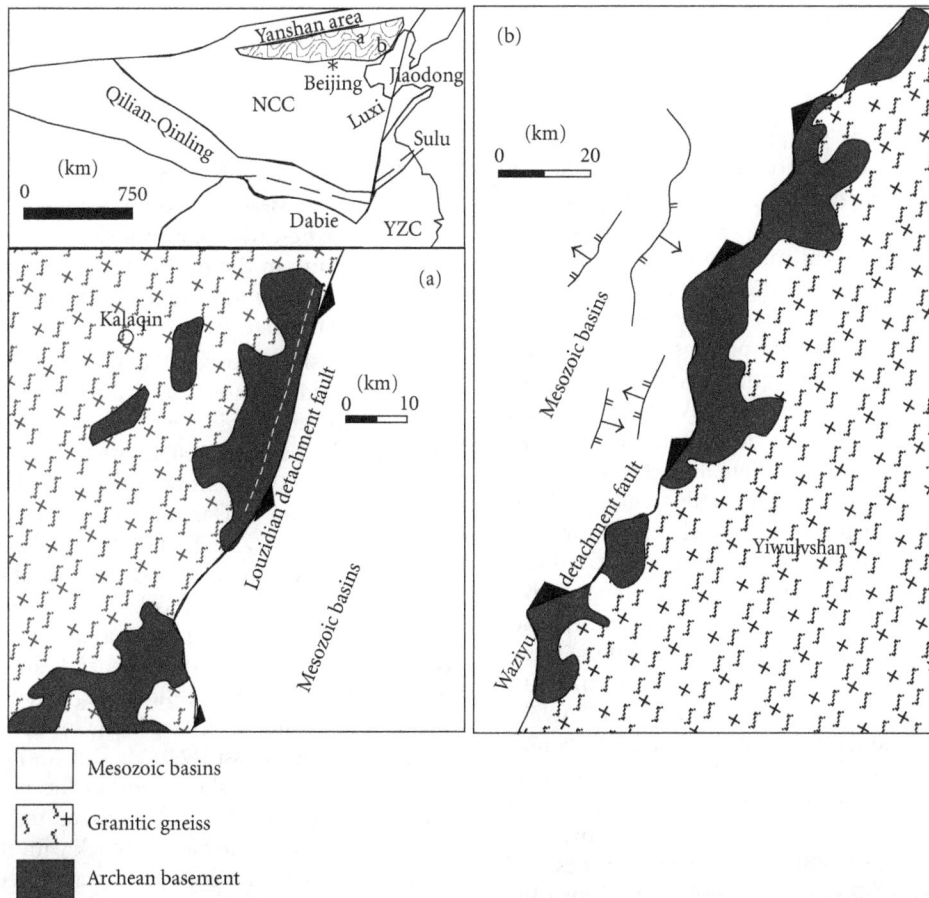

FIGURE 6: Geological map of the Eastern Yanshan Area. Simplified geological map showing major tectonic units of the Yanshan belt and its surrounding areas, with the locations of Kalaqin and Yiwulüshan metamorphic core complexes indicated by a rectangle. (a) Sketch geological map of the Kalaqin metamorphic core complex (modified after [29]); (b) Sketch geological map of the Yiwulüshan metamorphic core complex (adopted from [32]).

shearing, whereas the NNE trending ones are genetically related to sinistral strike slip and extensional faulting. The $^{40}Ar/^{39}Ar$ geochronology indicates that the top-to-the-southeast oblique thrusting event recorded by the EW-trending ductile shear zones mainly occurred in Late Triassic (Yanshanian, too) [24]. Recent studies have clarified a series of major Mesozoic contractional events in the Yanshan area [21, 25]. The Huairou thrust north of Beijing has overridden the Cambrian carbonates above the Middle Jurassic Tiaojishan formation ($J_2$) at Late Jurassic with a southern displacement direction. The Gubeikou-Pingquan fault features both south-vergent thrusting and dextral strike slipping with an age span from 148 to 132 Ma [21]. In summary, the activity period for contraction and thrusting at the Yanshan tectonic belt are mainly concentrated at Late Triassic and Late Jurassic.

Similar to most orogens, localized regions of large-magnitude extensional strain typified by metamorphic core complexes are also prominent geologic and physiographic features of the Yanshan tectonic belt. The examples include the Yunmengshan metamorphic core complex north of Beijing [26–28], the Kalaqin metamorphic core complex in the

Inner Mongolia [29–31], and the Yiwulüshan metamorphic core complex in western Liaoning [31, 32].

As described by Han et al. [29] and Shao et al. [30], the Kalaqin metamorphic core complex is a major structural component of the Ma'anshan uplift with a length of about 100 km and a width of about 35 km (Figures 6(a) and 6(b)). The core of the uplift consists of Archean gneisses of epidote-amphibolite facies (>2.5 Ga), marbles and TTG series, Early Proterozoic schists and marbles of low green schist facies, Late Proterozoic Minganshan group slates, Cambrian carbonates, Early Permian-Middle Jurassic plutonic rocks, and Early Cretaceous granitic rocks. The eastern and western flanks of the uplift consist mainly of Jurassic-Cretaceous terrestrial volcanic and sedimentary strata. The Louzidian detachment fault, which consists of the Louzidian ductile shear zone and the related Louzidian normal fault, separates metamorphosed complexes in the footwall to the west from the unmetamorphosed Mesozoic and Cenozoic sedimentary covers in the hanging wall. Field observations and structural analyses reveal that the ductile shear zone was genetically related to sinistral strike slip and extensional faulting. $^{40}Ar/^{39}Ar$ geochronology established an Early Cretaceous age for its formation [31].

As first reported by Ma et al. [33], geochronologically constrained by Zhang et al. [31] and further defined structurally by Darby et al. [32], the Yiwulüshan of western Liaoning displays many of the structural, metamorphic, and igneous features that characterize Cordilleran metamorphic core complexes. Situated along the eastern segment of the Yanshan area, the Yiwulüshan metamorphic core complex has an east-west dimension of 25 km and a north-south dimension of 60 km (Figure 6(c)). The core consists predominantly of Mesozoic granitoid plutons and their country rocks, which include migmatite, epidote-amphibolite facies paragneiss, and orthogneiss of the Archean crystalline basement. Plutons are granodioritic to granitic and grade outward into augen gneiss or migmatite. At structurally higher levels, it is the well-developed NNE-trending Waziyu ductile shear which, together with the related normal fault above it along the eastern margin of the Fuxin basin, separates the core of the dome from the hanging wall of the unmetamorphosed to low-grade metamorphosed Middle-Late Proterozoic Changcheng System and the Mesozoic sedimentary and volcanic rocks. Three synkinematic biotites from the Waziyu ductile shear zone give a $^{40}Ar/^{39}Ar$ plateau age span of 130–116 Ma and thus establish its main activity period as Early Cretaceous [31, 32]. Zircon U-Pb geochronology indicates that three representative batholiths (Lüshan, Jianlazishan, and Hengshan) at the core were mainly emplaced during Middle-to-Late Jurassic times, with ages of 164, 153, and 153 Ma, respectively [34]. Four other mineral samples from the core and the eastern flank give $^{40}Ar/^{39}Ar$ ages of 133 Ma (muscovite), 140 Ma (biotite), 156 Ma (K-feldspar), and 165 Ma (hornblende), respectively, (unpublished data), indicting the extensional exhumation of the metamorphic core complex range from east to west.

As a case with the Basin and Range Province of the Western United States, the isolated metamorphic core complexes within the Yanshan area, such as the Yunmengshan, Kalaqin, and Yiwulüshan metamorphic core complexes, represent the high-strain extension regions within a broad region of more distributed extensional deformation during Early Cretaceous [25–27, 30, 33, 35–38]. Their formation are contributes to gravitational collapse of an orogenically thickened crust that was facilitated by thermal weakening due to localized deep-seated plutonism and paleo-Pacific plate boundary reorganization [25, 31, 32].

# 5. Discussion and Conclusion

The statistics indicate that the main Mesozoic magmatic, metamorphic, and tectonic events on the Dabie, Luxi, Jiaodong, and Yanshan areas of the eastern NCB from south to north are consistent in time and genesis (Table 3, [39]). In the eastern NCB, the nearly S-N compression in Early Mesozoic dominated the formation of a series of thrusts trending almost E-W. In Cretaceous, the extension dominated the formation of a series of extensional detachment structures and metamorphic core complex. Many scholars had studied

the tectonic setting of the tectonic regime inversion in eastern NCB and gotten different opinions [7, 15, 40–43]. On the base of our studies and former researches, this paper discusses this issue briefly.

*5.1. The Early Mesozoic Compression Setting.* In the south of the NCB, the Mesozoic compressions occurred in Early-Middle Triassic-early Late Jurassic, such as the Dabie and Luxi areas, but, in the north, occurred a little later, such as the Yanshan area which experienced two periods: in the Late Triassic and the Late Jurassic.

The collision between the Yangtze block and the NCB began at Middle-Late Triassic, the Dabie-Sulu ultrahigh pressure metamorphic belt was formed at 242–224 Ma and cooled down to 300°C at 180 Ma [8, 9], which is approximately consistent with the early deformation age of ductile shear zone in Dabie-Sulu orogenic belt (Ar-Ar age of ~190 Ma, Tables 1 and 2, [11, 18]). Perhaps it was the geological record of orogenic relaxation at about 180–190 Ma, indicating the ending of collision orogenesis. The visual polar migration curve of NCB and Yangtze Block became consistency at 159 Ma [10]. In general, the collision between the NCB and the Yangtze Block might cause the nearly S-N strong compression thrusting in the southern NCB in Middle Triassic-Middle Jurassic (240 Ma–180 Ma) [20, 44].

The compressive shortening in the northern NCB was mainly affected by the collisions of the Mongolia block with the NCB and the Siberia block with the North China-Mongolia block. The collision time between the Mongolia block and the NCB was at about 250 Ma–200 Ma. This collision caused nearly S-N compression and formed a series of E-W striking thrust nappes of 250–200 Ma in Yanshan area, such as the thrust-nappes thrusting southward at 219 Ma in Yiwulüshan. The Mongolia-Okhotsk paleoocean gradually closed eastward in Early-Middle Jurassic (190–170 Ma) and caused collision between the Siberia block and the North China-Mongolia block. The collision might keep on until Late Jurassic (160–150 Ma) and even later [42, 45–48], which formed the Late Jurassic thrust-nappes thrusting southward in the Yanshan tectonic belt, such as the thrust nappes of the Cambrian overriding the Tiaojishan formation ($J_{2-3}$) in the Huaibei town of the Huairou area and the Gubeikou-Pingquan thrust fault in northern Beijing. During the relaxation period of compression, however, episodic extensional stress fields resulted in a series of Jurassic volcanic rocks.

*5.2. Late Mesozoic Extensional Tectonic Setting.* In the Late Mesozoic, the tectonic regime was inversed into extensional setting in the NCB. The extensions in the shallow crust were mainly displayed by extensional detachment faults and metamorphic core complex. The deformation age of extensional detachment shear zones in Yanshan tectonic belt is 140–100 Ma with a peak period of 120–110 Ma, while in the Dabie and the Luxi area it is 130–110 Ma. The intrusion ages of the magmatic rocks in the metamorphic core complex mostly are 150–110 Ma, only a few reached 170–160 Ma, a little earlier than the detachment shear deformation on the

TABLE 3: Time-range index of main Mesozoic structure events in the eastern North China Block [39].

| Structure belt | Dabie | | | West Shandong | Yanshan | |
|---|---|---|---|---|---|---|
| Rock type | (Ultra) mafic rocks | Granite/shear zone | Metamorphic rock (migmatite) | Intermediate-mafic rock | Intrusive | Shear zone |
| Isotopic age of compression (Ma) | 220–240 | 190 | 230–240 | 230–250 | 165–180 | ~150 ~219 |
| Isotopic age of extension (Ma) | 120–130 | ~120 | 117–130 | 115–130 | 111–130 | 116–140 |
| Structural index of compression | The oldest deposit in foreland molasse basin is Mid-Triassic and the youngest one is upper Jurassic | | | $\gamma_5^3$ undeformed | J$_2$ or J$_3$ (Tiaojishan group) deformed | |
| Structural index of extension | J$_3$, $\gamma_5^3$ deformed | | J$_3$ or K$_1$ deformed | K$_1$ deformed | | |

whole. Though some areas such as Yanshan tectonic belt still show compression in shallow crust where thrust napping was developing, underplating and delamination had developed deeper.

It usually takes about tens to hundreds million years for deep underplating and delamination to exhibit extensional detachment structure in the shallow crust. Chronology and deformed rock strain data (such as the Dabie extensional detachment belt (Figure 3)) showed that magma intrusion directly caused extensional detachment. Some geologists related the extension to the NNW or NW subduction of the west Pacific plate into the Eurasia plate. Many evidences indicate that East-Asia circum-Pacific active continental margin began at Middle Jurassic (180 Ma), peaking at Late Jurassic-Early Cretaceous, equivalent to the B episode of the movement in the Yanshan area [42].

In Jurassic, the NCB was acted by three plates, such as the Yangtze, Siberian, and west Pacific plates, and with the orogenesis going on, the crust gradually thickened, potential energy difference increased, and the main compression stress gradually changed from horizontal to vertical. Especially with the thickening of the NCB crust, the lithosphere became unbalanced and the asthenosphere upwelled. Lithosphere underplating and delamination could cause lithosphere rebound; then partial melting, magma intrusion and volcanism occurred, which resulted in the uplift extension detachment of the crust. But whether the underplating and delamination in the eastern North China, beginning at Late Jurassic-Cretaceous, was related to the west Pacific plate subduction still needs to be further studied.

Some researchers suggested that the Late Mesozoic extensional tectonic setting in the eastern North China were mainly resulted from mantle branch structures [3–5, 49, 50]. This is a good way to interpret the Late Mesozoic thermal dome-extensional tectonic setting of the eastern North China According to the study of the PGE distributions, it was suggested that the mantle was rich in PGE at 120 Ma, which is apparently different from the previous PGE-deplete mantle. Therefore, we speculate that the materials from the earth core was added to the Late Mesozoic mantle (at least 8‰, [15]), which supported the opinion of mantle branch structure to some degree.

In the Jiaodong peninsula, however, the sinistral strike-slip compressions dominated in Late Mesozoic, forming a series of thrust-nappes thrusting northward, and the deformation age of ~120 Ma (Table 2, [20]). The sinistral strike slipping of Jiaodong and the Tan-Lu faults could be related to the NE-ward subduction of the west Pacific plate. Therefore, sinistral strike-slip-compression stress field was generated in the east front of the Tan-Lu Faults zone, resulting in the sinistral strike-slip thrust faults and the clockwise rotation (at about 25°) of the blocks bounded by the thrust faults in the Jiaodong peninsula, similar to the Korea peninsula. Then deep structures could not be consistent with shallow ones and deep underplating and delamination; even magma intrusion and volcanism could have happened, but they were not immediately displayed in the shallow. It also suggested that the Tan-Lu Faults zone could be an important tectonic boundary, which caused significant difference in Mesozoic tectonic setting on the both sides of the Tan-Lu Faults.

## Acknowledgments

The authors would like to thank Professor Jiliang Li, Professor Wenjiao Xiao, Professor Mingguo Zhai, and Professor Zhihong Wang from Institute of Geology and Geophysics, Chinese Academy of Sciences, and Dr. Tianshan Gao from university of Science and Technology of China for their kind help in field work and discussion. This work has been supported by the following grants: the National Natural Science Foundation of China (Grants 41030422; 90714003).

## References

[1] S. Hu, Y. Zao, Z. Hu, J. Guo, and B. Xu, "Evolution and development of tectonics and magmatism at the active continental margin of east China (E 106°) during Mesozoic and Cenozoic," *Acta Petrologica Sinica*, vol. 10, no. 4, pp. 370–381, 1994 (Chinese).

[2] J. S. Ren, B. G. Niu, and J. Zheng, "Tectonic frame and geodynamic evolution of eastern China," *Geological Research*, vol. 29, no. 30, pp. 43–55, 1997 (Chinese).

[3] J. F. Deng, H. L. Zhao, and X. X. Mo, *Continental Roots-Plume Tectonics of China*, Beijing, China, Geological Publishing House, 1996.

[4] J. Deng, S. Su, Y. Niu et al., "A possible model for the lithospheric thinning of North China Craton: evidence from the Yanshanian (Jura-Cretaceous) magmatism and tectonism," *Lithos*, vol. 96, no. 1-2, pp. 22–35, 2007.

[5] S. V. Lysak, "Thermal history, geodynamics, and current thermal activity of lithosphere in China," *Russian Geology and Geophysics*, vol. 50, no. 9, pp. 815–825, 2009.

[6] M. Zhai, J. Yang, and W. Liu, "The large-scale metallogenesis and large-scale cluster of gold-deposits in eastern Shandong, China," *Science in China D*, vol. 44, no. 8, pp. 758–768, 2001.

[7] M. Zhai, R. Zhu, J. Liu et al., "Time range of Mesozoic tectonic regime inversion in eastern North China Block," *Science in China D*, vol. 47, no. 2, pp. 151–159, 2004.

[8] B. R. Hacker, L. Ratschbacher, L. Webb, T. Ireland, D. Walker, and D. Shuwen, "U/Pb zircon ages constrain the architecture of the ultrahigh-pressure Qinling-Dabie Orogen, China," *Earth and Planetary Science Letters*, vol. 161, no. 1—4, pp. 215–230, 1998.

[9] L. Ratschbacher, B. R. Hacker, A. Calvert et al., "Tectonics of the Qinling (Central China): tectonostratigraphy, geochronology, and deformation history," *Tectonophysics*, vol. 366, no. 1-2, pp. 1–53, 2003.

[10] S. A. Gilder, P. H. Leloup, V. Courtillot et al., "Tectonic evolution of the Tancheng-Lujiang (Tan-Lu) fault via Middle Triassic to Early Cenozoic paleomagnetic data," *Journal of Geophysical Research B*, vol. 104, no. 7, pp. 15365–15390, 1999.

[11] Q. Hou, Q. Liu, J. Li, and H. Zhang, "Late Mesozoic shear zones and its chronology in the Dabie Mountains, Central China," *Scientia Geologica Sinica*, vol. 42, no. 1, pp. 114–123, 2007.

[12] S. W. Dong, J. M. Hu, S. Z. Li et al., "The Jurassic deformation in the Dabie Mountains and its tectonic significances," *Acta Petrologica Sinica*, vol. 21, no. 4, pp. 1189–1194, 2005.

[13] S. Suo, Z. Zhong, and Z. You, "Extensional deformation of post ultrahigh-pressure metamorphism and exhumation process of ultrahigh-pressure metamorphic rocks in the Dabie massif, China," *Science in China D*, vol. 43, no. 3, pp. 225–236, 2000.

[14] Q. Liu, *Study on the distribution of Platinum group elements in Dabie (ultra-) mafic rocks and Fuxin volcanic rocks [Ph.D. dissertation]*, Graduate University of the Chinese Academy of Sciences, Beijing, China, 2005.

[15] Q. Liu, Q. L. Hou, X. H. Zhou, and L. W. Xie, "The distribution of platinum-group elements in gabbros from Zhujiapu, Dabie orogen," *Acta Petrologica Sinica*, vol. 21, no. 1, pp. 227–239, 2005.

[16] M. Zhai, J. Guo, Q. Wang, K. Ye, B. Cong, and W. Liu, "Division of petrological-tectonic units in the Northern Sulu ultra high pressure zone: an example of thick—skin thrust of crystalline units," *Scientia Geologica Sinica*, vol. 35, no. 1, pp. 16–26, 2000 (Chinese).

[17] M. Yang and G. Lü, *The Geology-Geochemistry of Gold Deposits of the Greenstone Belt in Jiaodong District, China*, Geological Publishing House, Beijing, China, 1996.

[18] H. Y. Zhang, Q. L. Hou, and D. Y. Cao, "Study of thrust and nappe tectonics in the eastern Jiaodong Peninsula, China," *Science in China D*, vol. 50, no. 2, pp. 161–171, 2007.

[19] X. X. Zhao, R. S. Coe, H. K. Chang et al., "Clockwise rotations recorded in Early Cretaceous rocks of South Korea: Implications for tectonic affinity between the Korea Peninsula and North China," *Geophysical Journal International*, vol. 139, no. 2, pp. 447–463, 1999.

[20] H. Y. Zhang, Q. L. Hou, and D. Y. Cao, "Tectono-chronologic constraints on a Mesozoic slip and thrust belt in the eastern Jiaodong Peninsula," *Science in China D*, vol. 50, no. 1, pp. 25–32, 2007.

[21] C. Zhang, G. Wu, D. Xu, G. Wang, and W. Sun, "Mesozoic tectonic framework and evolution in the central segment of the intraplate Yanshan orogenic belt," *Geological Bulletin of China*, vol. 23, no. 9-10, pp. 864–875, 2004 (Chinese).

[22] J. Hu, X. Liu, Y. Zhao, G. Xu, J. Liu, and S. Zhang, "On Yanshan intraplate orogene: an example from Taiyanggou area, Lingyuan, Western Liaoning Province, Northeast China," *Earth Science Frontiers*, vol. 11, no. 3, pp. 255–271, 2004 (Chinese).

[23] W. H. Wong, "Crustal movements and igneous activities in eastern China since Mesozoic time," *Bulletin of the Geological Society of China*, vol. 6, no. 1, pp. 9–37, 1927.

[24] X. Zhang, T. Li, and Z. Pu, "$^{40}$Ar/$^{39}$Ar thermochronology of two ductile shear zones from Yiwulüshan, West Liaoning Region: age constraints on the Mesozoic tectonic events," *Chinese Science Bulletin*, vol. 47, no. 13, pp. 1113–1118, 2002.

[25] G. A. Davis, Y. Zheng, C. Wang, B. J. Darby, C. Zhang, and G. E. Gehrels, "Mesozoic tectonic evolution of the Yanshan fold and thrust belt, with emphasis on Hebei and Liaoning provinces, northern China," in *Paleozoic and Mesozoic Tectonic Evolution of Central and Eastern Asia: From Continental Assembly to Intracontinental Deformation*, M. S. Hendrix and G. A. Davis, Eds., vol. 194 of *Geological Society of America Memoir*, pp. 171–197, 2001.

[26] G. A. Davis, X. Qian, Y. Zheng et al., "The Huairou (Shuiyu) ductile shear zone, Yunmengshan Mts.," in *Proceedings of the 30th international Geological Congress Field Trip Guide*, Beijing Geological Publishing House, Beijing, China, 1996.

[27] G. A. Davis, X. Qian, Y. Zheng et al., "Mesozoic deformation and plutonism in the Yunmengshan: a Chinese metamorphic core complex north of Beijing, China," in *The Tectonic Evolution of Asia*, A. Yin and T. M. Harrison, Eds., pp. 253–280, Cambridge University Press, Cambridge, UK, 1996.

[28] G. A. Davis, Y. Zheng, C. Wang, B. J. Darby, C. Zhang, and G. E. Gehrels, "Geometry and geochronology of Yanshan belt tectonics," in *Proceedings of the 100th Anniversary Celebration of Peking University: Collected Works of International Symposium on Geological Science*, pp. 275–292, Department of Geology, Peking University, Beijing, China, 1998.

[29] B. F. Han, Y. Zheng, J. Gan, and Z. Chang, "The Louzidian normal fault near Chifeng, inner Mongolia: master fault of a quasi-metamorphic core complex," *International Geology Review*, vol. 43, no. 3, pp. 254–264, 2001.

[30] J. A. Shao, L. Q. Zhang, W. Jia, and P. Y. Wang, "Harkin metamorphic core complex in Inner Mongolia and its upwelling mechanism," *Acta Petrologica Sinica*, vol. 17, no. 2, pp. 283–290, 2001 (Chinese).

[31] X. Zhang, H. Wang, and Y. Ma, "$^{40}$Ar/$^{39}$Ar age constraints on two NNE-trending ductile shear zones, Yanshan intraplate Orogen, North China Craton," *International Geology Review*, vol. 45, no. 10, pp. 936–947, 2003.

[32] B. J. Darby, G. A. Davis, X. Zhang, F. Wu, S. Wild, and J. Yang, "The newly discovered Wuziyu metamorphic core complex, Yiwulü Shan, western Liaoning province, North China," *Earth Science Frontiers*, vol. 11, no. 3, pp. 145–155, 2004.

[33] Y. Ma, S. Cui, G. Wu et al., "The structural feature of metamorphic core complex in Yiwulüshan mountains, western Liaoning," *Acta Geoscientia Sinica*, vol. 20, pp. 385–391, 1999 (Chinese).

[34] X. H. Zhang, Q. Mao, H. F. Zhang, and S. A. Wilde, "A Jurassic peraluminous leucogranite from Yiwulüshan, western Liaoning, North China craton: age, origin and tectonic significance," *Geological Magazine*, vol. 145, no. 3, pp. 305–320, 2008.

[35] G. A. Davis, B. J. Darby, Y. Zheng, and T. L. Spell, "geometric and temporal evolution of an extensional detachment fault, Hohhot metamorphic core complex, Inner Mongolia, China," *Geology*, vol. 30, pp. 1003–1006, 2002.

[36] L. E. Webb, S. A. Graham, C. L. Johnson, G. Badarch, and M. S. Hendrix, "Occurrence, age, and implications of the Yagan-Onch Hayrhan metamorphic core complex, southern Mongolia," *Geology*, vol. 27, no. 2, pp. 143–146, 1999.

[37] C. L. Johnson, L. E. Webb, S. A. Graham, M. S. Hendrix, and G. Badarch, "Sedimentary and structural records of Late Mesozoic high-strain extension and strain partitioning, east Gobi basin, Southern Mongolia," in *Paleozoic and Mesozoic Tectonic Evolution of Central and Eastern Asia: From Continental Assembly to Intracontinental Deformation*, M. S. Hendrix and G. A. Davis, Eds., vol. 194 of *Geological Society of America Memoir*, pp. 413–433, 2001.

[38] J. Ren, K. Tamaki, S. Li, and Z. Junxia, "Late Mesozoic and Cenozoic rifting and its dynamic setting in Eastern China and adjacent areas," *Tectonophysics*, vol. 344, no. 3-4, pp. 175–205, 2002.

[39] Q. L. Hou, Y. D. Wu, F. Y. Wu, M. G. Zhai, J. H. Guo, and Z. Li, "Possible tectonic manifestations of the Dabie-Sulu orogenic belt on the Korean Peninsula," *Geological Bulletin of China*, vol. 27, no. 10, pp. 1659–1666, 2008.

[40] M. Zhai, R. Zhu, J. Liu et al., "Time range of Mesozoic tectonic regime inversion in eastern North China Block," *Science in China D*, vol. 33, no. 10, pp. 913–920, 2003.

[41] Y. Zhao, Z. Yang, and X. Ma, "Geotectonic transition from Paleoasian system and Paleotethyan system to Paleopacific active continental margin in eastern Asia," *Scientia Geologica Sinica*, vol. 29, no. 2, pp. 105–119, 1994.

[42] Y. Zhao, G. Xu, S.-H. Zhang et al., "Yanshanian movement and conversion oftectonic regimes in East Asia," *Earth Sciences Frontiers*, vol. 11, no. 3, pp. 319–328, 2004.

[43] Z. Yadong, G. A. Davis, W. Cong, B. J. Darby, and Z. Changhou, "Major Mesozoic tectonic events in the Yanshan belt and the plate tectonic setting," *Acta Geologica Sinica*, vol. 74, no. 4, pp. 289–302, 2000.

[44] R. Y. Zhang, J. G. Liou, and W. G. Ernst, "The Dabie-Sulu continental collision zone: a comprehensive review," *Gondwana Research*, vol. 16, no. 1, pp. 1–26, 2009.

[45] A. M. Ziegler, P. M. Ree, D. B. Rowley, A. Bekker, Q. Li, and M. Hulver, "Mesozoic assembly of Asia: constraints from fossil floras. Tectonics, and paleomagnetism," in *Tectonic Evolution of Asia*, T. M. A. Harrison, Ed., pp. 371–400, Cambridge University Press, Cambridge, UK, 1996.

[46] A. Yin and S. Nie, "A Phanerozoic palinspastic reconstruction of China and its neighboring regions," in *Tectonic Evolution of Asia*, A. Yin and T. M. Harrison, Eds., pp. 442–485, Cambridge University Press, Cambridge, UK, 1996.

[47] L. Zonenshain, M. Kuzmin, and L. Natapov, *Geology of RSSR: A Plate Tectonic Synthesis*, vol. 21 of *Geodynamic Series*, American Geophysical Union, Washington, DC, USA, 1990.

[48] Q. R. Meng, "What drove late Mesozoic extension of the northern China-Mongolia tract?" *Tectonophysics*, vol. 369, no. 3-4, pp. 155–174, 2003.

[49] S. Niu, Z. Hou, and A. Sun, "The anti-gtavityu migration of metallogenic fluid from core and mantle," *Earth Science Frontiers*, vol. 8, no. 3, pp. 95–101, 2001.

[50] J. Zheng, S. Y. O'Reilly, W. L. Griffin, F. Lu, M. Zhang, and N. J. Pearson, "Relict refractory mantle beneath the eastern North China block: significance for lithosphere evolution," *Lithos*, vol. 57, no. 1, pp. 43–66, 2001.

# 5

# Fold-to-Fault Progression of a Major Thrust Zone Revealed in Horses of the North Mountain Fault Zone, Virginia and West Virginia, USA

**Randall C. Orndorff**

*U.S. Geological Survey, 926A National Center, Reston, VA 20192, USA*

Correspondence should be addressed to Randall C. Orndorff, rorndorf@usgs.gov

Academic Editor: David T. A. Symons

The method of emplacement and sequential deformation of major thrust zones may be deciphered by detailed geologic mapping of these important structures. Thrust fault zones may have added complexity when horse blocks are contained within them. However, these horses can be an important indicator of the fault development holding information on fault-propagation folding or fold-to-fault progression. The North Mountain fault zone of the Central Appalachians, USA, was studied in order to better understand the relationships of horse blocks to hanging wall and footwall structures. The North Mountain fault zone in northwestern Virginia and eastern panhandle of West Virginia is the Late Mississippian to Permian Alleghanian structure that developed after regional-scale folding. Evidence for this deformation sequence is a consistent progression of right-side up to overturned strata in horses within the fault zone. Rocks on the southeast side (hinterland) of the zone are almost exclusively right-side up, whereas rocks on the northwest side (foreland) of the zone are almost exclusively overturned. This suggests that the fault zone developed along the overturned southeast limb of a syncline to the northwest and the adjacent upright limb of a faulted anticline to the southeast.

## 1. Introduction

The North Mountain fault is a major structure that extends about 150 mi (240 km) from Rockbridge Co., central Virginia, northward to south-central Pennsylvania in the eastern U.S. Along the fault, Cambrian and Ordovician dominantly carbonate rocks occur in the hanging wall at the present erosion level and Silurian and Devonian clastic rocks occur in the footwall. The fault is a zone, which is as much as 1-mile wide, that contains rocks from Cambrian to Devonian age in a series of fault slices or horses. In the study area, the present leading edge lies on a footwall frontal ramp that rises from a lower décollement in Cambrian rocks. To the south, the North Mountain fault loses displacement in Rockbridge Co., where horizontal shortening is transferred to the Staunton-Pulaski thrust system located to the southeast [1].

Rader and Perry Jr. [2], Kulander and Dean [1, 3], and Dean et al. [4] interpreted fault genesis to a fault-propagation fold. In this model, the overturned limb of the fault-propagation fold is preserved in the footwall and total displacement is less than 10 mi (16 km). An alternative interpretation [5] based on seismic reflection data describes the North Mountain fault as the leading edge of a large thrust-sheet complex that duplicates Cambrian and Ordovician rocks in the Great Valley. Evans [5] estimated more than 35 mi (56 km) of displacement.

The North Mountain thrust sheet occupies an area between the Blue Ridge Mountains on the east and the Valley and Ridge province on the west. This area of the Central Appalachians is physiographically defined by the Great Valley or Shenandoah Valley (Figure 1). The thrust sheet is composed of approximately 10,000 ft (3,050 m) of Cambrian and Ordovician carbonate rocks. Within the eastern portion of

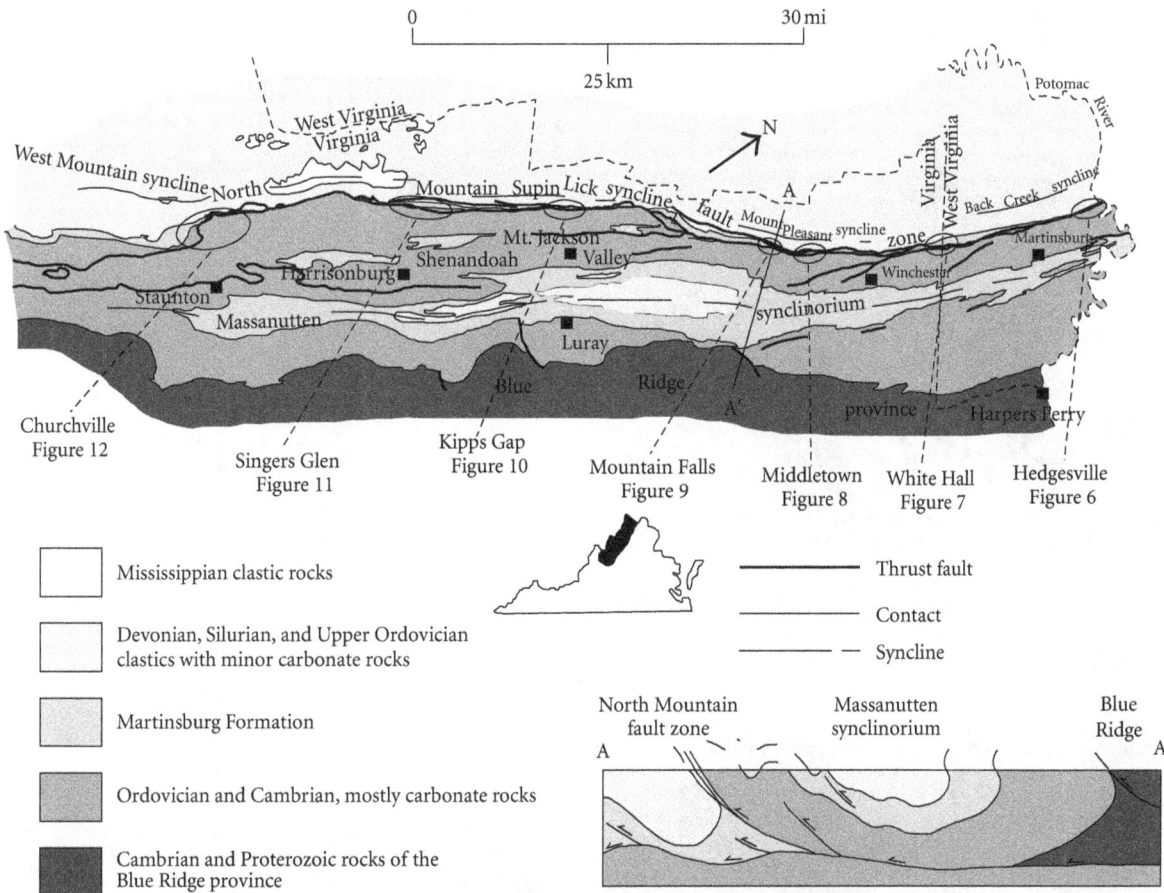

FIGURE 1: Generalized geologic map and cross section of part of the Central Appalachians showing the North Mountain fault zone and areas of detailed geologic mapping.

the North Mountain thrust sheet, Paleozoic carbonate rocks are typically steeply dipping and overturned. Within the western portion of the thrust sheet, the Paleozoic carbonate rocks are typically right-side up and southeast dipping. The major structure on the North Mountain thrust sheet is the doubly plunging Massanutten synclinorium, which extends from central Virginia to south-central Pennsylvania (Figure 1). Just west of the leading edge of the thrust sheet, a discontinuous ridge (Little North Mountain) is composed of the Silurian Tuscarora Quartzite. In places where the Tuscarora does not occur as isolated horses or as a continuous stratigraphic horizon on the footwall syncline, a ridge is not present (Figure 2).

Giles [6] was the first to recognize and describe a thrust fault along Little North Mountain in northern Virginia. He described the feature as a discontinuity between the Cambrian Elbrook Formation and the overturned and overridden Ordovician Martinsburg Formation. Butts and Edmundson [7] found little evidence of complex faulting. They suggested that thinning of stratigraphic units was due to depositional processes. Later, Butts and Edmundson [8] interpreted the rocks near North Mountain as a continuous sequence of previously unrecognized unconformities cut by a single fault. These early workers did not recognize the Ordovician rocks exposed as horses within the fault zone.

Their mapping showed the Elbrook Formation occurs in the hanging wall and rocks of the Martinsburg Formation and Tuscarora Quartzite occur in the footwall.

## 2. Lithotectonic Units

Rocks of the Valley and Ridge province of the North Mountain thrust sheet and the footwall of the thrust sheet in northern Virginia and West Virginia can be divided into three lithotectonic units bounded by detachments (Figure 3). The lower unit is comprised of the Middle Cambrian Elbrook Formation through the Middle Ordovician Edinburg Limestone (or Chambersburg Limestone in the West Virginia Panhandle). This unit is dominantly carbonate rock with medium- to thick-bedded dolomite and limestone in the lower part and thinner bedded carbonate rocks in the upper part. Structurally, large-amplitude folds occur in the basal part of the section and tighter folds occur in the upper part of the section.

The middle lithotectonic unit is comprised of shale, siltstone, sandstone, and shaly limestone of the Middle and Upper Ordovician Martinsburg Formation. This unit is 5,000 ft (1,524 m) thick and is complexly folded, faulted, and cleaved. The Martinsburg is exposed in the Massanutten synclinorium and within the North Mountain fault zone.

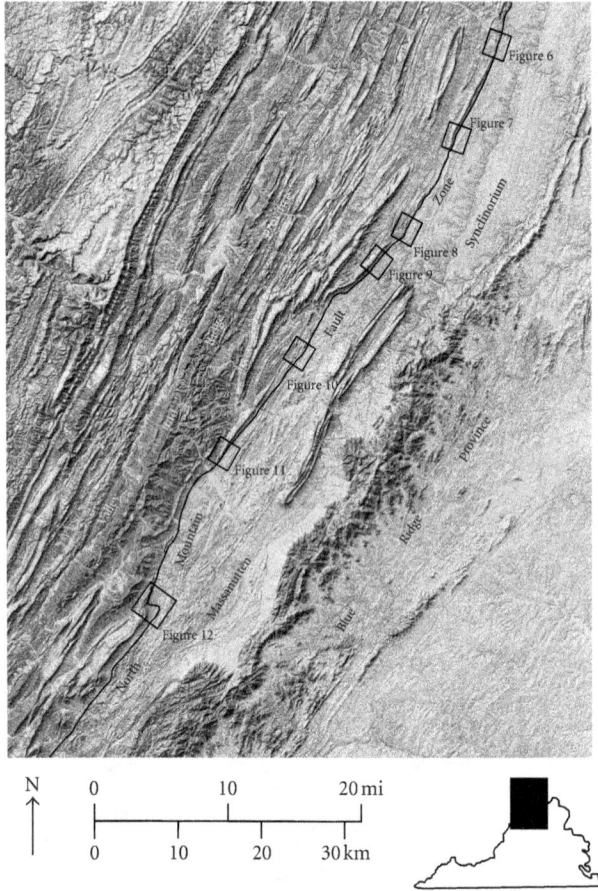

FIGURE 2: Digital elevation model (developed from U.S. Geological Survey National Elevation Dataset; http://ned.usgs.gov/) of part of the Central Appalachians showing the North Mountain fault zone and Massanutten synclinorium.

The upper lithotectonic unit consists of dominantly clastic rocks of the uppermost Ordovician through Devonian. These rocks are the youngest exposed in the Massanutten synclinorium and are also the footwall of the North Mountain fault zone. Evans [5] discussed the lithotectonic units in terms of competent and incompetent units, where incompetent units are detachments. In these terms, the lower lithotectonic unit is bounded by the Waynesboro Formation or Waynesboro detachment at the base and the Martinsburg detachment or middle lithotectonic unit above.

## 3. Detailed Geology along the North Mountain Fault Zone

Detailed geologic mapping holds the key to understanding the development of the North Mountain fault zone (Figure 4). 1 : 24,000-scale geologic mapping along the fault shows the typical map patterns and fault geometry of horses along a 70-mile (113-km) stretch from central Virginia to the eastern West Virginia panhandle (Figure 1).

In the areas mapped, the hanging wall of the North Mountain fault zone consists of southeast-dipping Cambrian

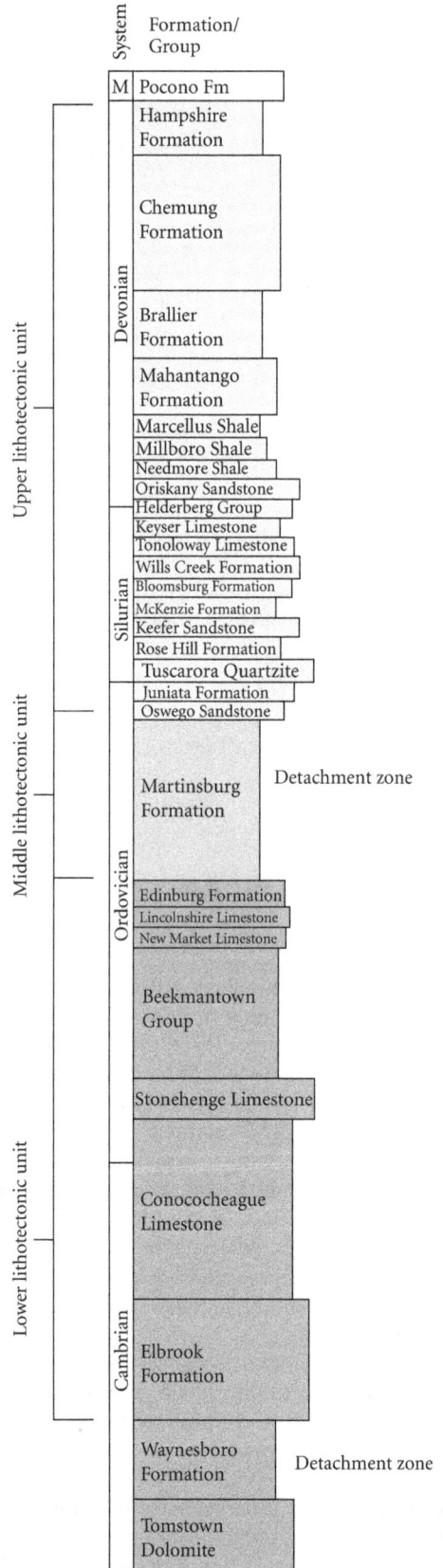

FIGURE 3: Stratigraphic column of geologic units exposed in the Great Valley and Valley and Ridge provinces showing lithotectonic units and detachment zones. Thicknesses are relative and not to scale. Modified from [5].

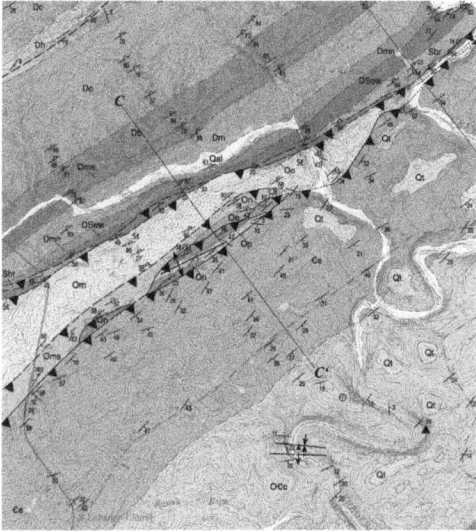

FIGURE 4: Section of the Middletown, Virginia, 7-1/2 minute geologic map [13] showing typical 1 : 24,000-scale geologic mapping used for this study. Figures 6–12 are from, or modified from, geologic maps at this scale.

Elbrook Formation rocks on the upright western limb of the Massanutten synclinorium (Figures 5–12). Although subsidiary folds occur on the western limb of the synclinorium, the beds along the fault zone consistently dip southeast. In the footwall of the North Mountain fault zone is a series of overturned synclines. In the Hedgesville quadrangle, Berkeley County, West Virginia (Figure 6), the fault zone consists of a series of upright anticlines and synclines that make up smaller 2nd-order folds on the west limb of the Massanutten synclinorium. The footwall is the overturned Back Creek syncline [9]. To the south in the White Hall quadrangle, Frederick County, Virginia (Figure 7), the hanging wall is made up of an anticline-syncline pair of the Welltown syncline and Pumpkin Ridge anticline, and the footwall is the overturned Mount Pleasant syncline composed of Devonian Mahantango Formation [10, 11]. The overturned Mount Pleasant syncline also occurs in the footwall in the Middletown and Mountain Falls quadrangles, Frederick and Shenandoah Counties, Virginia (Figures 8 and 9). Footwall structures are well exposed at Baldwin Gap in the Middletown quadrangle. There, Silurian and Devonian rocks of the overturned eastern limb of the Mount Pleasant syncline are exposed [12]. The geologic formations are thinner than in typical outcrop belts to the west in the Valley and Ridge province. Bedding-parallel décollements and several splays also occur. These faults are reverse faults with southeastward displacement. At Kipps Gap in Shenandoah County, Virginia, the footwall consists of overturned southeast-dipping Marcellus Shale of the Supin Lick syncline (Figure 10). The footwall in the Singers Glen quadrangle, Rockingham County, Virginia exposes Devonian shale as the overturned limb of the West Mountain syncline (Figure 11) that also makes up the footwall in the Churchville quadrangle, Augusta County, Virginia (Figure 12).

3.1. *Geology of Horses within the Fault Zone.* Between the hanging wall of the North Mountain fault zone and footwall synclines many horses occur. Although these horses have moved within the fault zone, the orientation of bedding remains consistent with Central Appalachian trends of about N30°–40°E. The fault zone ranges from 1/4-mile (0.4-km) wide across strike in the Hedgesville quadrangle, West Virginia (Figure 6) to 1/2-mile (0.8-km) wide in the Middletown quadrangle, Virginia (Figure 8) to 1-mile (1.6-km), in the Singers Glen quadrangle, Virginia (Figure 11). However, the number of exposed horses across the fault zone varies between one (Figures 7 and 12) and six (Figure 11) along this 70-mile stretch (113-km). For much of the length of the fault, multiple horses are stacked within the fault zone.

Along the entire trace of the fault zone, rocks in horses of the east side of the zone (hinterland) are right-side up and rocks on the west side (foreland) are almost exclusively overturned. In the Hedgesville quadrangle, the only right-side up rocks in horses in this stretch of the North Mountain fault zone are from the Beekmantown Group. The horses westward in the zone are made up of overturned rocks of the upper part of the Chambersburg Limestone (equivalent to the upper part of the Edinburg Formation in Virginia) and the lowermost rocks of the Martinsburg Formation. Thin horses of overturned Juniata Formation and Tuscarora Quartzite occur westward throughout the fault zone and form linear ridges.

In the White Hall quadrangle along the Virginia-West Virginia border, the largest horse in the fault zone is a steep north-plunging, upright anticline cored by the Rockdale Run Formation. This horse is thrust on a smaller horse of east-dipping, right-side up Pinesburg Station Dolomite through Chambersburg Limestone. Although the west side of the fault zone thrusts overturned Ordovician Martinsburg Formation over Upper Ordovician, Silurian, and Devonian rocks in the footwall, it appears that the fault dies out to the north where there is continuous exposure of Martinsburg through Lower Devonian Oriskany Sandstone [10].

The North Mountain fault zone in the Middletown quadrangle consists of many horses of older rock units over younger rock units [13]. The change from right-side up to overturned beds occurs within the uppermost part of the Edinburg Formation (equivalent to the upper part of the Chambersburg Limestone in West Virginia) and the lowest part of the highly deformed Martinsburg Formation. As many as nine horses of various Ordovician and Silurian units, with as many as five across strike, occur in the 1/2-mile (0.8-km) wide fault zone in this area (Figure 8). One horse is an upright anticline cored by Beekmantown Group. Another fault is recognized by the thin exposure of the Martinsburg Formation where the lowermost member, the Stickley Run Member, is thrust over uppermost Martinsburg and the Oswego Sandstone. The Martinsburg in this area of the Central Appalachians has been reported as much as 6,500 ft (1,981 m) thick [14], and the outcrop thickness along the North Mountain fault is less than 1,500 ft (457 m). Horses of the resistant Silurian Tuscarora Quartzite make up ridges along the fault zone west of the Martinsburg Formation exposures.

Map units

| | |
|---|---|
| Du/DSu | Devonian and Silurian units, undivided |
| Do | Oriskany Sandstone |
| Dh | Helderberg Group |
| St | Tuscarora Quartzite |
| Oj | Juniata Formation |
| Oo | Oswego Sandstone |
| Om | Martinsburg Formation |
| Oms | Stickley Run Member of Martinsburg Formation |
| Oe/Oc | Edinburg Formation (Va)/Chambersburg Limestone (WVa) |
| Ol | Lincolnshire Limestone |
| On | New Market Limestone |
| Ob | Beekmantown Group/Formation |
| O∈c | Conococheague Limestone |
| ∈e | Elbrook Formation |

Map symbols

▼     ▼     Thrust fault; dashed where approximately located

———————     Contact; dashed where approximately located

⊀₃₂     Strike and dip of bedding

⤬₇₁     Strike and dip of overturned bedding

FIGURE 5: Map units and map symbols used in Figures 6–12.

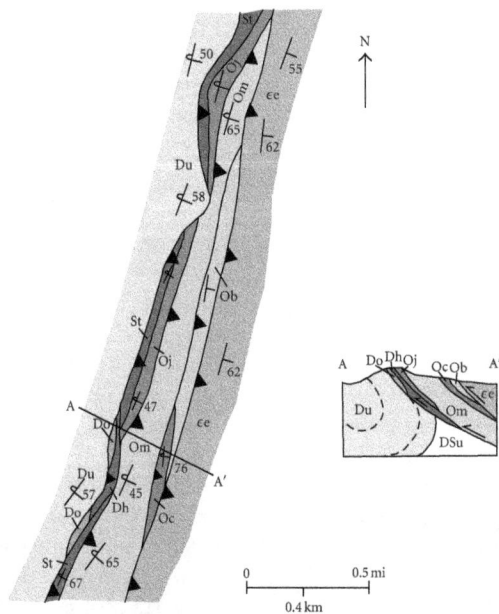

FIGURE 6: Geologic map of the North Mountain fault zone in the Hedgesville quadrangle, Berkeley County, West Virginia, modified from [9].

Fold-to-Fault Progression of a Major Thrust Zone Revealed in Horses of the North Mountain Fault Zone, Virginia and West Virginia, USA

67

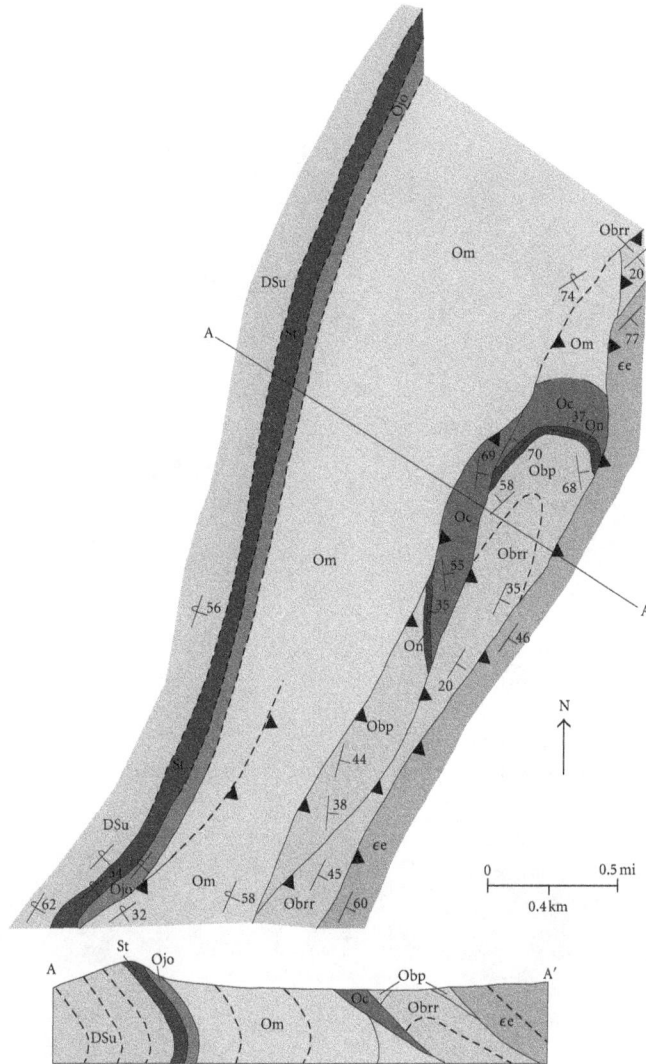

FIGURE 7: Geologic map of the North Mountain fault zone in the White Hall quadrangle, Frederick County, Virginia, modified from [10]. Obrr: Rockdale Run Formation of the Beekmantown Group; Obps: Pinesburg Station Dolomite of the Beekmantown Group.

Horses of Cambrian and Lower Ordovician units in the Mountain Falls quadrangle are thin, but still show the same relationships as those in other areas along the fault zone (Figure 9). Various horses mapped by McDowell [15] consist of right-side up Upper Cambrian and Lower Ordovician Conococheague Limestone, right-side up Middle Ordovician Beekmantown Formation (equivalent to the Rockdale Run Formation to the north), right-side up Middle Ordovician New Market and Lincolnshire Limestones, and overturned Stickley Run Member of the Martinsburg Formation. The west side of the fault zone consists of two linear, large horse blocks. One of these blocks consists of overturned beds of the upper part of the Martinsburg Formation and Oswego Sandstone, and the other overturned sandstone of the Tuscarora Quartzite.

In Kipps Gap in Shenandoah County, Virginia (Timberville and Orkney Springs quadrangles), various Cambrian through Devonian units occur as horses along the North Mountain fault zone (Figure 10). A quarry on the east side

of the fault zone exposes highly fractured limestone and dolostone of the Beekmantown Formation with shallow-dipping New Market Limestone just to the west delineating a west-dipping limb of a decapitated anticline. Westward, thin horses of the overturned Edinburg Formation and the Stickley Run Member of the Martinsburg Formation are thrust on upper units of the Martinsburg Formation that are thrust on Tuscarora Quartzite and limestone of the Silurian and Devonian Helderberg Group. The westernmost fault places the Tuscarora and Helderberg onto the Devonian Marcellus Shale, which makes up the eastern limb of the overturned Supin Lick syncline.

Horses along the North Mountain fault zone in the Singers Glen quadrangle, Rockingham County, Virginia, tend to be long and thin (Figure 11). Brent [16] mapped the various upright Ordovician units along Little North Mountain; however, they contain a reversed stacking order without faulting. Orndorff [17] recognized seven horses across a nearly 1-mile (1.6-km) wide section of the North

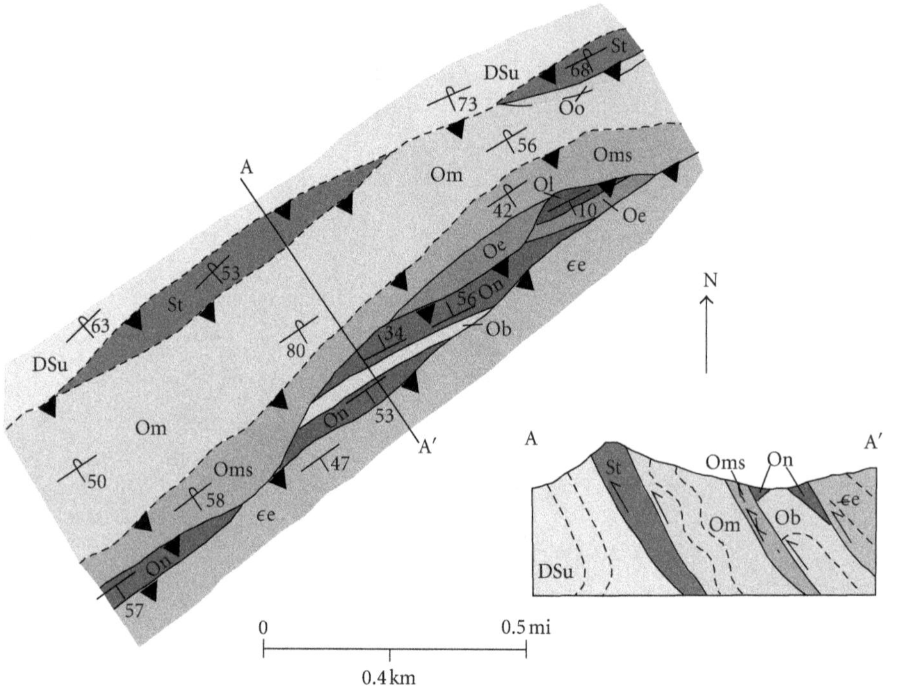

FIGURE 8: Geologic map of the North Mountain fault zone in the Middletown quadrangle, Shenandoah County, Virginia, from [13].

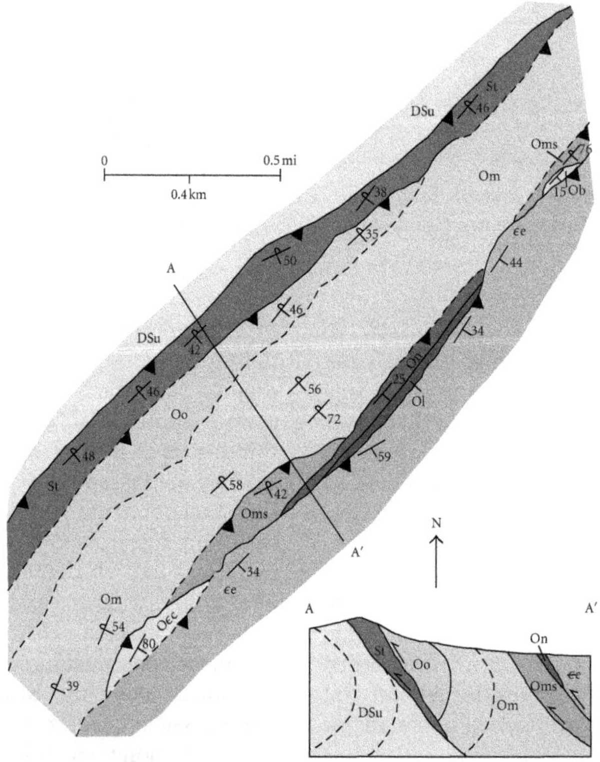

FIGURE 9: Geologic map of the North Mountain fault zone in the Mountain Falls quadrangle, Shenandoah County, Virginia, modified from [15].

Fold-to-Fault Progression of a Major Thrust Zone Revealed in Horses of the North Mountain Fault Zone, Virginia and West Virginia, USA

69

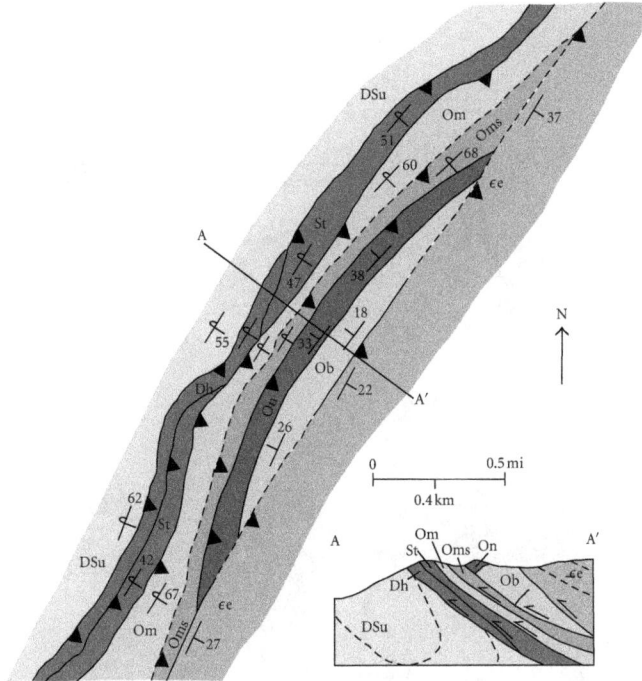

FIGURE 10: Geologic map of the North Mountain fault zone in the Kipps Gap area, Shenandoah County, Virginia (unpublished mapping by author).

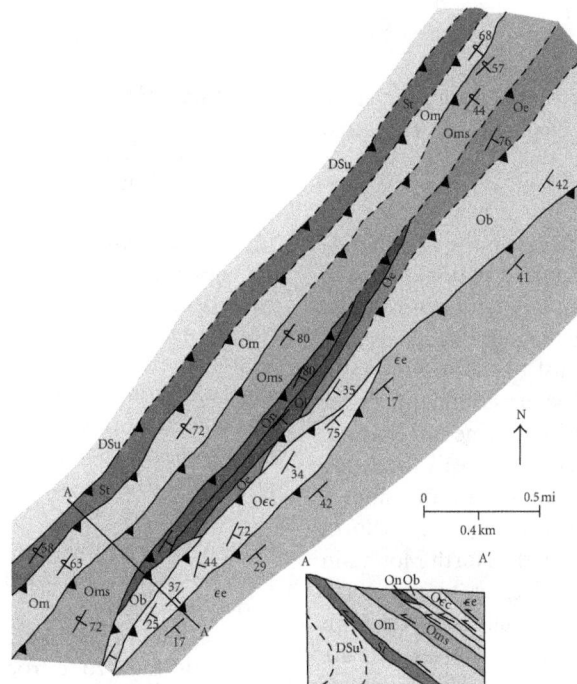

FIGURE 11: Geologic map of the North Mountain fault zone in the Singers Glen quadrangle, Rockingham County Virginia, from [17].

Mountain fault zone, which explained the order of the units mapped by Brent and their upright and overturned nature. Horses of right-side up stratigraphic units in the east part of the fault zone include the Upper Cambrian and Lower Ordovician Conococheague Limestone, two horses of Lower Ordovician Beekmantown Formation, and one with a section of Middle Ordovician New Market Limestone through Edinburg Formation. The first horse exposing overturned strata contains the Stickley Run Member of the Martinsburg Formation. This is continued southwestward

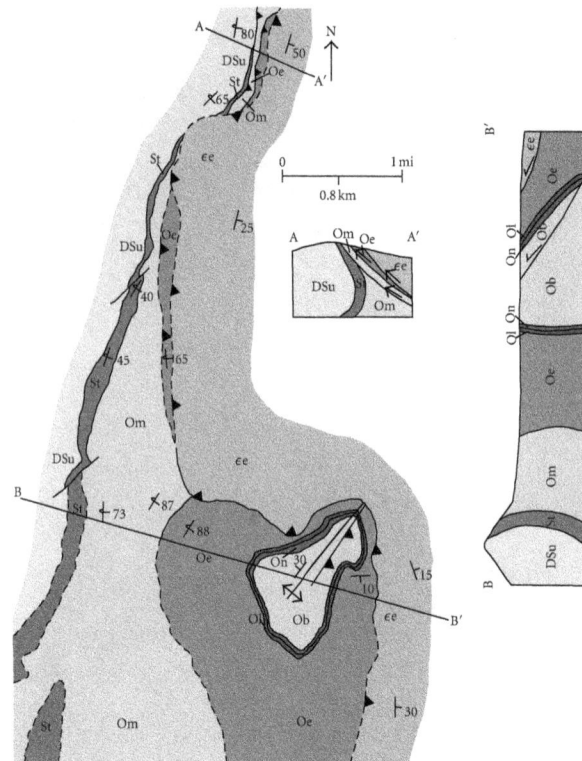

FIGURE 12: Geologic map of the North Mountain fault zone in the Churchville quadrangle, Augusta County, Virginia, modified from [19].

with horses of the upper part of the Martinsburg Formation and the Tuscarora Quartzite that is topographically expressed as Little North Mountain.

*3.2. Change in Fault Zone Characteristics.* In the Churchville quadrangle in Augusta County, Virginia, a major change in the characteristics of the North Mountain fault zone occurs. Displacement decreases and shortening of the Staunton-Pulaski fault system increases [1]. Just to the south of this area, the Pulaski fault becomes the predominant thrust system [18], and there is a transition zone in Augusta County. Geologic mapping by Rader [19] reveals flattening of the North Mountain fault and exposes an anticline of Middle Ordovician rocks in the footwall (Figure 12). North of this southeastward trend of the fault, the North Mountain fault zone is consistent with mapped areas to the north where rocks in the hanging wall are right-side up with lower Middle Ordovician units exposed in horses that are right-side up. Units stratigraphically above the Edinburg Formation exposed in horses are overturned including the Silurian Tuscarora Quartzite that makes up the east limb of the West Mountain syncline. The anticline on the footwall to the south of the southeastern jog of the fault becomes progressively overturned to the west. This structure may be the anticline that has been decapitated by the westward displacement of the North Mountain thrust sheet to the north.

# 4. Discussion

Boyer and Elliot [20] defined a horse as a pod of rock completely bound by two or more fault surfaces. Horses are derived from either the hanging wall or footwall of major thrusts and provide stratigraphic information from beneath a major thrust that cannot be obtained otherwise. Schultz [21] demonstrated the usefulness of the geology within horses in understanding the deformation of the Pulaski fault in southwestern Virginia. Here, rocks in the horses were the same age as those in the North Mountain fault zone; however, the Pulaski horses are complexly deformed with irregular fault contacts, the original stratigraphic sequences missing, and contain a predominance of inverted stratigraphy. By contrast, in the North Mountain fault zone, horses are not complexly deformed, original stratigraphic sequences are preserved, and there is a transition from right-side up to inverted sequences.

Analysis of bedding orientations within the North Mountain fault zone from detailed geologic mapping consistently parallel the strike of the Central Appalachians. Also, major folds on either side of the fault zone have consistent structural relationships (Figure 13). Horses within the fault zone document the deformation sequence. A key to the development of the fault zone as a fold-to-fault progression is the orientation of the horses along the leading edge. Anticlines preserved in horses have been identified along the 70-mile (113-km) stretch of the fault zone (Figure 1). The North Mountain fault zone in the Churchville quadrangle

Fold-to-Fault Progression of a Major Thrust Zone Revealed in Horses of the North Mountain Fault Zone, Virginia and West Virginia, USA

71

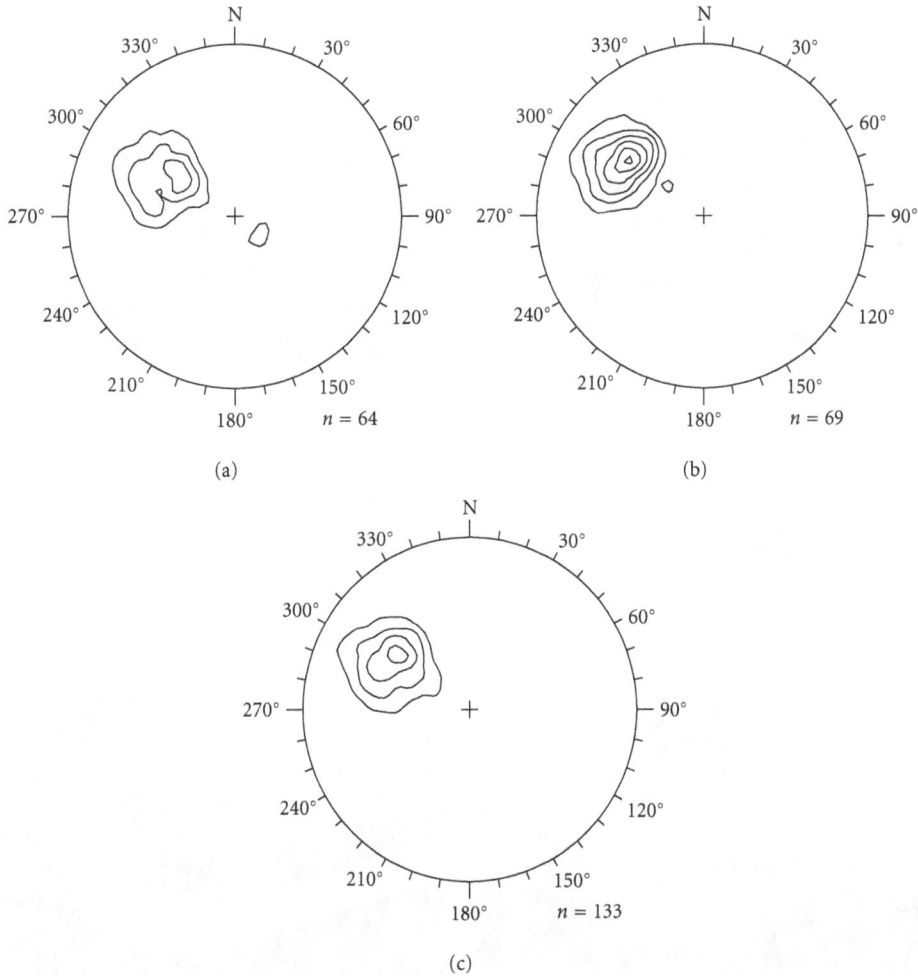

FIGURE 13: Lower-hemisphere equal-area stereographic plots of poles to bedding from detailed geologic mapping along the North Mountain fault zone. (a) Poles to bedding of upright Cambrian and Ordovician rocks. (b) Poles to bedding of overturned Ordovician to Devonian rocks. (c) Poles to bedding of all units in the North Mountain fault zone.

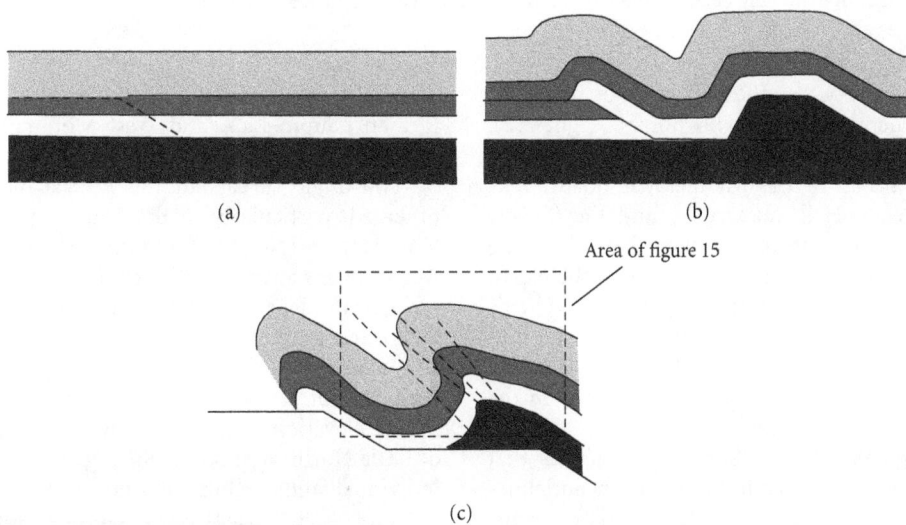

FIGURE 14: Conceptual model of detachment and folds showing progression of overturned folds and zones of shearing. (a) Initial compression develops ramps in the stratigraphic section. (b) Folds develop as shortening continues. (c) As folding continues, shear zones and faults develop along and oblique to the axial plane of the folds.

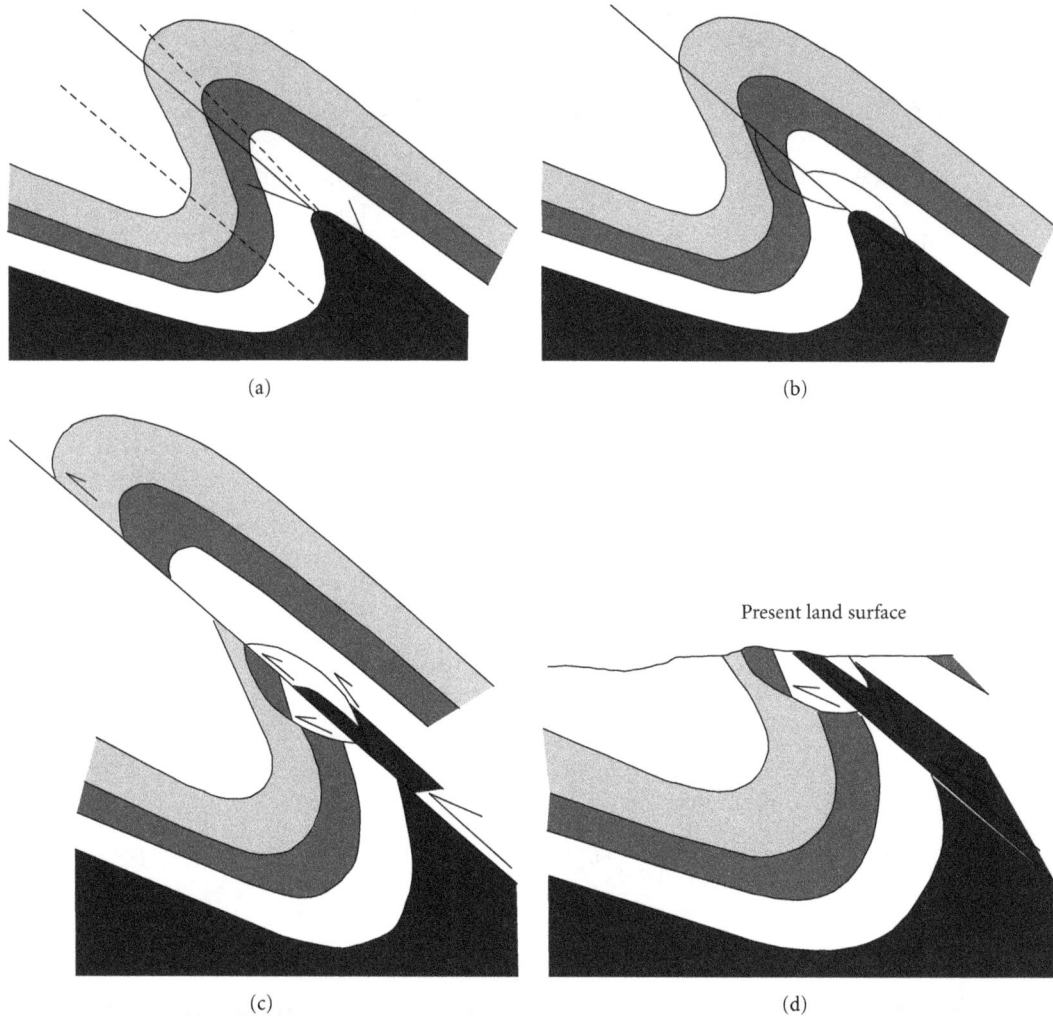

FIGURE 15: Conceptual model showing fold-to-fault progression and relationship of upright and overturned horses in the fault zone. (a) Overturned anticline-syncline pair with axial planes (dashed lines) and location of oblique and splay faults (heavy line). (b) Thrust faulting with origin of upright horse from upright limb of anticline. (c) Thrust faulting with origin of overturned horse from overturned limb of anticline. (d) Thrust fault zone showing current land surface and decapitated overturned anticline.

resides on a lateral ramp where the anticline was not over-ridden and consequently there are no horses (Figure 12). However, upright horses derived from this anticline where it was overridden occur in the Middletown quadrangle, Kipps Gap area, White Hall quadrangle, and the Singers Glen quadrangle as well as in many other areas along the fault zone. The transition from intact anticline to the North Mountain fault zone with horses occurs in the Churchville quadrangle. Also, this anticline documents the transition to the overturned western limb. Along the fault system to the north, rocks of the Edinburg Formation and higher in the stratigraphic section are overturned.

Willis [22] originally distinguished break thrusts that formed by thrusting the connecting limb of an anticline-syncline pair, where the hanging wall anticline was overthrust and, thus, preserved the footwall syncline. Erosion has removed the hanging wall cutoff [23]; therefore, the major synclinal structure of the North Mountain thrust sheet, Massanutten synclinorium, is in thrust contact with the large

regional overturned synclines such as the Back Creek, Mount Pleasant, Supin Lick, and West Mountain synclines (from north to south) (Figure 1).

Low-angle thrusts in the Appalachians are well documented, particularly in the Southern Appalachians [24]. However, the North Mountain fault zone in the Central Appalachians appears to be generally of moderate angle. For most of the length of the North Mountain fault zone, the fault is recognized topographically by Little North Mountain. The geometry of the various faults within the zone suggests fault dips of at least 30 to 40 degrees. This is determined by the elevation difference between the faults and the top of Little North Mountain, which gives a minimum angle for individual faults within the zone.

Suppe [25] recognized fault-propagation folds that form as stratigraphic layers are folded during propagation of a thrust through a sedimentary sequence and act similar to break thrusts. In other words, the fold forms as the fault propagates through the sedimentary sequence with

Fold-to-Fault Progression of a Major Thrust Zone Revealed in Horses of the North Mountain Fault Zone, Virginia and West Virginia, USA

73

progressive transfer of slip from the fault to the fold. Fault-propagation folds are the result of flexural bending of a layered sequence of rock in advance of the actual rupture and development of the fault plane. Though this model has been applied to thrust faults in the fold and thrust belt of the Appalachians, evidence from the North Mountain fault zone suggests that the folding occurred early in Alleghanian deformation and was not initiated by a propagating thrust fault (Figure 14).

Mitra [26] described faulted detachment folds that form in rock units with high competency contrasts and transition from folding to fault propagation as shortening increases. One characteristic of faulted detachment folds is that the fold-fault relationship shows a transition from folding to faulting with footwall synclines and decapitated anticlinal fold geometries [26]. The North Mountain fault zone occurs in rocks that have high competency contrasts between the Cambrian and Ordovician carbonates and the shaly units of the upper Middle Ordovician (Figure 3). This change in competency occurs in a transition in the Chambersburg/Edinburg Formation where dominant limestone and dolostone lithologies give way to calcareous shale and shaly limestone, and subsequently transitions into shale of the Martinsburg Formation, which is considered a detachment zone in the Central Appalachians [5]. Note that the change from right-side up to overturned horses in the North Mountain fault zone occurs in the transition from competent to less competent Middle Ordovician units.

Erslev [27] and Erslev and Mayborn [28] proposed trishear fault propagation through folds where the deformation zone is bounded by the axial planes of anticlinal hinges and the corresponding synclinal hinges. In this model, shear is distributed from the tip of the fault to a widening zone on the limb of steep and overturned beds. This distribution of shear may be the origin of subsidiary faults that led to the development of horses within the North Mountain fault zone.

## 5. Conclusions

Detailed examination of horses in the North Mountain fault zone document a fold-to-fault progression along the overturned southeast limb of a syncline to the northwest and the adjacent upright anticline to the southeast (Figure 15). Horses on the eastern part of the fault zone were derived from the upright limb or a hanging wall anticline. They structurally overlie overturned beds in the west part of the zone that were derived from the overturned limb of the subthrust syncline. Faulting occurred following the majority of the folding. The North Mountain fault plane is on a frontal ramp and is noncoaxial planar or oblique to an overturned anticline. This anticline is inferred since it is no longer exposed and the hanging wall cutoff has been removed by erosion. Evidence of this anticline exists in horses exposed along the present leading edge of the fault zone from central Virginia northward to the eastern panhandle of West Virginia. The moderate dip of the fault zone and consistent nature of the horses suggest that the North Mountain fault probably has less than 10 miles (16 kilometers) of horizontal displacement. Detailed geologic mapping is key to understanding complex structures in fold-and-thrust belts. Understanding the relationships of structures, transitions, and competencies of rock units and right-side up to overturned nature of rocks in horses aids in determination of relative fold and fault timing in major thrust zones.

## Acknowledgments

The author thanks Arthur P. Schultz and Robert C. Milici, U.S. Geological Survey, A.G. Leslie, British Geological Survey, and Robert Hatcher, University of Tennessee, for their suggestions and critical review of the paper. Also, the author appreciates the in-depth discussions and outcrop reviews on the structural geology of the Central Appalachians and North Mountain fault zone with C. Scott Southworth, David J. Weary, Jack B. Epstein, Robert C. McDowell, Daniel H. Doctor, E. William Decker, E. Brett Waller, and Kent D. Campbell.

## References

[1] B. R. Kulander and S. L. Dean, "Structure and tectonics of central and southern Appalachian Valley and Ridge and Plateau provinces, West Virginia and Virginia," *American Association of Petroleum Geologists Bulletin*, vol. 70, no. 11, pp. 1674–1684, 1986.

[2] E. K. Rader and W. J. Perry Jr., "Reinterpretation of the geology of Brocks Gap, Rockingham County, Virginia," *Virginia Division of Mineral Resources, Virginia Minerals*, vol. 22, pp. 37–45, 1976.

[3] B. R. Kulander and S. L. Dean, "The North Mountain-Pulaski fault system and related thrust sheet structure," in *Geometrics and Mechanisms of Thrusting, with Special Reference to the Appalachians*, G. Mitra and S. Wojtal, Eds., vol. 222, pp. 107–118, Geological Society of America, 1988.

[4] S. L. Dean, B. R. Kulander, and P. Lessing, "The structural geometry and evolution of foreland thrust systems, northern Virginia: Alternative interpretation and reply," *Geological Society of America Bulletin*, vol. 102, pp. 1442–1445, 1990.

[5] M. A. Evans, "The structural geometry and evolution of foreland thrust systems, northern Virginia," *Geological Society of America Bulletin*, vol. 101, no. 3, pp. 339–354, 1989.

[6] A. W. Giles, "Geology of Little North Mountain in northern Virginia and West Virginia," *Journal of Geology*, vol. 35, pp. 32–57, 1927.

[7] C. Butts and R.S. Edmundson, "Geology of little North Mountain in contributions to Virginia geology—part II," *Virginia Division of Mineral Resources Bulletin*, vol. 51, pp. 163–179, 1939.

[8] C. Butts and R. S. Edmundson, "Geology and mineral resources of Frederick County," *Virginia Division of Mineral Resources Bulletin*, vol. 80, 142 pages, 1966.

[9] S. L. Dean, B. R. Kulander, and P. Lessing, "Geology of the Hedgesville, Keedysville, Shepherdstown, and Williamsport quadrangles, Berkeley and Jefferson Counties," West Virginia Geological Economic Survey Map-WV31, scale 1:24,000, 1987.

[10] D. H. Doctor, R. C. Orndorff, R. A. Parker, D. J. Weary, and J. E. Repetski, "Geologic map of the White Hall quadrangle, Frederick County, Virginia and Berkeley County," West

Virginia. U.S. Geological Survey Open-file Report 2010-1265, scale 1:24,000, 2010.

[11] E. K. Rader, R. C. McDowell, T. M. Gathright III, and R. C. Orndorff, "Geologic map of Clarke, Frederick, Page, Shenandoah, and Warren Counties, Virginia," Lord Fairfax Planning District. Virginia Division of Mineral Resources Publication 143, scale 1:100,000, 1996.

[12] R. C. Orndorff and J. B. Epstein, "A structural and stratigraphic excursion through the Shenandoah Valley, Virginia," U.S. Geological Survey Open-file Report 94-573, 1994.

[13] R. C. Orndorff, J. B. Epstein, and R. C. McDowell, "Geologic map of the Middletown quadrangle, Frederick, Shenandoah, and Warren Counties, Virginia," U.S. Geological Survey Geologic Quadrangle Map GQ-1803, scale 1:24,000, 1999.

[14] J. B. Epstein, R. C. Orndorff, and E. K. Rader, "Middle Ordovician Stickley Run Member (new name) of the Martinsburg Formation, Shenandoah Valley, northern Virginia," *US Geological Survey Bulletin*, vol. 2135, pp. 1–13, 1995.

[15] R. C. McDowell, "Preliminary geologic map of the Mountain Falls quadrangle, Frederick and Shenandoah Counties, Virginia, and Hampshire County, West Virginia," U.S. Geological Survey Open-file Report 95-620, scale 1:24,000, 1995.

[16] W. R. Brent, "Geology and mineral resources of Rockingham County," *Virginia Division of Mineral Resources Bulletin*, vol. 76, p. 174, 1960.

[17] R. C. Orndorff, "Geologic map of the Muddy Creek drainage basin, Singers Glen quadrangle, Rockingham County, Virginia," U.S. Geological Survey Open-file Report 95-629, scale 1:24,000, 1995.

[18] M. J. Bartholomew, "Structural evolution of the Pulaski thrust system, southwestern Virginia," *Geological Society of America Bulletin*, vol. 99, pp. 491–510, 1987.

[19] E. K. Rader, "Geology of the Staunton, Churchville, Greenville, and Stuarts Draft quadrangles, Virginia," Virginia Division of Mineral Resources Report of Investigations 12, scale 1:24,000, 1967.

[20] S. E. Boyer and D. Elliott, "Thrust systems.," *American Association of Petroleum Geologists Bulletin*, vol. 66, no. 9, pp. 1196–1230, 1982.

[21] A. P. Schultz, "Horses in fensters of the Pulaski thrust sheet, southwestern Virginia: structure, kinematics, and implications for hydrocarbon potential of the Eastern Overthrust Belt," *US Geological Survey Bulletin*, vol. 1839, pp. A1–A13, 1988.

[22] B. Willis, "Mechanics of Appalachian structures," U.S. Geological Survey Annual Report 13, 1893.

[23] R. D. Hatcher Jr., W. A. Thomas, P. A. Geiser, A. W. Snoke, S. Mosher, and D. W. Wiltschko, "Alleghanian orogin," in *The Appalachian-Ouachita orogen in the United States*, R. D. Hatcher Jr., W. A. Thomas, and G. W. Viele, Eds., pp. 233–318, Geological Society of America, The Geology of North America F-2, 1989.

[24] L. D. Harris and R. C. Milici, "Characteristics of thin-skinned style of deformation in the Southern Appalachians, and potential hydrocarbon traps," *U.S. Geological Survey Professional Paper*, no. 1018, 1977.

[25] J. Suppe, *Principles of Structural Geology*, Prentice-Hall, Englewood, NJ, USA, 1985.

[26] S. Mitra, "Structural models of faulted detachment folds," *American Association of Petroleum Geologists Bulletin*, vol. 86, no. 9, pp. 1673–1694, 2002.

[27] E. A. Erslev, "Trishear fault-propagation folding," *Geology*, vol. 19, no. 6, pp. 617–620, 1991.

[28] E. A. Erslev and K. R. Mayborn, "Multiple geometries and modes of fault-propagation folding in the Canadian thrust belt," *Journal of Structural Geology*, vol. 19, no. 3-4, pp. 321–335, 1997.

# 6

# Fluoride in the Serra Geral Aquifer System: Source Evaluation Using Stable Isotopes and Principal Component Analysis

**Arthur Schmidt Nanni,**[1] **Ari Roisenberg,**[2] **Maria Helena Bezerra Maia de Hollanda,**[3] **Maria Paula Casagrande Marimon,**[4] **Antonio Pedro Viero,**[2] **and Luiz Fernando Scheibe**[1]

[1] *Universidade Federal de Santa Catarina (UFSC), Campus Universitário, Trindade, 88.010-970 Florianópolis, SC, Brazil*
[2] *Universidade Federal do Rio Grande do Sul (UFRGS), Avenida Bento Gonçalves 9500, prédio 43126/103, 91501-970 Porto Alegre, RS, Brazil*
[3] *Centro de Pesquisas Geocronológicas (CPGeo), Instituto de Geociências (USP), Universidade de São Paulo (USP), Rua do Lago 562, Cidade Universitária, 05508-080 São Paulo, SP, Brazil*
[4] *Universidade do Estado de Santa Catarina, UDESC-FAED, 88.035-001 Florianópolis, SC, Brazil*

Correspondence should be addressed to Arthur Schmidt Nanni; arthur.nanni@ufsc.br

Academic Editor: Umberta Tinivella

Groundwater with anomalous fluoride content and water mixture patterns were studied in the fractured Serra Geral Aquifer System, a basaltic to rhyolitic geological unit, using a principal component analysis interpretation of groundwater chemical data from 309 deep wells distributed in the Rio Grande do Sul State, Southern Brazil. A four-component model that explains 81% of the total variance in the Principal Component Analysis is suggested. Six hydrochemical groups were identified. $\delta^{18}O$ and $\delta^2H$ were analyzed in 28 Serra Geral Aquifer System samples in order to identify stable isotopes patterns and make comparisons with data from the Guarani Aquifer System and meteoric waters. The results demonstrated a complex water mixture between the Serra Geral Aquifer System and the Guarani Aquifer System, with meteoric recharge and ascending water infiltration through an intensive tectonic fracturing.

## 1. Introduction

In the last several decades, the global water consumption has dramatically increased, especially for agriculture, water supply, and industrial uses. This paper examines the fluoride content in water from the southernmost region of the fractured subcontinental Serra Geral Aquifer System (SGAS), an important aquifer that supplies a large amount of water for an economic developed region in Southern Brazil.

Fluoride content in water is beneficial to human health and in a moderate concentration (0.7–1.2 mg/L) prevents dental cavities. When used in excess can be toxic causing human and animal dental and skeletal fluorosis, which has been detected in China [1, 2], India [3], Kenya [4], and Israel [5], among other countries. The drinking water limit recommended by the World Health Organization for fluoride

is 1.5 mg/L [6]. In the study area, the SGAS fluoride average concentrations are around 0.24 mg/L, with a minimum value of 0.02 m/L and the highest at 3.03 mg/L.

The SGAS overlies the Guarani Aquifer System (GAS; [7]), which has been the focus of several recent studies due to its spatial extent and storage potential as a transboundary aquifer [8]. The area covered by the SGAS in Brazil, Uruguay, Argentina, and Paraguay is equivalent to 1,200,000 km$^2$ (Figure 1). In Brazil, these groundwater resources should be efficiently managed to protect its water potential and quality. The climate in the study area ranges from subtropical to temperate, with precipitation average of 1,550 mm/year.

The SGAS reaches its maximum thickness of about 1,800 m along the central axis of the Parana Basin, located in the Sao Paulo State. In the Rio Grande do Sul State, this aquifer has a wedge shape, ranging from 1,000 m in the eastern

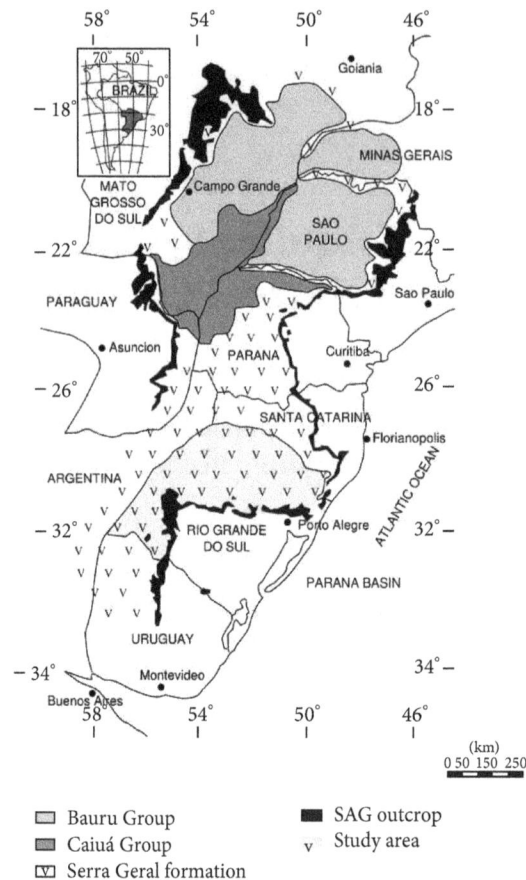

FIGURE 1: Location map showing the Serra Geral Formation (SGF) in South America (modified from [9]).

area bordering the Santa Catarina State to thickness of 200 m in the western area adjacent to Argentina Border.

The aim of this study is to understand the spatial distribution and geologic control of fluoride content in the SGAS, identifying hydrochemical groups with high fluoride Physicochemical analysis, geostatistics, and stable isotopes approaches were used along with a Geographic Information System (GIS) in order to identify the spatial distribution of the different water types.

1.1. Geologic and Hydrogeologic Framework. The study area is located in the Rio Grande do Sul State, Southern Brazil, and underlies 164,200 km$^2$, extending from 27°S to 31°S and from 50°W to 57°W. The volcanic sequence has an average thickness of about 550 m and consists of basaltic to rhyolitic units of Mesozoic age; the last ones occur on the Serra Geral Formation (SGF) top sequence. The clayish soil thickness in the SGF ranges from few decimeters in the eastern area to several meters in the west and northwest. This soil controls water infiltration and, as a consequence, the SGAS recharge by meteoric waters [10].

Studies have suggested groundwater mixing processes between the SGAS and older confined sedimentary aquifer systems from the Parana Basin, including the GAS, through ascending groundwater infiltration [11–16].

Distinct tectonic fracturing occurs in the study area, controlling geomorphological features and groundwater circulation. These structures are directly related to the opening of the South Atlantic Ocean, which is reflected in the major NE and NW fault and fracture directions. Neotectonics is considered to play an important hydrogeological role in the eastern region. Major tectonic systems define individual hydrochemical sectors, which are considered as individual hydrogeological blocks [15, 17, 18]. The tectonic block limited by the Terra de Areia-Posadas and Mata-Jaguari Fault Systems (Figure 2) is uplifted with respect to the neighboring blocks. In the north, the adjacent block is divided into smaller units with a gradual terrain lowering from east to west, influenced by NE normal faults, parallel to the Leao and Perimpo Fault Systems.

The identification of hydrochemical groups in the SGAS is an important tool for the definition of tectonic structures, supporting the fluoride ascending circulation hypothesis in the aquifer [22]. The meteoric recharge pattern as $Ca^{2+}HCO_3^-$ waters and also ascending infiltrations by $Na^+HCO_3^-$ waters with $SO_4^{2-}$ and $Cl^-$ in the GAS, resulting in a complex water mixture [15]. Several authors have pointed

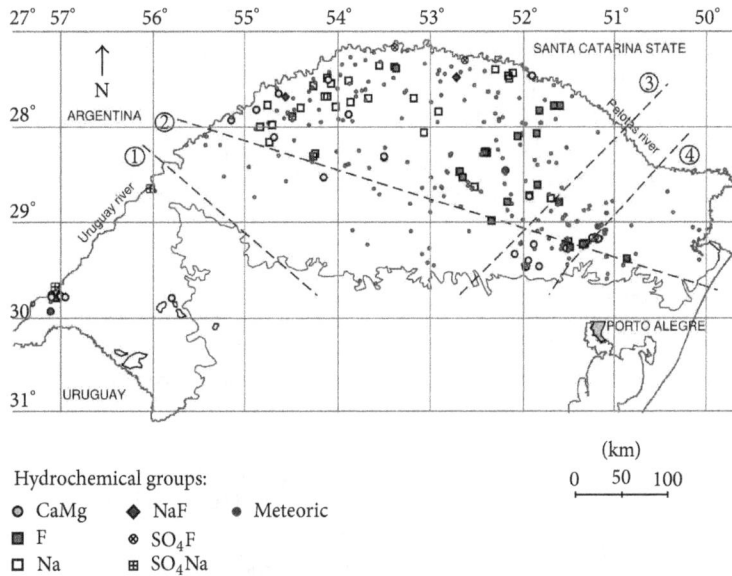

FIGURE 2: Sample points, hydrochemical groups distribution on SGAS, and major tectonic fault systems: (1), Mata-Jaguari, (2) Terra de Areia-Posadas, (3) Perimpo, and (4) Leao.

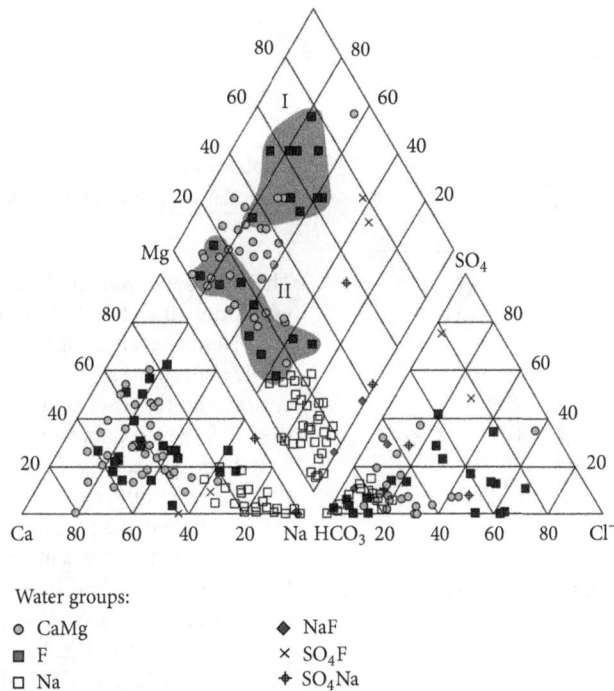

FIGURE 3: Hydrochemical groups distribution in the piper diagram, showing two fluoride hydrochemical types (I—sulfated and II—bicarbonated, [19]).

out that high fluoride content in groundwater can be related to $Na^+HCO_3^-$ ascending infiltrations waters [11, 15, 22–24]. Such mixing and hydrochemical groups can both result from $Ca^{2+}HCO_3^-$ and $Na^+HCO_3^-$ waters [25].

## 2. Materials and Methods

### 2.1. Multivariate Statistics. Principal components analysis (PCA) was used to discriminate hydrochemical groups in

the 309 physicochemical analyses of the SGAS in order to systematize the interpretation of a large number of variables through the maximization of differences between them [26].

The PCA was performed to 2, 3, and 4 principal components and normalized to a value of "1," minimum eigenvalue accepted [27]. Subsequently, the outliers were cut by visual analysis in boxplots and dispersion charts for $F^-$, $Na^+$, $Ca^{2+}$, $Mg^{2+}$, $HCO_3^-$, $SO_4^{2-}$, and $Cl^-$, considering water classification. This statistical procedure restricts the analysis

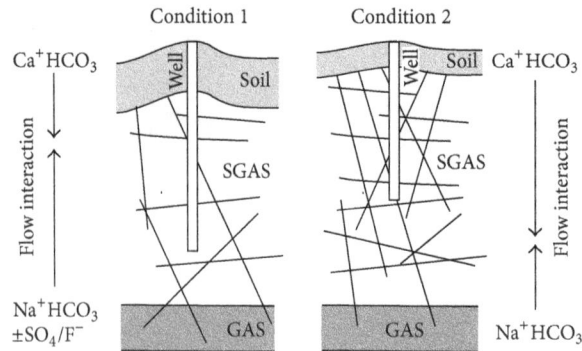

FIGURE 4: Water flow interactions to different geological conditions.

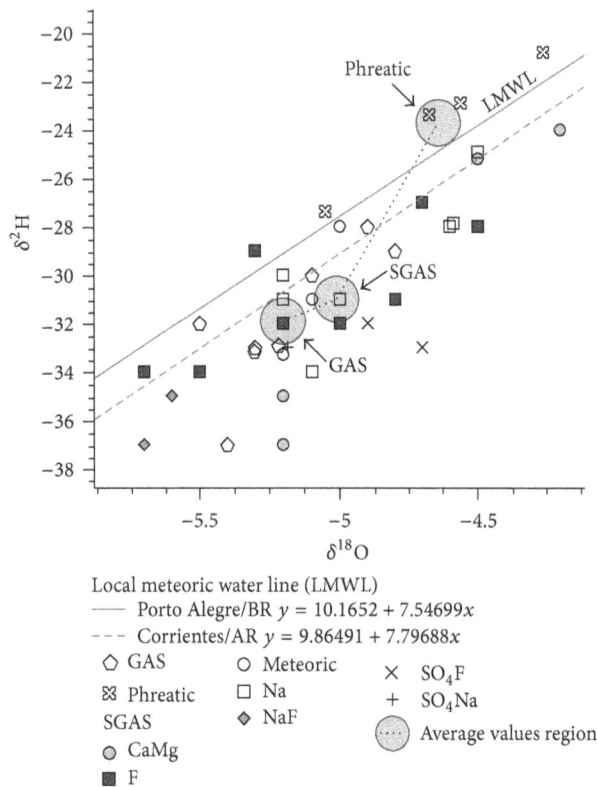

FIGURE 5: $\delta^{18}O$ (‰) versus $\delta^2H$ (‰) in GAS, SGAS, and phreatic aquifer.

TABLE 1: Rotated matrix with four components and characteristic parameters. Data from [21].

| Parameters | Components | | | |
|---|---|---|---|---|
| | 1 | 2 | 3 | 4 |
| $Mg^{2+}$ | **.874** | .063 | .072 | .139 |
| $Ca^{2+}$ | **.811** | .231 | −.045 | −.043 |
| $Na^+$ | −.143 | .188 | **.870** | .271 |
| $Cl^-$ | .353 | **.707** | .206 | −.051 |
| $HCO_3^-$ | **.572** | .095 | **.699** | −.091 |
| $SO_4^{2-}$ | .030 | **.872** | .074 | .237 |
| $F^-$ | .067 | **.145** | **.150** | **.956** |
| Explained variance | 37.433 | 20.260 | 12.694 | 10.770 |
| Cumulative % of variance | 37.433 | 57.693 | 70.386 | 81.156 |

The hydrogen was analyzed through the reaction of water with pure zinc at 500°C for liberation of the $H_2$ gas.

Twenty seven samples were selected in order to establish stable isotope patterns in the SGAS. These results were considered together with eight additional ones belonging to the GAS (seven data from [28] and one from this study) and four from phreatic waters from an adjacent area [28]. This procedure is appropriate for understanding water mixture features between aquifers and to estimate their relative ages.

Unfiltered water samples were collected and stored in 100 mL polypropylene bottles to provide adequate volume for separate analyses of $\delta^{18}O$ and $\delta^2H$. All data considered in this paper are expressed in ‰ with respect to Vienna Standard Mean Ocean Water (VSMOW) on the Delta scale:

$$\delta = \left[ \frac{(1000 + R_{sample})}{(1000 + R_{standard})} - 1 \right] * 1000, \quad (1)$$

where $R_{sample}$ is the $^{18}O/^{16}O$ or $^2H/^1H$ ratio of the sample and $R_{standard}$ is the corresponding ratio in VSMOW. Analytical precisions were better than ±0.5‰ for $\delta^{18}O$ and ±2‰ for $\delta^2H$.

The altitude and the distance from the ocean can influence stable isotopes patterns. The average altitude of the SGAS in the area is 500 m, and the spatial position ranges from the coast in the eastern sector to 800 km inland in the western sector.

towards the fluoride focus, avoiding interferences from other parameters. The spatial distribution and interpretation of principal components defined by PCA were performed with a GIS environment.

## 2.2. Stable Isotopes.

Stable isotopes were analyzed in the Geochronological Research Center of the Sao Paulo University. The analysis for oxygen was accomplished through the balanced isotopic reaction between water and $CO_2$ over 24 hours.

The $CO_2$ extraction was done in a vacuous line off-line and subsequently was read in a Thermo Finnigan Delta Plus/Advantage spectrometer with dual-inlet installed.

TABLE 2: Cluster central scores, dominant components and hydrochemical groups. Data from [21].

| Component | Cluster | | | | | | | | |
|---|---|---|---|---|---|---|---|---|---|
| | 1 | 2 | 3 | 4 | 5 | 6 | 7 | 8 | 9 |
| CaMg | 1.614 | −0.613 | −0.228 | 0.511 | −0.152 | −0.724 | 2.042 | −0.722 | −1.720 |
| $SO_4$ | 0.987 | 4.411 | −0.377 | −1.198 | −0.113 | 9.303 | −0.236 | −0.260 | −0.348 |
| Na | −0.892 | 3.032 | −0.881 | −0.667 | −0.305 | −0.845 | 1.092 | 1.653 | 2.691 |
| F | −0.330 | −1.357 | 1.244 | 6.986 | −0.250 | 2.849 | 0.285 | −0.034 | 4.786 |
| Cases | 22 | 2 | 19 | 3 | 212 | 2 | 11 | 36 | 2 |
| Hydrochemical group | CaMg | $SO_4$Na | F | F | Meteoric | $SO_4$F | CaMg | Na | NaF |

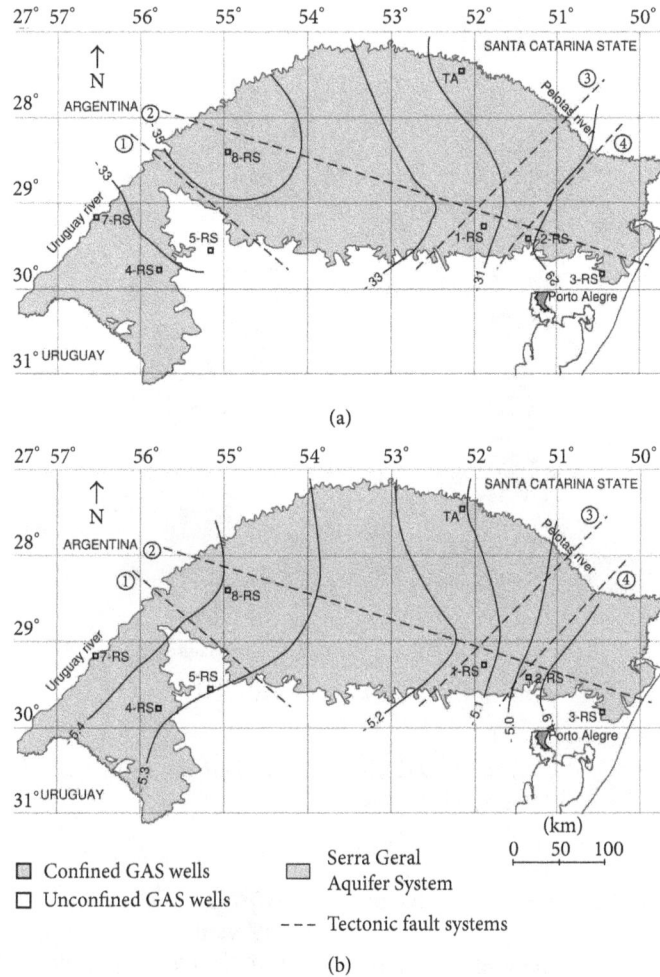

(a)

(b)

FIGURE 6: Hydrogeological spatial trends of $\delta^2$H (a) and $\delta^{18}$O (b) in the GAS (data from [20]) and major tectonic fault systems: (1) Mata-Jaguari, (2) Terra de Areia-Posadas, (3) Perimpo, and (4) Leao.

Due to this, ISOHIS [29] isotope data from the precipitation sampling stations were selected both in Porto Alegre, Southern Brazil, located in coastal and low altitude condition (GNIP code 8396700) and Corrientes, Argentina, located in interior and low altitude (GNIP code 8716600) area showed to compare data obtained in this research.

## 3. Results and Discussion

*3.1. Principal Components Analysis.* The PCA results show the correspondence of each component to the analyzed parameters. The solution using the eigenvalue criterion results in four components that explain 81% of the total variance. Component 4 is highly weighted by fluoride (Table 1).

Component 1 contains $HCO_3^-$, $Ca^{2+}$, and $Mg^{2+}$, typical for calcium and magnesium bicarbonated waters with short to medium residence time. In Component 2, $SO_4^{2-}$ and $Cl^-$ prevail, indicating calcium and sodium-chlorinated and sulfate waters. Component 3 is distinguished by $HCO_3^-$ and $Na^+$, representing sodium bicarbonated waters. The $F^-$ followed by $Na^+$ and $SO_4^{2-}$ defines Component 4, which

(a)

(b)

Water groups:
- ⊙ CaMg
- ■ F
- □ Na
- ⊞ SO₄Na
- ◆ NaF
- ⊗ SO₄F
- ○ Meteoric

▨ Serra Geral Aquifer system

--- Tectonic fault systems

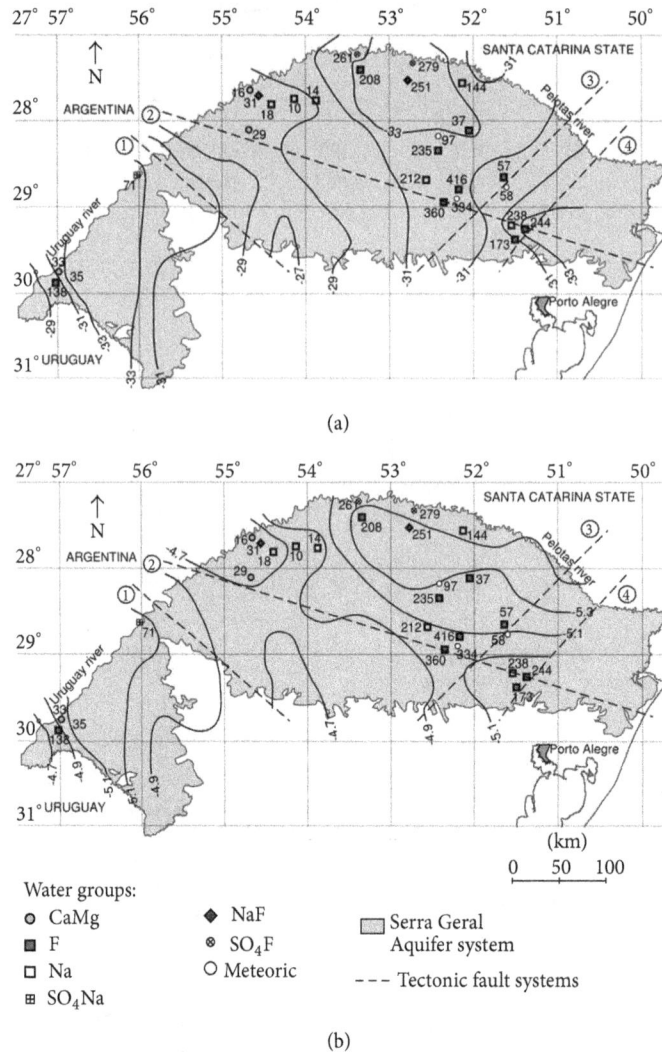

FIGURE 7: Hydrogeological spatial trends of $\delta^2 H$ (a) and $\delta^{18}O$ (b), hydrochemical groups in the SGAS and major tectonic fault systems: (1), Mata-Jaguari, (2) Terra de Areia-Posadas, (3) Perimpo, and (4) Leao.

comprises the high $F^-$ group linked to sodium-sulfated waters.

Based on these components, the solution for nine clusters was selected, and it shows the best approach to the hydrochemical groups. Subsequently, a cluster interpretation was performed based on central scores of each cluster. In order to assist the interpretation, each cluster was renamed according to the predominant ion (Table 2), defining hydrochemical groups. Thus, Component 1 has been renamed to CaMg, Component 2 to SO₄, Component 3 to Na, and Component 4 to F.

Clusters 3 and 4 are composed predominantly by the F group. Clusters 1 and 7 comprise the CaMg group. Cluster 8 is composed by the Na group. Other clusters show prevalence of two components, resulting in the groups SO₄Na, SO₄F, and NaF. Cluster 5 comprises the Meteoric waters and other samples with no relationship to the four principal components.

3.2. Hydrochemistry. The geochemical data based on PCA results were plotted in a piper diagram (Figure 3), with the four renamed clusters defined by the PCA. The majority of wells (212 out of 309 wells) have a composition close to *meteoric waters* with $Ca^{2+}HCO_3^-$ nature and do not include high fluoride waters. Due to this, 212 samples are not represented in the diagrams in order to obtain a better fit by the other groups.

The CaMg group represents a predominant $HCO_3^-$ water type where $Mg^{+2}$ appears in more than 50% of the wells (Figure 3).

The Na group shows their typical distribution in the piper diagram, representing $Na^+HCO_3^-$ waters.

$SO_4Na$, $SO_4F$, and NaF groups appear only in two wells each. Both wells compose distinct groups as shown in Figure 3.

F distribution in the piper diagram maintains an association with sulfated and bicarbonated groundwaters. These

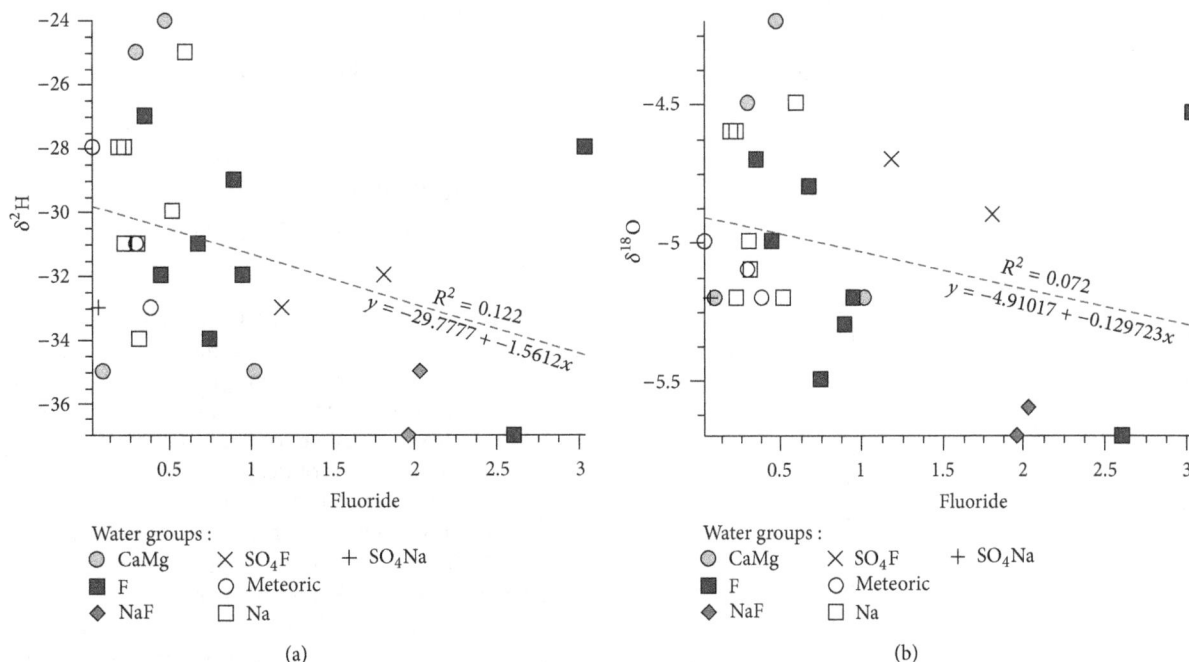

FIGURE 8: Fluoride versus $\delta^{18}O$ and $\delta^2H$ and the SGAS hydrochemical groups.

hydrochemical characteristics are also related to different groundwater sources or point out to mixture features during groundwater ascending infiltration, where ionic enrichment including increasing fluoride content can result from the interaction of ascending/descending water recharge due to longer residence time in the Na and CaMg hydrochemical groups [19]. Thus, the $SO_4F$, NaF, F, and $SO_4Na$ hydrochemical groups have a closed relationship interaction with two or more fracture directions (Figure 4), where $SO_4^{2-}$, $Na^+$, and fluoride can ascend.

3.3. Stable Isotopes. The 27 analyzed samples from SGAS are distributed among the six water groups defined in the PCA. For GAS waters, one sample was analyzed and aggregated other 7 results [20]. Another 4 results to phreatic waters [28] were aggregated as mentioned. The $\delta^{18}O$ (‰) ranges from −5.0 to −4.2 (phreatic waters), from −5.7 to −4.2 (SGAS), and from −5.5 to −4.8 (GAS), while the $\delta^2H$ (‰) varies from −27.4 to −20.8 (phreatic waters), from −37 to −24 (SGAS), and from −37 to −28 (GAS; Table 3). The SGAS shows a scattered distribution compared to the GAS (Figure 5). Nevertheless, the average of $\delta^{18}O$ and $\delta^2H$ values in both aquifer systems is similar, and the SGAS is located in an intermediate position but nearer to GAS when compared to phreatic waters. This reinforces the hypothesis of water circulation through the SGAS to the GAS, with minor influence of phreatic/meteoric waters in the sampled field.

The GAS data plotted in a map (Figure 6) show decreasing $\delta^2H$ and $\delta^{18}O$ values from east to west, reflecting the distance from the ocean and the groundwater flow direction through the GAS. The lowest $\delta^{18}O$ and $\delta^2H$ values in the GAS are related to the highest confinement and distance from direct recharge areas. In this sense, tectonic fractures can provide

meteoric to medium residence time waters to the GAS across the Serra Geral Formation, indicating descending water recharge processes, corresponding to Condition 2 described into Figure 4.

The SGAS $\delta^{18}O$ and $\delta^2H$ results plotted in the same map show a different arrangement from the GAS, with the lowest values in the northern, western, and eastern portions (Figure 7). These differences are related to the influence of tectonic fractures and, as a consequence, to the hydrochemical groups distribution [19].

In the northern region, the lowest $\delta^{18}O$ and $\delta^2H$ values coincide with a greater volcanic sequence thickness and thicker soils, providing for an increase in groundwater residence time and reduced meteoric water recharge. The prevalence of $SO_4F$, NaF, and F hydrochemical groups in this portion confirms the long residence time and older relative age to this waters.

In the western sector of Figure 6, samples with the lowest $\delta^{18}O$ and $\delta^2H$ values are located along the Uruguay River Fault System in an NE-SW direction. In this area, the SGAS is thinner, and the intense tectonic fracturing favors to the groundwater ascension through the confined GAS.

In the eastern area, the lowest $\delta^{18}O$ and $\delta^2H$ values (samples 238 and 244) are related to the Leao and Terra de Areia-Posadas Fault Systems, which allow for groundwater ascension through the confined GAS.

The scatter diagram of fluoride versus $\delta^{18}O$ and $\delta^2H$ (Figure 8) reveals, in a broad sense, that lower isotope ratios (most negative values) are associated with higher fluoride content, reflecting water-rock interaction during a long residence time. Similar conditions were also described by [24] in the GAS in the Sao Paulo State, Southern Brazil. This behavior can be associated with ascending infiltration GAS waters.

TABLE 3: $\delta^{18}O$ and $\delta^2H$ values of phreatic waters (a), GAS (a), and SGAS (b). Data from [28] (*) and [20] (**). TA is a code to Tres Arroios City (sampled by author). Fluoride content is exclusive to SGAS.

(a)

| Sample code | Aquifer | $\delta^{18}O$ (‰) | $\delta^2H$ (‰) |
|---|---|---|---|
| PMT1* | | −4.67 | −23.4 |
| PMT2* | Phreatic | −4.26 | −20.8 |
| PMCG3* | | −5.05 | −27.4 |
| PMBF7* | | −4.56 | −22.9 |
| Averages | | −4.64 | −23.7 |
| TA | | −5.1 | −30.0 |
| 1-RS** | | −5.2 | −33.0 |
| 2-RS** | | −4.9 | −28.0 |
| 3-RS** | GAS | −4.8 | −29.0 |
| 4-RS** | | −5.3 | −33.0 |
| 5-RS** | | −5.3 | −33.0 |
| 7-RS** | | −5.5 | −32.0 |
| 8-RS** | | −5.4 | −37.0 |
| Averages | | −5.28 | −33.2 |

(b)

| Sample code | SGAS water group | $F^-$ (mg/L) | $\delta^{18}O$ (‰) |
|---|---|---|---|
| 35 | CaMg | 0.1 | −5.2 |
| 33 | CaMg | 1.02 | −5.2 |
| 29 | CaMg | 0.3 | −4.5 |
| 16 | CaMg | 0.47 | −4.2 |
| 416 | F | 0.45 | −5.0 |
| 360 | F | 0.68 | −4.8 |
| 235 | F | 0.95 | −5.2 |
| 208 | F | 0.75 | −5.5 |
| 173 | F | 0.9 | −5.3 |
| 138 | F | 3.03 | −4.5 |
| 57 | F | 0.35 | −4.7 |
| 37 | F | 2.6 | −5.7 |
| 244 | Na | 0.32 | −5.1 |
| 238 | Na | 0.52 | −5.2 |
| 212 | Na | 0.31 | −5.0 |
| 144 | Na | 0.23 | −5.2 |
| 18 | Na | 0.23 | −4.6 |
| 14 | Na | 0.19 | −4.6 |
| 10 | Na | 0.6 | −4.5 |
| 251 | NaF | 1.96 | −5.7 |
| 31 | NaF | 2.03 | −5.6 |
| 279 | SO$_4$F | 1.19 | −4.7 |
| 261 | SO$_4$F | 1.81 | −4.9 |
| 71 | SO$_4$Na | 0.07 | −5.2 |
| 334 | Meteoric | 0.39 | −5.2 |
| 97 | Meteoric | 0.3 | −5.1 |
| 58 | Meteoric | 0.03 | −5.0 |
| Averages | | 0.81 | −5.01 |

## 4. Conclusions

In the Rio Grande do Sul State, extensive groundwater mixing processes operate between the SGAS and GAS through defined and intense tectonic fractures. Six hydrochemical groups were distinguished in the SGAS, three of them containing substantial fluoride contents in combination with $Na^+$ and $SO_4^{2-}$. Ascending recharge processes are considered to define ionic enrichment, especially in $Na^+$, $F^-$, $SO_4^{2-}$, and $Cl^-$.

The comparison between $\delta^{18}O$ and $\delta^2H$ spatial distribution in the SGAS demonstrates a different scenario concerning the GAS. The differences are probably explained by the interference of different tectonic structures than those in the GAS which are responsible for the defined hydrogeological blocks. The combination of deep fault zones and high pressure and temperature conditions in a confined GAS supports the hypothesis of ascending groundwater infiltration.

The depleted $\delta^{18}O$ and $\delta^2H$ values suggest that fluoride enrichment in the SGAS is related to percolation through the GAS in confinement and long residence time conditions. The similar values of $\delta^{18}O$ and $\delta^2H$ in the two aquifer systems suggest, in some regions, that ascending recharge processes are more intense than meteoric recharge. In the Rio Grande do Sul State, this behavior increases from east to west, which is the groundwater flow direction in the GAS. Similar behavior is observed in the SGAS, except in some scarps and valleys, as is demonstrated by the distribution of the CaMg water group. The fluoride input to the groundwater is probably related to water-rock interaction in the GAS and in older aquifer systems below it.

## Acknowledgments

The funding for this study was provided by the Brazilian National Research Council (CNPq), Rio Grande do Sul Research Foundation (Fapergs), and Rio Grande do Sul Environmental Agency (Fepam).

## References

[1] N. Lin, J. Tang, and J. Bian, "Geochemical environment and health problems in China," *Environmental Geochemistry and Health*, vol. 26, no. 1, pp. 81–88, 2004.

[2] W. Genxu and C. Guodong, "Fluoride distribution in water and the governing factors of environment in arid north-west China," *Journal of Arid Environments*, vol. 49, no. 3, pp. 601–614, 2001.

[3] V. Vijaya Kumar, C. S. T. Sai, P. L. K. M. Rao, and C. S. Rao, "Studies on the distribution of fluoride in drinking water sources in Medchal Block, Ranga Reddy District, Andhra Pradesh, India," *Journal of Fluorine Chemistry*, vol. 55, no. 3, pp. 229–236, 1991.

[4] W. K. Moturi, M. P. Tole, and T. C. Davies, "The contribution of drinking water towards dental fluorosis: a case study of Njoro division, Nakuru district, Kenya," *Environmental Geochemistry and Health*, vol. 24, no. 2, pp. 123–130, 2002.

[5] U. Kafri, A. Arad, and L. Halicz, "Fluorine occurrence in groundwater in Israel and its significance," *Journal of Hydrology*, vol. 106, no. 1-2, pp. 109–129, 1989.

[6] WHO, *Fluorides*, Environmental Health Criteria no. 227, United Nations Environmental Programme, International Labour Organization, World Health Organization, Geneva, Switzerland, 2002.

[7] H. C. N. S. Campos, "Modelación conceptual y matemática del acuífero guaraní, Cono Sur," *Acta Geológica Leopoldensia*, vol. 23, no. 4, pp. 3–50, 2000.

[8] L. M. Aráujo, A. B. França, and P. E. Potter, "Hydrogeology of the Mercosul aquifer system in the Paraná and Chaco-Paraná Basins, South America, and comparison with the Navajo-Nugget aquifer system, USA," *Hydrogeology Journal*, vol. 7, no. 3, pp. 317–336, 1999.

[9] A. Batezelli, A. R. Saad, V. J. Fulfaro, A. C. Corsi, P. M. B. Landim, and J. A. J. Perinotto, "Análise de bacia aplicada às unidades mesozoicas do triângulo mineiro (sudeste do Brasil): uma estratégia na prospecção de recursos hídricos subterrâneos," *Águas Subterrâneas*, vol. 19, no. 1, pp. 61–73, 2005.

[10] P. A. R. Reginato and A. J. Strieder, "Integração de dados geológicos na prospecção de aqüíferos fraturados na Formação Serra Geral," *Revista Águas Subterrâneas*, vol. 20, no. 1, pp. 1–14, 2006.

[11] M. Szikszay, J. Teissedre, U. Barner, and E. Matsui, "Geochemical and isotopic characteristics of spring and groundwater in the State of São Paulo, Brazil," *Journal of Hydrology*, vol. 54, no. 1–3, pp. 23–32, 1981.

[12] G. Fraga, *Origem de fluoreto em águas subterrâneas dos Sistemas Aqüíferos Botucatu e Serra Geral da Bacia do Paraná [Ph.D. thesis]*, Programa de Pós-Graduação em Recursos Minerais e Hidrogeologia, Universidade de São Paulo, 1992.

[13] C. V. Portela Filho, F. J. F. Ferreira, E. F. Rosa Filho, A. C. Buchmann, and S. P. Rostirolla, "Estudo preliminar da conexão entre os aquíferos Serra Geral e Guarani com base em dados aeromagnetométricos e hidroquímicos," in *XII Congresso Brasileiro de Águas Subterrâneas*, vol. 1, pp. 74–86, Editora da ABAS, São Paulo, Brazil, 2002.

[14] C. V. Portela Filho, F. J. F. Ferreira, E. F. Rosa Filho, A. C. Buchmann, and S. Rostirolla, "Compartimentação magnética-estrutural do sistema aquífero Serra Geral e sua conectividade com o Sistema Aqüífero Guarani na região central do arco de Ponta Grossa (Bacia do Paraná)," *Revista Brasileira de Geociências*, vol. 3, pp. 369–381, 2005.

[15] J. L. F. Machado, *Compartimentação espacial e arcabouço hidro-estratigráfico do Sistema Aqüífero Guarani no Rio Grande do Sul [Ph.D. thesis]*, Programa de Pós-Graduação em Geologia—Área de Concentração em Geologia Sedimentar, Universidade do Vale do Rio Dos Sinos, São Leopoldo, Brazil, 2005.

[16] L. F. Scheibe and R. Hirata, "O contexto tectônico dos sistemas aqüíferos guarani e serra geral em santa catarina: uma revisão," in *XV Congresso Brasileiro de Águas Subterrâneas*, pp. 1–14, Associação Brasileira de Águas Subterrâneas, Curitiba, Brazil, 2008.

[17] N. A. Lisboa, *Fácies, estratificações hidrogeoquímicas e seus controladores geológicos, em unidades hidrogeológicas do Sistema Aqüífero Serra Geral, na Bacia do Paraná, Rio Grande do Sul [Ph.D. thesis]*, Curso de Pós-Graduação em Geociências, Universidade Federal do Rio Grande do Sul, 1996.

[18] N. A. Lisboa and E. Menegotto, "Diferenciações hidrogeoquímicas no sistema aqüífero Serra Geral no Rio Grande do Sul," in *XII Simpósio Brasileiro de Recursos Hídricos*, pp. 489–496, 1997.

[19] A. Nanni, A. Roisenberg, M. P. C. Marimon, and A. P. Viero, "The hydrochemical facies and anomalous fluoride content in the Serra Geral aquifer system, southern Brazil: a GIS approach with tectonic and principal component analysis," *Environmental Geology*, vol. 58, no. 6, pp. 1247–1255, 2009.

[20] A. A. Kimmelmann e Silva, A. C. Reboucas, and M. M. F. Santiago, "$^{14}$C analyses of groundwater from the Botucatu Aquifer System in Brazil," *Radiocarbon*, vol. 31, no. 3, pp. 926–933, 1989.

[21] A. Nanni, A. Roisenberg, J. M. G. Fachel, G. Mesquita, and C. Danieli, "Fluoride characterization by principal component analysis in the hydrochemical facies of Serra Geral Aquifer System in Southern Brazil," *The Anais da Academia Brasileira de Ciências*, vol. 80, no. 4, pp. 693–701, 2008.

[22] O. A. B. Licht, *A geoquímica multielementar na gestão ambiental—identificação e caracterização de províncias geoquímicas naturais, alterações antrópicas da paisagem, áreas favoráveis à prospecção mineral e regiões de risco para a saúde no estado do Paraná, Brasil [Ph.D. thesis]*, Pós-Graduação em Geologia, Área de Concentração—Geologia Ambiental, Setor de Ciências da Terra, Universidade Federal do Paraná, Curitiba, Brazil, 2001.

[23] A. V. L. Bittencourt, E. F. Rosa Filho, E. C. Hindi, and A. C. Buchmann Filho, "A influência dos basaltos e de misturas com águas de aquíferos sotopostos nas águas subterrâneas do Sistema Aqüífero Serra- Geral na bacia do rio piquiri, Paraná—BR," *Revista Águas Subterrâneas*, vol. 17, pp. 67–75, 2003.

[24] O. Sracek and R. Hirata, "Geochemical and stable isotopic evolution of the Guarani Aquifer System in the state of São Paulo, Brazil," *Hydrogeology Journal*, vol. 10, no. 6, pp. 643–655, 2002.

[25] E. F. Rosa Filho, J. X. Montano, and U. Duarte, "Estudo do movimento das águas subterrâneas do Sistema Aqüífero Guarani (SAG) através de isótopos, no Paraná, São Paulo e no Uruguai," *Revista Latino Americana de Hidrogeologia*, pp. 109–121, 2007.

[26] A. O. Invernizzi and S. M. B. Oliveira, "Hydrochemical characterization of a watershed through factor analysis," *Revista de Águas Subterrâneas*, vol. 18, pp. 67–77, 2004.

[27] H. F. Kaiser, "The varimax criterion for analytic rotation in factor analysis," *Psychometrika*, vol. 23, no. 3, pp. 187–200, 1958.

[28] M. P. Marimom, *Origem das Anomalias de Fluoreto nas Águas Subterrâneas utilizadas para abastecimento público da Formação Santa Maria, Estado do Rio Grande do Sul, Brasil [Ph.D. thesis]*, Curso de Pós-graduação em Geociências, Universidade Federal do Rio Grande do Sul, Porto Alegre, Brazil, 2006.

[29] IAEA—International Atomic Energy Agency, Isotope Hydrology Information System, the ISOHIS Database, 2004, http://isohis.iaea.org.

# 7

# A Fast Interpretation Method for Inverse Modeling of Residual Gravity Anomalies Caused by Simple Geometry

## Khalid S. Essa

*Geophysics Department, Faculty of Science, Cairo University, P.O. 12613, Giza, Egypt*

Correspondence should be addressed to Khalid S. Essa, khalid_sa_essa@yahoo.com

Academic Editor: Steven L. Forman

An inversion technique using a fast method is developed to estimate, successively, the depth, the shape factor, and the amplitude coefficient of a buried structure using residual gravity anomalies. By defining the anomaly value at the origin and the anomaly value at different points on the profile, the problem of depth estimation is transformed into a problem of solving a nonlinear equation of the form $f(z) = 0$. Knowing the depth, the shape factor can be estimated and finally the amplitude coefficient can be estimated. This technique is applicable for a class of geometrically simple anomalous bodies, including the semiinfinite vertical cylinder, the infinitely long horizontal cylinder, and the sphere. The efficiency of this technique is demonstrated with gravity anomaly due to a theoretical model, in each case with and without random errors. Finally, the applicability is illustrated using the residual gravity anomaly of Mobrun ore body, situated near Noranda, QC, Canada. The interpreted depth and the other model parameters are in good agreement with the known actual values.

## 1. Introduction

Inversion of gravity data is nonunique in the sense that the observed gravity anomalies in the plane of observation can be explained by a variety of density distributions. One way to solve this ambiguity is to assign a suitable geometry to the anomalous body with a known density followed by inversion of gravity anomalies [1]. Although simple models may not be geologically realistic, they are usually are sufficient to analyze sources of many isolated anomalies [2]. The interpretation of such an anomaly aims essentially to estimate the body parameters such as shape, depth, and radius. Several graphical and numerical methods have been developed for interpreting gravity anomalies caused by simple bodies. The simplest formula used to approximate depth to a causative body from the residual gravity data is the half-$g_{max}$ rule [3, 4]. However, the drawback with this approach is that it is highly subjective and can lead to large errors [5]. Gupta [6] presented a numerical approach to determine the depth to cylindrical and spherical models from the residual gravity data. Abdelrahman [7] argued that inserting the maximum

gravity value $g_{max}$ as a known parameter in Gupta's formulation may lead to large error in the calculation of depth in the existence of noise. Recently, several computer-based methods of inverting gravity data to determine model parameters have been presented with various levels of success [8–10]. A simple method proposed by Essa [11] is used to determine the depth and shape factor of simple shapes from residual gravity anomalies along the profile. Another automatic method, the least-squares method, was proposed by Asfahani and Tlas [12], by which the depth and amplitude coefficient can be determined. The principal difficulty with the inversion methods is the inherent nonuniqueness of the solution [13]. Therefore, there is still a need for an interpretation technique that is robust, rapid, and can provide parameters of the bodies in field situations.

In this paper, an inversion technique based on nonlinear equation $z = f(z)$ to analyze gravity anomalies due to simple structures. The inversion technique simultaneously estimates the depth, the nature of the source (shape factor), and the amplitude coefficient of the buried structures. The accuracy of the result obtained by this procedure depends upon the accuracy to which the residual anomaly can be separated from the Bouguer anomaly. In most cases, graphical methods [5] or standard numerical methods [14–16] can be used to separate the residual gravity anomaly attributable to the buried structure from the Bouguer data. Also, the accuracy of the result of the present method depends on the extent to which the source body conforms to one of the assumed geometries. The methodology is illustrated with theoretical models, in each case with and without random errors, and tested by the gravity anomaly of the Mobrun ore body, situated in the mining district of Noranda, QC, Canada.

## 2. The Method

The general vertical component of the gravity anomaly expression produced by a sphere (3D), an infinite long horizontal cylinder (2D), and a semiinfinite vertical cylinder (3D) is shown in Figure 1 and given in Abdelrahman et al. [17] as

$$g(x_i, z, q) = A \frac{z^m}{(x_i^2 + z^2)^q}, \tag{1}$$

where

$$A = \begin{cases} \frac{4}{3} \pi G \sigma R^3, \\ 2\pi G \sigma R^2, \\ \pi G \sigma R^2, \end{cases} \quad m = \begin{cases} 1, \\ 1, \\ 0, \end{cases} \tag{2}$$

$$q = \begin{cases} \dfrac{3}{2} & \text{for a sphere} \\ 1 & \text{for a horizontal cylinder} \\ \dfrac{1}{2} & \text{for a vertical cylinder } R \ll z. \end{cases}$$

In (1), $z$ is the depth, $q$ is the shape factor, for example, the shape factors for the semiinfinite vertical cylinder (3D) (the gravity response in case of the semiinfinite vertical cylinder is only applicable when the radius of the cylinder is much smaller than the distance from observation position to the top of the cylinder. This is called the "thin vertical rod approximation." For the general case of the right vertical cylinder, the gravity response is much more complicated), horizontal cylinder (2D), and sphere (3D) are 0.5, 1.0, and 1.5, respectively. Also, the shape factor for the finite vertical cylinder is approximately 1 [18]. The shape factor ($q$) approaches zero as the structure becomes a nearly horizontal bed, and approaches 1.5 as the structure becomes a perfect sphere (point mass). $x_i$ is the position coordinate, $\sigma$ is the density contrast, $G$ is the universal gravitational constant, and $R$ is the radius.

At the origin ($x_i = 0$), (1) gives the following relationship:

$$g(0) = \frac{A}{z^{2q-m}}. \tag{3}$$

Using (1), we obtain the following normalized equation at $x_i = \pm N$ and $x_i = \pm M$ where $N = 1, 2, 3, \dots$ and $M = 1, 2, 3, \dots$

$$\frac{g(N)}{g(0)} = \left( \frac{z^2}{N^2 + z^2} \right)^q, \qquad \frac{g(M)}{g(0)} = \left( \frac{z^2}{M^2 + z^2} \right)^q. \tag{4}$$

Let $F = (g(N)/g(0))$ and $T = (g(M)/g(0))$ then from (4), we get:

$$z = e^{([(\ln F/\ln T)*(\ln(z^2/(M^2+z^2)))]+\ln(N^2+z^2))/2}, \quad M \neq N. \tag{5}$$

Equation (5) can be solved for $z$ using the standard methods for solving nonlinear equations (e.g., [19]), and its iteration form can be expressed as:

$$z_f = f(z_j), \tag{6}$$

where $z_j$ is the initial depth, and $z_f$ is the revised depth; $z_f$ will be used as the $z_j$ for the next iteration. The iteration stops when $|z_f - z_j| \leq e$, where $e$ is a small predetermined real number close to zero. The source depth is determined by solving one nonlinear equation in $z$. Any initial guess for $z$ works well because there is always one global minimum. Theoretically, two different values of $N$ and $M$ are enough to determine the depth. In practice, more than two values of $N$ and $M$ are preferable because of the presence of noise in the data.

Once, the depth ($z$) is known, the shape factor ($q$) can be estimated from the following form:

$$q_c = \frac{\ln F}{\ln(z_c^2/(N^2 + z_c^2))}, \tag{7}$$

where $z_c$ is the estimated depth. Finally, knowing the shape factor ($q$), the amplitude coefficient ($A$) can be estimated from the following form:

$$A_c = g(0)z_c^{2q_c - m}, \tag{8}$$

where $q_c$ is the estimated shape factor.

For each $N$ and $M$ value, we compute the values of the model parameters ($z$, $q$, and $A$) from (5), (7), and (8), respectively. Theoretically, the anomaly values at the origin and any two $N$ and $M$ distances are just enough to determine the model parameters. However, in practice, it is recommended to use all possible combinations of $N$ and $M$ values to determine the most appropriate source parameters solutions from all gravity data. We then measure the goodness of fit between the observed and computed gravity data for each set of solutions. The simplest way to compare two gravity profiles is to compute the standard error ($\mu$) between the observed values and the values computed from estimated values of $z$, $q$, and $A$. The model parameters which give the least root mean sum squares differences are the best. In this way, we can select the best-fit source parameters solutions from all gravity data.

## 3. Synthetic Examples

We computed three different residual gravity anomalies, each consisting of the effect of local structure (semiinfinite

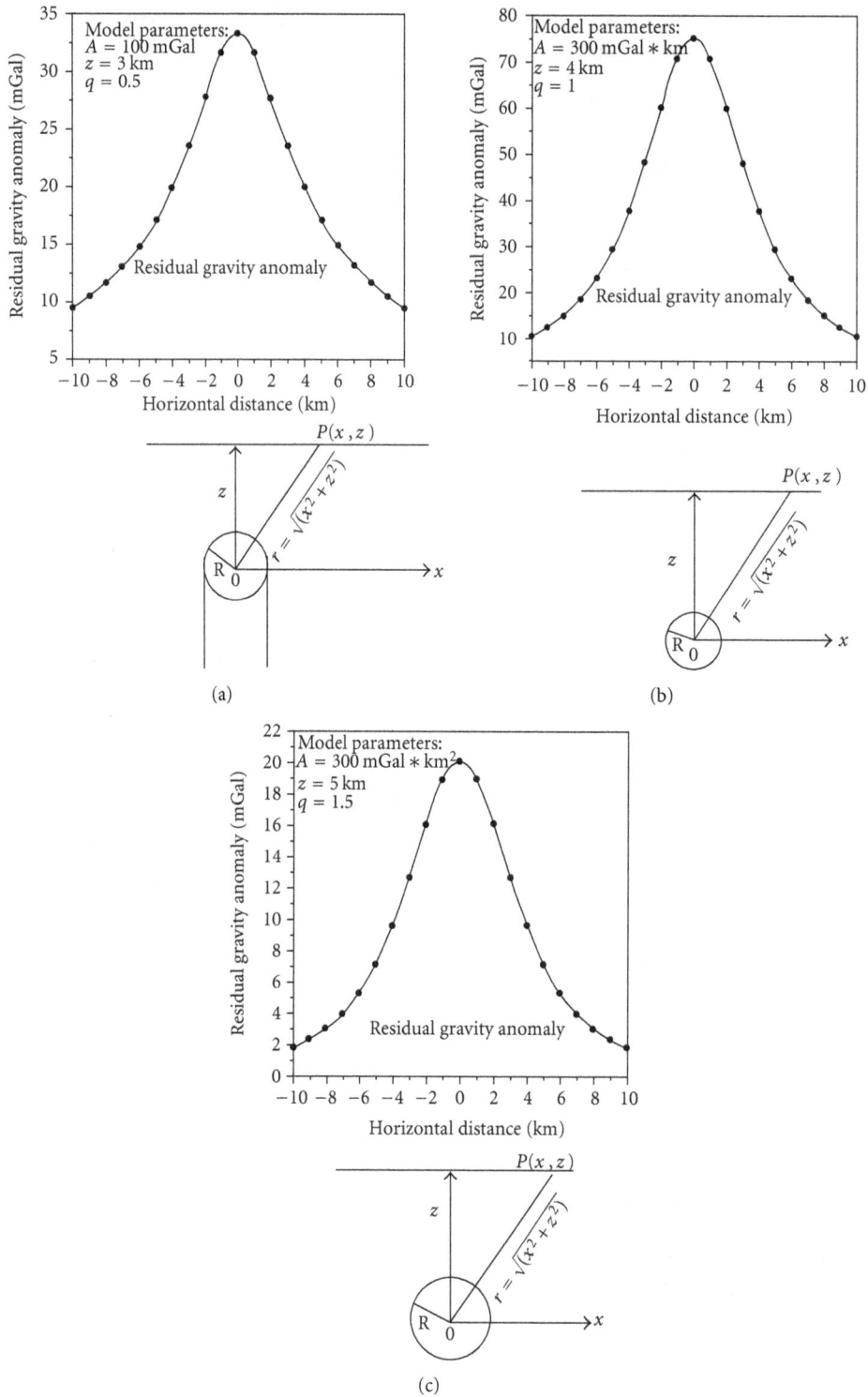

FIGURE 1: Residual gravity anomalies and schematic diagrams for various simple geometrical structures: (a) vertical cylinder, (b) horizontal cylinder, and (c) sphere.

vertical cylinder, horizontal cylinder, and sphere). The model equations representing the models are

$$cg_1(x_i) = \frac{100}{\left(x_i^2 + 3^2\right)^{0.5}},$$

$$g_2(x_i) = \frac{1200}{\left(x_i^2 + 4^2\right)},$$

$$g_3(x_i) = \frac{2500}{\left(x_i^2 + 5^2\right)^{1.5}}.$$

$$(9)$$

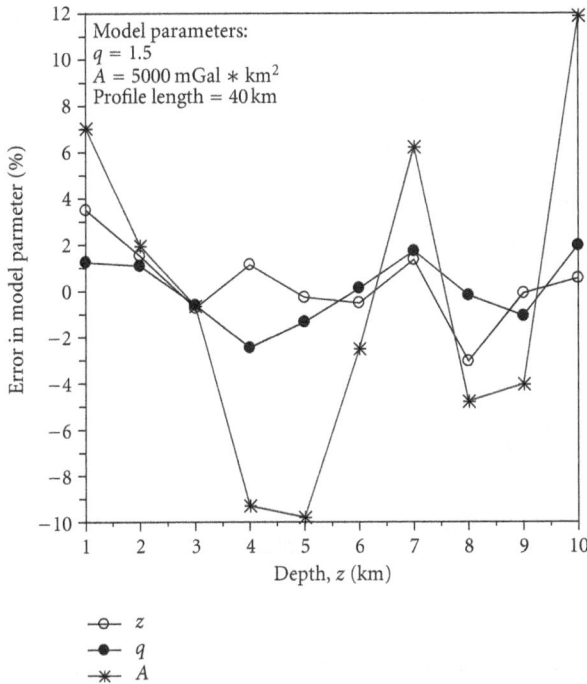

FIGURE 2: The effect of the depth of burial of a sphere model on the inverted parameters for synthetic data contaminated by 10% random noise.

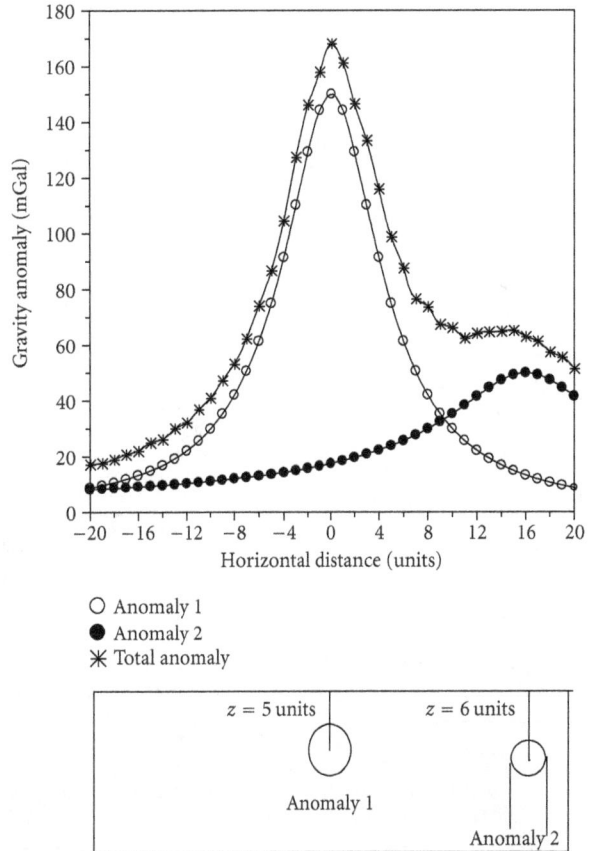

FIGURE 3: Noisy composite gravity anomaly consisting of the combined effects of an intermediate structure (horizontal cylinder with $A = 750$ mGal $*$ unit and $z = 5$ units) [anomaly 1] and interference from neighboring structure (semiinfinite vertical cylinder with $A = 300$ mGal and $z = 6$ units) [anomaly 2].

The three gravity anomalies are shown in Figure 1. Equations (5), (7), and (8) were applied to the residual anomaly profiles, yielding the model parameters: the depth, the shape factor, and the amplitude coefficient, respectively, solutions for all possible $N$ and $M$ points. The computed model parameters for the three models (a semiinfinite vertical cylinder model, a horizontal cylinder model and a sphere model) are summarized in Tables 1–3, respectively. In order to examine the influence of the noise on this approach, a 10% random noise has been added to the synthetic data using the following expression:

$$g_{rand}(x_i) = g(x_i)[1 + (RND(i) - 0.5) * 0.1], \quad (10)$$

where $g_{rand}(x_i)$ is the contaminated anomaly value at $x_i$, and RND $(i)$ is a pseudorandom number whose range is $(0,1)$. The interval of the pseudorandom number is an open interval, that is, it does not include the extremes 0 and 1.

Table 1 shows the model parameters ($z$, $q$, and $A$) in the case of using a semiinfinite vertical cylinder model results are the same when using synthetic data. The depth is within 1%, the shape factor is within 2% and the amplitude coefficient is within 5.9%. Table 2 shows the model parameters in the case of using a horizontal cylinder model results are the same when using synthetic data. After adding 10% random errors in the synthetic data, the depth is within 4.2%, the shape factor is within 7%, and the amplitude coefficient is within 13.2%. Table 3 shows the model parameters in the case of using a sphere model results are the same when using synthetic data. After adding 10% random errors in the

synthetic data, the depth is within 8.8%, the shape factor is within 4.6%, and the amplitude coefficient is within 3.3%.

In all cases examined, the exact values of the depth ($z$), the shape factor ($q$), and the amplitude coefficient ($A$) were obtained when using synthetic data without random errors. However, in studying the error response of the present method, synthetic examples contaminated with 10% random errors were considered. Good results are obtained by using the present algorithm—particularly for shape and depth estimation, which is a primary concern in gravity prospecting and other geophysical work.

3.1. Effect of Random Noise. We compute a gravity anomaly due to a sphere model (profile length = 40 km, $q = 1.5$, and $A = 5000$ mGal $*$ km$^2$; station separation interval = 1 km) buried at different depths. The computed gravity anomaly $g(x_i)$ was contaminated with random errors with a noise level of 10 mGal using the following equation:

$$\Delta g_{rand}(x_i) = g(x_i) + 10(RND(i) - 0.5), \quad (11)$$

where $\Delta g_{rand}(x_i)$ is the contaminated anomaly value at $x_i$.

Following the interpretation method, (5), (7), and (8) were used to determine depth, shape factor, and amplitude

TABLE 1: Numerical results of the present method applied to the semiinfinite vertical cylinder synthetic example ($q = 0.5$, $z = 3$ km, and $A = 100$ mGal; profile length = 20 km; sampling interval = 1 km) without and with 10% random noise.

| $M$ (km) | Using synthetic data | | | Using data with 10% random errors | | |
|---|---|---|---|---|---|---|
| | Computed depth (km) | Computed shape factor | Computed amplitude factor (mGal) | Computed depth (km) | Computed shape factor | Computed amplitude factor (mGal) |
| 2.00 | 3.00 | 0.50 | 100.00 | 2.97 | 0.47 | 94.33 |
| 3.00 | 3.00 | 0.50 | 100.00 | 2.90 | 0.54 | 104.01 |
| 4.00 | 3.00 | 0.50 | 100.00 | 2.42 | 0.38 | 65.53 |
| 5.00 | 3.00 | 0.50 | 100.00 | 2.97 | 0.47 | 94.40 |
| 6.00 | 3.00 | 0.50 | 100.00 | 3.31 | 0.58 | 134.60 |
| 7.00 | 3.00 | 0.50 | 100.00 | 3.15 | 0.53 | 112.76 |
| 8.00 | 3.00 | 0.50 | 100.00 | 3.26 | 0.56 | 126.73 |
| 9.00 | 3.00 | 0.50 | 100.00 | 3.19 | 0.54 | 118.41 |
| 10.00 | 3.00 | 0.50 | 100.00 | 3.06 | 0.50 | 102.89 |
| Average (km) | 3.00 | 0.50 | 100.00 | 3.03 | 0.51 | 105.96 |
| Percent of error | 0.00 | 0.00 | 0.00 | 1.00 | 2.00 | 5.96 |

TABLE 2: Numerical results of the present method applied to the horizontal cylinder synthetic example ($q = 1.0$, $z = 4$ km, and $A = 300$ mGal $*$ km; profile length = 20 km; sampling interval = 1 km) without and with 10% random noise.

| $M$ (km) | Using synthetic data | | | Using data with 10% random errors | | |
|---|---|---|---|---|---|---|
| | Computed depth (km) | Computed shape factor | Computed amplitude factor (mGal) | Computed depth (km) | Computed shape factor | Computed amplitude factor (mGal) |
| 2.00 | 4.00 | 1.00 | 300.00 | 4.66 | 1.19 | 350.25 |
| 3.00 | 4.00 | 1.00 | 300.00 | 4.18 | 1.12 | 349.38 |
| 4.00 | 4.00 | 1.00 | 300.00 | 4.15 | 1.11 | 327.14 |
| 5.00 | 4.00 | 1.00 | 300.00 | 4.00 | 1.03 | 331.93 |
| 6.00 | 4.00 | 1.00 | 300.00 | 4.05 | 1.06 | 362.68 |
| 7.00 | 4.00 | 1.00 | 300.00 | 3.78 | 0.93 | 336.27 |
| 8.00 | 4.00 | 1.00 | 300.00 | 4.00 | 1.03 | 329.87 |
| 9.00 | 4.00 | 1.00 | 300.00 | 4.28 | 1.01 | 318.85 |
| 10.00 | 4.00 | 1.00 | 300.00 | 4.49 | 1.11 | 350.93 |
| Average (km) | 4.00 | 1.00 | 300.00 | 4.17 | 1.07 | 339.69 |
| Percent of error | 0.00 | 0.00 | 0.00 | 4.25 | 7.00 | 13.23 |

coefficient, respectively. The percentage of error in model parameters was plotted against the model depth for comparison. Numerical results are shown in Figure 2. We verified numerically that the depth, shape factor, and amplitude coefficient are within 3.5%, 2%, and 11.9%, respectively. Good results are obtained by using the present algorithm because our technique is robust in the presence of noise.

*3.2. Application to Composite Anomalies.* The composite gravity anomaly in mGal, shown in Figure 3, consists of the combined effects of an intermediate structure of interest (a 2D horizontal cylinder with $z = 5$ units, $A = 750$ mGal $*$ unit, and a station separation of 1 depth unit) and an interference from neighboring rocks (a 3D semiinfinite vertical cylinder with $z = 6$ units, and $A = 300$

mGal; a station separation of 1 depth unit) The anomaly was computed using the following expression:

$$\Delta g(x_i) = \frac{3750}{(x_i^2 + 25)} \text{ (2D horizontal cylinder)}$$
$$+ \frac{300}{(x_i^2 + 36)^{0.5}} \text{ (3D vertical cylinder).} \tag{12}$$

In Figure 3, anomaly 1 is the anomaly due to the intermediate structure of interest, and anomaly 2 is the anomalies due to the interference from neighboring structures represented by a horizontal cylinder and a vertical cylinder, respectively. The composite gravity anomaly $\Delta g(x_i)$ is also contaminated with a 5% random error. Following the same interpretation method, the result is shown in Table 4. The

TABLE 3: Numerical results of the present method applied to the sphere synthetic example ($q = 1.5$, $z = 5$ km, and $A = 500$ mGal $*$ km$^2$; profile length = 20 km; sampling interval = 1 km) without and with 10% random noise.

| M (km) | Using synthetic data | | | Using data with 10% random errors | | |
|---|---|---|---|---|---|---|
| | Computed depth (km) | Computed shape factor | Computed amplitude factor (mGal) | Computed depth (km) | Computed shape factor | Computed amplitude factor (mGal) |
| 2.00 | 5.00 | 1.50 | 500.00 | 2.79 | 1.39 | 334.34 |
| 3.00 | 5.00 | 1.50 | 500.00 | 4.48 | 1.17 | 146.71 |
| 4.00 | 5.00 | 1.50 | 500.00 | 4.99 | 1.44 | 407.92 |
| 5.00 | 5.00 | 1.50 | 500.00 | 3.00 | 1.20 | 271.76 |
| 6.00 | 5.00 | 1.50 | 500.00 | 5.07 | 1.49 | 493.73 |
| 7.00 | 5.00 | 1.50 | 500.00 | 5.07 | 1.49 | 496.48 |
| 8.00 | 5.00 | 1.50 | 500.00 | 5.30 | 1.62 | 844.65 |
| 9.00 | 5.00 | 1.50 | 500.00 | 5.25 | 1.59 | 750.98 |
| 10.00 | 5.00 | 1.50 | 500.00 | 5.16 | 1.54 | 602.08 |
| Average (km) | 5.00 | 1.50 | 500.00 | 4.56 | 1.44 | 483.18 |
| Percent of error | 0.00 | 0.00 | 0.00 | 8.80 | 4.66 | 3.36 |

TABLE 4: Numerical results of the present method applied to the horizontal cylinder synthetic example ($q = 1$, $z = 5$ km, and $A = 750$ mGal $*$ km; profile length = 40 km; sampling interval = 1 km) without and with 5% random noise.

| M (km) | Using synthetic data | | | Using data with 5% random errors | | |
|---|---|---|---|---|---|---|
| | Computed depth (km) | Computed shape factor | Computed amplitude factor (mGal) | Computed depth (km) | Computed shape factor | Computed amplitude factor (mGal) |
| 3.00 | 7.29 | 1.54 | 2252.78 | 5.38 | 0.76 | 393.01 |
| 4.00 | 5.87 | 1.01 | 1812.75 | 6.91 | 1.22 | 2686.23 |
| 5.00 | 5.30 | 0.82 | 590.16 | 5.93 | 0.91 | 706.47 |
| 6.00 | 4.94 | 0.72 | 435.48 | 5.73 | 0.85 | 561.92 |
| 7.00 | 4.67 | 0.64 | 359.25 | 5.46 | 0.78 | 423.99 |
| 8.00 | 4.43 | 0.58 | 211.87 | 5.02 | 0.67 | 280.06 |
| 9.00 | 4.20 | 0.52 | 179.18 | 4.50 | 0.54 | 187.71 |
| 10.00 | 4.00 | 0.47 | 156.33 | 4.18 | 0.47 | 152.74 |
| Average (km) | 5.09 | 0.79 | 749.72 | 5.39 | 0.78 | 674.02 |

result is generally in very good agreement with the model parameters ($q = 1$, $z = 5$ units; $A = 750$ mGal $*$ unit). Good results are obtained by using the present algorithm for model parameters which are of primary concern in gravity prospecting and other geophysical work. It is also emphasized that the method works well when dealing with gravity data having interference from neighboring anomalies and noise.

*3.3. The Effect of the Offset in the Origin Point of the Gravity Profile.* When interpreting real gravity data, inaccurate selection of the origin point of the gravity profile can lead to errors in estimating the gravity parameters. In order to examine this effect, we have introduced some successive errors ($\delta x$) of 0, $\pm 0.2$, $\pm 0.4$, ..., $\pm 1.5$ units to the horizontal coordinate $x$ of (1) of the gravity forward modeling calculations. The corresponding gravity dataset of a semiinfinite vertical cylinder model ($q = 0.5$, $z = 5$ units, and $A = 1200$ mGal;

profile length = 20 units, $N = 2$ units, and $M = 3$ units) has been inverted following the same procedures. Maximum absolute errors in the gravity inverse parameters ($z$, $q$, and $A$) are found to be 53%, 106%, and 180%, respectively (Figure 4(a)). This figure shows that the algorithm proposed has recovered almost the true values of $z$, $q$, and $A$ for each anomalous body at zero offset ($\delta x = 0$ unit). They also show that as $\delta x$ increases, the absolute error in $z$, $q$, and $A$ in general increases too, and the sign of the error in these parameters can change. In order to examine the accuracy and stability of the introduced algorithm on the combined effects of inaccurate origin selection and noise contamination, 5% random noise has been added to the various gravity dataset described in the previous paragraph, and then inverted. The maximum absolute errors in the gravity parameters ($z$, $q$, and $A$) obtained from inversion for the vertical cylinder models are found to be 71%, 99%, and 108%, respectively (Figure 4(b)). The analysis introduced demonstrates that

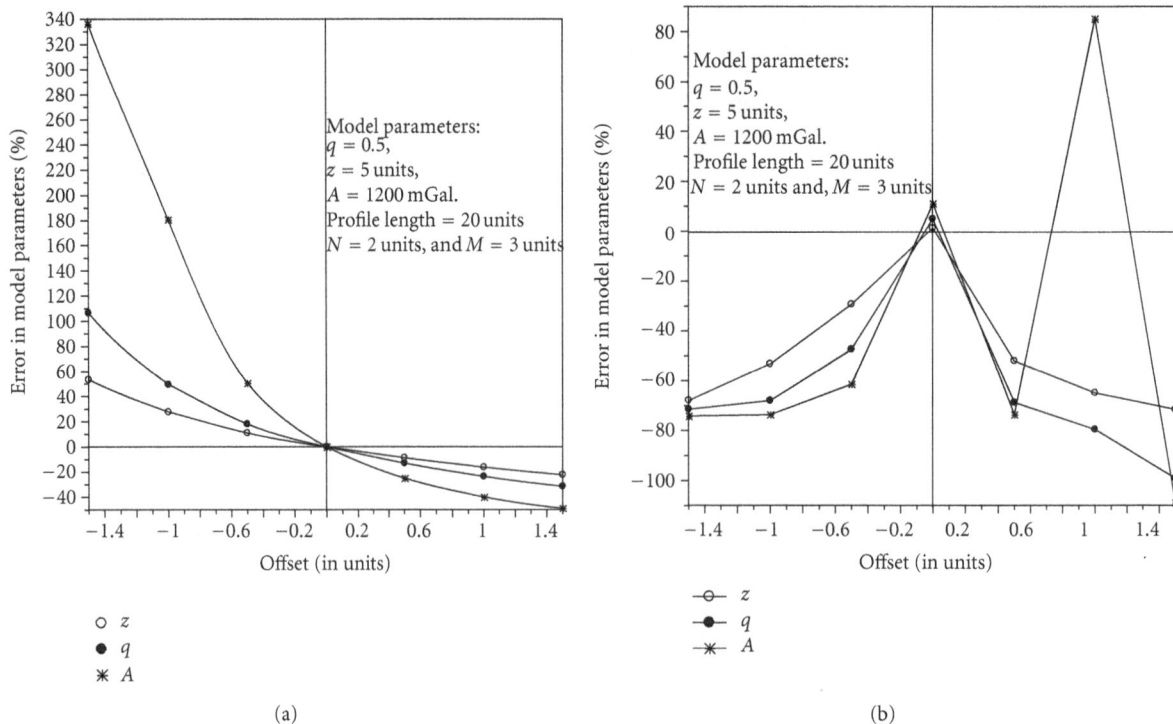

FIGURE 4: The effect of the offset in the origin point of a semiinfinite vertical cylinder model: (a) on the inverted model parameters for noise free synthetic data and (b) on the inverted model parameters for synthetic data contaminated by 5% random noise.

the present method is stable and can estimate the gravity parameters with a reasonable accuracy depending upon the embedded noise level and the offset value in the origin point.

*3.4. The Sensitivity of the Removing Regional Field.* The composite gravity anomaly (in mGal), shown in Figure 5, which consists of a local structure (sphere with $z = 3$ units and $A = 750$ mGal $*$ units$^2$; profile length = 20 units) and a 2nd-order polynomial of regional structure. The model equation is

$$\Delta g(x_i) = \frac{2250}{\left(x_i^2 + 9\right)^{1.5}} \text{ (A spherical model)}$$

$$- 0.02x_i^2 + 2x_i + 10 \text{ (2nd-order regional field).} \tag{13}$$

Using a separation technique to remove the effect of the regional structure (graphical method; [5]) and reestimate the model parameters ($z$, $q$, and $A$) (Table 5). Table 5 shows that the results are still incorrect because the field is still contaminated by remaining regional field. By using different separation methods [14–16], the residual separated and inverted the residual anomaly (Table 5).

## 4. Field Example

The fast algorithm has been adapted for interpreting residual gravity anomalies related to three different types of structures, for example, a sphere, a vertical cylinder, and a horizontal cylinder. The standard error ($\mu$) is used in this

paper as statistical preference criterion in order to compare the observed and calculated values. This $\mu$ is given by the following mathematical relationship:

$$\mu = \sqrt{\frac{\sum_{i=1}^{N} \left[g(x_i) - g_c(x_i)\right]^2}{N}}, \tag{14}$$

where $g(x_i)$ is the observed gravity values, and $g_c(x_i)$ is the calculated gravity values. A residual gravity field anomaly taken from Canada has been interpreted using the proposed technique in order to examine its applicability and stability.

*4.1. Mobrun Sulfide Body.* The residual gravity anomaly profile (Figure 6) over the Mobrun sulfide body in Noranda, QC, Canada (after [20]) was digitized at an interval of 33.5 m. The method was applied to the anomaly profile using a sampling interval of 33.5 m to determine the model parameters of the buried structure using all successful combinations of $N$ and $M$ values. Then, we computed the standard error ($\mu$) between the observed values and the values computed from estimated parameters $z$, $q$, and $A$ for each $N$ and $M$ value. The results are shown in Table 6 for the cases of $N$ and $M$ values where the $\mu$ difference between the modelled and observed data is less than 1.1 mGal. Also we computed the set of mean values. The set of mean values of the model parameters is rejected because it has a larger $\mu$ value (0.06 mGal) than the $\mu$-value of the optimum set (0.02 mGal). Also we computed the set of mean values. The set of mean values of the model parameters is rejected because it has a larger $\sigma$ value (0.055 mGal) than the $\mu$-value

TABLE 5: Numerical results of the present method applied to composite gravity anomaly (in mGal), which consists of a local structure (sphere with $z$ = 3 units and $A$ = 750 mGal $*$ units$^2$; profile length = 20 units) and a 2nd-order polynomial of regional structure.

| M (units) | Composite synthetic gravity data inverted | | | Using graphical separation techniques, remaining residual data inverted | | | Using analytical method, the remaining residual data inverted | | |
|---|---|---|---|---|---|---|---|---|---|
| | $z$ (units) | $q$ | $A$ (mGal $*$ unit$^2$) | $z$ (units) | $q$ | $A$ (mGal $*$ unit$^2$) | $z$ (units) | $q$ | $A$ (mGal $*$ unit$^2$) |
| 3.00 | 2.29 | 0.73 | 135.65 | 2.54 | 0.99 | 230.87 | 3.00 | 1.50 | 750.00 |
| 4.00 | 2.04 | 0.61 | 109.29 | 2.42 | 0.92 | 195.12 | 3.00 | 1.50 | 750.00 |
| 5.00 | 1.81 | 0.51 | 94.78 | 2.30 | 0.85 | 166.59 | 3.00 | 1.50 | 750.00 |
| 6.00 | 1.59 | 0.43 | 87.68 | 2.19 | 0.78 | 145.15 | 3.00 | 1.50 | 750.00 |
| 7.00 | 1.40 | 0.37 | 85.45 | 2.08 | 0.72 | 129.56 | 3.00 | 1.50 | 750.00 |
| 8.00 | 1.22 | 0.32 | 86.62 | 1.98 | 0.67 | 118.34 | 3.00 | 1.50 | 750.00 |
| 9.00 | 1.07 | 0.27 | 90.42 | 1.89 | 0.63 | 110.23 | 3.00 | 1.50 | 750.00 |
| 10.00 | 0.94 | 0.24 | 96.49 | 1.81 | 0.59 | 104.35 | 3.00 | 1.50 | 750.00 |
| Average (unit) | 1.55 | 0.43 | 98.30 | 2.15 | 0.77 | 150.02 | **3.00** | **1.50** | **750.00** |

TABLE 6: Numerical results of the present method applied to the Mobrun field example, Canada (best-fit in bold).

| N | M (m) | Depth $z$ (m) | Shape factor $q$ | Amplitude coefficient $A$ (mGal) | Standard error $\mu$ (mGal) |
|---|---|---|---|---|---|
| 33.50 | 67.00 | 25.68 | 0.61 | 55.75 | 0.49 |
| 33.50 | 100.50 | 26.54 | 0.64 | 55.48 | 0.34 |
| 33.50 | 134.00 | 28.66 | 0.71 | 55.51 | 0.13 |
| 67.00 | 100.50 | 28.64 | 0.68 | 56.07 | 0.16 |
| **67.00** | **134.00** | **33.34** | **0.78** | **59.10** | **0.02** |
| 100.50 | 134.00 | 42.36 | 0.93 | 72.38 | 0.21 |
| −33.50 | −67.00 | 24.89 | 0.52 | 58.71 | 1.01 |
| −33.50 | −100.50 | 29.53 | 0.64 | 57.14 | 0.16 |
| −33.50 | −134.00 | 30.74 | 0.68 | 57.45 | 0.08 |
| −67.00 | −100.50 | 41.30 | 0.84 | 68.43 | 0.14 |
| −67.00 | −134.00 | 40.14 | 0.82 | 66.46 | 0.11 |
| −100.50 | −134.00 | 38.15 | 0.79 | 63.86 | 0.06 |
| Average (m) | | 32.50 | 0.72 | 60.53 | 0.06 |

TABLE 7: Comparative results of Mobrun field example, Canada.

| | UsingGrant and West [20] method | Using Roy et al. [21] method | Using present method |
|---|---|---|---|
| $z$ (m) | 30 | 29.44 | 33.3 |
| $q$ | — | 0.77 | 0.78 |
| $A$ (mGal) | — | — | 59.1 |

of the optimum set (0.02 mGal). The optimum set is given at $N$ = 67 m and $M$ = 134 m. The best-fit model parameters are $z$ = 33.34 m, $q$ = 0.72, and $A$ = 59.1 mGal (Figure 6). This suggests that the shape of the ore body resembles a semiinfinite vertical cylinder, probably with a large radius. This is because the shape factor computed by the present method (0.72) is located between the shape factor of a perfect semiinfinite vertical cylinder ($q$ = 0.5) and the shape factor of an infinite horizontal cylinder ($q$ = 1.0). It is evident from the field example that our method gives good insight from gravity data of a short profile length concerning the nature of the source body. This is because the geologic situation is not complicated. The present method may not be applied to real data in complex geologic situations to obtain reliable or detailed information about the different shallow sources. This is true because each gravity measurement determines, at the station location, the sum of all effects from the surface downward. In complex geologic situations, the gravity profile is seldom a simple picture of a single isolated disturbance but always is a combination of two or more anomalies of shallow origin and very broad anomalies of regional nature, which may have their origin below the section within which the geologic interest lies. The shape and the depth to the top of the ore body obtained by the present method agree very well with those obtained from Roy et al. [21] and drilling information [20] (Table 7).

## 5. Conclusion

The problem of determining the appropriate depth, shape factor, and amplitude coefficient of a buried structure from

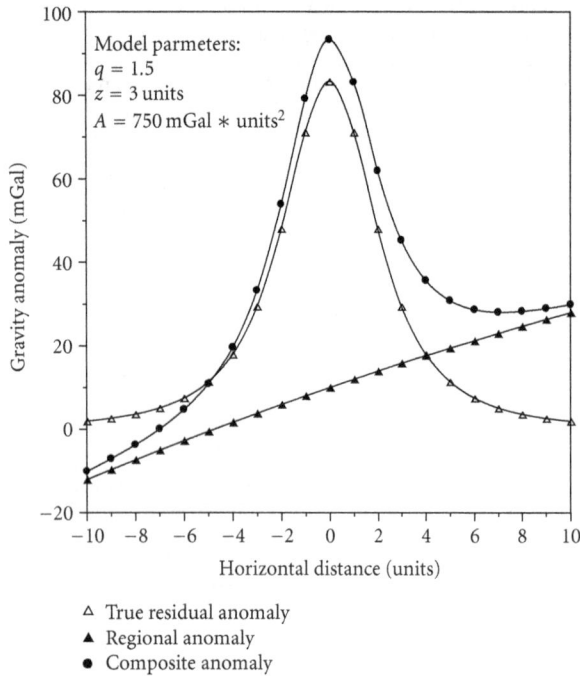

FIGURE 5: Composite gravity anomaly of a buried sphere model ($q = 1.5$, $A = 750$ mGal $*$ units$^2$ and $z = 3$ units; profile length = 20 units) and second-order regional.

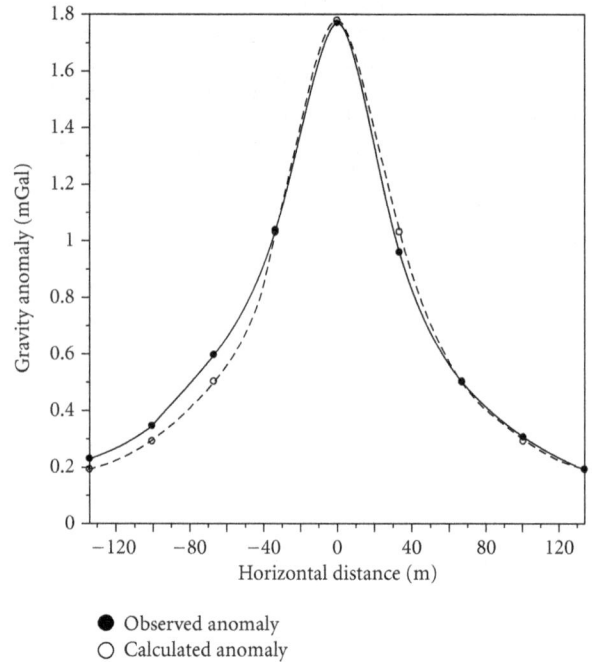

FIGURE 6: The measured gravity response (black circles) over a Mobrun sulfide ore body in Noranda, QC, Canada, and the predicted response (white circles) computed from the present inversion method.

the residual gravity data of a short or a long profile length can be solved using the present method. A simple and rapid inversion approach is formulated to use the anomaly values at the origin and two pairs of measured data points ($\pm N$ and $\pm M$). The repetition of the method using all possible combinations of such pairs of measured points will lead to the best-fitting model. This happens when these two pairs of points contain the least amount of noise in the entire set of measured data. It is also emphasized that the calculated gravity anomaly of a set of mean values of the model parameters obtained by the present method does not necessarily guarantee it matches the observed anomaly values when the data contain measurement errors. The advantages of this method over previous graphical and numerical techniques used to interpret gravity data are (1) all the three model parameters can be obtained from all observed data, (2) the method is automatic, and (3) the method works well even when gravity data was noisy. Moreover, the advantage of the present method over the least-squares method is that the method does not require computation of analytical or numerical derivatives with respect to the model parameters. Also, the disadvantages of this technique are very difficult to interpret more complicated structures and the accuracy of the results depending upon the removing unwanted field (regional). Finally, in view of the above facts, we envisage the application of this method in solving various problems related to potential field data interpretation in the future.

## Appendix

$$g(x_i, z, q) = A \frac{z^m}{(x_i^2 + z^2)^q}. \tag{A.1}$$

At the origin ($x_i = 0$), (A.1) gives the following relationship:

$$g(0) = \frac{A}{z^{2q-m}}. \tag{A.2}$$

Using (A.1), we obtain the following normalized equation at $x_i = \pm N$ and $x_i = \pm M$, where $N = 1, 2, 3, \ldots$ and $M = 1, 2, 3, \ldots$

$$\frac{g(N)}{g(0)} = \left( \frac{z^2}{N^2 + z^2} \right)^q,$$

$$\frac{g(M)}{g(0)} = \left( \frac{z^2}{M^2 + z^2} \right)^q. \tag{A.3}$$

Let $F = (g(N)/g(0))$ and $T = (g(M)/g(0))$ by taking algorithm to both sides:

$$\ln F = \ln \left( \frac{g(N)}{g(0)} \right) = \ln \left( \frac{z^2}{N^2 + z^2} \right)^q = q \ln \left( \frac{z^2}{N^2 + z^2} \right). \tag{A.4a}$$

$$\ln T = \ln \left( \frac{g(M)}{g(0)} \right) = \ln \left( \frac{z^2}{M^2 + z^2} \right)^q = q \ln \left( \frac{z^2}{M^2 + z^2} \right), \tag{A.4b}$$

by dividing (A.4a) and (A.4b), we get

$$\frac{\ln F}{\ln T} = \frac{\ln(z^2/(N^2+z^2))}{\ln(z^2/(M^2+z^2))}.$$ (A.4c)

From (A.4c), we get

$$\frac{[(\ln F/\ln T) * (\ln(z^2/(M^2+z^2)))] + \ln(N^2+z^2)}{2} = \ln z.$$ (A.4d)

By taking an exponential to both sides, we get

$$z = e^{([(\ln F/\ln T)*(\ln(z^2/(M^2+z^2)))]+\ln(N^2+z^2))/2}, \qquad M \neq N.$$ (A.5)

Equation (A.5) can be solved for $z$ using the standard methods for solving nonlinear equations (e.g., [19]), and its iteration form can be expressed as

$$z_f = f(z_j).$$ (A.6)

## Acknowledgments

The author wishes to express his sincere thanks to Professor Steven Forman, the editor, Professor Michel Chouteau, and the reviewer for their excellent suggestions, keen interests and thorough review that improved the paper. Many thanks to Professor El-Sayed M. Abdelrahman, Geophysics Department, Faculty of Science, Cairo University, for his constant help and encouragement.

## References

[1] V. Chakravarthi and N. Sundararajan, "Ridge-regression algorithm for gravity inversion of fault structures with variable density," *Geophysics*, vol. 69, no. 6, pp. 1394–1404, 2004.

[2] E. S. M. Abdelrahman and T. M. El-Araby, "A least-squares minimization approach to depth determination from moving average residual gravity anomalies," *Geophysics*, vol. 58, no. 12, pp. 1779–1784, 1993.

[3] L. L. Nettleton, "Gravity and magnetic calculation," *Geophysics*, vol. 7, pp. 293–310, 1942.

[4] W. M. Telford, L. P. Geldart, R. E. Sheriff, and D. A. Key, *Applied Geophysics*, Cambridge University Press, London, UK, 1976.

[5] L. L. Nettleton, *Gravity and Magnetic in Oil Prospecting*, McGraw-Hill Book, 1976.

[6] O. P. Gupta, "A least-squares approach to depth determination from gravity data," *Geophysics*, vol. 48, no. 3, pp. 357–360, 1983.

[7] E. M. Abdelrahman, "Discussion on: a least-squares approach to depth determination from gravity data by O. P. Gupta," *Geophysics*, vol. 55, no. 3, pp. 376–378, 1990.

[8] Y. Li and D. W. Oldenburg, "3-D inversion of gravity data," *Geophysics*, vol. 63, no. 1, pp. 109–119, 1998.

[9] X. Li and M. Chouteau, "Three-dimensional gravity modeling in all space," *Surveys in Geophysics*, vol. 19, no. 4, pp. 339–368, 1998.

[10] O. Boulanger and M. Chouteau, "Constraints in 3D gravity inversion," *Geophysical Prospecting*, vol. 49, no. 2, pp. 265–280, 2001.

[11] K. S. Essa, "A simple formula for shape and depth determination from residual gravity anomalies," *Acta Geophysica*, vol. 55, no. 2, pp. 182–190, 2007.

[12] J. Asfahani and M. Tlas, "An automatic method of direct interpretation of residual gravity anomaly profiles due to spheres and cylinders," *Pure and Applied Geophysics*, vol. 165, no. 5, pp. 981–994, 2008.

[13] Y. Li and D. W. Oldenburg, "3-D inversion of magnetic data," *Geophysics*, vol. 61, no. 2, pp. 394–408, 1996.

[14] E. M. Abdelrahman, S. Riad, E. Refai, and Y. Amin, "On the least-squares residual anomaly determinations," *Geophysics*, vol. 50, no. 3, pp. 473–480, 1985.

[15] B. N. P. Agarwal and C. Sivaji, "Separation of regional and residual anomalies by least-squares orthogonal polynomial and relaxation techniques: a performance evaluation," *Geophysical Prospecting*, vol. 40, no. 2, pp. 143–156, 1992.

[16] K. S. Essa, "Gravity data interpretation using the s-curves method," *Journal of Geophysics and Engineering*, vol. 4, no. 2, pp. 204–213, 2007.

[17] E. M. Abdelrahman, A. I. Bayoumi, Y. E. Abdelhady, M. M. Gobashy, and H. M. El-Araby, "Gravity interpretation using correlation factors between successive least-squares residual anomalies," *Geophysics*, vol. 54, no. 12, pp. 1614–1621, 1989.

[18] E. S. M. Abdelrahman and H. M. El-Araby, "Shape and depth solutions from gravity data using correlation factors between successive least-squares residuals," *Geophysics*, vol. 58, no. 12, pp. 1785–1791, 1993.

[19] W. H. Press, B. P. Flannery, S. A. Teukolsky, and W. T. Vetterling, *Numerical Recipes, The Art of Scientific Computing*, Cambridge University Press, 1986.

[20] F. S. Grant and G. F. West, *Interpretation Theory in Applied Geophysics*, McGraw-Hill Book, 1965.

[21] L. Roy, B. N. P. Agarwal, and R. K. Shaw, "A new concept in Euler deconvolution of isolated gravity anomalies," *Geophysical Prospecting*, vol. 48, no. 3, pp. 559–575, 2000.

# 8

# The Dabie Extensional Tectonic System: Structural Framework

**Quanlin Hou,[1] Hongyuan Zhang,[2] Qing Liu,[1] Jun Li,[3] and Yudong Wu[4]**

[1] *Graduate University of the Chinese Academy of Sciences, Beijing 100049, China*
[2] *School of Earth Sciences and Resources, China University of Geosciences, Beijing 100083, China*
[3] *Institute of Geology and Geophysics, The Chinese Academy of Sciences, Beijing 100029, China*
[4] *MLR Key Laboratory of Metallogeny and Mineral Assessment, Institute of Mineral Resources, CAGS, Beijing 100037, China*

Correspondence should be addressed to Quanlin Hou, quhou@gucas.ac.cn

Academic Editor: Yi-Wen Ju

A previous study of the Dabie area has been supposed that a strong extensional event happened between the Yangtze and North China blocks. The entire extensional system is divided into the Northern Dabie metamorphic complex belt and the south extensional tectonic System according to geological and geochemical characteristics in our study. The Xiaotian-Mozitan shear zone in the north boundary of the north system is a thrust detachment, showing upper block sliding to the NNE, with a displacement of more than 56 km. However, in the south system, the shearing direction along the Shuihou-Wuhe and Taihu-Mamiao shear zones is tending towards SSE, whereas that along the Susong-Qingshuihe shear zone tending towards SW, with a displacement of about 12 km. Flinn index results of both the north and south extensional systems indicate that there is a shear mechanism transition from pure to simple, implying that the extensional event in the south tectonic system could be related to a magma intrusion in the Northern Dabie metamorphic complex belt. Two $^{40}Ar$-$^{39}Ar$ ages of mylonite rocks in the above mentioned shear zones yielded, separately, $\sim$190 Ma and $\sim$124 Ma, referring to a cooling age of ultrahigh-pressure rocks and an extensional era later.

## 1. Introduction

Dabie area is well known of owning one largest area of ultrahigh-pressure metamorphic belt (UHPB) in the world, located, as a tectonic zone, between the North China Block and Yangtze Block (Figure 1).

The Dabie area experienced a complicated tectonic evolution during the Mesozoic and resulted in thrust-nappe, extensional detachment, and strike-slip structures [1, 2]. Much attention has been paid to the tectonic evolution of the Dabie Mountains, with most tectonic models proposing although compressional tectonism for the formation of the orogen. However, Mesozoic extensional structures in the Dabie Mountains are also obvious and important for understanding the Mesozoic tectonic regime inversion from compression to extension occurred throughout the Dabie Mountains and even in the eastern North China Block [3].

The purpose of this paper is to figure out the structural framework of the Dabie Late Mesozoic extensional detachment system by analyzing deformation, tectonic styles, and physical conditions, to constrain the time of the extensional tectonics, and finally to discuss the tectonic implications.

## 2. Tectonic Background

The general geology of the Dabie area has been described in multiple publications [5–13]. Briefly, from north to south, the Dabie Mountains can be divided into four major tectonic units: (1) the North Huaiyang Flysch belt (NHMB) mainly composed of the Foziling Group (Pt$_2$); (2) the Northern Dabie metamorphic complex belt (NDMCB), composed dominantly of granitic gneisses of TTG composition, with the Xiaotian-Mozitan and the Shuihou-Wuhe shear zones defining the northern and southern boundaries, respectively; (3) the ultrahigh-pressure metamorphic belt (UHPB) refers to the Central Dabie ultrahigh pressure metamorphic complex and is bounded in the south by the Taihu-Mamiao shear zone; (4) the high-pressure belt (HPB) refers to the Southern Dabie high-pressure blueschist/greenschist terrane [10],

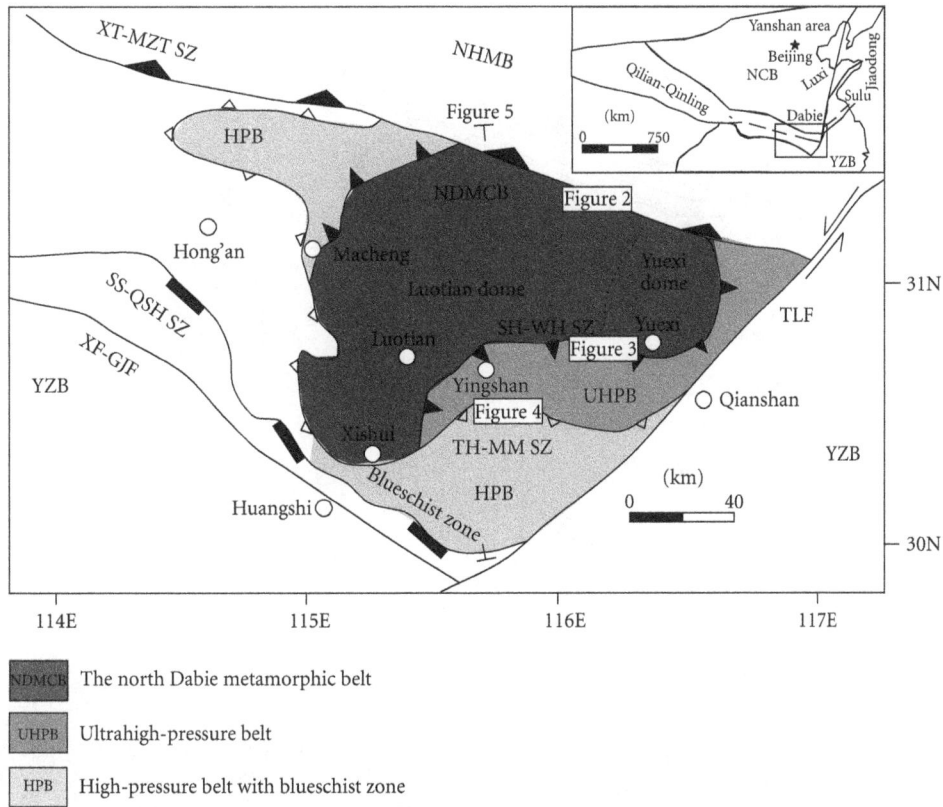

FIGURE 1: Simplified map of late Mesozoic main shear zones in eastern Dabie Mountain, Central China (after Suo et al., 2000 [4]). NCC: Northern China Block; YZC: Yangzi Block; NHMB: North Huaiyang metamorphic belt; NDMCD: North Dabie metamorphic complex belt; UHPB: Ultrahigh-pressure metamorphic belt; HPB: high-pressure metamorphic belt; SH-WH SZ: Shuihou-Wuhe shear zone; TH-MM SZ: Taihu-Mamiao shear zone; SS-QSH SZ: Susong-Qingshuihe shear zone; XT-MZT SZ: Xiaotian-Mozitan shear zone; TLF: Tancheng-Lujiang Fault (Tanlu Fault); XF-GJF: Xiangfan-Guangji Fault.

whose southern margin is defined by the Susong-Qingshuihe shear zone (Figure 1).

Several contrasting tectonic models have been proposed for the Dabie orogenic belt. Suo et al. [4, 14] identified extensional tectonism in the Dabie orogen during the middle-late Mesozoic. Song [15] defined extensional structures in metamorphic rocks with different ages that formed prior to the continental collision that produced the Qinling-Dabie orogen. Based on geochemical analyses of the Wangmuguan pluton in Xinxian, Zhang et al. [16] proposed that the pluton formed within extensional conditions after the formation of the orogen and was related to lithosphere detachment. A two-stage extension model proposed by Jin et al. [17] describes extension during the early Caledonian period, which resulted in the local sea and the beginnings of the Dabie-Qinling metamorphic core complex; extension during the Yanshan period coincided with the formation of the core complex and the intrusion of granite. Wang and Yang [18] investigated the extensional domes in the Dabie Mountains. Li [19] studied the late Mesozoic extension in the eastern part of the Dabie Mountains. Among these studies, there is a hot debate about the timing of the onset of extension, its direction, and tectonic styles. In this paper, we describe the results of our field observations and geochemical analysis, which provide some

new insights into understanding the evolution of the late Mesozoic extensional detachment zones in the Dabie area.

## 3. Deformation History of Extensional Detachment Zones

The NDMCB is the crystalline core of the late Mesozoic extensional structures. These features can be divided into two tectonic systems, the north extensional tectonic system and the south extensional tectonic system (Figure 1).

*3.1. The North Extensional Tectonic System.* The principal detachment shear zone is the XT-MZT SZ, which is located between the NHMB and the NDMCB. To the east (i.e., along the Huoshan-Zhujiapu Road), the brittle-ductile extensional shear zone dips to the NE (60°) at an angle of 30°–40° to the horizontal. To the west (i.e., in the Jinzhai-Qingshan area), the shear zone dips towards the NNE-NE (30°–50°). From south to the north, the dip angle changes from steep (about 70°) to gentle. Hence, it is shaped like a shovel and is locally sinuate (orientation 220° and dip 20°). The displacement orientation on the shear zone is to the NE or NNE. Noticeably, the extensional shear zone locally in the

(a)

(b)

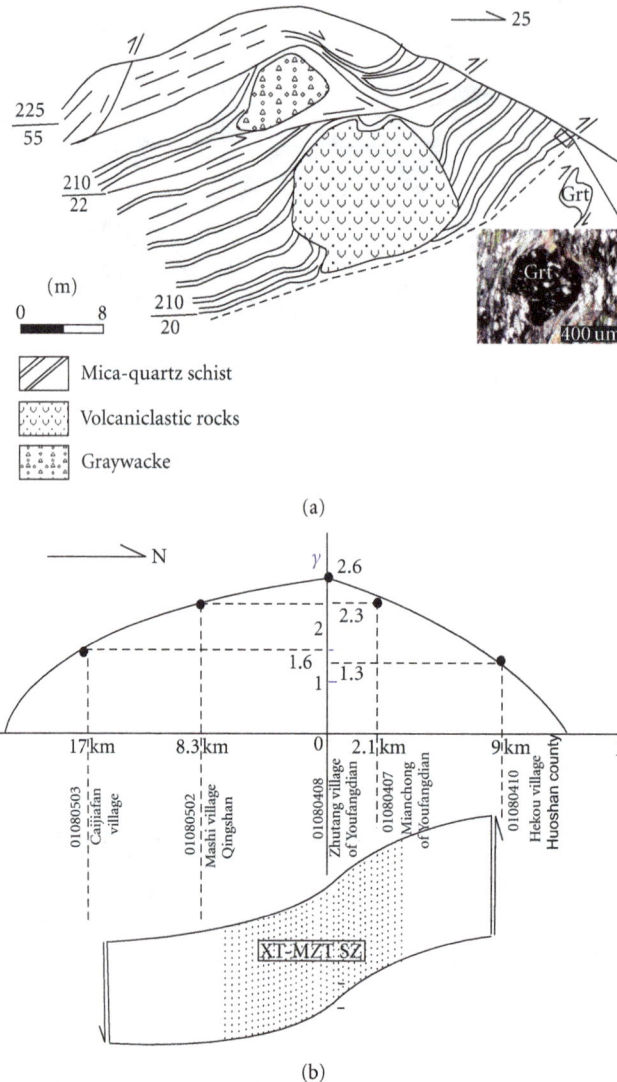

FIGURE 2: (a) The north extensional detachment zone suggests extension shearing after late Jurassic from the XT-MZT SZ. Late Jurassic volcaniclastic tectonite blocks are hosted by mica-quartz schist (Pt$_2$). The photo right below shows snow ball structure observed by microscopy. (b) The shear length of the XT-MZT SZ is more than 56 km.

NHMB has developed into an inconsecutive extensional crenulation cleavage (C′, the same as S$_3$ in some places) with NE-dip. The extensional crenulation cleavage is an important character of the north extensional tectonic system. The foliations (S$_1$ or S$_2$) dipping SW in quartz schist (Pt$_2$) may represent earlier overshear deformation in the NHMB. Asymmetric augen in amphibole gneiss in the Huoshan area indicates shearing towards the NE. Using the optical microscope, garnet in quartz schist has a "snowball" texture and also confirms that NE-directed shearing (Figure 2(a)).

According to a systematical strain-measurement analysis of the snowball textures in garnet and deformed quartz grains, the shear strain ($\gamma$) in the central part of the north detachment zone is up to 2.6 and gradually decreases outward. Measurements of a rock finite strain in the detachment zone indicated that the shear displacement is at least 56 km.

3.2. The South Extensional Tectonic System. The south extensional tectonic system is composed mainly of one deep ductile shear zone and ltwo brittle-ductile shear zones.

3.2.1. One Deep Ductile Shear Zone (Shuihou-Wuhe). The Shuihou-Wuhe ductile shear zone (SH-WH SZ) consists of feldspar mylonite and formed under lower crustal conditions; the shear zone was named the "first southern detachment zone" by Zhong et al. [20]. The shear zone dips towards the S, SSW, and SSE at an angle of 40–60°. The S-C shear fabrics, sheath folds, and shear folds in the shear zone consistently dip towards the south (Figures 3(a) and 3(b)). Much of these extensional shear deformations superpose on earlier thrust-shearing deformation and strike-slip deformation. The shear strain ($\gamma$) in the zone is up to 5, such as observations in the north of Yingshan (Figure 3(a)).

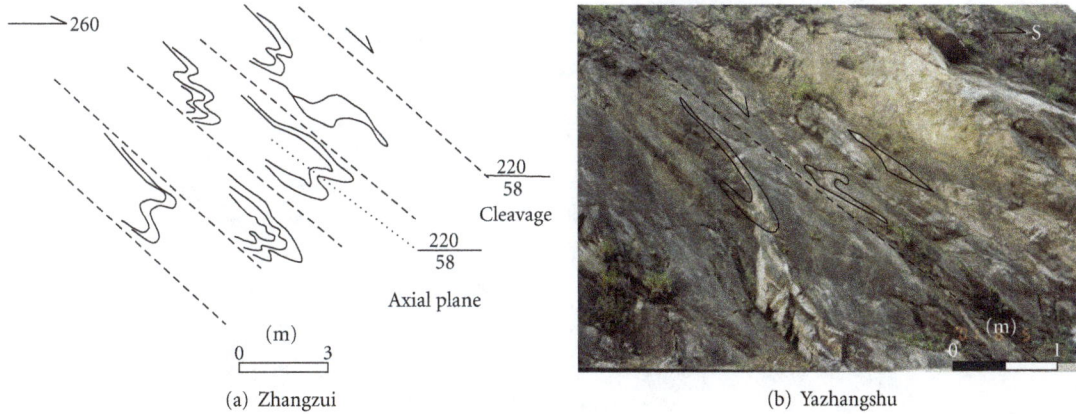

(a) Zhangzui

(b) Yazhangshu

FIGURE 3: Quartz bands in gneiss in the Shuihou-Wuhe shear zone, the shear zone dips SSW; (a) and (b) are, respectively, from places the Zhangzui village and the Yazhangshu primary school of the Yingshan County.

Based on strain-measurement analysis, the strain ellipsoid of the shear zone displays the Flinn parameter ($K$) range of 0.01–0.1, corresponding with pancake-shape flattening and reflecting intense compression and pure shear deformation.

According to the fabric analysis of deformed quartz, the small-circle belt at the east segment of the SH-WH SZ reflects high-temperature deformations conditions (>700°C) and relatively slow strain-rate velocity ($10^{-7} s^{-1}$). The middle segment was asymmetric with point belonging to a rhomb slide system, which reflects middle-high-temperature deformational conditions (650–700°C). The quartz fabric analysis also shows that the principal stress direction was oriented NE-SW, which is consistent with the principal stress direction determined from the preferred orientation of amphibole long axes. The paleodifferential stress was calculated at about 92 Mpa based on quartz dislocation density.

### 3.2.2. Two Brittle-Ductile Shear Zones (Taihu-Mamiao, Susong-Qingshuihe)

*(a) The Taihu-Mamiao Shear Zone.* The Taihu-Mamiao shear zone (TH-MM SZ) lies to the south of the SH-WH SZ and is bounded by the southern Dabie and Susong metamorphic zones in the north and south, respectively (Figure 1). The TH-MM SZ dips mainly towards the S or SSE; the dip angle changes from steep to gentle with increasing depth. A series of recumbent folds, S-C shear fabrics, asymmetric augen, extensional lineations in some profiles indicate that extension was to the SSE (Figures 4(a) and 4(b)). Based on strain-measurement analysis, the Flinn parameter for the shear zone is about 1, indicating a planar strain.

*(b) The Susong-Qingshuihe Shear Zone.* The Susong-Qingshuihe shear zone (SS-QSH SZ) lies to the south of the TH-MM SZ and to the north of the Xiangfan-Guangji fault (XF-GJ F.). The SS-QSH SZ marks the interface between high-pressure eclogite and blueschist units (Figure 1). The stretch lineation of monzogranitic mylonite in Qingshuihe plunges towards the SSW. According to strain-measurement analysis,

the Flinn parameter of the shear zone is more than 7, which suggests cigar-shaped, extensional deformation. The shear displacement along the detachment zone is more than 12 km based on strain measurement analysis. Deformed quartz displays asymmetic point coarctation, belonging to a rhomb slide system, and suggesting middle-to-high temperatures during deformation. Based on the dislocation density of quartz, we calculated the paleodifferential stress 70–84 Mpa, which is smaller than that of the SH-WH SZ.

Based on the structural analysis above, the three shear zones in the south extensional tectonic system probably represent deeper, middle, and upper detachment systems, ranging from north to south, respectively. Deformation temperatures and differential stresses decrease from the northern Dabie complex to the south. Remarkably, strain changes from north to south; from flattened (i.e., $K \ll 1$) in the northern the SH-WH SZ, to planar (i.e., $K \approx 1$), and then to extensional strain (i.e., $K \gg 1$) in the southern SS-QSH SZ. This transition of strain reflects the influence of magmatic intrusion during the extensional detachment.

## 4. Chronological Constraints on the Extensional Detachment

$^{40}Ar/^{39}Ar$ dating of biotite and hornblende from the four extensional detachment zones in Dabie area are shown in Table 1. The data can be divided into two groups, ca. 124 Ma and ca. 190 Ma, respectively. The north extensional tectonic system and the SS-QSH SZ were active at ca. 124 Ma, representing the time of extensional detachment to this period. In contrast, the SH-WH and TH-MM SZs, on the north and south sides of ultrahigh-pressure metamorphic zone, respectively, have ages of ca. 190 Ma, representing the time of exhumation of ultrahigh-pressure metamorphic rocks.

## 5. Discussion

The north extensional detachment zone hosts a shear zone that cuts the Late Proterozoic Xinyang Group (Pt₂n) mica-quartz schist. The shear zone contains allochthonous blocks

(a)

(b)

FIGURE 4: Extensional flat-ramp structure in the Taihu-Mamiao shear zone, indicating S-extension (near Luoxi village along the road between Liuyang and Hualiangting in the Taihu County). The occurrences of gneiss cleavage, flat and ramp, and extensional lineation are marked.

TABLE 1: Summary of mineral $^{40}$Ar-$^{39}$Ar data in the main shear zone of eastern Dabie Mountains.

| Shear zone | Sample locations | Mineral | Weighted mean plateau age (Ma) |
| --- | --- | --- | --- |
| XT-MZT SZ | Qingshan town | Biotite | 124.17 ± 0.25 |
| | Zhangchong village | Biotite | 126.91 ± 0.30 |
| SH-WH SZ | Shuihou village | Biotite | 190.59 ± 0.42 |
| TH-MM SZ | Luoxi village | Hornblende | 197.41 ± 0.46 |
| | Luoxi village | Biotite | 189.42 ± 0.29 |
| SS-QSH SZ | Qingshuihe village | Biotite | 124.87 ± 0.21 |
| | Chenhan village | White mica | 194.01 ± 0.36 |
| | Qingshuihe village | Biotite | 127.96 ± 0.30 |

of Late Jurassic volcaniclastic rocks and tuff that are up to several meters in size (Figure 2).

Large-scale, harmonic, recumbent folds also occur in the Late Jurassic volcaniclastic rocks and the Late Proterozoic mica schist. In the south extensional detachment system, the Hong'an Group (Pt) and Yanshanian granite (J$_3$—K$_1$) experienced extensional shear deformation at the same time. These features suggest that the extensional detachment of both the south and north detachment systems took place after the Late Jurassic.

The age group ~200 Ma is coincide with the U-Pb ages, Sm-Nd ages, and Rb-Sr ages of UHP rocks [22–26], which reflect protracted cooling or partial resetting by Jurassic or Cretaceous magmatism.

In addition, massive granitic intrusions and numerous ultrabasic plutons were emplaced in the NDMCB during

TABLE 2: The comparison of deformation ages from southern Dabie belt of China.

| Age groups | Data of XT-MZT SZ [21] | | Our data from other shear zones |
| --- | --- | --- | --- |
| ~200 Ma | | | 229.8 ± 70.97 |
| | | | 219.52 ± 1.57 |
| 150~190 Ma | 156.7 ± 1.5 | | 197.41 ± 0.46 |
| | | | 194.01 ± 0.36 |
| | | | 190.59 ± 0.42 |
| | | | 189.42 ± 0.29 |
| | | | 156.5 ± 7.15 |
| 145~110 Ma | 136.5 ± 1.3 | 132.0 ± 0.8 | |
| | 129.8 ± 0.6 | 141.9 ± 1.2 | |
| | 129.2 ± 1.0 | 127.1 ± 0.7 | 127.96 ± 0.30 |
| | 133.1 ± 0.9 | 128.3 ± 0.9 | 124.87 ± 0.21 |
| | 134.1 ± 0.9 | 127.0 ± 0.8 | 124.56 ± 2.4 |
| | 136.8 ± 1.1 | 122.5 ± 0.6 | |
| | 130.7 ± 1.0 | 120.7 ± 0.7 | |

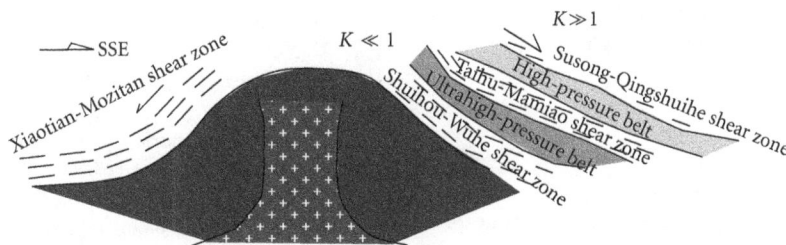

FIGURE 5: Extensional detachment model for the late Mesozoic in the Dabie Mountains.

the extensional phase at about 120 Ma. The Ar-Ar and Rb-Sr isotopic analyses of rocks in the Dabie group also record this time at about 110–145 Ma (Table 2). All these lines of evidence suggest that they were closely related to extensional detachment. The Cretaceous lacustrine, fluvial, and piedmont facies (several kilometers thick in the Hefei basin) also reflect a regional extensional setting and intense mountain-basin movement. Therefore, voluminous magmatic emplacement and lithosphere delamination during the early Cretaceous induced the rapid uprising of the North Dabie central area, with extensional detachment on both sides. The intense denudation resulted in the thick sediment deposits in the Hefei basin. The high- and ultrahigh-pressure eclogites were probably emplaced during this event (Figure 5). Large-scale extension in the Dabie Mountains during the late Mesozoic is representative of the tectonic regime inversion that affects the eastern North China Block.

Lastly, the deposition of platinum group elements (PGEs) in ultramafic and/or mafic rocks with two ages, ca. 120 Ma and ca. 230 Ma, in the North Dabie complex core suggests that their source region was the upper mantle. The PGE data indicate that the late Mesozoic upper mantle (ca. 120 Ma) enriched in PGE, whereas depleted PGE before 120 Ma. The late Mesozoic upper mantle with PGE enrichment in the Dabie region is contaminated with about 8% Earth core materials, as the PGE contents in the Earth's core are much higher than in the upper mantle [27, 28]. If this is correct,

then the PGE mantle enrichment must be related to the late Mesozoic extensional detachment in the Dabie area.

## 6. Conclusions

(1) The main shear zones in the Dabie area are characterized by extensional detachment during late Mesozoic era. The XT-MZT SZ is detached to the NNE, the SH-WH SZ, and the TH-MM SZ are displaced to the SSE, while the SS-QSH SZ is displaced to the SW.

(2) The shear length of the XT-MZT SZ is more than 56 km and that of the SS-QSH SZ is more than 12 km. The Flinn parameter of the Shuihou-Wuhe shear zone is much smaller than 1 (i.e., 0.01–0.1), which suggests that the shear zone was flattened when it formed. The Flinn parameter of the Taihu-Mamiao shear zone is about 1 (i.e., 1.1), whereas the Susong-Qingshuihe shear zone is much more than 1 (i.e., 7.6), which suggests that they were formed during extension. From north to south in the south extensional tectonic system, these Flinn parameter values display the transition from pure shear to simple shear, possibly reflecting the active intrusion of magma during the extensional detachment.

(3) Two deformation ages, ~190 Ma and ~124 Ma based on mineral $^{40}$Ar-$^{39}$Ar data, are concluded from the main shear zones in the Dabie area. The early age (~190 Ma) could be related to the UHP cooling and reversion during orogenesis, whereas the later one (~124 Ma) could represent

the extensional detachment age after the formation of the orogeny.

(4) The strain analysis, chronology, and the mantle enrichment in platinum group elements suggests that magmatic intrusion in the north Dabie complex core is the main cause for extensional detachment structures during the late Mesozoic.

## Acknowledgments

The authors thank Professors Jiliang Li, Wenjiao Xiao, Mingguo Zhai, and Zhihong Wang from the Institute of Geology and Geophysics, Chinese Academy of Sciences, and Tianshan Gao from the University of Science and Technology of China for the excellent field and inner laboratory discussions; they also thank Dr. Paul Duuring from The University of British Columbia, Canada for much help. This work is supported by the National Natural Science Foundation of China (Grant no. 41030422), the Major Project of Chinese Academy of Sciences (Grants KZCX1-07), and the Key project of the National Natural Science Foundation of China (Grant no. 40234050).

## References

[1] S. Li, T. M. Kusky, X. Liu et al., "Two-stage collision-related extrusion of the western Dabie HP-UHP metamorphic terranes, central China: evidence from quartz c-axis fabrics and structures," *Gondwana Research*, vol. 16, no. 2, pp. 294–309, 2009.

[2] S. Z. Li, X. Liu, Y. H. Suo et al., "Triassic folding and thrusting in the Eastern Block of the North China Craton and the Dabie-Sulu orogen and its geodynamics," *Acta Petrologica Sinica*, vol. 25, pp. 2031–2049, 2009.

[3] S. Z. Li, T. M. Kusky, G. Zhao et al., "Mesozoic tectonics in the Eastern Block of the North China Craton: implications for subduction of the Pacific plate beneath the Eurasian plate," *Geological Society Special Publication*, no. 280, pp. 171–188, 2007.

[4] S. Suo, Z. Zhong, and Z. You, "Extensional deformation of post ultrahigh-pressure metamorphism and exhumation process of ultrahigh-pressure metamorphic rocks in the Dabie massif, China," *Science in China, Series D*, vol. 43, no. 3, pp. 225–236, 2000.

[5] E. Eide, "A model for the tectonic history of HP and UHPM regions in east central China," in *Ultrahigh-Pressure Metamorphism*, R. G. Coleman and X. M. Wang, Eds., pp. 391–426, Cambridge Press, 1995.

[6] X. M. Wang, R. Y. Zhang, and J. G. Liou, "UHPM terrane in east central China," in *Ultrahigh Pressure Metamorphism*, R. G. Coleman and X. M. Wang, Eds., pp. 356–390, Cambridge University Press, Cambridge, UK, 1995.

[7] B. L. Cong, *Ultrahigh-Pressure Metamorphic Rocks in the Dabieshan-Sulu Region of China*, Kluwer Academic Publishers, Science Press Beijing, 1996.

[8] B. R. Hacker, X. Wang, E. A. Eide, and L. Ratschbacher, "The qinling-dabie ultrahigh-pressure collisional orogen," in *The Tectonic Evolution of Asia*, A. Yin and T. M. Harrison, Eds., pp. 345–370, Cambridge University Press, Cambridge, UK, 1996.

[9] J. G. Liou, R. Y. Zhang, X. M. Wang, E. A. Eide, W. G. Ernst, and S. Maruyama, "Metamorphism and tectonics of highpressure and ultrahigh-pressure belts in the Dabie-Su-Lu region, China," in *The Tectonic Evolution of Asia*, A. Yin and T. M. Harrison, Eds., pp. 300–344, Cambridge University Press, Cambridge, UK, 1996.

[10] B. M. Jahn, F. Wu, C. H. Lo, and C. H. Tsai, "Crust-mantle interaction induced by deep subduction of the continental crust: geochemical and Sr-Nd isotopic evidence from post-collisional mafic-ultramafic intrusions of the northern Dabie complex, central China," *Chemical Geology*, vol. 157, no. 1-2, pp. 119–146, 1999.

[11] S. Li, T. M. Kusky, G. Zhao et al., "Two-stage Triassic exhumation of HP-UHP terranes in the western Dabie orogen of China: constraints from structural geology," *Tectonophysics*, vol. 490, no. 3-4, pp. 267–293, 2010.

[12] S. Z. Li, G. C. Zhao, G. W. Zhang et al., "Not all folds and thrusts in the Yangtze foreland thrust belt are related to the Dabie Orogen: insights from Mesozoic deformation south of the Yangtze River," *Geological Journal*, vol. 45, no. 5-6, pp. 650–663, 2010.

[13] S. Li, T. M. Kusky, G. Zhao et al., "Thermochronological constraints on two-stage extrusion of HP/UHP terranes in the Dabie-Sulu orogen, east-central China," *Tectonophysics*, vol. 504, no. 1–4, pp. 25–42, 2011.

[14] S. Suo, L. Sang, Y. Han et al., *The Petrology and Tectonics in Dabie Precambrian Metamorphic Terranes, Central China*, China University of Geosciences Press, Wuhan, China, 1993.

[15] H. L. Song, "Early extensional tectonics in Qinglin-Dabie orogen," in *Extensional Tectonics Research*, X. L. Qian, Ed., pp. 12–21, Geology Press, Beijing, China, 1994.

[16] Q. Zhang, W. P. Ma, J. W. Jin et al., "Geochemistry and tectonic significance of post-tectonic gabbro from Wangmuguan of Xinxian county, Henan province," *Chinese Journal of Geochemistry*, vol. 24, 4, pp. 341–350, 1995.

[17] W. Jin, H. Song, and W. Ma, "Extensional tectonics in tongbai-west Dabie mountain," *Scientia Geologica Sinica*, vol. 32, no. 2, pp. 156–164, 1997.

[18] G. C. Wang and W. R. Yang, "Uplift evolution during Mesozoic-Cenozoic of the Dabie orogenic belt: evidence from the tectono-chronology," *Earth Science—Journal of China University of Geosciences*, vol. 23, no. 5, pp. 461–467, 1998.

[19] J. Li, *The Dabie Shan main shear zone of Late Mesozoic and extension structure [M.S. thesis]*, Institute of Geology and Geophysics, Chinese Academy of Sciences, 2003.

[20] Z. Q. Zhong, S. Suo, and Z. D. You, "Extensional tectonic framework of post high and ultrahigh pressure metamorphism in Dabieshan, China," *Earth Science—Journal of China University of Geosciences*, vol. 23, no. 3, pp. 225–229, 1998.

[21] Y. S. Wang, B. W. Xiang, G. Zhu et al., "$^{40}$Ar-$^{39}$Ar geochronology records for post-orogenic extension of the Xiaotian-Mozitan fault," *Geochimica*, vol. 38, pp. 458–471, 2009.

[22] L. E. Webb, L. Ratschbacher, B. R. Hacker et al., "Kinematics of exhumation of high-and ultrahigh-pressure rocks in the Hong'an and Tongbai Shan of the Qinling-Dabie collisional orogen, eastern China," *GSA Memoirs*, vol. 194, pp. 231–245, 2001.

[23] J. C. Grimmer, L. Ratschbacher, M. McWilliams et al., "When did the ultrahigh-pressure rocks reach the surface? A $^{207}$Pb/$^{206}$Pb zircon, $^{40}$Ar-$^{39}$Ar white mica, Si-in-white mica, single-grain provenance study of Dabie Shan synorogenic foreland sediments," *Chemical Geology*, vol. 197, no. 1–4, pp. 87–110, 2003.

[24] B. R. Hacker, S. R. Wallis, L. Ratschbacher, M. Grove, and G. Gehrels, "High-temperature geochronology constraints on the tectonic history and architecture of the ultrahigh-pressure Dabie-Sulu orogen," *Tectonics*, vol. 25, no. 5, Article ID TC5006, 2006.

[25] B. R. Hacker, S. R. Wallis, M. O. Mcwilliams, and P. B. Gans, "$^{40}$Ar-$^{39}$Ar constraints on the tectonic history and architecture of the ultrahigh-pressure Sulu orogen," *Journal of Metamorphic Geology*, vol. 27, no. 9, pp. 827–844, 2009.

[26] B. R. Hacker, L. Ratschbacher, and J. G. Liou, "Subduction, collision and exhumation in the ultrahigh-pressure Qinling-Dabie orogen," *Geological Society Special Publication*, no. 226, pp. 157–175, 2004.

[27] Q. Liu, *Study on the distribution of Platinum group elements in Dabie (ultra-) mafic rocks and Fuxin volcanic rocks [Ph.D. thesis]*, Graduate University of the Chinese Academy of Sciences, 2005.

[28] Q. Liu, Q. L. Hou, X. H. Zhou, and L. W. Xie, "The distribution of platinum-group elements in gabbros from Zhujiapu, Dabie orogen," *Acta Petrologica Sinica*, vol. 21, no. 1, pp. 227–239, 2005.

# 9

# Tectonic and Hydrothermal Activities in Debagh, Guelma Basin (Algeria)

Said Maouche,[1] Abdeslam Abtout,[1] Nacer-Eddine Merabet,[1] Tahar Aïfa,[2] Atmane Lamali,[1] Boualem Bouyahiaoui,[1] Sofiane Bougchiche,[1] and Mohamed Ayache[1]

[1] CRAAG, BP 63, Bouzareah, 16340 Alger, Algeria
[2] Geosciences-Rennes, CNRS UMR 6118, Université de Rennes 1, Bat.15, Campus de Beaulieu, 35042 Rennes Cedex, France

Correspondence should be addressed to Said Maouche; said_maouche@yahoo.fr

Academic Editor: Salvatore Gambino

Quaternary and Pliocene travertines, deposited from hot springs, can reveal much about neotectonic and hydrothermal activity. The aim of this work is the understanding of the actual tectonic activity in the Guelma Basin and in one of its spa structures. Gravity data were collected during a field study in the Hammam Debagh (HD) area and then analyzed to better highlight the architecture of its subsurface underlying structures. This analysis was performed by means of a Bouguer anomaly, upward continuations, and residual and derivative maps. Comparison of gravity maps, field geology, geomorphic observations, and structural maps allowed us to identify the major structural features in the Hammam Debagh. As a result, we confirm the position of the Hammam Debagh active fault which is superimposed to the hydrothermal active source in the NW-SE direction characterized by a negative gravity anomaly.

## 1. Introduction

Plio-Quaternary travertine deposits from hot springs can reveal much about the neotectonic history as demonstrated by many examples worldwide [1]. Several open cracks parallel to active normal faults were identified in deposits at Pamukkale, Turkey [2]. In New Guinea, travertine deposits are controlled by fracturing [3]. Numerous studies show how the deposition of travertine has ceased in relation to tectonic activity (i.e, Hula Valley, Israel [4]). Other examples of tectonic activity on a normal fault generating hydrothermal systems are known at Mammoth Hot Springs [5] in Slovakia [6]; Aveyron, France [7], and Shelsley, United Kingdom [8].

Tilting, warping, and faulting of the Algerian crust were followed by travertine formation [9]. Several hydrothermal systems and associated travertine deposits are identified in the Tellian Atlas (e.g., Hammam Debagh (HD) in the eastern part and Hammam Boughrara in the western part).

Along the Tellian Atlas (Figure 1), most of the hydrothermal sources are located along active faults, activated or reactivated during major seismic events such as Bouchegouf and Hammam N'baïlis faults [10, 11]. Hammam Debagh (HD) hydrothermal source is located in the western part of the Guelma Basin and is one of the famous spas in northeastern Algeria [10, 12, 13]. The HD area is situated in the western limit of the Guelma Plio-Quaternary pull-apart basin created between two overlapping east-west dextral strike slip faults [10, 14]. This active hydrothermal source is responsible for several meters' thickness of travertine deposits. The presence of active hydrothermal source indicates the presence of fault not observed so far. This tectonic feature represents an important element to be clarified since it also represents the structural emplacement of the western limit of the Guelma Neogene basin (Figure 2). This area experienced several seismic events [11, 15, 16]. The seismic activity seems to be related to an NW-SE unknown active fault as shown by focal mechanisms particularly that of 2003 HD earthquake [15, 17, 18].

In this study, an attempt is made for understanding the recent tectonic activity that generates the seismicity in the vicinity of HD. For that purpose, we used a multidisciplinary

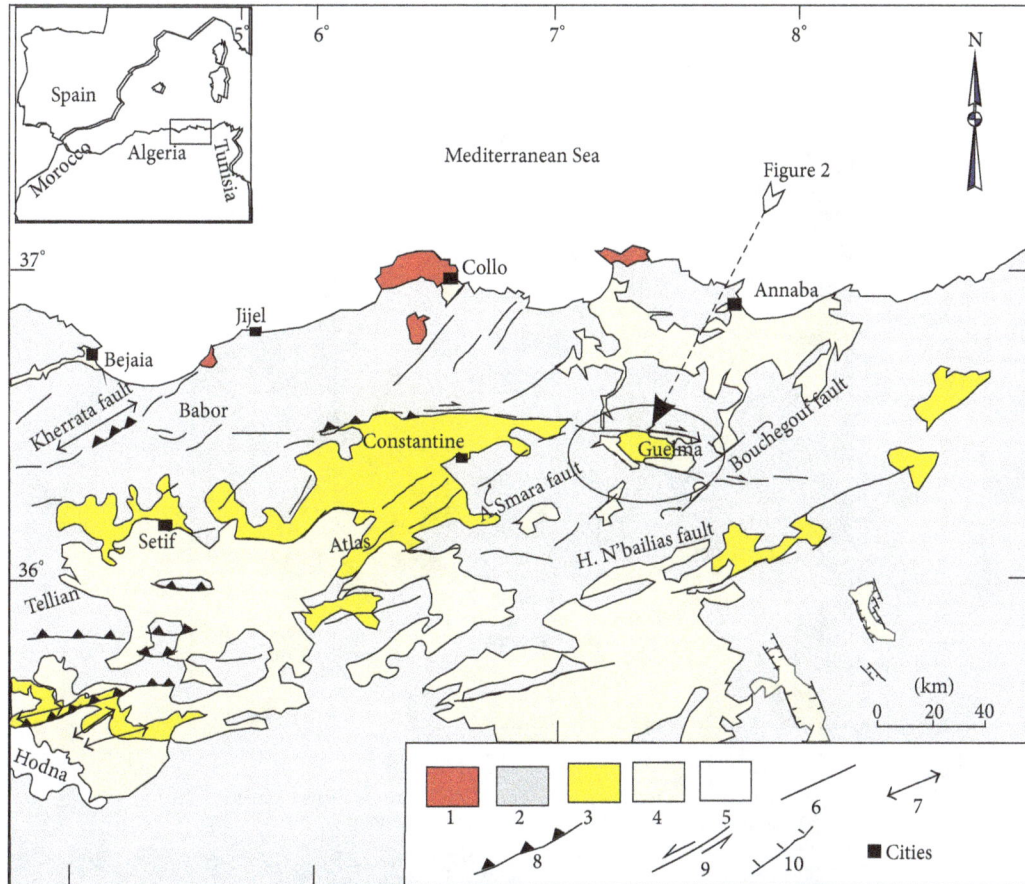

FIGURE 1: Tectonic map of the Eastern Tellian Atlas of Algeria [11] and the Guelma seismogenic localization: (1) volcanism, (2) Jurassic-Cretaceous and lower Cenozoic basement, (3) Neogene post-nappe deposits, (4) Plio-Quaternary deposits, (5) Quaternary deposits, (6) fault, (7) anticline, (8) reverse fault, (9) strike-slip fault, and (10) normal fault.

approach consisting in the combination of geology (tectonics and neotectonics), geomorphology, and geophysics (gravimetry) with the objective of recognizing and detecting the seismogenic source likely responsible for the hydrothermal activity in the HD-Roknia lineament. The main important result we obtained consists of the localization of the Plio-Quaternary hydrothermal primary source along the structural western limit of the Guelma seismogenic basin.

## 2. Geological and Morphological Outline

*2.1. Geological Outline.* The Tellian Atlas of Algeria underwent significant tectonics during the Neogene time. Neotectonic features correspond to E-W to NE-SW trending folds, reverse faults affecting Quaternary deposits ([17, 19] and references therein). Along the folded mountains, intermountain Neogene post-nappe Basins are characterized by the presence of Miocene and Plio-quaternary sediments (e.g., Soummam, Hodna, Constantine, and Guelma basins [11, 17]). It is worthwhile noting that the Guelma basin (considered in this study) is rather particular since, as a pull-apart basin, it is not very common in Algeria. This seismogenic

basin is situated at the eastern end of the Constantine E-W trending dextral strike slip (Figure 1). This typically pull-apart basin formed during the post-nappe period is filled by late tertiary to quaternary sediments. The Mio-Pliocene "post-nappe" continental deposits consist of two sets: (a) sandstone interbedded with marl representing the post-Tortonian period [10]; (b) at the top, particularly in the HD area, we may find clays and conglomerates, red tuffs, and lacustrine limestone succession. According to Glacon and Rouvier 1972 [20], this series is equivalent to the Messinian observed in Tunisia. The alluvial terraces deposited by Bou Hamdane and Seybous rivers constitute the quaternary cycle of the Guelma seismogenic basin.

*2.2. The Hammam Debagh Hydrothermal Source.* The HD group of springs is one of the bathing places still in use since the Roman's era. Several studies [21, 22] point out their main chemical water composition (Table 1). The output temperature ranges from 88.5°C to 97.8°C. The geothermometric study indicates that the maximum temperature is around 300°C at depth. The boreholes drilled in the region evidenced significant abnormalities which range as follows: 1°C—16 m, 1°C—11 m, and even 1°C—7 m [12].

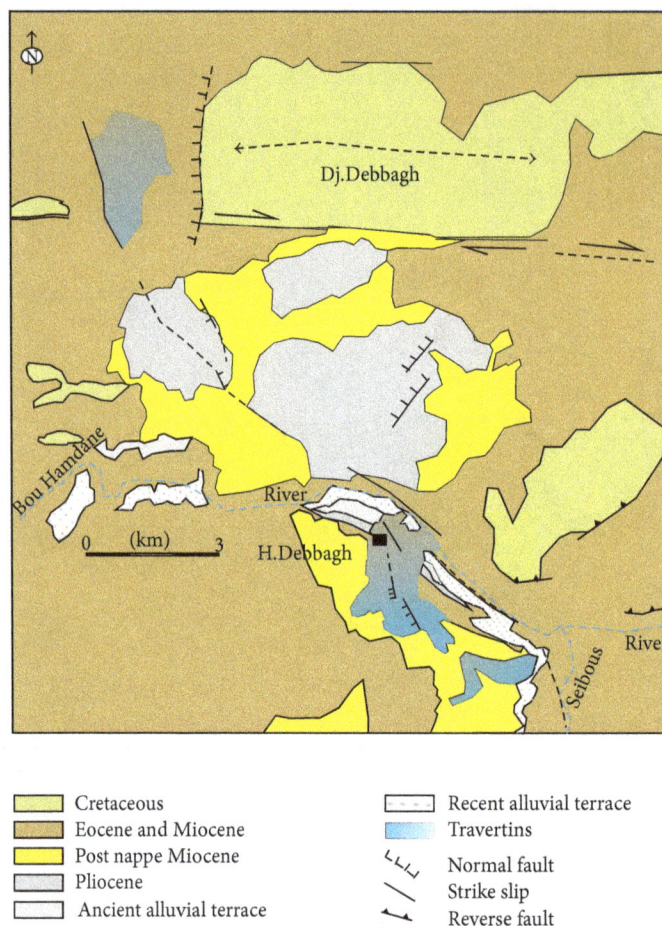

FIGURE 2: Geological map of the western part of the Guelma basin (see position in Figure 1), showing the study area; full line: observed normal fault; dashed line: interpreted normal fault.

TABLE 1: The Hammam Debbagh thermal springs, principal chemical constituents and physical parameters [13–25].

| Site | Temperature of water (°C) | Flow (L/mn) | Total dissolve solids (ppm) | Principal chemical constituents | Associated rocks | Remarks |
|---|---|---|---|---|---|---|
| Hammam Debbagh | 88.5–98 | 6000 | 1.466 | $MgCO_3$ = 257, $MgSO_4$ = 176, $MgCl_2$ = 416, $NaCl$ = 416, $KCl$ = 79, Gaz = 97%, $CO_2$ = 2.5%, Na = 0.5%, $H_2S$ | Faulted Miocene, Pliocene and Quaternary | (i) Large deposits of travertines containing Pisolite of Aragonite (ii) Cloud streams (iii) Roman careers (iv) Traces arsenic (As) 6.5 ppm |

In addition to the temperature parameter, this water is rich in sodium chloride and calcium sulfate. It contains radon (0.1 and 10 millimicrocuries/L) and arsenic (0.5 mg/L) and emits hydrogen sulphide [23]. These chemical parameters show that this source can be linked to a magmatic system.

2.3. The Travertine Deposits and Morphological Structures. Travertine deposits are calcareous, limestone accumulations formed in lakes, rivers, springs, and caves. They consist of crystalline deposits or incrustations formed by rapid

precipitation of calcium carbonate from cold or thermal waters at surface; they show very restricted lateral extent. Travertines of the study area are thermal or hydrothermal calcareous deposits related to HD source, represented by two main facies: the porous biodetrical facies and a constructed one, or the waterfall deposits related to plentiful and regular drainages of waters loaded with calcium, magnesium bicarbonates, and sulphides (Table 1, Figure 3(e)). These deposits, are of laminar type and were deposited in sheets during Pliocene-Quaternary. Southward the village, these facies are mined in active quarries. Locally, the travertine presents

FIGURE 3: Photographs of the main active site of Hammam Debagh hot springs (HD): (a) travertine accumulation on Quaternary soil; (b) clogged (filling or sealing) alluvial deposits (ancient elevated alluvial terrace); (c) well-developed cone-shaped accumulation, approximately 7 m high. Note the near vertically dipping layers and the stair-step topography that characterizes terraced mound accumulations; (d) steeply dipping to near-vertical laminae, transverse section through a fissure ridge accumulation displaying the internal structure along the fissure and the downward flow over both sides; (e) present travertine deposition (active thermal spring); (f) the more recent alluvial terrace close to Bou Hamdane river.

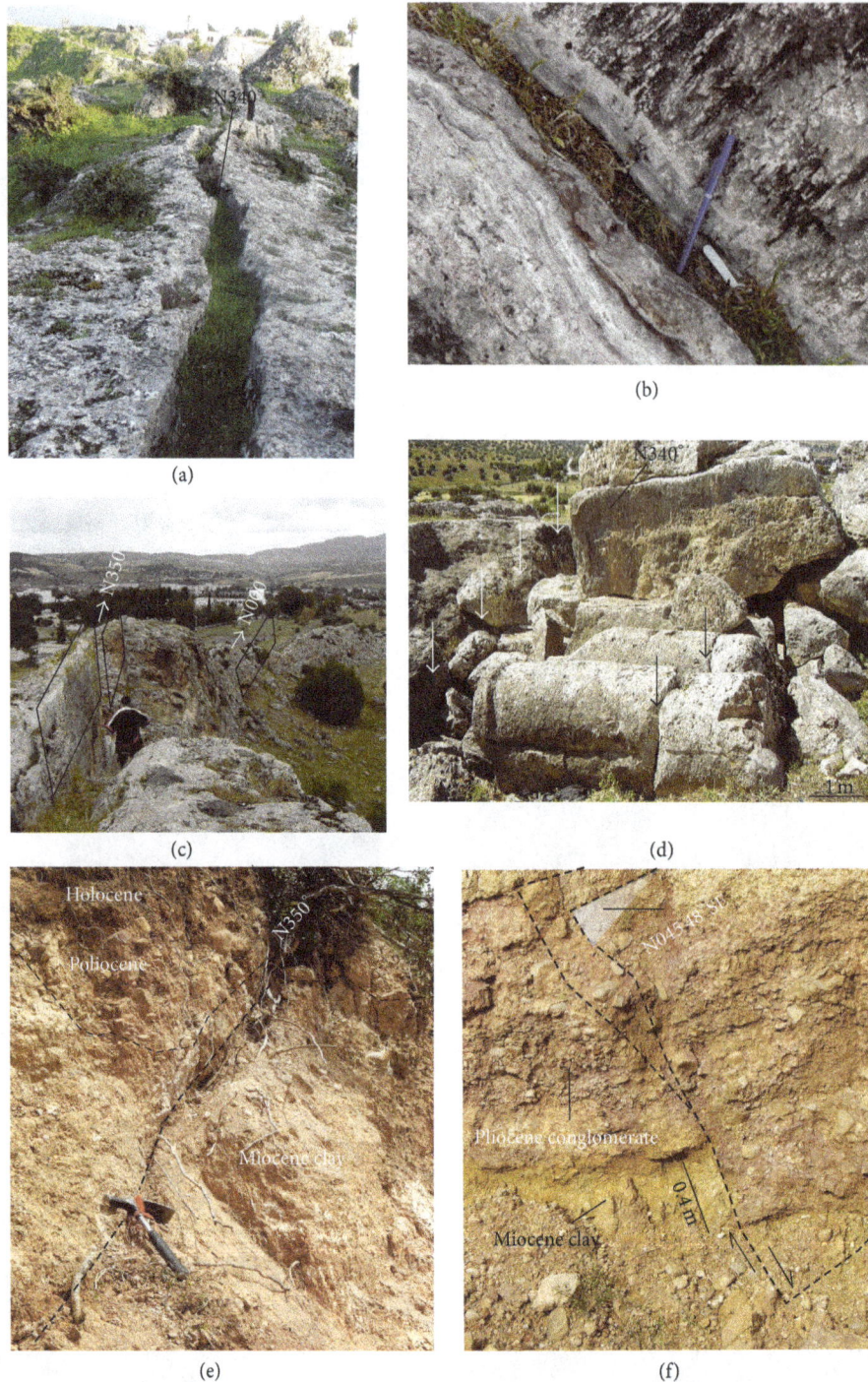

FIGURE 4: Examples of observed recent tectonic evidences: (a, b) fissured ridge at HD. Travertine precipitates from hot waters that emerge along the axial ridge making up the cone. They are part of a radically different internal structure than the terraced mound deposits; (c) vertical transverse fracture; (d) picture showing surface rupture and broken columns in roman career probably related to an earthquake occurrence; the arrow indicates the position of fracture; (e, f) recent normal faults affecting the Pliocene deposits.

a remarkable feature consisting in low porosity laminates alternating with compact levels. These layers are observed in the coal face of the Roman quarries.

The travertine deposits systematically show conglomerates on their bottom, attesting the presence of a series of alluvial terraces clogged by hydrothermal deposits (Figure 3(b)).

In several cases, they show a quaternary soil on their bottom (Figure 3(a)) that confirms the young age of the HD travertine. Their average thickness is from 10 to 20 m, and they present a little dip. The hydrographic network individualizes at least three levels of alluvial terraces; one of them is represented in Figure 3(f). The dip value variations observed

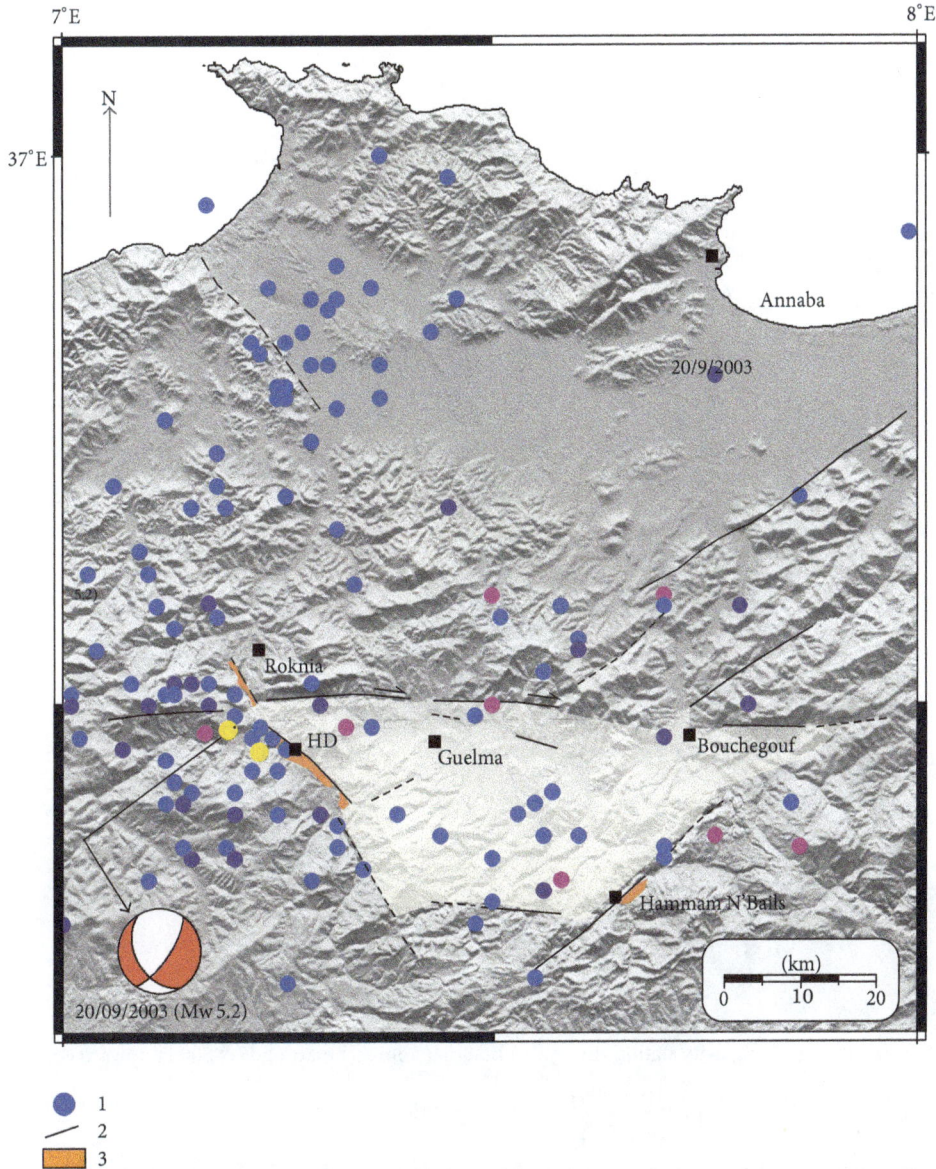

FIGURE 5: Seismicity distribution in the Guelma basin and surroundings: (1) earthquakes, (2) active and potentially active fault, and (3) travertine deposits; yellow circle corresponds to the largest earthquakes that occurred in the region with the focal mechanism of the 20/09/2003 HD earthquake: seismicity data are from [16].

from a terrace level to another one indicate that the lower terrace, close to the Bou Hamdane River, is more recent. The abundance of these alluvial deposits (mainly composed of conglomerates) evidences the existence of strong streaming of the Bou Hamdane River during the Quaternary before it changes its river path, probably in response to tectonic uplift and travertine deposits. In this area, different shapes of positive morphological travertine structures described as dykes, domes, and terraces (Figures 3(c), 3(d), and 3(e)) are injected in the system of alluvial terraces. A regular cone presents a simple cross section with a central orifice and the "parasitic cone" developed after the original cone has become inactive. In the south of the HD, a second facies of travertine outcrops forming a series of small flaps or tatters.

As known, travertine buildups around spring orifices along fractures such as joints of faults form elongate ridges. At this site, one of the ridges' wall exceeds 7 m height and is elongated in the N 340° direction (NW-SE) (Figures 4(a) and 4(c)). In some cases, we observe precipitations from hot waters emerging along the axial ridge (Figure 4(b)). At the site of the Roman quarry, located south of the village, we observe a significant normal movement on the NW-SE fault (Figure 4(d)). This recent tectonic movement is also observed outside the travertine area (Figures 4(e) and 4(f)) and shows normal faults affecting the Pliocene deposits.

*2.4. Neotectonics and Seismotectonics.* The Tellian Atlas structure is a fold and thrust fault system [17]. Along this domain

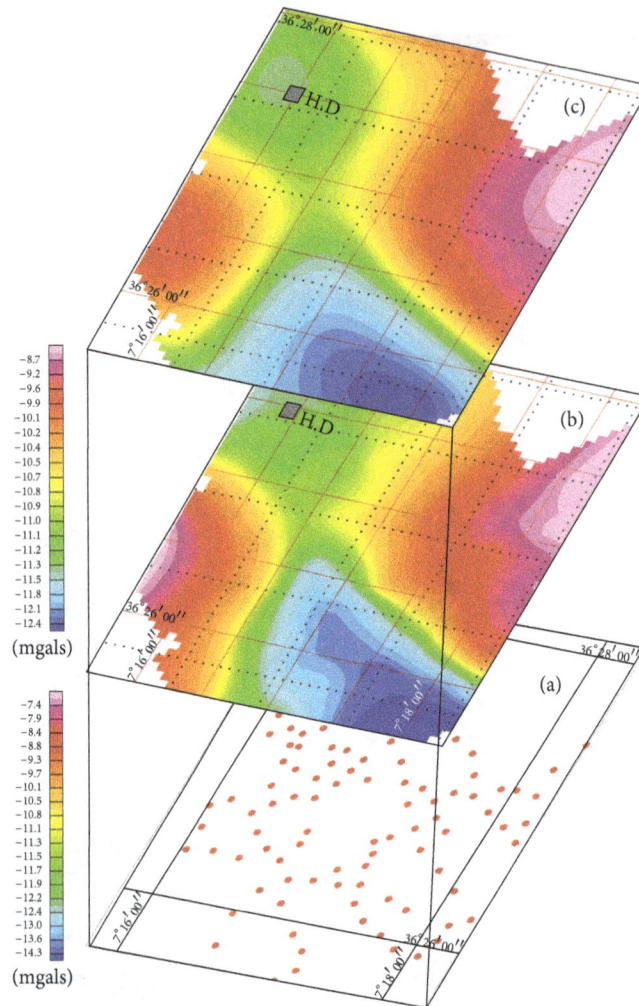

FIGURE 6: Gravity map: (a) gravity station survey, (b) Bouguer anomaly map, and (c) 500 m upward continuation.

(Figure 1), the recent tectonic activity responsible for the moderate seismicity is related to the 5-6 mm/y convergent movement of the African plate towards the Eurasian plate, with an NNW direction [24–26]. The stress field acting along northern Algeria generates tectonic structures in the NE-SW and E-W direction. According to several studies [14, 17, 19, 22], the structures of the postthrusting intermountain Neogene basins are characterized by reverse faulting in the western and central part of the Tellian Atlas domain, as attested by the rupture mechanism highlighted by the focal mechanism of significant earthquakes, whereas strike slip mechanism is predominant in the eastern part [11, 16, 17]. In the Guelma seismogenic basin, earthquakes exhibit focal mechanisms showing normal faulting regime. The size of the pull-apart basin is quite important, and this is linked to the amount of slip and distance between overlapping segments. In this basin (Figure 5), two particular faults are observed. The Bouchegouf and Hammam N'Bailis faults, along which crop out the Quaternary deposits related to hydrothermal springs [10, 11], are potentially active and could be in relation to the damaging earthquakes of the Guelma seismogenic zone

(Figure 5) [16, 27]. Several earthquakes struck this region: Héliopolis on December 17, 1850 ($I_0$ = VI EMS); Guelma on February 10, 1937 ($I_0$ = VIII EMS, mb = 5.4); and recently HD on September 20, 2003 (mb = 5.2) with a normal fault mechanism (a small strike slip component can also be observed on the focal mechanism of the 2003 HD earthquake in Figure 5).

According to several authors [10, 12, 18] the HD-Roknia source alignment has neotectonic signification and is related to an NW-SE fault. The field work we carried out reveals fault structures in the same direction, with normal apparent movement (Figures 2 and 5). Such observations are visible in the recent HD travertine deposits. Moreover, juxtaposed alluvial terraces, present in this zone, attest on the vertical movement related to the recent activity on NW-SE to NNW-SSE fault. We notice and confirm a good correlation between the fault-plane orientation observed in the field (cracks in the travertine), from the 2003 focal mechanism and the trends of the geomorphic structures related to the hydrothermal sources (i.e., dykes, ramparts, and cones). The NW-SE to NNW-SSE normal faults are also in agreement

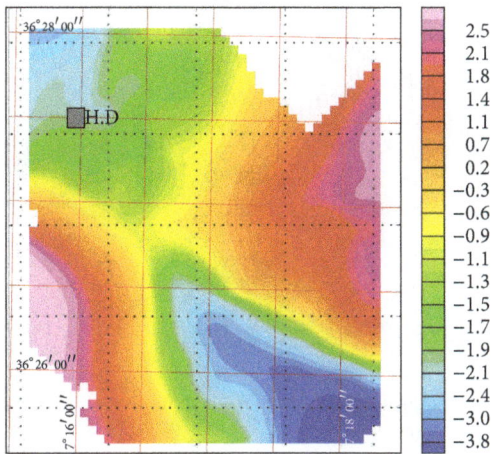

(a)

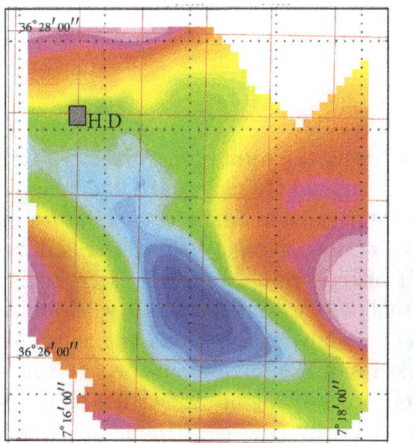

(b)

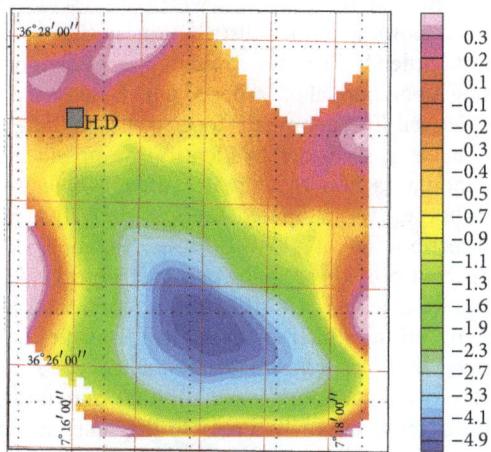

(c)

FIGURE 7: Residual anomaly maps: (a) order 1, (b) order 2, and (c) order 3.

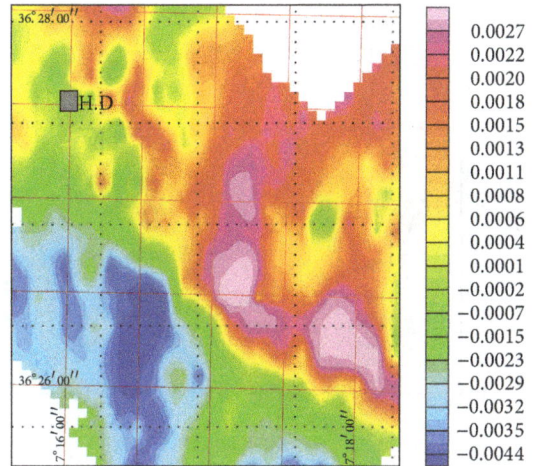

(a)

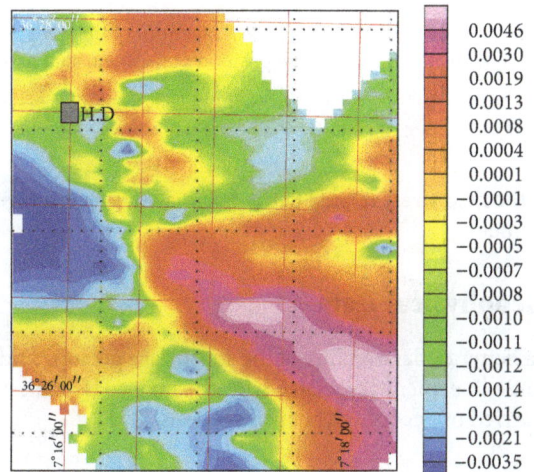

(b)

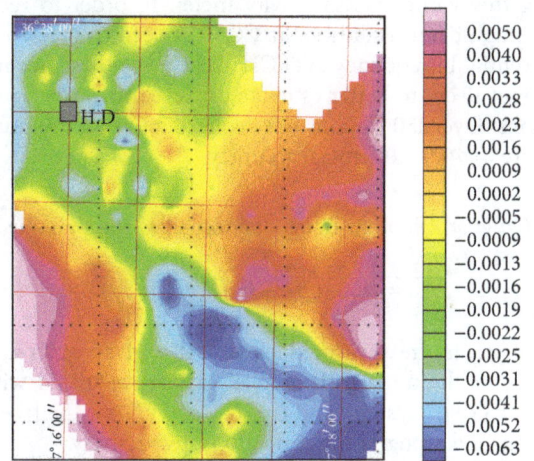

(c)

FIGURE 8: First derivative maps along the three axes $(X, Y, Z)$ ($drvx$, $drvy$ and $drvz$, resp.).

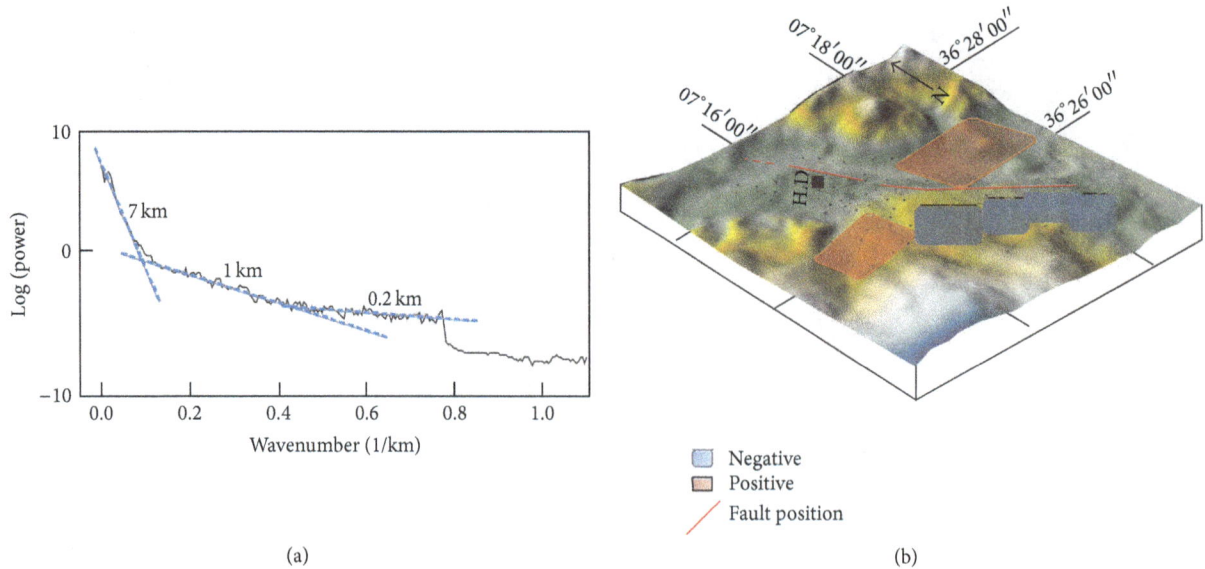

FIGURE 9: (a) Radially averaged power spectrum of the Bouguer anomaly showing 3 sources at 0.2 km, 1 km, and 7 km depths; (b) location of the main fault (red line) with the position of the main source depths deduced from the power spectrum.

with the field stress related to the Eurasia-Africa convergence movement.

## 3. Geophysical Data

For a better understanding of the subsurface geology and the tectonic lineament associated with the major HD fault zone, we carried out geophysical investigations.

*3.1. Gravity Data.* A geophysical prospection was carried out in the vicinity of HD, where essentially travertine deposits crop out. The gravity field values have been measured using the Scintrex CG3 gravity meter. In order to reduce topographic uncertainties, the CG3 has been connected to a differential bifrequency HTECH Z12 GPS with an accuracy of about 10 cm to better estimate the levelling of the gravity stations. Over 150 stations were collected during this survey (Figure 6) with 0.015 mGal accuracy.

*The Bouguer Anomaly.* Bouguer anomaly is computed using

$$A_B = g_m - \left( g_{th} - C_{al} + C_p - C_T \right), \qquad (1)$$

where $g_m$ is the measured gravity at a given station, $g_{th}$ represents the normal theoretical gravity calculated within the IUGG1967 system, and $C_{al}$, $C_p$, and $C_T$ are the free-air, plateau, and topographic corrections, respectively.

$C_{al}$ and $C_p$ are computed taking into account the determined GPS altitudes and with a correcting density value of $2.4\,\text{g/cm}^3$ obtained from the measurements of representative specimen of the geological formations [28].

The Bouguer anomaly map (Figure 6) is drawn from a regular grid of the Bouguer anomaly values obtained

by interpolation using the minimum curvature. This map, with amplitudes ranging between −7 and −16 mGal, mainly shows a significant negative anomaly located southeastwards, bordered by two positive anomalies northwards. It includes all the vertical effects from surface to deep sources. To determine the shallow density distribution, we eliminated the "regional" effect by using the polynomial method at different orders. The first residual anomaly obtained by subtracting a polynomial of "order 1" mainly shows the same anomaly as the Bouguer anomaly. The negative anomaly shows an amplitude value of 5 mGal while both positive anomalies reach a value of 3 mGal (Figure 7).

The second residual anomaly map obtained by subtracting a polynomial of "order 2," which represents shallow anomalies, shows a similar distribution as that of a polynomial of "order 1." We also computed a residual anomaly map using a polynomial of "order 3" which shows a negative anomaly as well. This anomaly is located all over the southern part of the map. In order to identify the possible origins of the anomalies and to evidence the possible limits in between bodies, we computed the first derivative maps in three directions (X, Y, and Z, resp., Figure 8). On the southern centre of these maps, we may observe a gravity discontinuity-oriented NW-SE. Upward continuation maps at different altitudes filtered the high frequencies and thus highlighted the long wavelength anomalies. For increasing values of upward continuations at altitudes from 0 to 500 m, the main negative anomaly moves to the southeastern part of the map (Figure 6). The anomaly depths, as a function of their wavelengths, are obtained by using the calculation of the Bouguer anomaly average radial spectrum. Three main solutions have been evidenced (Figure 9). The deeper one is located at more than 7 km, while two other particular bodies, shallower, are located at ~1 km and ~0.2 km. The deeper body

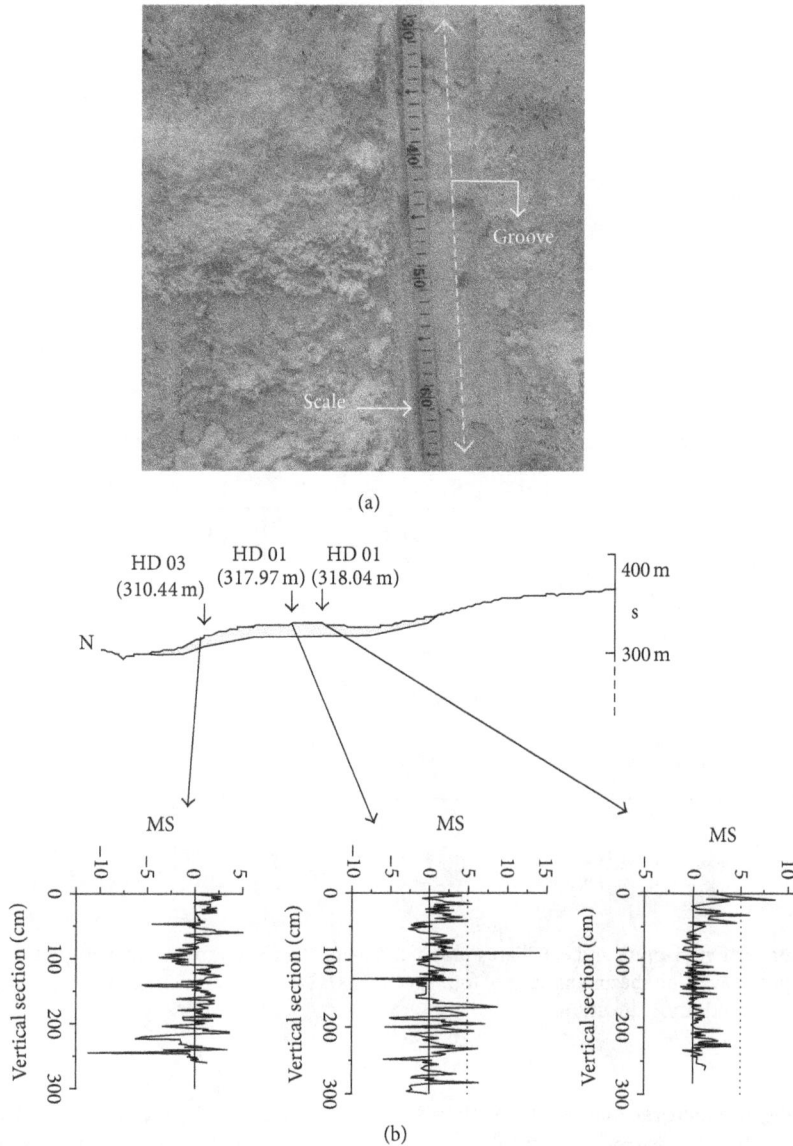

FIGURE 10: (a) Magnetic susceptibility (MS) measurement along a groove opened for acquisition; (b) MS profiles along N-S topographic cross section on the travertine deposits.

should be a regional structure related to the basement of the Guelma Neogene Basin.

### 3.2. Travertine Magnetic Susceptibility Measurements.

The preliminary rockmagnetic data show that the magnetic signature associated with travertine is generally characterized by low susceptibility values due to the carbonate nature. In order to better understand the process of incrustation, mineralogical enrichment, and paleocurrent flow, magnetic susceptibility signature is used. Magnetic susceptibility (MS) profiles were carried out in several sites. To reduce the local alteration effect, grooves were opened before getting accurate measurement using MS2 sensors. The location and the altitude of the different profiles we measured are calculated using the same differential bifrequency HTECH Z12 GPS. In

fact, the MS2 Bartington instrument enables us to use both sensors: MS2D for calibration and the MS2F which is adapted for data acquisition along opened grooves (Figure 10(a)). The distance between measured points is equivalent to the width of the sensor, that is, ~1 cm in the case of MS2F.

Series of bulk magnetic susceptibility (MS) curves were drawn for each profile to determine MS variation versus depth. If we suppose that the precipitation of travertine takes place in horizontal or subhorizontal topographic conditions, lateral correlation between profiles can reveal postdeposit deformation. As example, to highlight lateral MS variation on successive profiles, even if they are taken at different altitudes, we compare three of them (Figure 10(b)). Preliminary results show that MS values are the strongest at the lower part of some profiles indicating the basement of the travertine series while other profiles show slightly elevated MS values at the

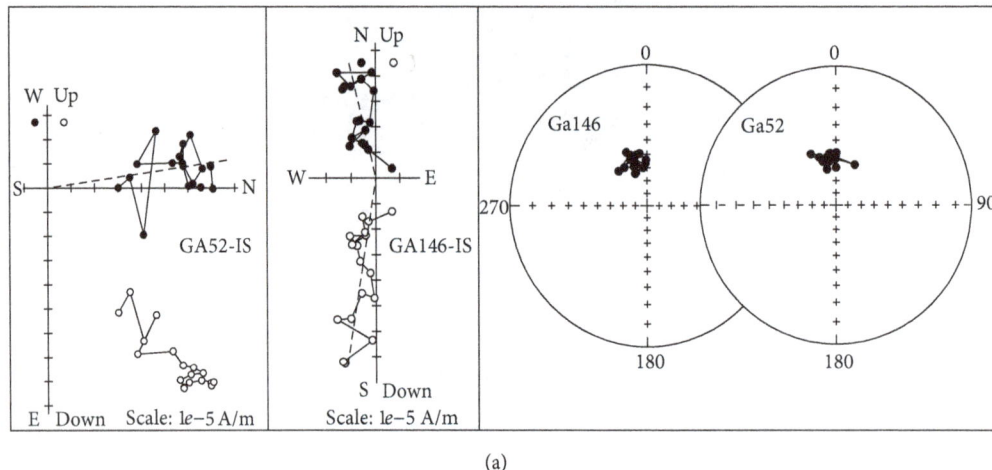

(a)

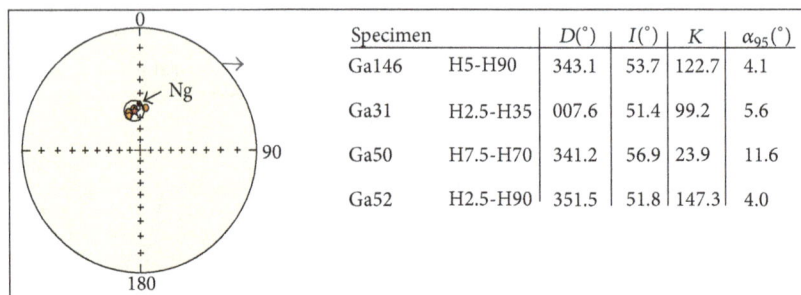

| Specimen | | $D(°)$ | $I(°)$ | $K$ | $\alpha_{95}(°)$ |
|---|---|---|---|---|---|
| Ga146 | H5-H90 | 343.1 | 53.7 | 122.7 | 4.1 |
| Ga31 | H2.5-H35 | 007.6 | 51.4 | 99.2 | 5.6 |
| Ga50 | H7.5-H70 | 341.2 | 56.9 | 23.9 | 11.6 |
| Ga52 | H2.5-H90 | 351.5 | 51.8 | 147.3 | 4.0 |

★ Mean magnetization direction $D = 351.2$, $I = 53.9$, $K = 113.9$, $\alpha_{95} = 8.6$

(b)

FIGURE 11: (a) Orthogonal projection diagram and equal area plot of the magnetization of two specimens of the pilot series; (b) preliminary paleomagnetic analysis on pilot travertine specimens: case of four specimens (Ga31, Ga50, Ga52, and Ga146) and the different paleomagnetic parameters including mean magnetization direction.

top, suggesting anthropogenic factors such as atmospheric deposition of magnetic spherules formed as combustion products (see HD 2 in Figure 10(b)). All sections show more lateral variation indicating some differences in the sedimentary processes within the travertine. In the case of profiles 1, 2, and 3 (Figure 10(b)) performed on travertine wall, it yields the probable paleocurrent flow. In that case, it evidences flow direction from south to north. Hence, the travertine deposits are growing northward (Figure 10(b)). Another hypothesis is that the MS profiles' variations can be mainly related to chemical origins, as the magnetic enrichment increases this magnetic parameter. Indeed, outside the anthropogenic enrichment origin observed on the top of profiles (example in HD 2), the little increase of the MS is probably due to the spas activity which mostly grows with tectonic activity (active faulting). Most indications are given by our preliminary paleomagnetic study carried out on the sampled travertine. A pilot series of eight standard paleomagnetic specimens was demagnetized under an alternating field (AF) at the cryogenic 2G magnetometer at CEREGE (France). The results show that most specimens exhibit more or less a good AF demagnetization where the total natural remnant magnetization (NRM) is deleted at around 2 mT or at maximum of 4 mT,

with weak NRM around $1.5 \times 10^{-6}$ Am$^2$, which attest of specific magnetite bearing mineralogy. In Figure 11(a), the orthogonal diagrams highlight that the NRM exhibits only one component despite the weakness of the NRM intensities. Only few specimens (e.g., Ga50 and Ga52) show hard AF demagnetization due to the presence of magnetite spherules. On the other hand, the demagnetization of the pilot series of specimens gives a good indication about the magnetic direction which suggests a Plio-Quaternary age (Figure 11(b)) [29]. Finally, this preliminary paleomagnetic approach, combining the travertine magnetic susceptibility and preliminary paleomagnetic analysis, can be used to demonstrate the spas activity. As suggested previously, using the MS profile signals, one can determine different levels of deposits related to high and/or low spas activity phases which can be attributed to instantaneous tectonic movement.

## 4. Discussion

*4.1. Geological and Morphological Contribution.* The travertine deposit of HD has been accumulated along an active normal fault which generated earthquakes in the past. The existence of extensional fractures out of the area attests that this structure somehow contributes to the pull-apart system of the Guelma basin. It is clear that travertine becomes deposited upon the inner and the outer walls as well and progressive widening of the fracture takes place. This is due to the continuation of the crustal tension, or high hydrostatic pressure within the ridge. The local buildups will result in a series of small interconnected mounds gradually increasing in height. This morphology attests that the water flowing from the fracture is often discontinuous, leading to a linear array of individual springs. Smaller movements occurred along the major fault or downslope movements and follow the fracture during earthquake activity (Figure 10). Good examples of tensional fractures are known at HD, and the travertine landslide near the village was probably initiated by this process. Therefore, the HD fault could be responsible for the local seismic activity. This is corroborated by the focal mechanisms of several earthquakes which occurred along the western limit of the Guelma basin that exhibit an NNW-SSE fault plane.

The presence of hydrothermal sources at a given site can change the organization of drainage, so as they profoundly influence the fluvial systems and their development. In HD, the principal drainage pattern is represented by Bou Hamdane River which crosses this zone in its northern side (Figure 2). As shown in Section 2, the overall morphology of the site of HD shows that the river was deviated from its primary flow position. This is evidenced, at the base of the front of travertine, by the presence of detrital material abandoned by the river as a terrace. The cemented alluvium modifies the Bou Hamdane River flow. This scenario is shown by the river bend that bypasses the northern boundary of the travertine area and the northern direction on the travertine flow deposit as suggested by the MS profiles. Hence, this river has been deviated ~100 m to the north by the impinging travertine deposits during their development. After the initial incision of a primary flow of hydrothermal water from a shallow volcanic source (as shown by the gravity data), the evolution of the travertine deposits could be interpreted as an active hydrothermal feature related to active tectonics.

*4.2. Geophysical Data.* The gravity data show that the HD hydrothermal source is shallow and located at the south of the village. These data provide a picture of the hydrothermal deposits formation at depth surrounding the position of the heat source (7 km) to the surface. The chemical and physical parameters can explain the volcanic or plutonic nature of the origin source. The major gravimetric axe is superimposed to the position of HD and the numerous normal faults observed in the vicinity of HD. The preliminary paleomagnetic analysis corroborates a recent age of the travertine deposits and suggests that the surface deformation is of Quaternary age such as the surface rupture observed in the Roman career. The MS profiles highlight the S to N flow direction and attest of the probable increase in water mineralogical drift. The simple typical scenario proposed all over the world of such deposits (heated seat deep, subvertical fault, and hydrothermal steam) can be applied to the site of HD and surroundings.

*4.3. Structural Implication.* Previous geological studies in the Guelma basin did not reveal the existence of the tectonic lineament of HD-Roknia. This is due in part to the absence of structural investigation. This work has revealed minors fault structures in the NNW-SSE direction, with apparent normal movement outcropping in Miocene, Pliocene sandstones, and travertine. These young NNW-SSE fracture systems are juxtaposed to HD and Roknia spas. The recent activity of this fault is attested by the moderate earthquakes that occurred in this zone. The HD major fault is similar to the Bouchegouf, Hammam N'Bailis (Guelma region), and Sigus (to the west in Constantine region) active faults which are also juxtaposed to hydrothermal sources.

## 5. Conclusion

The different field observations and measurements evidenced an active tectonic structure with normal faulting along which a system of travertine was deposited. From tectonic point of view, the numerous fissures observed are parallel to the active extensional fault responsible for the local seismicity. As mentioned before, the HD case is similar to the travertine deposits at Pamukkale, Turkey [2], and at La Roche Fontaine, France [30] and also the fracturing of a Madang Province (New Guinea) travertine Pleistocene to Holocene deposit [3]. The 2003 HD earthquake seems to be clearly related to the fault newly identified in this study (namely, HD Fault) which is seismogenic structure similar to the already known H. N'Bailis, Bouchegouf, Sigus, and Ain Smara active faults.

## Acknowledgments

This paper was supported by (Centre de Recherche en Astronomie, Astrophysique et Géophysique) CRAAG, Project code G002/08 (Contribution géophysique à l'étude de l'évolution géodynamique du bassin sédimentaire de Guelma (Algérie)). The authors are grateful to the Magnetic Laboratory of CEREGE (France) for the preliminary paleomagnetic pilot measurements. They thank Dr. A. Harbi for the numerous and fruitful discussions on the seismicity of the region. They are grateful to the Editor Dr. Gambino and anonymous reviewers for their comments, remarks, and suggestions.

## References

[1] P. L. Hancock, R. M. L. Chalmers, E. Altunel, and Z. Çakir, "Travitonics: using travertines in active fault studies," *Journal of Structural Geology*, vol. 21, no. 8-9, pp. 903–916, 1999.

[2] E. Altunel and P. L. Hancock, "Morphology and structural setting of Quaternary travertines at Pamukkale, Turkey," *Geological Journal*, vol. 28, no. 3-4, pp. 335–346, 1993.

[3] W. F. Humphreys, S. M. Awramik, and M. H. P. Jebb, "Freshwater biogenic tufa dams in Madang Province, Papua New Guinea," *Journal of the Royal Society of Western Australia*, vol. 78, no. 2, pp. 43–54, 1995.

[4] A. Heimann and E. Sass, "Travertines in the Northern Hula Valley, Israel," *Sedimentology*, vol. 36, no. 1, pp. 95–108, 1989.

[5] K. E. Bargar, "Geology and thermal history of Mammoth Hot Springs, Yellowstone National Park, Wyoming," *United States Geological Survey Bulletin*, vol. 1444, pp. 1–55, 1978.

[6] V. Lozek, "Molluscan characteristics of the Pleistocene warm periods with particular reference to the last interglacial," *Berichte der Deutschen Gesellschaft fur Geologische Wissenschaft Reihe A*, vol. 14, pp. 439–469, 1969.

[7] P. Ambert and A. Tavoso, "Les formations quaternaires de la vallée du Tarn entre Millau et Saint-Rome de Tarn," *Paleobiologie-Continentale*, vol. 12, pp. 185–193, 1981.

[8] A. Pentecost, H. A. Viles, A. S. Goudie, and D. H. Keen, "The travertine deposit at Shelsley Walsh, Hereford & Worcestershire," *Transactions of the Woolhope Naturalists Field Club*, vol. 50, pp. 25–36, 2000.

[9] M. Julian and J. Martin, "Signification géomorphologique des tufs et des travertins," *Bulletin de l'Association Géographique de France*, no. 479-480, pp. 219–233, 1981.

[10] J. M. Vila, *La chaine alpine d'Algérie orientale et des confins algéro-tunisiens [Thèse de Doctorat]*, Université de Pierre et Marie Curie, Paris, France, 1980.

[11] A. Harbi, S. Maouche, and A. Ayadi, "Neotectonics and associate seismicity in the Eastern Tellian Atlas of Algeria," *Journal of Seismology*, vol. 3, no. 1, pp. 95–104, 1999.

[12] P. Deleau, "Etude géologique des régions de Jemmaps, Hammam Maskoutine et du col des oliviers. Thèse," *Bulletin: Service de la Carte Géologique de l'Algérie*, vol. 2, no. 8, p. 583, 1938.

[13] H. Dib, "Guide pratique des sources thermales de l'Est Algérien," *Mémoires du Service Géologique National*, no. 15, p. 106, 2008.

[14] M. Meghraoui, *Géologie des zones sismiques du Nord de l'Algérie. Paléosismologie, Tectonique Active et Synthèse sismotectonique [Thèse de Doctorat d'Etat]*, Université de Paris-Sud Orsay, 1988.

[15] D. Benouar, "Materials for the investigation of the seismicity of Algeria and adjacent regions during the twentieth century," *Annali di Geofisica*, vol. 37, no. 4, p. 862, 1994.

[16] A. Harbi, A. Peresan, and G. F. Panza, "Seismicity of Eastern Algeria: a revised and extended earthquake catalogue," *Natural Hazards*, vol. 54, no. 3, pp. 725–747, 2010.

[17] M. Meghraoui, J. L. Morel, J. Andrieux, and M. Dahmani, "Néotectonique de la chaîne Tello-Rifaine et de la Mer d'Alboran: une zone complexe de convergence continent-continent," *Bulletin de la Société Géologique de France*, vol. 167, pp. 143–159, 1996.

[18] A. Harbi, S. Maouche, and H. Benhallou, "Re-appraisal of seismicity and seismotectonics in the north-eastern Algeria part II: 20th century seismicity and seismotectonics analysis," *Journal of Seismology*, vol. 7, no. 2, pp. 221–234, 2003.

[19] S. Maouche, M. Meghraoui, C. Morhange, S. Belabbes, Y. Bouhadad, and H. Haddoum, "Active coastal thrusting and folding, and uplift rate of the Sahel Anticline and Zemmouri earthquake area (Tell Atlas, Algeria)," *Tectonophysics*, vol. 509, no. 1-2, pp. 69–80, 2011.

[20] G. Glacon and H. Rouvier, "L'unité du col d'Adissa (confins Algéro-tunisiens septentrionaux): lithologie et stratigraphie; conséquences structurale et paléogéographique de son individualisation," *Bulletin de la Société Géologique de France*, vol. 7-13, pp. 100–110, 1972.

[21] G. A. Waring, "Thermal springs of the United States and other countries of the world, a summary," *United States Geological Survey*, vol. 492, pp. 1–383, 1965.

[22] A. Deschamps, M. Bezzeghoud, and A. Bounif, "Seismological study of the Constantine (Algeria) earthquake (27 October 1985)," in *Seismicity, Seismotectonics and Seismic Risk of the Ibero-Maghrebian Region*, J. Mezcua and A. Udias, Eds., vol. 8, pp. 163–173, Instituto Geográfico Nacional, Madrid, Spain, 1991.

[23] S. Guigue, Les sources thermominérales de l'Algérie. Tome I et II, Serv. Carte Géol. de l'Algérie, 3ème série, 5ème et 9ème fasc, 1940.

[24] D. McKenzie, "Active tectonics of the Mediterranean region," *Geophysical Journal. Royal Astronomical Society*, vol. 30, pp. 109–185, 1972.

[25] J. M. Nocquet and E. Calais, "Geodetic measurements of crustal deformation in the Western Mediterranean and Europe," *Pure and Applied Geophysics*, vol. 161, no. 3, pp. 661–681, 2004.

[26] E. Serpelloni, G. Vannucci, S. Pondrelli et al., "Kinematics of the Western Africa-Eurasia plate boundary from focal mechanisms and GPS data," *Geophysical Journal International*, vol. 169, no. 3, pp. 1180–1200, 2007.

[27] A. Harbi and S. Maouche, "Les principaux séismes du Nord-Est de l'Algérie," *Mémoires du Service Géologique National*, no. 16, p. 106, 2009.

[28] A. Abtout, L. Hamai, and B. Bouyahiaoui, Etude gravimétrique de la région de Guelat Bousbaa-Nador et du bassin d'Oued Zenati.(Guelma): reconnaissance des aquifères. Rapport ANRH (Agence National des Ressources Hydriques), 2005.

[29] J. D. A. Zijderveld, "AC demagnetization of rocks: analysis of results," in *Method in Paleomagnetism*, D. W. Collinson, K. M. Creer, and S. K. Runcorn, Eds., pp. 254–286, Elsevier, Amsterdam, The Netherlands, 1967.

[30] C. Pareyn and H. Salimeh, "Dislocation catastrophique d'un tuf lapidifie dans l'arriere-pays d'Honfleur (Calvados, France)," *Bulletin Centre de Géomorphologie du Caen*, vol. 38, pp. 151–159, 1990.

# 10

# Quarry Waste: Chances of a Possible Economic and Environmental Valorisation of the Montorfano and Baveno Granite Disposal Sites

**G. A. Dino, M. Fornaro, and A. Trentin**

*DST, Università degli Studi di Torino, Via Valperga Caluso, 35, 10125 Torino, Italy*

Correspondence should be addressed to G. A. Dino, giovanna.dino@unito.it

Academic Editor: Atle Nesje

The Montorfano and Baveno granite ore bodies are located in the Lake District (VCO-NE Piedmont). They were and are still quarried as dimension stones, with a consequent production of a huge volume of "waste." In 1995, an Italian company (GMM S.p.A) decided to invest in the valorisation of granite quarry waste as a secondary raw material. An in situ geological prospecting was carried out in order to evaluate the characteristics of the material and the thickness and volume of the useful disposal sites that could be used (by means of geophysical surveys). As a consequence of the field work, the amount of quarry waste was estimated as nearly 2 Mm$^3$. Chemical analysis was carried out on 75 granite samples, in order to individuate the Fe$_2$O$_3$ grade: from 1.321–2.593% of the original waste to 0.160–0.228% after the "dry process" treatment. Three different detailed maps that show the typology, the locations, and the quality distribution of the material in the dumping areas have been drawn up.

## 1. Geological Setting

The main quarry disposal sites, the subject of the present research, are located in two different areas: the Sengio and Ciana Tane-Pilastretto dumps, which are located on the southern slope of the Montorfano Massif, while the Braghini site pertains to Monte Camoscio, on the right side of the Toce River (Ossola Valley) (Figure 1).

The two granitic plutons belong to the Hercynian Magmatic belt, which is located between the *Ivrea-Verbano* zone and the *Serie dei Laghi* unit. These units pertain to the South Alpine system basement and are situated to the south of the Insubric line. The Permian granites, *Graniti dei Laghi*, consist of nonmetamorphosed to weakly metamorphosed rocks [1, 2] and constitute an elongated SE-NW batholith, between the Biella-Valsessera and Val d'Ossola zones. This batholite is formed by singular plutons which develop a narrow contact aureole [3] in the *Scisti dei Laghi* and the *Strona-Ceneri* formation. The batholith is composed of five plutons: Biella-Valsessera, Alzo-Roccapietra, Quarna, Mottarone-Baveno, and Montorfano. Their age has been estimated, using a radiometric test, as 275 My [4]. The

study has focused on the Mottarone-Baveno and Montorfano plutons: these granitic bodies, NE-SW elongated, at present separated by quaternary deposits, occupy an area of about 30 km$^2$ and intrude awkwardly into the *Scisti dei Laghi* unit.

The Mottarone-Baveno pluton is mainly formed by two different varieties of granite: a *pink granite* (called "Rosa Baveno," Table 1), a historically famous ornamental stone, which has been quarried in the district over the last centuries, and a *white granite*, which represents the prevalent volumetric variety in the pluton, and which is no longer quarried. The miarolitic granite, with a small percentage of even red granite, constitutes the upper layer of the pluton (this layer dips about 15° to the E): the transition from the white to the pink granite is mainly chromatic (Table 1) and is probably caused by chemical diadochy of the Fe$^{3+}$, which substitutes the Al in the K-feldspar structure [5]. The pink granite is characterized by the presence of intergranular albite and fluorite.

The Montorfano pluton is formed by a medium-grained white granite and a small amount of a "green granite." It is actually an episyenite consisting mainly of albite and chlorite. This so-called Verde Mergozzo occurs in the northern slope

TABLE 1: Petrographic analysis of the Montorfano white granite and the Baveno pink granite.

| | | Chemical composition | | | | | | |
|---|---|---|---|---|---|---|---|---|
| | Mineralogy | $Al_2O_3$ | $Fe_2O_3$ | $TiO_2$ | CaO | MgO | $K_2O$ | $Na_2O$ |
| Baveno pink granite | Quartz, K-feldspar, plagioclase, biotite, zircon, allanite and fluorite | 13,98 | 2,301 | 0,26 | 1,19 | 0,29 | 4,97 | 3,18 |
| Montorfano white granite | Plagioclase, quartz, K-feldspar, biotite, apatite, zircon, allanite | 13,75 | 2,2895 | 0,25 | 1,25 | 0,265 | 4,875 | 3,225 |

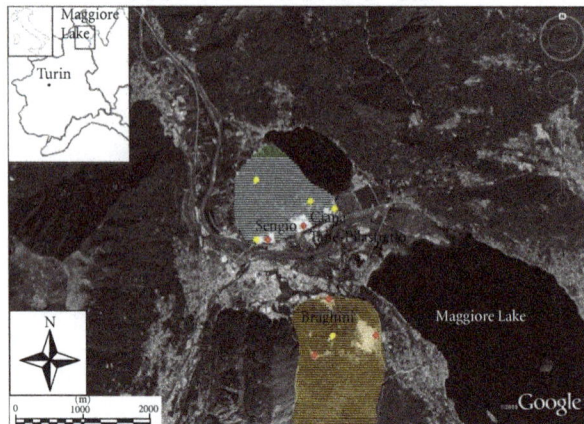

FIGURE 1: Geographic context of the studied area. The main image shows the satellite orthophoto of a portion of Massiccio dei Laghi batholith. In the northern part of the map (Montorfano area), it is possible to individuate both the white granite (represented by grey stripes) and the green granite portion (green Mergozzo type, represented by green stripes). The orange area defines the Baveno-Mottarone pluton. Furthermore in the image the exploited quarry dumps (red points) are represented and the ancient quarry waste areas, potentially useful for a future valorisation (yellow points).

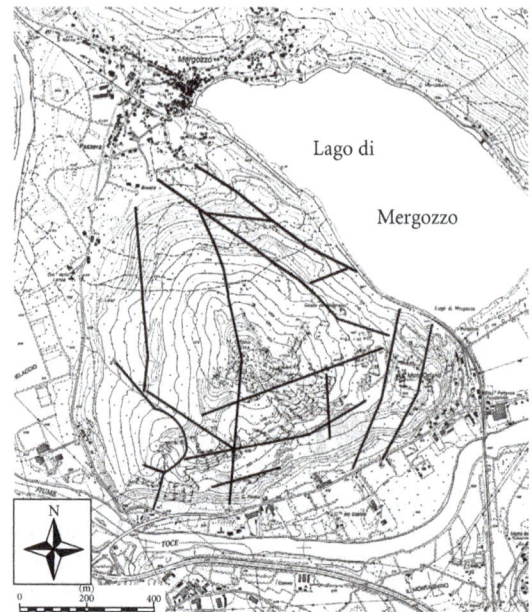

FIGURE 2: Structural sketch map of the Montorfano pluton, showing the main directions of the principal fracture systems and the main faults in the granite massif. These directions depend on the regional faults system (Gravellona-Fondotoce-Bieno, Candoglia-Mergozzo-Fondotoce, and Pogallo-Lago d'Orta faults) [6].

of the pluton, where the unconformity with the *Serie dei Laghi* unit is evident (Figure 1).

The Montorfano massif is delimited by a complex structural fault system (Figure 2) [6]:

(i) the Gravellona-Fondotoce-Bieno fault, WSW-ENE oriented, divides the Montorfano from Mottarone-Baveno massif,

(ii) the Candoglia-Mergozzo-Fondotoce fault, NNW-SSE, corresponds to the major axes of the Mergozzo lake,

(iii) the Pogallo-Lago d'Orta fault, in the N-S direction, is in the western part of the pluton.

## 2. Mining Context

The studied granite resource was only quarried in the past for a local use: in the XVI century, production grew and the material was exported and used for the construction of important buildings and by manufactures in the neighbouring regions (above all the pink granite for Lazzaretto in Milan). The production increases even more during the XVII

century, thanks to the modernization of the exploitation technology and innovative quarry techniques (such as the extraction of megalithic blocks for use as columns for historical palaces and important churches).

In the XVIII and XIX centuries, the granitic structures and manufactures could be found throughout the North of Italy and even reached Rome; in 1830, 82 columns of Montorfano White Granite were sent to Rome, by means of river transport, for the realization of the famous *Basilica of San Paolo Fuori le Mura*. Baveno pink granite was also used for the realization of Christopher Columbus monument in New York and the King's palace in Bangkok.

The quarries were exploited using explosives (large charge rock blasting (large charge rock blasting, fired in small adits or tunnels, driven in the face of the quarry (at the level of the floor). Coyote shooting), [7]: the dimension stones, sectioned after the explosion, were placed in the quarry yard for the subsequent working phase. In the past, this excavation technique produced, during the past, a huge amount of waste, which were stocked on the lower side of the hill of the

FIGURE 3: Southern side of Montorfano massif.

massifs (Montorfano and Baveno), thus forming differently shaped "waste" dumps ("*ravaneti*").

The volumes of quarry waste are a clear example of the problems connected to mining activities: the exploitation works in this territory have caused and are causing an evident hazard for the population, as well as significant environmental and landscape impacts on this rather touristic area (*Lago Maggiore* territory) (Figure 3).

Thanks to the evolutions in quarrying techniques, the quantity of waste has been reduced over the last few decades, but the problems connected to the management of such a kind of materials have remained unsolved. A possible way of solving the problem, with contemporary economic benefits, should be the new exploitation of the materials, stocked in dumps, for the production, by specific mineral dressing treatment, of secondary raw materials (SRMs) for industry (mainly for the production of Grès Porcelain stoneware) through specific mineral dressing treatments [8].

According to the regulations, the raw materials for the Grès Porcelain stoneware had to include pure quartz and feldspar; for this reason the price of the final product is expensive. The introduction of secondary raw materials from the exploitation of quarry waste from the Montorfano and Baveno granites, whose mineral composition (33% quartz, 62% feldspar) is quite similar to the desired mixture, represents the basis for a new waste valorisation which nowadays represents an interesting economic reality which has produced a subsequent decrease in the costs of the raw materials.

The feldspar market, as an industrial mineral, has increased year after year in Italy: from 200.000 t in 1975 to 2,5 Mt in recent years (2005 data); the imported natural resources (about 2,0 Mt) have to be added to these amounts. Some industrial mineral factories, such as Ecomin S.p.A. (Ecomin is one of the companies of *Gruppo Minerali Maffei S.p.A* Group) in Verbania, decided to produce SRM from granite quarry waste.

This treatment plant, which has operated in the field of dressing of raw material since 1995, has the concession to exploit the three granite quarry waste disposal sites: the Sengio and Ciana Tane-Pilastretto areas (white granite) and Braghini area (pink granite exploitation).

In order to guarantee the highest level of safety in the overhanging yards and the stability of the quarry fronts, the material layer is exploited from the top to the bottom

of the whole volume: the quarried material is then loaded into dumpers and transported to the treatment plant. The large-sized blocks that are stocked in the areas are reduced using blasting and breaking techniques, or are used, as such, as riprap or armour stones. The total recovery of the granite waste has exposed the underlying bedrock, in order to minimize the hydrogeologic hazard in the area.

The ore, conveyed from the quarry waste areas to the plant, is treated by crushers, roller mills, and so forth, in order to reduce each grain size class and to obtain 1.25 mm as the maximum grain size dimension. It is then sieved to obtain different grain size materials and to separate the powder granite from the other products. Finally this material passes through electromagnetic separators, which select the ferromagnetic minerals from the final product, characterized by appropriate physical-chemical properties. The waste produced during the enrichment phase (powder granite, ferromagnetic minerals, mud) is also treated to obtain byproducts which are used in other applications.

In particular, the main product is commercially known as F60P (quartz feldspar mixture: 60% of feldspar, mostly K-feldspar), whose production is about *140.000 t/year*. Different byproducts, commercially known as SNS-sand (as a premix for building uses), NGA-coarse black sand (used for industrial sandblasting), SF-wet feldspar (for the ceramic industry), and SF100 and SF200 (used as fillers in cement industries), have to be added to the F60P production. The total amount of byproducts is about *70.000 t/year*.

The importance of such a treatment is that, at first, it is possible to valorise quarry waste as secondary raw material, and then it is possible to achieve the goal of a zero-waste-volume production, with a consequent reduction ins costs for quarry enterprises and indisputable environmental advantages for the territory.

The Gruppo Minerali Maffei S.p.A has carried out different geo-mechanic stability controls on the slopes and has characterised the three exploited areas, in order to establish the possibility of rock detachment, both from the mountain faces and from the debris landfill (debris flow and/or rock collapse).

As far as the modelling of the potential gravitational dynamics is concerned, an ILA programme, which allows the safety factor (Fs) values to be calculated, on the basis of common stability models (Bishop, Jambu, etc.), was adopted. The parameters used to calculate the Fs are summarised in Table 2.

On the basis of the reported data, it was possible to calculate the Fs (December 2004) for the three areas and to foresee the Fs in 5 (December 2009) and 10 years (December 2014), see Table 3.

The Fs value was obtained considering the different heights of the rock face (granitic basement), considering a hypothetical rock collapse from different altitudes, and also considering the "in-progress" mining activity (on the granite debris).

The evaluated Fs values, connected to geomechanic stability of the debris deposits, are very close to the equilibrium limit. Moreover, it is possible to underline that the Fs pertaining to the mining exploitation of the quarry waste

TABLE 2: Quarry dumps characteristics (granite substratum and debris landfill).

| | Sengio area | | Ciana-Tane Pilastretto area | | Braghini area | | |
| --- | --- | --- | --- | --- | --- | --- | --- |
| | Granite substratum | Debris landfill | Granite substratum | Debris landfill | Granite substratum | Debris landfill | Eluvio-colluvial and fluvio-glacial deposits |
| Saturated volume weight (kN/m$^3$) | 25 | 18.5 | 25 | 18.5 | 25 | 18.5 | 20 |
| Not drained cohesion (kN/m$^3$) | 0 | 0 | 0 | 0 | 0 | 0 | 0.1 |
| Internal friction angle | 45° | 38° | 45° | 38° | 45° | 38° | 28° |

deposits (in the studied areas) increases more and more over the years.

The data reported in Table 3 confirm, once again, that mining exploitation of the granite quarry disposal sites (new mines) significantly improves the geomechanical stability of the investigated areas.

What emerges from the shown data (Table 3) is that in December 2004, and not considering regular mining exploitation and contextual slope recovery, the safety factor (Fs) was very close to the limit imposed by law (Fs = 1.3; Ministerial Decree 11/03/1988). It is possible to underline that the current configuration of the debris body is suitable for a stable equilibrium over the long term. Thanks to the mining exploitation, the stability of the slope (rock with a very little part of debris material) will improve more and more (see Table 3).

## 3. Waste Valorisation

*3.1. Geological Survey.* The present study illustrates a detailed geological and geomineral survey of the three mining areas, which was focused mainly on the granite waste deposits and on quaternary formations (fluvial and glacial deposits), in order to delineate their geometric surface.

Furthermore, the research described the structural, granulometric, and stability properties of the waste deposits and, at the same time, conducted a petrologic characterization of the bedrock, with the purpose of underlining the general characteristics (petrology, size distribution, morphology of the bedrock, and stability of the slopes) of the three deposits. This data collection is useful to describe the general characteristic of the waste deposits, compared to the adjacent areas, in order to evaluate the volume of the ore bodies, using surface and thickness data (see Section 3.2). One of the goals of the study was to evaluate the past environmental rehabilitation measures conducted in the past in some dumps and the development of the future rehabilitation measures in the three investigated areas.

The granite waste (Figure 4) is composed of >30 mm material (70–75%), <30 mm (20%), and metric rocks (5 to 10%): it is classifiable as a granitic gravel sand, with a small percentage of silt and the absence of clay. The presence of the silt fraction is caused by weathering of the granitic sand and gravel. Only metric boulders are formed, above all, by white and/or pink granites, in the Braghini area.

The Sengio white granite deposit (Figure 1) has a particular characteristic, which has been observed only in this massif sector: in the rock volume, some millimetric femic

FIGURE 4: Granite quarry waste detail. The different, casual grain size distribution compromise the stability of the coarse volume.

concentrations (max 2 cm), characterized by an oxidized ring around the femic masses, are recognizable in the rock volume. This globular femic concentration has been described by some authors as a globular dissemination of arsenopyrite mineral [9] which causes an important decrease in the final quality of the product.

The femic masses are frequently the cause of the lower quality of the dimension stone; therefore, a large quantity of waste was produced in the Sengio area during exploitation. When the orebody became uneconomic, because of the decrease in quality, the mine conductor abandoned the quarry and transferred the exploitation to the SE area.

The Ciana Tane-Pilastretto (Figure 1) name originates from the presence of Ciana and Tane-Pilastretto quarries, at the top of the granitic waste dumps. A huge quantity of waste (more than in the Sengio area) covered the area, because the Ciana and Tane-Pilastretto quarries were exploited for a longer period, thanks to the better quality of the granitic orebody (characterized by lower femic disseminations in the inner plutonic position).

The Ciana Tane-Pilastretto granitic waste is formed by metric blocks in the superficial layer and sandy to gravel materials in underlying volume, which can be observed from the road that cuts the middle sector of the dump. The granulometric fractioning is caused by the mobilization of the metric boulders that have been left on the superficial layer, above all from the top of the granite waste exploitation.

The Braghini dump (Figure 1) is instead formed by pink granite clasts coming from the Monte Camoscio quarries: the main minerals are quartz, pink to red K-feldspars, plagioclase, and biotite. The materials occupy the basal part of the slope, partially blocking the flow of a stream.

Quarry Waste: Chances of a Possible Economic and Environmental Valorisation of the Montorfano and Baveno Granite Disposal Sites

119

TABLE 3: Geotechnical stability checks.

| | Sengio area | | Ciana-Tane Pilastretto area | | Braghini area | |
|---|---|---|---|---|---|---|
| | Slope face within 290–350 m (sliding surfaces characterising the contact between debris and basement) | Slope face within 260–350 m (sliding surfaces characterising the contact between debris and basement) | General stability of the quarrying face (sliding surfaces passing through a point) | Local stability of the quarrying face (sliding surfaces passing through a point) | Slope face within 590–680 m (sliding surfaces characterising the contact between debris and basement) | Slope face within 540–585 m (sliding surfaces characterising the contact between debris and basement) |
| Fs (December 2004) | 1.25–1.37 | 1.28–1.30 | 1.39–1.59 | 1.52–1.83 | 1.11 | 1.21–1.34 |
| Fs (Evolution within 5 years, December 2009) | 1.33–1.67 | 1.36–1.41[1] | 2.27–3 | >3 | 1.45–1.86 | 1.32–2.01 |
| Fs (Evolution within 10 years, December 2014) | > 2.10–3[2] | 1.7–1.95[3] | — | — | — | 2.04–3 |

[1] Circular surfaces tangent to the contact: rock basement-debris
[2] Slope face within 265–290 m (sliding surfaces characterising the contact between debris and basement)
[3] Slope face within 235–290 m (sliding surfaces characterising the contact between debris and basement).

Model resistivity with topography
iteration 5 RMS error = 3.2
ws48_Sengio_passo5.bin

Unit electrode spacing = 5 m

(a)

Model resistivity with topography
iteration 4 RMS error = 20.8
ws48_Tane_passo5.bin

Unit electrode spacing = 5 m

(b)

Model resistivity with topography
iteration 3 RMS error = 9.4
ws48_Braghini_passo2,5.bin

Unit electrode spacing = 2.5 m

Low resistivity values; it represents the finest quarry dump material with presence of water (aquifer).

Medium resistivity values; it represents the most part of the granite quarry waste.

High resistivity values; it materializes the granite bedrock and the granite boulder.

(c)

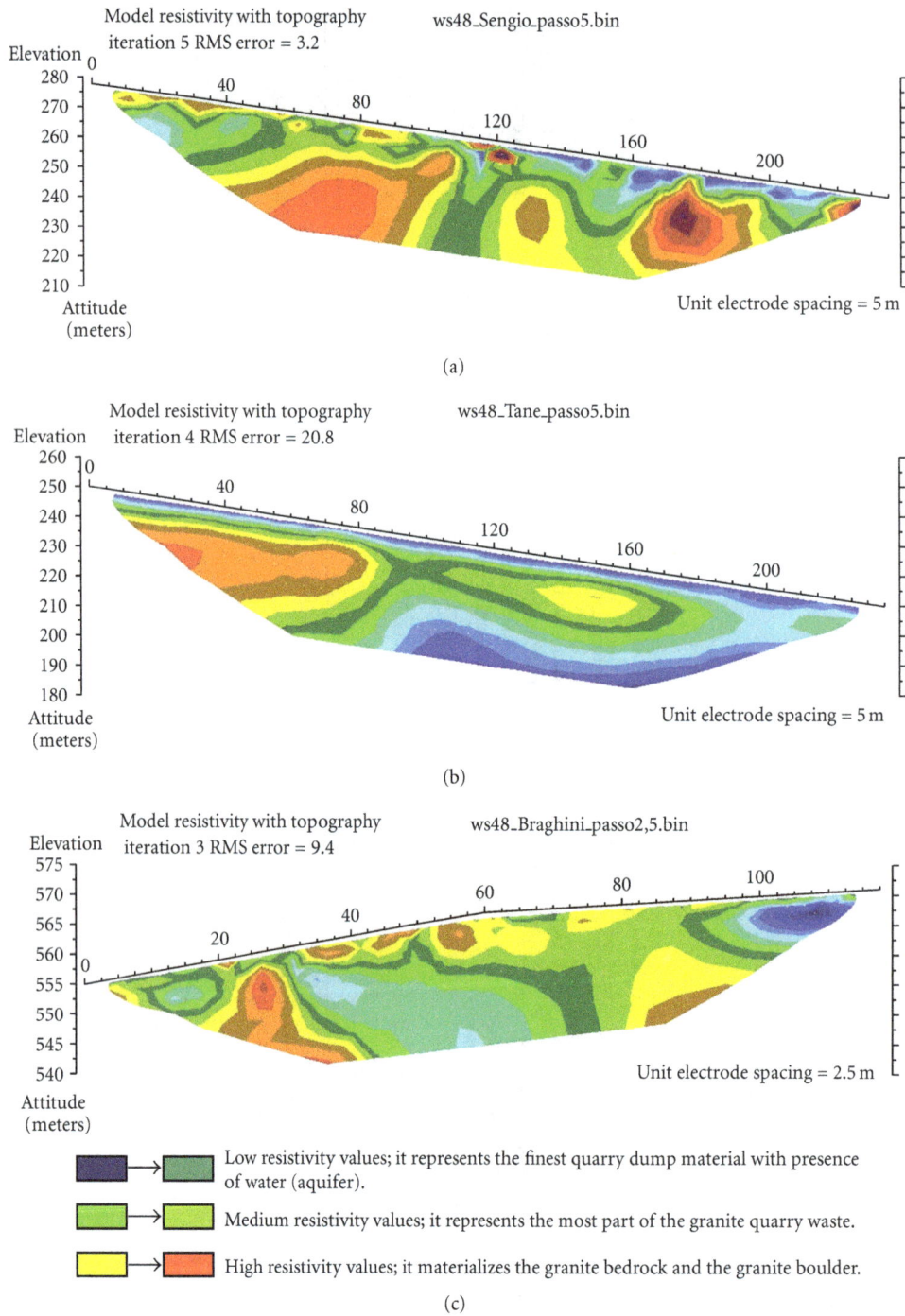

FIGURE 5: (a) Sengio area. (b) Tane Pilastretto area. (c) Braghini area.

An important characteristic of the waste, when considering it as raw materials for ceramic industries, is the different weathering alteration of the sandy fractions, caused by the different steps of the granite waste during the centenarian activity: the vegetation is also composed of different plants because of the different periods of deposition. This factor is very important when deciding, projecting, and programming the best way to valorize the materials in a treatment plant.

3.2. *Volumes Determination.* Another goal of the study was to determine the volumes of the ore bodies: a geophysical survey—an electrical tomography investigation—was carried out to estimate the thickness of the waste deposits. Thanks to a processing phase, the collected field data (resistivity data) were processed to obtain a resistivity model of the investigated area and a 2D resistivity cross-section.

In the Sengio area (Figure 5(a)), the tomography track is positioned in the median sector of the dump, through

Quarry Waste: Chances of a Possible Economic and Environmental Valorisation of the Montorfano and Baveno
Granite Disposal Sites

121

TABLE 4: Thickness and volumes for the investigate areas.

| Quarry waste dumps | Average thickness calculated thanks to tomography cross-sections (m) | Ore body volume (m³) |
|---|---|---|
| Sengio | 20 m | 361.600 |
| Ciana Tane-Pilastretto | 25 m | 1.489.000 |
| Braghini | 15 m (tomography limit) | 158.000 |

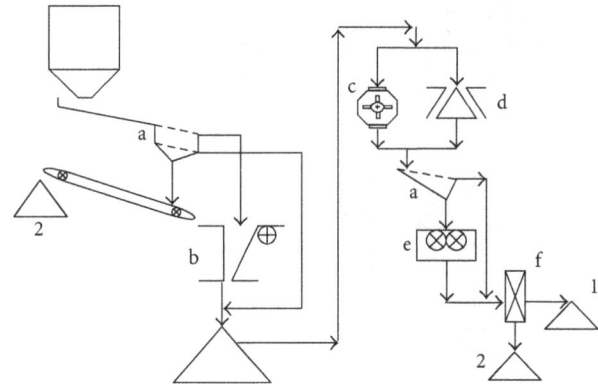

a: sieves
b, c, d: crushers
e: mill crushers
f: magnetic separation (different steps)
1: main product (F60P)
2: by products

FIGURE 6: Simplified flow-chart of Ecomin transformation plant.

an internal working road that crosses the whole volume. The distinctive parameters for the geophysical survey were 5 meters of electrode distance to obtain a length of 240 meters and a sampling rate of 250 ms.

The same configuration as that of the Sengio area was used in the Ciana Tane-Pilastretto (Figure 5(b)) area. In this area the position of the cross-section was set in the basal sector of the waste volume.

Finally, in the Braghini dump (Figure 5(c)), the tomography cross section was set at a N-S direction, crossing the waste area in the lowest sector of the deposit; in this case the configuration of the survey was 48 electrodes, at the distance of 2.5 meters for every electrode, with a total section of about 117.5 meters.

It is possible to show the tomographic profiles of the three studied dumps (Figure 5). The considerable grain size distribution of the material conditioned the electrical signal answer. Even so, the configuration and the border between the waste deposit and the substratum of rock are clearly visible.

It is possible to estimate the deepness of the granitic bedrock in the cross section, thanks to the use of different colours, which correspond to different resistivity values. The resistivity difference can be related to the unconformity of the granitic waste and bedrock.

A comparison between the tomography cross sections and the geological surveys data made it possible to estimate the thickness of the waste deposits.

Elaborating the parameter reported in Table 4, it is possible to estimate a total volume of the dumps as about 2 Mm³ (about 3.6 Mt). About 20% has to be subtracted from this volume as it was selected in the treatment plant as not being directly utilizable as second raw material for ceramic. This material has in fact to be treated in another plant (as byproduct). Therefore, the resources directly treatable in the Ecomin plant should be about 2,9 Mtons. Considering a feeding to the plant of nearly 200 t/year, it is possible to estimate, at least, 15 years for the production lifetime.

It is important to underline that the waste amount in the old quarry dumps is periodically refilled by the flowing waste and that the research for "new ore bodies" is still in progress. Therefore a longer treatment plant activity can be hoped for.

*3.3. Laboratory Phase.* After a geological survey a representative sampling on the debris deposit was carried out. The material characterised by a size distribution of 30 to 150 mm was sampled: 30 mm is the lower limit of the "stone chips" (debris rocks) which can be treated directly in the transformation plant (Ecomin).

Each area of the Sengio and Ciana-Tane Pilastretto quarry dumps was split into a square net of about 50 m per side, while the Braghini quarry dump was divided into a net of about 30 m per side.

78 samples were taken in the three studied dumps: 26 samples from the Sengio quarry dumps, 30 from the Ciana Tane-Pilastretto area, and 22 from the Braghini area. The samples were treated and analysed in the Minerali Industriali Lab. (Cacciano—Masserano, BI).

The laboratory treatment involved a simulation of the ore dressing pilot plant to what normally happens in the Ecomin treatment plant (Figure 6); in particular the process in the pilot plant (Figure 7) can be summarised as a progressive crushing (by means of jaw crushers and roller mills) in order to guarantee a size cut. The thus obtained material (TQ) was then treated by means of two passages in a magnetic separator with permanent magnet cylinders, in order to obtain the final product (2SM). The 2SM samples were grinded and compressed to create tablets for the chemical analysis, using an X-ray spectrometer (Siemens SRS 303). Also the TQ samples were also analysed (chemical analysis), in order to individuate the characteristics of the granitic rocks.

Chemical analyses are fundamental to evaluate the $Fe_2O_3$ grade of each dumps; this parameter is necessary to calculate the quality of the expected products from the treatment plant. On the basis of the lab results, it was possible to draw up, for each investigated area, a $Fe_2O_3$ grade map in order to individuate the zones characterised by the best quality (Figure 8).

In Figure 8 there are three maps concerning the geo-ore deposit quality in each studied areas.

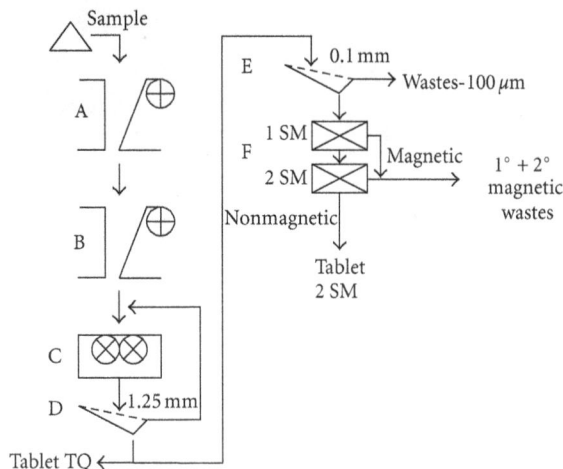

A, B: alternative crushers
C: mill crusher
D, E: laboratory sieves
F: magnetic separation (different steps)

FIGURE 7: Simplified flow-chart for the laboratory pilot plant.

TABLE 5: Fe$_2$O$_3$ average grade for the three studied areas.

| Quarry dumps | Average grade % Fe$_2$O$_3$ (TQ) | Average grade % Fe$_2$O$_3$ (2SM) |
|---|---|---|
| Sengio (white granite) | 2,317 | 0.205 |
| Tane-Pilastretto (white granite) | 2,262 | 0,184 |
| Braghini (pink granite) | 2,301 | 0.210 |

The Fe$_2$O$_3$ grade was quite different (Table 5); the decrease of the quality in the Sengio and Braghini samples, after the magnetic separation, is probably caused by the particular physical characteristics peculiar of the material stocked in the two deposits.

In the Sengio quarry dump, the presence of altered arsenopyrite concentrations in the rocks is probably the reason for the bare magnetic separation during the treatment. On the other hand, the decrease in the quality of the final products in the Braghini quarry dump is probably caused by the weathering alteration in the fine matrix of the granite waste deposit, which compromises the magnetic separation.

Furthermore, looking at the three maps, it is possible to see that there are different Fe$_2$O$_3$ grades, even in the same dump area, which depend on the progressive waste stockpiling in dumps over the years (from different exploited areas). The data regarding the Fe$_2$O$_3$ grade distribution in the three areas are essential to guarantee the integral exploitation of the "waste resources" and to assure a controlled and uniform feeding to the treatment plant.

The lab testing phase also regarded the search for the right method to decrease the Fe$_2$O$_3$ grade in the samples characterised by a very low quality; in these cases a deeper crushing and adding magnetic separations would be the correct ways to valorise the material. On the basis of the

test results, it is possible to notice a quality increase in the quality of treated material, in particular in the 0.1–0.5 mm size distribution classes.

As far as the efficiency of the magnetic separation in a future experimental phase is concerned, it is possible to notice that the Fe$_2$O$_3$ grade does not decrease substantially after the passages in the separator (3-4 passages): in fact the grade tendency is close to an asymptote. Such a result, directly imputable to the liberation grade of the single grains, is closely connected to the size distribution of the final product.

## 4. Quarry Dumps Rehabilitation

The old dumps can be considered as "new mines": they can be rehabilitated during the exploitation phase by the enterprise which owns the mining concession [10].

The valorisation of the material stocked in the dumps guarantees positive results, both for the company (ECOMIN) and the interests of the community at large. The systematic exploitation of the "waste" accumulated during the years and/or actually produced is the basic condition for

(i) the safety of the landfill areas, intervening with appropriate accommodation and environmental recovery,

(ii) the impacts reduction (dust, etc.), due to the presence of incoherent material placed on the dump slopes,

(iii) a positive economic return, due to the exploitation of the material placed in the dumps, which should not be quarried (with explosives and/or mechanical equipment), but only picked up and sent to the treatment plant.

Projects concerning "land use changes in areas characterised by hydrological constraints" have been approved, as far as the exploitation plans are concerned. As a result, the project envisages the exploitation of debris materials stocked on the slope, in order to reach the granite substrate (Figure 9). In parallel to the exploitation, an environmental restore of the areas has had to be carried out, at the same time as the exploitation. It offers a partial renaturalization of the site, in order to return it to its original morphological and landscape configuration, by reducing the visual and environmental impacts. In fact, thanks to appropriate naturalistic interventions, it is possible to recreate the typical environmental conditions typical of the quarried area before the exploitation. An additional and equally important objective of mitigating risk is to improve the stability of the dump slopes. Specific forced re-naturalisation interventions were carried out in some areas, which are completely free of debris material and characterised by high slopes: small terraces, built directly onto the granite substrate, on which top soil have been placed to encourage growth. Bush species were then implanted in these artificial inlets, with very good results (which should be verified in the future).

Other well-known forced "bio-engineering" re-vegetation methods were used for the reclamation of the areas [11]. The result is an rehabilitation that has been applied

Quarry Waste: Chances of a Possible Economic and Environmental Valorisation of the Montorfano and Baveno Granite Disposal Sites

123

(a)

(b)

■ Fe$_2$O$_3$ oxide <0.19%     ■ Fe$_2$O$_3$ oxide >0.22%
■ Fe$_2$O$_3$ oxide 0.19–0.22%   • Sampling points
▨ Service track

(c)

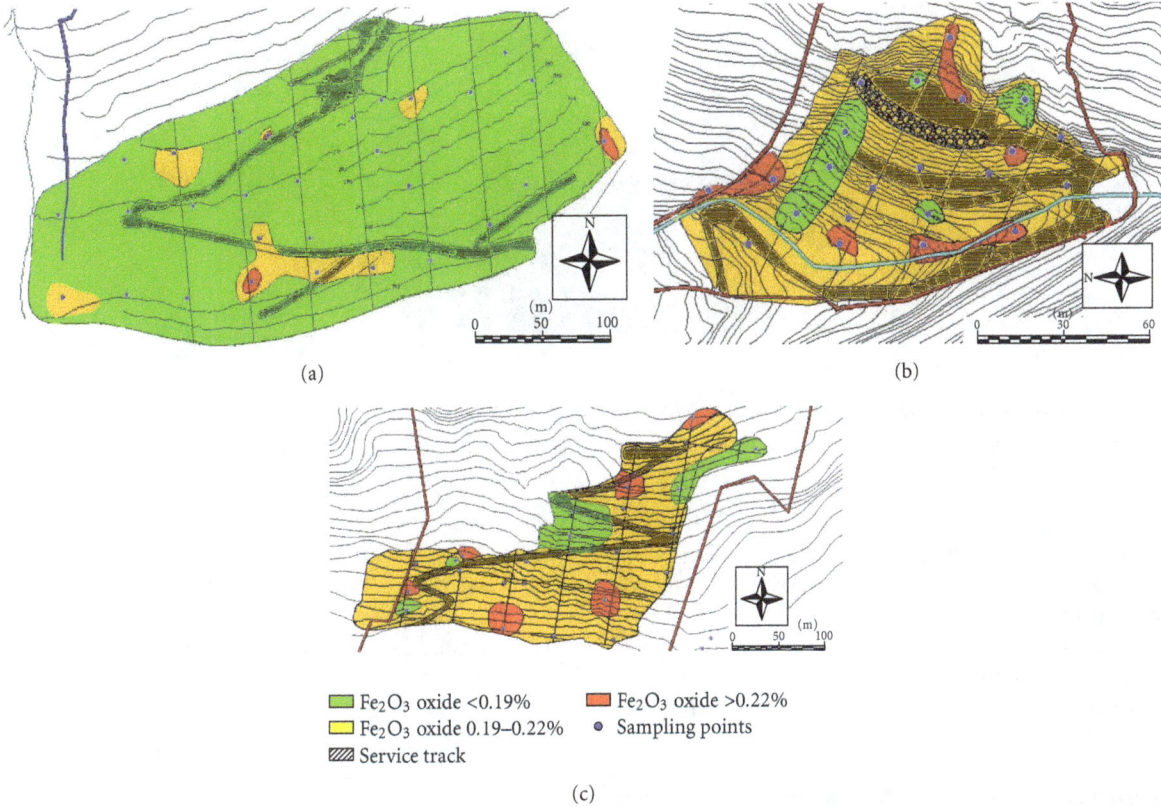

FIGURE 8: (a) Ciana-Tane Pilastretto Area. It is possible to notice the service track, in the middle of the quarry dump (black line), and a stream in the upper left side of the map (blue line). (b) Braghini Area. It is possible to notice, in the middle of the map, a rockfall protection embankments (wire netting area), the Rio Cavallaccio stream just above the rockfall protection embankments (blue line) and the concession area borders (red lines). (c) Sengio Area. Just in the middle of the investigated area it is represented a service track (black line). The borders of the concession area are portrayed as red lines.

FIGURE 9: The uncovering of granite substrate thanks to mechanical equipments. The material is transported to the plant, located just below the slope (on the other side of the street).

perfectly to the morphological and vegetation context if the surrounding areas (see Figure 10).

The key elements, necessary to achieve the total safety conditions, seem to be the "walls" that have been placed at the base of the excavation, and which have been made using the less valuable material produced during the exploitation. These artificial hills are frequently useful to create a "screen" or a visual impediment to the outside world, in harmony with the surrounding nature. They are also important as protection elements for the adjacent and very busy SS34 (State highway, Figure 2).

The frequent fall of large amounts of water, due to the heavy rainfall that periodically affects the area, constitutes a repeated and hard test for the stability of the faces, which are often bares and characterised by the presence of incoherent materials.

Therefore, it is possible to ask, what would have happen to the safety of roads, sides, and so forth, if the miner had not decided to intervene, and to exploit the Baveno and Montorfano "waste-deposits." The exploitation of these materials is truly favourable for slopes stability and for the safety of the roads and infrastructure near the quarries.

A wise and planned mining activity is also important for the defence of the territory, as well as for the production of secondary raw materials (MPSs), which are profitable for the company and for the country.

## 5. Conclusions

The paper emphasizes that the interest of private companies for the systematic recovery of quarry waste should be

FIGURE 10: (a) Quarry dump rehabilitation (Braghini area). (b) Quarry dump rehabilitation after the exploitation of the pink granite waste, in order to obtain secondary raw material for ceramic industries. Scala dei Ratti yard (Baveno area), quarried by Mineral Baveno s.r.l.

interpreted as a signal of the good will of private and public bodies to guarantee environment and territory protection and the safety of the quarries.

The research points out that it is possible to ensure sustainable development for mining activities, guaranteeing, at the same time, profit for the virtuous companies involved in the exploitation, valorisation, and recovery of the "new ore-bodies."

In particular, treating and valorising the waste stored in quarry sites, it is possible to guarantee

(i) fewer (with less impact) problems concerning environmental impact and hydrogeological problems connected to quarry dumps,

(ii) greater slope stability,

(iii) to bring to surface other parts of the granite ore body, which were hidden by waste piles,

(iv) potential revenues from the treatment of MPS exploited from the dump/ore body, and so forth.

Granite quarry wastes represent, therefore, an important alternative (integrating) source, as a substitute to the exploitation of "virgin" material from the primary quartz and feldspar mines.

As already mentioned, the exploitation of the quarry waste often ensures a correct environment recovery and the safety of slopes affected by the dumps.

In order to fully exploit the granite resource from the dumps it is still essential to conduct a complete investigation in order to estimate the volumes, size distribution and chemical and mineralogical characteristics of the material.

The research should involve both a field investigation (detecting geostructural remote land sensing, geophysical surveys, etc.) and a weighty part of lab analysis (mineral-chemical characterization of the raw material as such and of the products obtained from the treatment, ore enrichment in a pilot plant, etc.).

Thanks to a detailed study of the *Graniti dei Laghi* area (Baveno and Montorfano), it has been possible to

(i) calculate the quantity of material available for the treatment plant (2,9 Mtons),

(ii) assess the quality of these materials in order to select what to send to the treatment plant and what to keep separate, because of their poorer quality. This unusable fraction should also be used for environmental recovery and the rehabilitation of the landfill morphology,

(iii) evaluate the reserves available in dumps, added to a coming waste from the quarry, which will ensure the productivity of the treatment plant for not less than 15 years.

## Acknowledgments

The authors would like to thank the Gruppo Minerali Maffei S.p.A., Ecomin s.r.l, and Minerali Industriali S.p.A. for their fundamental help during the field surveys and the laboratory phase; a special thanks are due to Eng. G. Bozzola, Dr. T. Mestriner, Eng. E. Salvaia, Dr. A. Lorenzi, and Dr. S. Vegis. Thanks are also due to Prof. A. Godio and the Geophysics Lab at DITAG-Politecnico di Torino, for their help during the geophysical processing. Finally, they would like to thank Rag. D. Marchetti (Giacomini S.p.A. enterprise) for the kindness during the visit to the Mineral Baveno quarrying area and treatment plant.

## References

[1] A. Boriani, L. Burlini, V. Caironi, E. Giobbi Origoni, A. Sassi, and E. Sesana, "Geological and petrological studies on the Hercynian plutonism of Serie dei Laghi—geological Map of its occurrence between Val Sesia and Lago Maggione (N-Italy)," *Rendiconti—Societa Italiana di Mineralogia e Petrologia*, vol. 43, no. 2, pp. 367–384, 1988.

[2] E. G. Origoni, R. Bocchio, A. Boriani, M. Carmine, and L. De Capitani, "Late-Hercynian mafic and intermediate intrusives of Serie dei Laghi (N-Italy)," *Rendiconti—Societa Italiana di Mineralogia e Petrologia*, vol. 43, no. 2, pp. 395–409, 1988.

[3] A. Boriani, V. Caironi, M. Oddone, and R. Vannucci, "Some petrological and geochemical constraints on the genesis of the Baveno-Mottarone and Montorfano plutonic bodies," *Rendiconti—Societa Italiana di Mineralogia e Petrologia*, vol. 43, no. 2, pp. 385–393, 1988.

Quarry Waste: Chances of a Possible Economic and Environmental Valorisation of the Montorfano and Baveno
Granite Disposal Sites

125

[4] A. Boriani, E. G. Origoni, and L. Pinarelli, "Paleozoic evolution of southern Alpine crust (northern Italy) as indicated by contrasting granitoid suites," *Lithos*, vol. 35, no. 1-2, pp. 47–63, 1995.

[5] C. Poisiana, *I graniti minori dell'Ossola inferiore: caratterizzazione tecnico-petrografica e loro usi nella storia*, Tesi di Laurea in Scienze Geologiche, Università degli Studi di Milano, 2002.

[6] U. Zezza, "Il granito del Montorfano: coltivazione e sue caratteristiche tecniche," in *Atti del Convegno Internazionale Sulla Coltivazione di Pietre e Minerali Litoidi*, p. 36, Torino, Italy, 1974.

[7] M. Pieri, *Marmologia. Dizionario di marmi e graniti italiani ed esteri*, Hoepli, Milano, Italy, 1966.

[8] G. A. Dino, A. Gioia, M. Fornaro, and S. Bonetto, "Monte Bracco quartzite dumps: chance of recovery as second raw material for glass and ceramic industries," in *Proceedings of the 1st International Conference on the Geology of Tethys*, pp. 203–208, Cairo University, November 2005.

[9] B. Bigioggero, A. Boriani, G. V. Dal Piaz, and G. Martinotti, "Escursione: da Arona alla Val Formazza attraverso il Verbano, l' Ossola e le sue valli. Sosta 3.5.—Strada Montorfano-Mergozzo, granito di Montorfano con filoncelli aplitici," in dal Piaz G.V., Guida geologica. Vol. 3/1: Le Alpi dal Monte Bianco al lago Maggiore, Ed. BeMa. Collana: Guide Geologiche Regionali (reprinted), 2004.

[10] G. Bozzola, L. Garrone, L. Ramon, and D. Savoca, " Un esempio concreto di riutilizzo di prodotti di scarto: da granito da discarica a materia prima per ceramica e vetreria," *Ghaziabad Engines and Machines*, vol. 4, pp. 17–19, 1995.

[11] H. M. Schiechtl and G. Sauli, "Nuove tecniche di bioingegneria nei ripristini di cave e miniere," in *Atti Convegno Suolo e Sottosuolo*, vol. 3, pp. 1347–1356, Associazione Mineraria Subalpina, Torino, Italy, 1989.

# 11

# Applications of Vitrinite Anisotropy in the Tectonics: A Case Study of Huaibei Coalfield, Southern North China

**Yudong Wu,[1,2] Quanlin Hou,[2] Yiwen Ju,[2] Daiyong Cao,[3] Junjia Fan,[4] and Wei Wei[2]**

[1] MLR Key Laboratory of Metallogeny and Mineral Assessment, Institute of Mineral Resources, CAGS, Beijing 100037, China
[2] Key Lab of Computational Geodynamics of Chinese Academy of Sciences and College of Earth Science,
  Graduate University of the Chinese Academy of Sciences, Beijing 100049, China
[3] Key Laboratory of Coal Resources, China University of Mining and Technology, Beijing 100083, China
[4] Key laboratory of Basin Structure and Petroleum Accumulation, CNPC and PetroChina Research Institute
  of Petroleum Exploration and Development, Beijing 100083, China

Correspondence should be addressed to Quanlin Hou, quhou@gucas.ac.cn

Academic Editor: Hongyuan Zhang

29 oriented and 10 nonoriented coal samples are collected in the study from three different regions of the Huaibei coalfield, eastern China, and their vitrinite reflectance indicating surface (RIS) parameters are systematically calculated and analyzed. Using the available methods, Kilby's transformations and RIS triaxial orientations are obtained. The magnitudes and orientations of the RIS axes of the three regions were respectively projected on the horizontal planes and vertical sections. The results show that the samples in high deformed region have significant anisotropy magnitudes (higher $B_w/R_{max}$ and $R_{am}$ values) with a biaxial negative style, whereas the samples in the slightly deformed area have unimpressive anisotropy magnitudes with a biaxial negative style. Thermal metamorphism superposed might enhance the complication and variation of RIS style. RIS projection analysis deduced that the RIS orientation is mainly controlled by regional tectonic stress, and likely influenced by deformation mechanisms of coal.

## 1. Introduction

During the coalification, tectonic-thermal events could influence the vitrinite reflectance of coal [1, 2]. Effects of temperature and stress on modifying vitrinite reflectance anisotropy (VRA) were proved by high temperature-pressure experiments on bituminous anthracite [3–13]. Patterns of VRA [14, 15] in coal are usually represented by a three-dimensional graphical form resembling an ellipsoid and termed as RIS (reflectance indicating surface).

The investigation of the RIS for vitrinite in coals is performed by means of sections cut from oriented samples, and the magnitudes and orientations of the ellipsoid parameters are obtained by calculating the data measured from three mutually perpendicular surfaces. This method has been reported [14] and verified by high temperature-pressure experiments [16] and tectonic stress field analysis [11, 15, 17–20]. The anisotropy of RIS in this method is quantified

by a function of rank [15], including in terms of bireflectance ratio [21], anisotropy ratio [17] and bireflectance [22].

The RIS styles may also be obtained by nonoriented sections from crushed coal samples. The method requires measurement of the maximum and minimum apparent reflectance from a series of randomly oriented vitrinite particles, and the magnitudes of RIS principal reflectance axes were estimated by using the reflectance crossplots [23, 24]. A series of parameters named as Kilby's transformations in the method are proposed more effectively to estimate the anisotropy magnitude and shape of RIS ellipsoid, including RIS style ($R_{st}$), anisotropy magnitude ($R_{am}$), and reflectance of equivalent volume isotropic RIS ($R_{ev}$). Compared with Levine's method, basing on the application of vitrinite maximum reflectance versus vitrinite random reflectance [25], less time consuming is found during Kilby's data processing [13, 23, 26–30]. In the nonoriented sections, however, the orientations of three principal reflectance axes are missing.

Previous works suggested that the anisotropic strains developed by lithostatic pressure, especially by tectonic strains, including oriented pressures and shears. High rank coals and anthracites show evidence of the presence of biaxial negative materials [15, 19, 23, 31, 32]. As the result of tectonic dislocations, in many cases associated with fractures in the plane normal to the coal bedding, it is possible to find microtextures responsible for the biaxial positive character [29, 30]. It is found that minimum reflectance develops incrementally parallel to the direction of maximum compressive stress ($\sigma_1$) during coalification, and maximum reflectance develops incrementally parallel to the direction of minimum compressive stress ($\sigma_3$) [1, 14]. These viewpoints are verified by simple shear experiments in high temperature-pressure environment [28, 33, 34]. On this basis, the RIS may approximate to finite strain ellipsoid [35], and the orientations and magnitudes of RIS principal reflectance axes may correspondingly have further usages in the tectonic stress field analysis.

The Huaibei Coalfield in Anhui Province of China has abundant coal resources. Thermal model research showed that the coal experienced low temperature (<230°C) after being buried during the coalification process [36] and, therefore, thermal function was limited in the coalification history. Although the coalification process was influenced by other geological agencies such as tectonic stress, the RIS parameters may provide semiquantitative information about the magnitudes and orientations of tectonic deformation.

In this paper, 29 oriented coal samples from the working face of the underground mines and 10 nonoriented coal samples from the core drilling were collected. In order to estimate the magnitudes and orientations of the three principal reflectance axes, the RIS was built on the oriented coal sample blocks. The orientation of RIS was obtained by calculating the data measured from three mutually perpendicular surfaces. Meanwhile, the anisotropy magnitude and shape were analyzed with the parameters such as bireflectance ratio, Flinn's diagram, and Kilby's transformations. The effects of tectonic stress and thermal evolution on coal are therefore expected to be evaluated.

## 2. Geology of the Study Area

The Huaibei Coalfield in the southeast area of North China Craton was influenced by the Tancheng–Lujiang sinistral strike-slip fault in Mesozoic [39–42] and a related arcuate thrust system [37]. Many researchers believed that the arcuate thrust system was a west-thrust imbricate fan with the arc-top near to Huaibei City [37, 43]. A syncline formed in the margin of the thrust system had a certain influence on the burial depth of coal seams and the level of coal rank [44]. Coal-bearing strata mainly occur in Permo-Carboniferous age.

The Huaibei Coalfield is divided into three regions representing different stress-thermal environments. Vitrinite reflectance of coal samples collected from these regions was measured and sampling locations are shown in Figure 1(a).

(1) Region A, including the coal mines of Taoyuan (label 3), Qinan (label 4), Qidong (label 5), and Xutuan (label 9). The strata were likely influenced less by magmatism and tectonic agency and more by burial metamorphism.

(2) Region B, including the coal mines of Haizi (label 6), Linhuan (label 7), Tongting (label 8) and Baishan (label 10). Both Mesozoic deformation and magmatism affected deeply the strata and likely impacted on the samples collected in the region.

(3) Region C, including the coal mines of Luling (label 1), Zhuxianzhuang (label 2), Shitai (label 11) and Yangzhuang (label 12). The strata in the region strongly deformed, and samples are probably highly influenced by effects of tectonic deformation and thermal-shearing.

(4) The lithostratigraphic sequence of the study area is shown in Figure 1(b). The major coal seams occur in the lower Permian system ($P_1$) and minor coal seams locally occur in the upper Permian System ($P_2$). Southward marine regression during early Permian time led to deposition in an island-lagoon environment and followed by terrestrial delta-plain deposits that contained extensive and minable coal seams [44]. The deltaic environment became unstable in early Late Permian, and coal seams became thinned, discontinuous and unminable. The alluvial deposits in Late Permian marked the end of coal accumulation in the Huaibei coalfield [44]. All the oriented samples collected from the working face in coal seams of 3rd, 7th, 8th and 10th and the nonoriented samples collected from the well-drills belonging to the Permian strata.

## 3. Experimental Section

Coal macerals of samples were determined by spot checks before the reflectance measurement, and the reflectance was measured by means of the method described in GB/T 8899-1998 (equated with ISO7404/3 standard). At least 500 points for each sample were measured on the same polished specimen. The contents of vitrinite, inertinite, liptinite, and mineral matter were obtained and are shown in Table 1. Coal lithotype in the Huaibei coalfield is dominated by clarain, with secondary durian and vitrain, and coal maceral is dominated by vitrinite (40.6~92.5%), which is followed by inertinite and liptinite (Table 1).

*3.1. Sample Preparation.* The 29 oriented samples were prepared following the rules described in GB/T 16773-2008 (equated with ISO7404/2 standard) and cut into cubes with approximately 3-4 cm in dimensions (Figure 2). One pair of planes are paralleling to the horizontal plane and the other two mutually perpendicular pairs normal to the horizontal one. Each sample, therefore, has three pairs of mutually

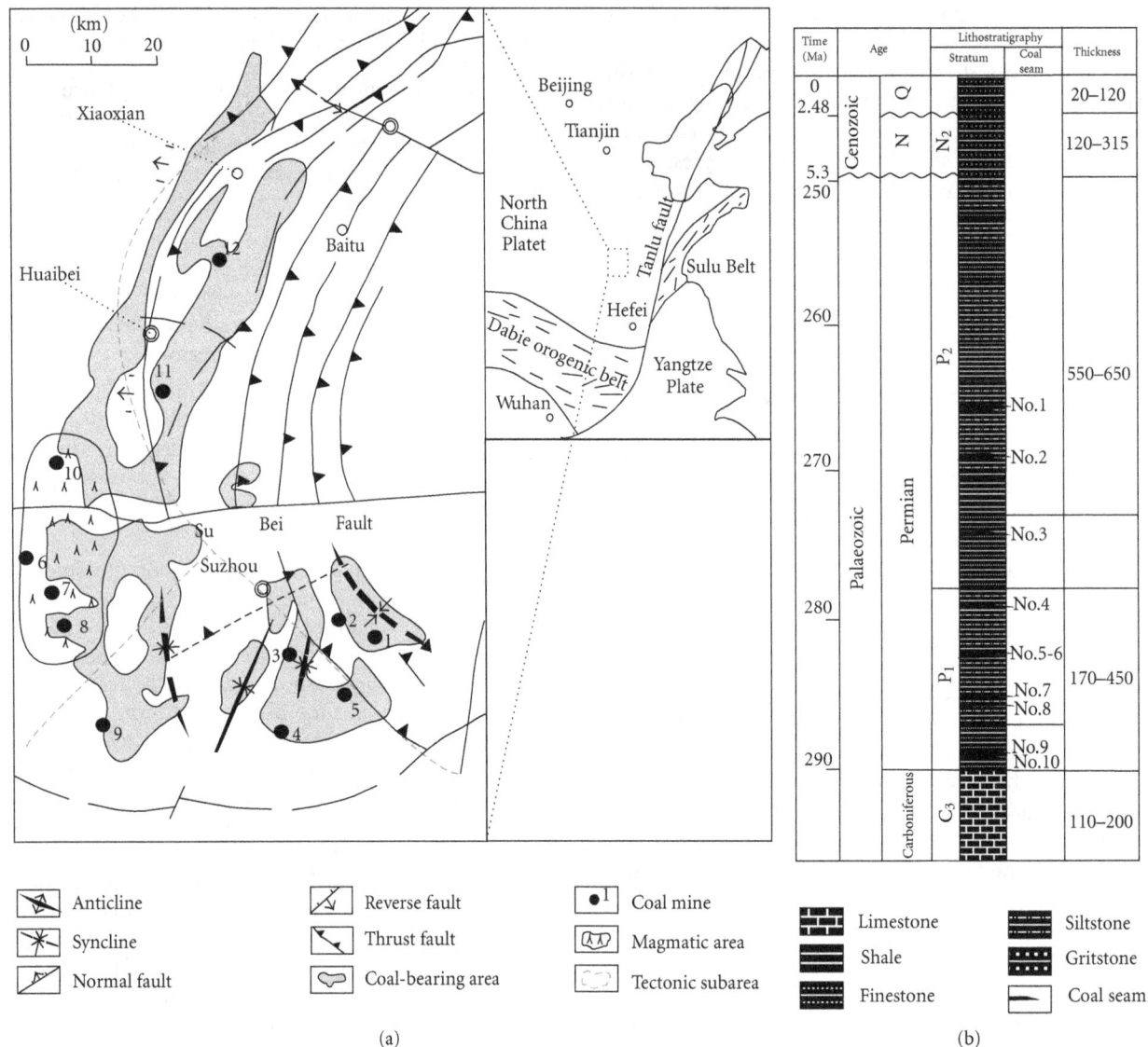

FIGURE 1: Simplified geologic map with late Paleozoic stratigraphic column of the Huaibei Coalfield. Coal mine label: (1) Luling mine, (2) Zhuxianzhuang mine, (3) Taoyuan mine, (4) Qinan mine, (5) Qidong mine, (6) Haizi mine, (7) Linhuan mine, (8) Tongting mine, (9) Xutuan mine, (10) Baishan mine, (11) Yangzhuang mine, (12) Shitai mine, The location is shown in the inset of Figure 1(a): Areas in gray where Upper Permian coal bearing strata are preserved in the Huaibei coalfield. Oriented coal samples were gathered from the Luling (LL), Taoyuan (TY), Haizi (HZ), Linhuan (LH), Xutuan (XT), Shitai (ST), and Yangzhuang (YZ) mines. The nonoriented coal samples were collected from bore holes in TY, XT, LH and Qidong (QD) coal mines. As shown in Figure 1(b), the Mesozoic and lower Cenozoic strata were suffered from erosion and missed. The Permian strata mainly consist of sandy mudstone and fine sandstone with 10 major coal seams (Nos. 1~10). Geological data are compiled based on [37, 38] the stratigraphic column is modified from China National Administration of coal Geology (1996).

perpendicular surfaces (a/a′, b/b′, and c/c′ as shown in Figure 2) under the process of polishing surface.

The 11 nonoriented samples were prepared by a modified procedure of GB/T 16773-2008 (equated with ISO 7404/02 standard). These samples were also used in our previous work for thermal simulation of maturation for coal [36].

### 3.2. Optical Microscopic Measurements.
The measurements were performed for each oriented sample using an MPV-3 Combi (Leitz) microscope, reflected light, and oil immersion objectives. The maximum and minimum apparent reflectance ($R_{max}$ and $R_{min}$) on each polished surface were measured in polarized light ($\lambda = 546\mu$) and by rotating the microscope stage through 360° and following the analytical method described in GB/T 6948-2008 (equated with ISO7404/5 standard). at least 20 points on each surface of each sample were measured to obtain the mean statistical values and azimuths of $R_{max}$ and $R_{min}$.

Table 1: Maceral composition of coal samples studied.

| Sample number | V (%) | I (%) | E (%) | MM (%) |
|---|---|---|---|---|
| TYM03 | 40.6 | 22.5 | 36.7 | 0.2 |
| TY02 | 66.4 | 18.4 | 14.4 | 0.8 |
| XTM08 | 89.2 | 9.2 | 0.8 | 0.8 |
| LHM03 | 67.4 | 31.7 | x | 0.9 |
| LHM02 | 81.9 | 17.4 | x | 0.7 |
| LHM10 | 76.8 | 22.8 | x | 0.4 |
| HZM03 | 92.5 | 7.0 | x | 0.5 |
| LLM05 | 58.2 | 24.2 | 15.2 | 2.4 |
| LLM08 | 65.0 | 17.9 | 16.3 | 0.8 |
| LLM01 | 67.8 | 18.3 | 12.4 | 1.5 |
| YZM12 | 67.3 | 32.6 | x | 0.1 |
| STM05 | 67.7 | 30.8 | x | 1.5 |

See Figure 1 for sample location. Abbreviations: $V \rightarrow$ vitrinite content (vol.%); $I \rightarrow$ inertinite content (vol.%); $E \rightarrow$ Exinite content (vol.%); $MM \rightarrow$ mineral matter content (vol.%). $x$ = content < 0.1%.

As for nonoriented samples, at least 20 points were measured with the same analytical method, and the mean statistical values of $R_{max}$ and $R_{min}$ were obtained.

### 3.3. Data Processing.
On the basis of measured mean values, the values and occurrences of maximum reflectance ($R_{max}$), intermediate reflectance ($R_{int}$), and minimum reflectance ($R_{min}$) axes may be obtained by solving ellipsoid equation in analytic geometry of space [35], which is the magnitudes and orientations of RIS. The computing method is in Figure 3 as follows:

$$\lambda'_x x^2 + \lambda'_y y^2 + \lambda'_z z^2 - 2\gamma'_{xy} xy - 2\gamma'_{yz} yz - 2\gamma'_{xz} zx = 1, \quad (1)$$

where

$$\lambda'_x = \frac{1}{\lambda_x}, \quad \lambda'_y = \frac{1}{\lambda_y}, \quad \lambda'_z = \frac{1}{\lambda_z}, \quad \gamma' = \frac{\gamma}{\lambda} = \gamma\lambda',$$

$$\lambda'_x = \frac{\cos^2\bar{\theta}_x}{\left(\bar{R}_{max}\right)^2_{xy}} + \frac{\sin^2\bar{\theta}_x}{\left(\bar{R}_{min}\right)^2_{xy}},$$

$$\lambda'_y = \frac{\sin^2\bar{\theta}_x}{\left(\bar{R}_{max}\right)^2_{xy}} + \frac{\cos^2\bar{\theta}_x}{\left(\bar{R}_{min}\right)^2_{xy}}$$

$$\gamma'_{xy} = \left[\frac{1}{\left(\bar{R}_{max}\right)^2_{xy}} - \frac{1}{\left(\bar{R}_{min}\right)^2_{xy}}\right] * \sin\bar{\theta}_x * \cos\bar{\theta}_x,$$

$$\lambda'_y = \frac{\cos^2\bar{\theta}_y}{\left(\bar{R}_{max}\right)^2_{yz}} + \frac{\sin^2\bar{\theta}_y}{\left(\bar{R}_{min}\right)^2_{yz}},$$

$$\lambda'_z = \frac{\sin^2\bar{\theta}_y}{\left(\bar{R}_{max}\right)^2_{yz}} + \frac{\cos^2\bar{\theta}_z}{\left(\bar{R}_{min}\right)^2_{yz}}$$

$$\gamma'_{yz} = \left[\frac{1}{\left(\bar{R}_{max}\right)^2_{yz}} - \frac{1}{\left(\bar{R}_{min}\right)^2_{yz}}\right] * \sin\bar{\theta}_y * \cos\bar{\theta}_y,$$

$$\lambda'_z = \frac{\cos^2\bar{\theta}_z}{\left(\bar{R}_{max}\right)^2_{zx}} + \frac{\sin^2\bar{\theta}_z}{\left(\bar{R}_{min}\right)^2_{zx}},$$

$$\lambda'_x = \frac{\sin^2\bar{\theta}_z}{\left(\bar{R}_{max}\right)^2_{zx}} + \frac{\cos^2\bar{\theta}_z}{\left(\bar{R}_{min}\right)^2_{zx}},$$

$$\gamma'_{zx} = \left[\frac{1}{\left(\bar{R}_{max}\right)^2_{zx}} - \frac{1}{\left(\bar{R}_{min}\right)^2_{zx}}\right] * \sin\bar{\theta}_z * \cos\bar{\theta}_z. \quad (2)$$

This ellipsoid equation has three eigenvalues, corresponding with the values of three RIS ellipsoid axes ($R_{max} > R_{int} > R_{min}$):

$$1 + e_i = \sqrt{\lambda_i}; \quad i = 1, 2, 3. \quad (3)$$

The occurrence of three RIS ellipsoid axis is calculated according to the formula

$$A_i = \frac{\arctan M_i}{L_i}, \qquad B_i = \arccos N_i, \quad i = 1, 2, 3, \quad (4)$$

where $A_i$ and $B_i$ represent the direction and dip angle of the axis in the coordinate system, respectively

$$L_i = \cos R_{ix} = \frac{X_i}{\sqrt{\lambda_i}}; \qquad M_i = \cos R_{iy} = \frac{Y_i}{\sqrt{\lambda_i}},$$

$$N_i = \cos R_{iz} = \frac{Z_i}{\sqrt{\lambda_i}}, \quad i = 1, 2, 3,$$

$$X_i + Y_i + Z_i = \lambda_i,$$

$$\frac{X_i}{\lambda'_y \lambda'_z - \lambda'_i \lambda'_y - \lambda'_i \lambda'_z + \lambda'^2_i - \gamma'^2_{yz}}$$

$$= \frac{-Y_i}{-\gamma'_{xy} \lambda'_z + \gamma'_{xy} \lambda'_i - \gamma'_{zx} \gamma'_{yz}}$$

$$= \frac{-Z_i}{\gamma'_{xy} \gamma'_{yz} + \gamma'_{zx} \lambda'_y - \gamma'_{zx} \lambda'_i}. \quad (5)$$

The geological occurrence of RIS ellipsoid axis ($Dir_i \angle Dip_i$), however, needs a data conversion as follows:

if $B_i \geq 0$, $Dir_i = C_i - A_i \pm 360°$, $Dip_i = B_i$,

if $B_i < 0$, $Dir_i = C_i - A_i \pm 180°$, $Dip_i = -B_i$, $\quad (6)$

where $0 < Dir_i < 360°$, $0 < Dip_i < 90°$; where $C_i$ represents the geological direction of arrowhead in Figure 2.

Therefore, the reflectance anisotropy (bireflectance $B_w$ and $R_{am}$ coefficient values) and RIS shapes (such as $R_{ev}$, $R_{st}$ coefficient value, Flinn's parameters, etc.) can be determined. The details of the parameters are given as follows.

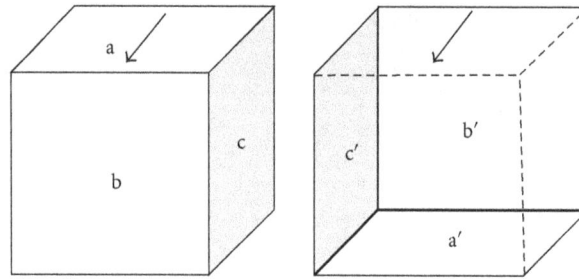

FIGURE 2: Diagram of three sections cut from oriented blocks. Arrowhead on the top surface of the block shows the spatial orientation. Surfaces marked with a/a′, b/b′, c/c′ were mutually perpendicular with the surfaces a and a′ horizontal and c and c′ vertical and parallel to the arrowhead.

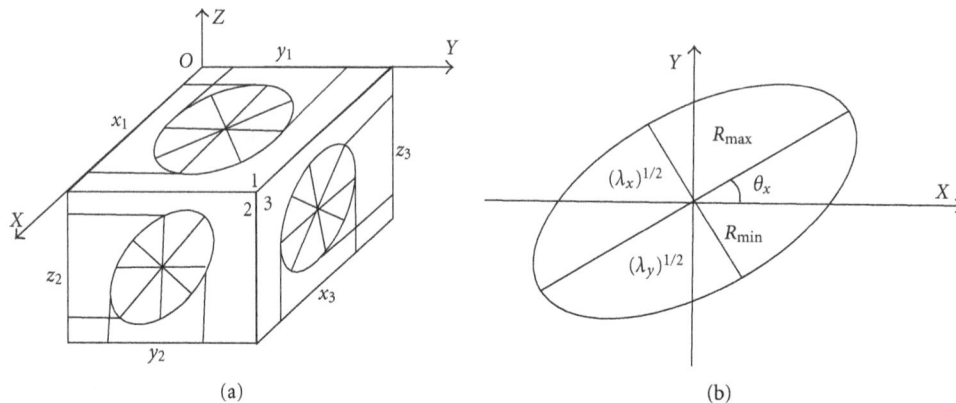

(a)                                                                  (b)

FIGURE 3: Coordinates and coal brick VR ellipses [11]. A coal sample brick was demonstrated in Figure 3(a), the measuring result were quantitatively represented in the $YXZ$ coordinate system. The value and orientation of $R_{max}/R_{min}$ on each polished surface were shown on three mutually perpendicular cube face, represented by the long/minor axis of ellipse. For example, as shown in Figure 3(b), the measured value ($R_{XOY \cdot max}$ and $R_{XOY \cdot min}$) and their orientations ($\theta_x$) are expressed as their coordinates and the RIS ellipsoid equation can be built.

(1) $B_w$ coefficient value [22], representing bireflectance value, is calculated according to the formula.

$$B_w = R_{max} - R_{min}. \tag{7}$$

$B_w/R_{max}$ [21] represents the bireflectance ratio.

(2) Flinn's parameter was firstly proposed for examining the geometry of three-dimensional homogeneous strain [35, 45]. Formula of Ramsay was chosen here

$$K = \frac{\varepsilon_1 - \varepsilon_2}{\varepsilon_2 - \varepsilon_3}, \tag{8}$$

where

$$\varepsilon_1 = \ln(R_{max}), \qquad \varepsilon_2 = \ln(R_{int}), \qquad \varepsilon_3 = \ln(R_{min}). \tag{9}$$

(3) $R_{am}$ coefficient value [23, 24], representing the RIS anisotropy magnitude, is calculated according to the formula

$$R_{am} = (x^2 - y^2)^{1/2}, \tag{10}$$

where

$$x = \frac{R_{int}}{R_{max} + R_{int} + R_{min}} + \left[y + \frac{1}{3}\cos 30°\right]\tan 30° - \frac{1}{2},$$

$$y = \left[\frac{R_{max}}{(R_{max} + R_{int} + R_{min})} - \frac{1}{3}\right]\cos 30°. \tag{11}$$

This value is the distance between the plotted position of RIS and the position of an isotropic RIS, a single measure of anisotropy for all styles of RIS [23]. Isotropic constituent of coal is characterized by $R_{am} = 0$. The higher coefficient value is, the stronger optical anisotropy will be. Typical $R_{am}$ values for raw vitrinites are from 0.03 to 0.05 and the value of 0.1 represents strong anisotropy of a constituent.

(4) $R_{st}$ coefficient value [23, 24], representing RIS style, is calculated according to the formula

$$R_{st} = 30 - \arctan\left(\frac{x}{y}\right), \tag{12}$$

where $x$ and $y$ are determined in the same way as shown above. $R_{st}$ values range from ($-30$) to ($+30$) and the RIS shape is described as

$R_{st} = (-30)$: uniaxial negative; $R_{st} = (+30)$: uniaxial positive; $-30 < R_{st} < 0$: biaxial negative; $0 < R_{st} < +30$: biaxial positive; $R_{st} = 0$: biaxial neutral.

(5) $R_{ev}$ coefficient value [23, 24], representing the reflectance of equivalent volume of isotropic RIS, is calculated according to the formula

$$R_{ev} = (R_{max} \times R_{int} \times R_{min})^{1/3}. \quad (13)$$

## 4. Results

*4.1. Anisotropic Properties of Nonoriented Samples.* $R_{max}$ and $R_{min}$ of 11 non-oriented samples were measured and the anisotropic properties of the material such as $B_w$ and $B_w/R_{max}$ were calculated as shown in Table 2.

*4.2. Magnitude and Orientation of RIS.* The magnitudes and orientations of RIS for 29 oriented samples were calculated by measuring the values and azimuths of the $R_{max}$ and $R_{min}$. Many of the coal samples suffered from tectonic deformation during coalification. The deformation types were identified by the proposed tectonic coals standard [38] and divided into three series: brittle-ductile transition, ductile shear, and brittle crush as shown in Table 3.

*4.3. Anisotropic Property of Oriented Samples.* Based on magnitudes and orientations of RIS, Levine's rank and Kilby's transformations are obtained as shown in Table 4.

## 5. Discussion

*5.1. Major Facts Affecting VRA.* Both temperature and stress have influences on modifying VRA. It is therefore necessary to estimate roughly the thermal effects during coalification before analyzing RIS.

In previous works, the mean random vitrinite reflectance ($R_m\%$) shows a strong correlation ($r^2 = 0.7$, $n > 600$) with maximum burial temperature ($T_{max}$ in $°C$). These data are modeled by the linear regression equation [46]:

$$\ln(R_m) = 0.0078 T_{max} - 1.2. \quad (14)$$

The conversion between $R_{max}$ and $R_m$ is following the analytical method described in GB/T 6948-2008 (equated with ISO7404/5 standard). On the basis of correlation given above, the maximum temperatures sustained during burial are estimated.

As shown in Table 5, the measured reflectance values range from 0.74% to 3.01, indicating conversion of medium-volatile bituminous coal to anthracite coal. Most of the studied samples experienced a maximum temperature between $120~20°C$, less than 290°C The available high temperature-pressure experiments on bituminous anthracite show that there is little change in vitrinite reflectance anisotropy under the temperature less than 400°C [47, 48]. It is reasonable to believe that the vitrinite reflectance anisotropy here was mainly caused by the tectonic differential stress.

TABLE 2: The vitrinite reflectance anisotropy of non-oriented samples, Huaibei coalfield.

| Coal mine | Sample number | Coal seam | $R_{max}$ | $R_{min}$ | $B_w$ | $B_w/R_{max}$ | $R_{max}/R_{min}$ |
|---|---|---|---|---|---|---|---|
| Linhuan | LH10 | 9 | 1.37 | 1.24 | 0.13 | 0.09 | 1.10 |
| Shitai | ST04 | 3 | 1.44 | 1.29 | 0.15 | 0.10 | 1.12 |
| Taoyuan | TY01 | 7 | 1.02 | 0.9 | 0.12 | 0.12 | 1.13 |
| Taoyuan | TY02 | 10 | 1.00 | 0.91 | 0.09 | 0.09 | 1.10 |
| Taoyuan | TY03 | 10 | 1.14 | 1.02 | 0.12 | 0.11 | 1.12 |
| Taoyuan | TY04 | 7 | 1.06 | 0.95 | 0.11 | 0.10 | 1.12 |
| Taoyuan | TY05 | 3 | 0.87 | 0.79 | 0.08 | 0.09 | 1.10 |
| Qidong | QD01 | 9 | 1.09 | 0.97 | 0.12 | 0.11 | 1.12 |
| Qidong | QD02 | 7 | 0.94 | 0.87 | 0.07 | 0.07 | 1.08 |
| Qidong | QD03 | 9 | 0.89 | 0.77 | 0.12 | 0.13 | 1.16 |
| Qidong | QD04 | 10 | 0.98 | 0.88 | 0.10 | 0.10 | 1.11 |

Abbreviations: $R_{max}$: maximum reflectance; $R_{int}$: intermediate reflectance; $R_{min}$: minimum reflectance; $B_w$: bireflectance, $R_{max} - R_{min}$.

*5.2. Magnitudes of RIS.* The three sampling regions represent different stress-thermal environments and, therefore, the corresponding coal samples experienced different effects of deformation and metamorphism. The parameters of reflectance anisotropy (bireflectance $B_w$ and $R_{am}$ coefficient values) and RIS shapes (such as $R_{ev}$ and $R_{st}$ coefficient values, Flinn's parameters) are analyzed as follows.

*5.2.1. Bireflectance.* The bireflectance increases progressively with upgrading of coal rank under the heat treatment without differential stress involved. In that case, the bireflectance for the coal ranking from medium-volatile bituminous to anthracite coal should be at the same level and the $B_w/R_{max}$ coefficient, representing bireflectance ratio, should be suitable for the bireflectance analysis among various coal ranks.

Relationships between $R_{max}$ and $B_w/R_{max}$ are shown in Figure 4. The range of $R_{max}$ value is 0.81~1.21% in Region A (Taoyuan mine, Qidong mine, and Xutuan mine) and 1.21~3.13% in Region B (Linhuan mine and Haizi mine). This obvious difference above indicates that samples from Region B was dominated by the magmatic thermal activity. However, compared with Region A (Luling mine, Shitai mine, and Yangzhuang mine), samples in Region C where experienced extra effects of tectonic deformation and shearing heat, are manifested by the various $R_{max}$ ranges of 0.83~1.72%.

In contrast to Region B, the $B_w/R_{max}$ value in Region C and Region A simultaneously increased with the $R_{max}$ and indicates that the reflectance in the two regions is dominated by the tectonic deformation and shearing heat.

The 95% confidence intervals of $B_w/R_{max}$ value are $0.13 \pm 0.0195$ in Region A and $0.23 \pm 0.0333$ in Region C. The difference between two regions represents the effects of tectonic stress. The same indictor in Region B is $0.17 \pm 0.0326$. The coal samples in the region were partly influenced by Mesozoic magmatism, and represent polarization of

TABLE 3: Magnitude and orientation of RIS, Huaibei coalfield.

| Region | Coal number | Coal seam | DT | $R_{max}$ axes | | | $R_{int}$ axes | | | $R_{min}$ axes | | |
|---|---|---|---|---|---|---|---|---|---|---|---|---|
| | | | | Value | Strike | Dip | Value | Strike | Dip | Value | Strike | Dip |
| Region A | TYM03 | 8 | BD | 1.01 | 41° | 81° | 0.95 | 177° | 21° | 0.77 | 266° | 6° |
| | TYM04 | 8 | BC | 1.00 | 323° | 69° | 0.90 | 153° | 55° | 0.81 | 37° | 44° |
| | TYM06 | 10 | BC | 0.87 | 234° | 89° | 0.85 | 28° | 49° | 0.74 | 160° | 16° |
| | TYM07 | 10 | BC | 0.79 | 164° | 4° | 0.75 | 291° | 72° | 0.72 | 82° | 89° |
| | XTM01 | 7 | BC | 1.11 | 151° | 69° | 0.99 | 239° | 36° | 0.89 | 37° | 63° |
| | XTM02 | 7 | BD | 1.11 | 106° | 87° | 1.01 | 157° | 1° | 0.97 | 248° | 77° |
| | XTM06 | 7 | DS | 1.13 | 248° | 85° | 1.07 | 107° | 40° | 0.98 | 25° | 63° |
| | XTM07 | 7 | BC | 1.11 | 242° | 64° | 1.05 | 126° | 4° | 0.98 | 60° | 45° |
| Region B | LHM02 | 7 | DS | 1.31 | 339° | 87° | 1.26 | 250° | 67° | 1.14 | 139° | 86° |
| | LHM03 | 7 | DS | 1.55 | 300° | 88° | 1.25 | 108° | 6° | 1.17 | 196° | 54° |
| | LHM04 | 9 | BD | 1.46 | 256° | 54° | 1.40 | 344° | 22° | 1.29 | 82° | 67° |
| | LHM05 | 9 | BD | 3.09 | 56° | 74° | 2.69 | 310° | 54° | 2.36 | 222° | 49° |
| | LHM09 | 9 | BD | 1.44 | 295° | 56° | 1.31 | 174° | 36° | 1.15 | 25° | 82° |
| | LHM12 | 10 | BD | 1.24 | 62° | 85° | 1.15 | 196° | 5° | 1.10 | 272° | 82° |
| | LHM14 | 10 | BD | 1.50 | 344° | 15° | 1.36 | 205° | 72° | 1.18 | 73° | 55° |
| | LHC01* | — | — | 1.37 | 337° | 41° | 1.15 | 246° | 1° | 1.13 | 156° | 41° |
| | HZM03 | 10 | BC | 1.92 | 121° | 32° | 1.70 | 47° | 89° | 1.63 | 222° | 35° |
| | HZM04 | 10 | BC | 2.01 | 98° | 27° | 1.98 | 333° | 82° | 1.82 | 186° | 89° |
| | HZM10 | 8 | DS | 2.66 | 332° | 33° | 2.54 | 233° | 16° | 2.51 | 151° | 84° |
| | HZC02* | — | — | 1.46 | 93° | 26° | 1.32 | 345° | 31° | 1.17 | 214° | 46° |
| | HZC03* | — | — | 1.45 | 3° | 36° | 1.26 | 101° | 11° | 1.13 | 205° | 51° |
| | BSC04* | — | — | 2.61 | 24° | 1° | 2.16 | 289° | 87° | 2.08 | 114° | 3° |
| | BSC05* | — | — | 2.54 | 249° | 8° | 2.21 | 343° | 26° | 1.96 | 149° | 62° |
| Region C | YZM11 | 3 | BC | 1.70 | 30° | 77° | 1.59 | 160° | 89° | 1.34 | 215° | 63° |
| | YZM14 | 3 | BC | 1.69 | 132° | 88° | 1.61 | 248° | 29° | 1.38 | 335° | 88° |
| | YZM16 | 3 | BC | 1.82 | 164° | 24° | 1.60 | 3° | 65° | 1.30 | 254° | 74° |
| | STM01 | 3 | BD | 1.52 | 128° | 84° | 1.47 | 4° | 11° | 1.21 | 275° | 75° |
| | STM02 | 3 | BC | 1.41 | 215° | 33° | 1.36 | 17° | 83° | 1.10 | 97° | 42° |
| | STM03 | 3 | BD | 1.72 | 94° | 2° | 1.25 | 209° | 64° | 1.18 | 16° | 2° |
| | STC06* | — | — | 1.52 | 186° | 22° | 1.37 | 277° | 3° | 1.19 | 15° | 67° |
| | LLM03 | 8 | BD | 1.06 | 47° | 50° | 0.99 | 261° | 38° | 0.88 | 171° | 77° |
| | LLM05 | 8 | DS | 0.84 | 322° | 74° | 0.76 | 247° | 17° | 0.59 | 118° | 63° |
| | LLM07 | 10 | BC | 0.95 | 156° | 45° | 0.86 | 345° | 47° | 0.74 | 256° | 57° |
| | LLM08 | 10 | BC | 0.83 | 146° | 18° | 0.80 | 357° | 79° | 0.75 | 235° | 83° |
| | LLM10 | 8 | DS | 1.11 | 307° | 67° | 0.92 | 35° | 42° | 0.83 | 168° | 68° |

*Data complied from [18].

Abbreviations: DT: deformation type; BD: brittle-ductile transition; DS: ductile shear; BC: brittle crush; $R_{max}$: maximum reflectance; $R_{int}$: intermediate reflectance; $R_{min}$: minimum reflectance.

$R_o$ value. Meanwhile, some strata underwent stronger deformation with increasing $B_w/R_{max}$ values.

5.2.2. Reflectance of Equivalent Volume Isotropic RIS. RIS parameter $R_{ev}$, the reflectance of the isotropic RIS of equivalent volume, as a result of the chemical structure ordering during heating, is suggested to characterize the basic structure unites (BSU) [49].

As shown in Figure 5, positive linear regression relationships between $R_{max}$ and $R_{ev}(r^2)$ for Regions A, B, and C are 0.956, 0.980, and 0.965, respectively. The correlations between $R_{ev}$ values and the hydrogen contents ($H^{daf}$) have

TABLE 4: Calculated results of RIS parameters, Huaibei coalfield.

| Region | Coal number | Coal seam | DT | $R_{st}$ | $R_{am}$ | $R_{ev}$ | $B_w$ | $B_w/R_{max}$ | Axis ratio $R_{max}/R_{int}$ | $R_{int}/R_{min}$ | $R_{max}/R_{min}$ | K |
|---|---|---|---|---|---|---|---|---|---|---|---|---|
| Region A | TYM03 | 8 | BD | −14.43 | 0.05 | 0.90 | 0.24 | 0.24 | 1.07 | 1.23 | 1.31 | 0.33 |
| | TYM04 | 8 | BC | 1.06 | 0.04 | 0.90 | 0.19 | 0.19 | 1.11 | 1.11 | 1.23 | 0.96 |
| | TYM06 | 10 | BC | −20.43 | 0.03 | 0.82 | 0.13 | 0.15 | 1.03 | 1.15 | 1.18 | 0.20 |
| | TYM07 | 10 | BC | −0.05 | 0.02 | 0.75 | 0.07 | 0.09 | 1.05 | 1.05 | 1.10 | 0.95 |
| | XTM01 | 7 | BC | 2.52 | 0.04 | 0.99 | 0.23 | 0.20 | 1.12 | 1.12 | 1.26 | 1.04 |
| | XTM02 | 7 | BD | 14.66 | 0.03 | 1.03 | 0.14 | 0.12 | 1.10 | 1.04 | 1.14 | 2.49 |
| | XTM06 | 7 | DS | −6.05 | 0.03 | 1.06 | 0.16 | 0.14 | 1.06 | 1.10 | 1.16 | 0.64 |
| | XTM07 | 7 | BC | −4.73 | 0.02 | 1.05 | 0.12 | 0.11 | 1.05 | 1.07 | 1.13 | 0.71 |
| Region B | LHM02 | 7 | DS | −12.79 | 0.03 | 1.23 | 0.18 | 0.13 | 1.04 | 1.11 | 1.16 | 0.41 |
| | LHM03 | 7 | DS | 17.42 | 0.06 | 1.31 | 0.38 | 0.24 | 1.23 | 1.07 | 1.32 | 2.93 |
| | LHM04 | 9 | BD | −9.73 | 0.02 | 1.38 | 0.17 | 0.12 | 1.04 | 1.08 | 1.13 | 0.51 |
| | LHM05 | 9 | BD | 2.89 | 0.05 | 2.70 | 0.72 | 0.23 | 1.15 | 1.14 | 1.31 | 1.04 |
| | LHM09 | 9 | BD | −3.89 | 0.04 | 1.29 | 0.30 | 0.21 | 1.10 | 1.15 | 1.26 | 0.70 |
| | LHM12 | 10 | BD | 12.26 | 0.02 | 1.16 | 0.14 | 0.11 | 1.08 | 1.04 | 1.13 | 2.08 |
| | LHM14 | 10 | BD | −4.52 | 0.03 | 1.34 | 0.33 | 0.22 | 1.10 | 1.16 | 1.28 | 0.67 |
| | LHC01* | — | — | 25.69 | 0.04 | 1.21 | 0.24 | 0.18 | 1.19 | 1.02 | 1.21 | 9.98 |
| | HZM03 | 10 | BC | 15.63 | 0.03 | 1.75 | 0.30 | 0.15 | 1.13 | 1.05 | 1.18 | 2.65 |
| | HZM04 | 10 | BC | −21.15 | 0.02 | 1.94 | 0.19 | 0.09 | 1.02 | 1.09 | 1.10 | 0.19 |
| | HZM10 | 8 | DS | 18.44 | 0.01 | 2.57 | 0.15 | 0.06 | 1.05 | 1.01 | 1.06 | 3.63 |
| | HZC02* | — | — | −1.14 | 0.04 | 1.31 | 0.29 | 0.20 | 1.11 | 1.13 | 1.25 | 0.84 |
| | HZC03* | — | — | 6.18 | 0.05 | 1.27 | 0.32 | 0.22 | 1.15 | 1.12 | 1.28 | 1.29 |
| | BSC04* | — | — | 21.95 | 0.05 | 2.27 | 0.53 | 0.20 | 1.21 | 1.04 | 1.25 | 5.01 |
| | BSC05* | — | — | 4.55 | 0.05 | 2.22 | 0.58 | 0.23 | 1.15 | 1.13 | 1.30 | 1.16 |
| Region C | LLM03 | 8 | BC | −7.01 | 0.04 | 0.98 | 0.18 | 0.17 | 1.07 | 1.12 | 1.21 | 0.59 |
| | LLM05 | 8 | BC | −11.51 | 0.07 | 0.72 | 0.24 | 0.29 | 1.10 | 1.28 | 1.41 | 0.40 |
| | LLM07 | 10 | BC | −6.37 | 0.05 | 0.85 | 0.21 | 0.22 | 1.10 | 1.17 | 1.28 | 0.60 |
| | LLM08 | 10 | BD | −8.05 | 0.02 | 0.79 | 0.08 | 0.10 | 1.04 | 1.07 | 1.11 | 0.58 |
| | LLM10 | 8 | BC | 11.71 | 0.06 | 0.94 | 0.27 | 0.25 | 1.20 | 1.11 | 1.33 | 1.84 |
| | YZM11 | 3 | BC | −12.87 | 0.05 | 1.54 | 0.37 | 0.21 | 1.07 | 1.19 | 1.27 | 0.38 |
| | YZM14 | 3 | BC | −15.35 | 0.04 | 1.55 | 0.32 | 0.19 | 1.05 | 1.17 | 1.23 | 0.32 |
| | YZM16 | 3 | BC | −4.68 | 0.06 | 1.56 | 0.53 | 0.29 | 1.14 | 1.23 | 1.41 | 0.63 |
| | STM01 | 3 | BD | −21.59 | 0.05 | 1.39 | 0.32 | 0.21 | 1.03 | 1.22 | 1.26 | 0.17 |
| | STM02 | 3 | BC | −20.95 | 0.05 | 1.28 | 0.31 | 0.22 | 1.04 | 1.24 | 1.29 | 0.18 |
| | STM03 | 3 | BD | 23.20 | 0.08 | 1.37 | 0.54 | 0.31 | 1.38 | 1.06 | 1.46 | 5.57 |
| | STC06* | — | — | −3.00 | 0.05 | 1.35 | 0.33 | 0.22 | 1.11 | 1.15 | 1.28 | 0.74 |

*Data complied from [18] except for the coal seam and deformation type of these four samples. Abbreviations: BD: brittle-ductile transition; DS: ductile shear; BC: brittle crush; DT: deformation type; $R_{max}$: maximum reflectance; $R_{int}$: intermediate reflectance; $R_{min}$: minimum reflectance; $B_w$: bireflectance, $R_{max} − R_{min}$; K: Flinn's parameter, $(\varepsilon_1 \text{-} \varepsilon_2)/(\varepsilon_2 \text{-} \varepsilon_3)$, where $\varepsilon_1 = \ln(R_{max})$, $\varepsilon_2 = \ln(R_{int})$, $\varepsilon_3 = \ln(R_{min})$; $R_{am}$: RIS anisotropy magnitude, $(x^2 − y^2)^{1/2}$; $R_{st} \rightarrow R_{IS}$ style, 30-arctan $(x/y)$; $R_{ev}$: reflectance of equivalent volume isotropic RIS, $(R_{max} \times R_{int} \times R_{min})^{1/3}$; where $x = R_{int}/(R_{max} + R_{int} + R_{min}) + [y + 1/3 \cos 30°] \times \tan 30° − 1/2$, $y = [R_{max}/(R_{max} + R_{int} + R_{min}) − 1/3] \times \cos 30°$.

TABLE 5: Maximum temperatures sustained during burial, Upper Permian strata, Huaibei coalfield.

| Region A | | | Region B | | | Region C | | |
|---|---|---|---|---|---|---|---|---|
| Number | $R_{max}$ | $T_{max}$ (°C) | Noumber | $R_{max}$ | $T_{max}$ (°C) | Number | $R_{max}$ | $T_{max}$ (°C) |
| TYM03 | 1.01 | 147 | LHM02 | 1.31 | 180 | STM01 | 1.52 | 200 |
| TYM04 | 0.94 | 138 | LHM03 | 1.52 | 200 | STM02 | 1.41 | 190 |
| TYM06 | 0.88 | 129 | LHM04 | 1.41 | 190 | STM03 | 1.72 | 215 |
| TYM07 | 0.81 | 119 | LHM05 | 3.13 | 283 | STC06* | 1.52 | 200 |
| TY01 | 1.02 | 148 | LHM09 | 1.34 | 183 | YZM11 | 1.70 | 214 |
| TY02 | 1.00 | 146 | LHM12 | 1.21 | 170 | YZM14 | 1.69 | 213 |
| TY03 | 1.14 | 163 | LHM14 | 1.37 | 186 | YZM16 | 1.82 | 223 |
| TY04 | 1.06 | 153 | LHC01* | 1.37 | 186 | ST01 | 1.44 | 193 |
| TY05 | 0.87 | 128 | LH01 | 1.37 | 186 | LLM03 | 1.11 | 159 |
| QD01 | 1.09 | 157 | HZM03 | 1.82 | 223 | LLM05 | 0.83 | 122 |
| QD02 | 0.94 | 138 | HZM04 | 1.99 | 234 | LLM07 | 0.96 | 141 |
| QD03 | 0.89 | 131 | HZM10 | 2.58 | 261 | LLM08 | 0.88 | 129 |
| QD04 | 0.98 | 143 | HZC02* | 1.46 | 194 | LLM10 | 1.04 | 151 |
| XTM01 | 1.13 | 162 | HZC03* | 1.45 | 193 | | | |
| XTM02 | 1.12 | 160 | BSC04* | 2.61 | 263 | | | |
| XTM06 | 1.13 | 162 | BSC05* | 2.54 | 260 | | | |
| XTM07 | 1.21 | 170 | | | | | | |

*Data complied from [18]. During the calculation, $R_{max}$ is converted into random vitrinite reflectance at first, and then the maximum buried temperature are obtained by the formula given above. Abbreviations: $R_{max}$: maximum reflectance; $T_{max}$: maximum buried temperature.

been verified that the $R_{ev}$ depends on the chemical composition of anthracites rather than the three-dimensional BSU [30]. Since the chemical composition and structure mainly evolved under the effect of temperature and coincided with the results, the linear- regression slope of Region B seems to be mainly influenced by the thermal effect.

*5.2.3. RIS Style and Anisotropy Magnitude.* The coefficient values of $R_{st}$ and $R_{am}$ construct polar scatter plots proposed by [23], and the three-dimensional style and anisotropy magnitude of RIS can be analyzed together. As shown in Figure 6(a), samples from Region A and C are characterized by substantial RIS style of biaxial negative ($R_{st}$ ranges between −30~15). The anisotropy magnitude in Region C ($R_{am}$ ranges between 0.04~0.08), however, is higher than that in Region A ($R_{am}$ ranges between 0.02~0.04), which coincides with the distribution of bireflectance ratio. Conversely, as shown in Figure 6(b), the samples from Region B are characterized by significant biaxial positive character or unconspicuous biaxial negative ($R_{st}$ ranges between −15~30) and the anisotropy magnitude ($R_{am}$) is various, ranging from 0.02 to 0.06.

*5.2.4. RIS-Logarithmic Flinn Diagrams.* For a more complete characterization of the optical properties of both raw and carbonized anthracites, the modified Flinn's $K$ parameters [35], related to the anisotropy as well as the optical character, are also calculated and shown in. The samples from Region C and A are characterized by evolvement of RIS style from constriction types to flattening types ($0 < k < 1$ shown in

Figure 7(a)). However, RIS styles of the samples from Region B are complicated and various (as shown in Figure 7(b)), corresponding to plane strain in finite strain analysis.

*5.2.5. RIS Anisotropy Evolution Stages .* The deformation path of finite strain analyses proposed by Ramsay suggested that the cleavage is developed by the route of "sphere types → uniaxial oblate types → uniaxial prolate types → uniaxial oblate types → flattening types". The anisotropic coal samples may have the same evolution process. A related evolutionary path of RIS style during coalification was reported by Levine [15] and the stages are given as follows.

The first stage was after the deposition. The peat was subjected to mild geothermal process due to broad regional subsidence and burial of overlying rocks. During this period, the ambient geologic stresses were nontectonic and only due to vertical static pressure loading by the overlying strata. Owing to the vertical downward lithostatic pressure, RIS would show the style of uniaxial negative. At the second stage, the anisotropic strain developed by tectonic differential stress (in a lateral direction) and lithostatic pressure (in the vertical direction) and the coal seams would suffer from extrusion in two directions and the anisotropic RIS represents the style of biaxial positive. At the third stage, with the increase of buried depth (more than 1000 m), the pressure on the coal seams from the overlying strata may approximately represent as isotropic hydrostatic pressure (Heim's hypothesis). The tectonic differential stress is upgraded with the enhancement of tectonic agency. All of these factors would make the anisotropic RIS style as biaxial negative.

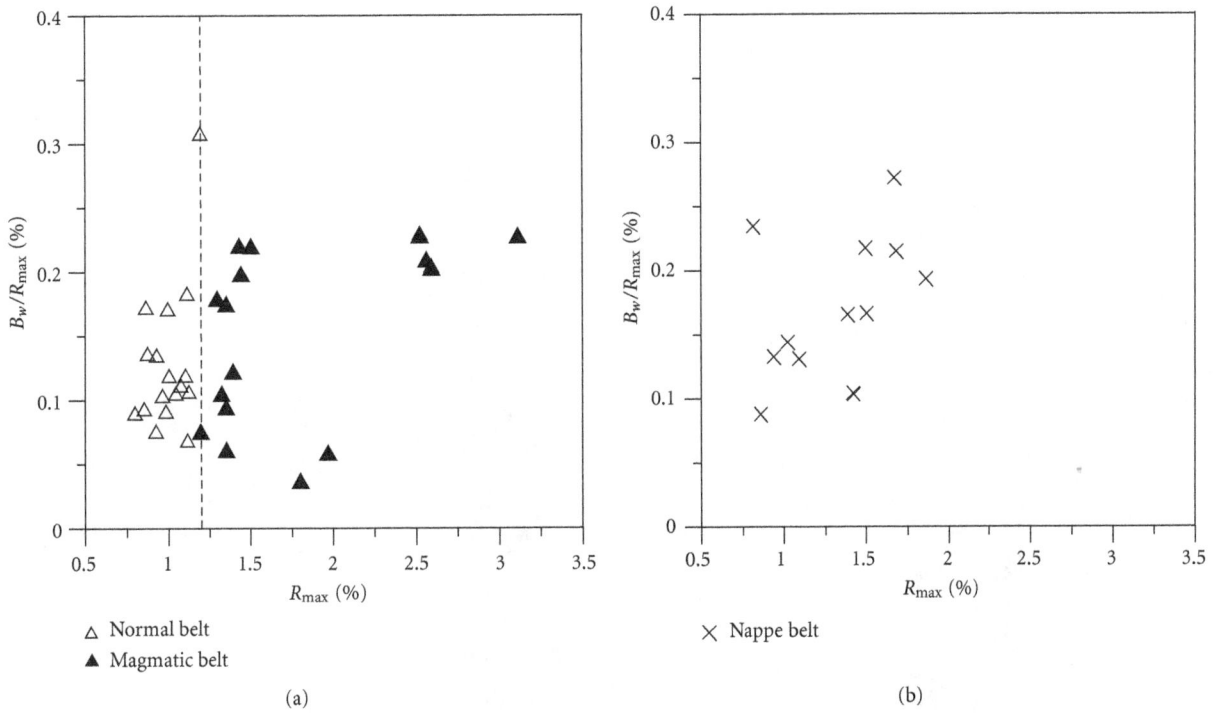

FIGURE 4: Correlations between $R_{max}$ and $B_w/R_{max}$ for the anthracites. $R_{max}$: maximum reflectance; $B_w/R_{max}$: bireflectance ratio.

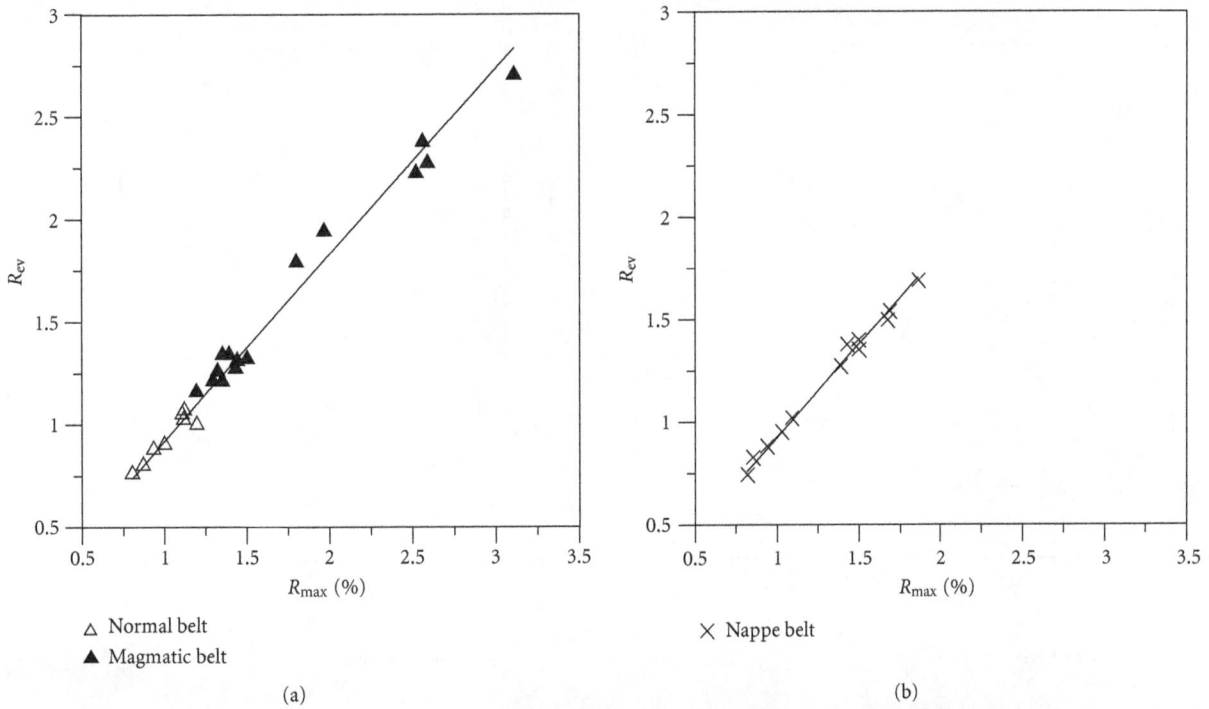

FIGURE 5: Correlation between $R_{max}$ and $R_{ev}$ for the anthracites. Region A: $y = (0.866 * x) + 0.057$ (coef. of determination, $r^2 = 0.956$); Region B: $y = 0.888x + 0.031$ (coef. of determination, $r^2 = 0.980$); Region C: $y = (0.838 * x) + 0.063$ (coef. of determination, $r^2 = 0.965$); $R_{max}$: maximum reflectance; $R_{ev}$: Reflectance of equivalent volume isotropic RIS, $(R_{max} \times R_{int} \times R_{min})^{1/3}$.

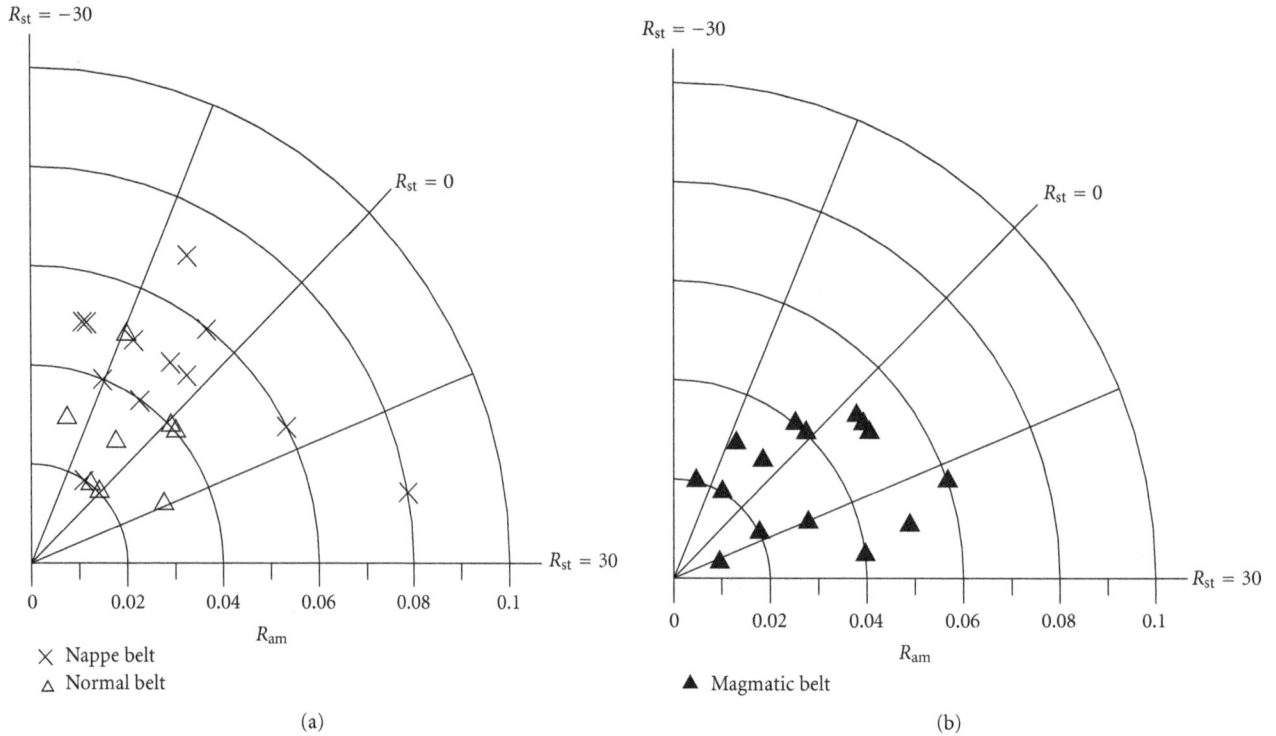

FIGURE 6: $R_{st}$-$R_{am}$ polar plots of anthracites. $R_{am}$-RIS anisotropy magnitude, $(x^2 - y^2)^{1/2}$; $R_{st}$-RIS style, 30-arc $\tan(x/y)$ where $x = R_{int}/(R_{max} + R_{int} + R_{min}) + [y + 1/3 \cos 30°] \times \tan 30° - 1/2, y = [R_{max}/(R_{max} + R_{int} + R_{min}) - 1/3] \times \cos 30°$; RIS shape described as $-30 < R_{st} < 0$-biaxial negative; $0 < R_{st} < +30$-biaxial positive; $R_{st} = (-30)$ uniaxial negative; $R_{st} = (+30)$ uniaxial positive, $R_{st} = 0$-biaxial neutral.

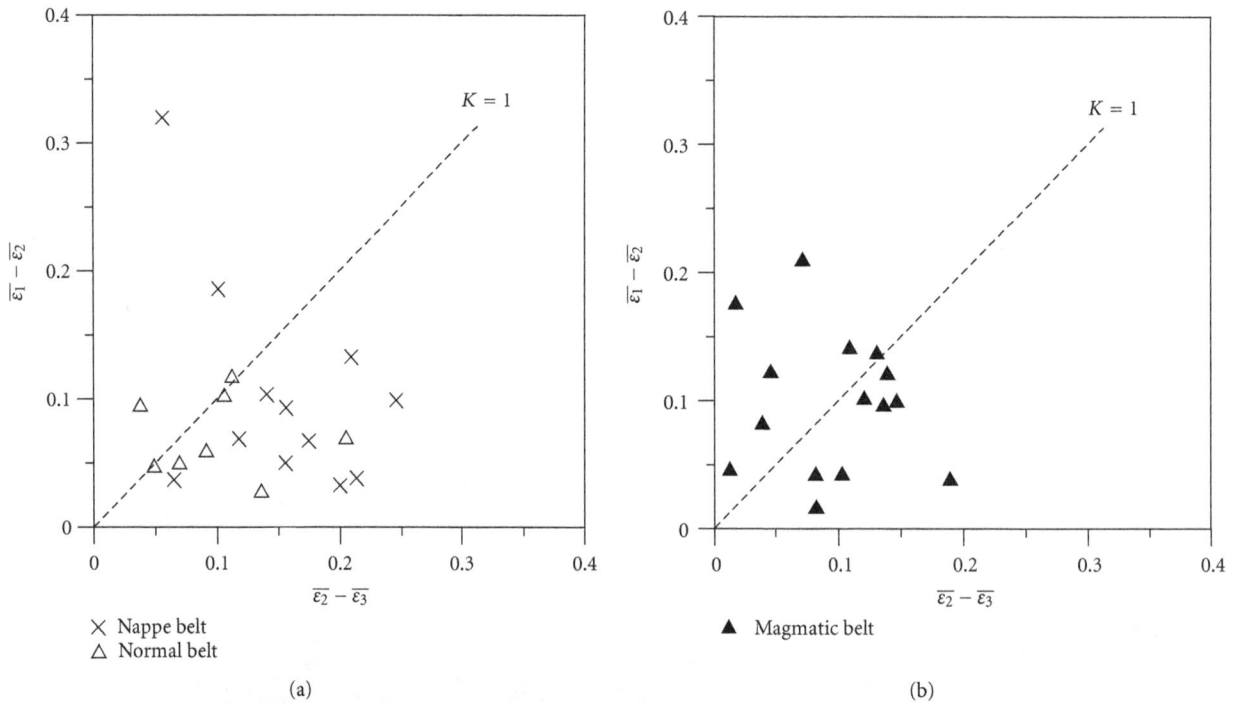

FIGURE 7: RIS-logarithmic Flinn diagrams. The arrowhead directs the evolutionary path of RIS style; $K$-Flinn'S parameter [35], $(\varepsilon_1 - \varepsilon_2)/(\varepsilon_2 - \varepsilon_3)$, where $\varepsilon_1 \ln(R_{max})$, $\varepsilon_2 = \ln(R_{int})$, $\varepsilon_3 = \ln(R_{min})$; $K = \infty$-uniaxial prolate types; $K = 0$-uniaxial oblate types; $K = 1$-plane strain types; $1 < K < \infty$-constriction types; $0 < K < 1$-flattening types.

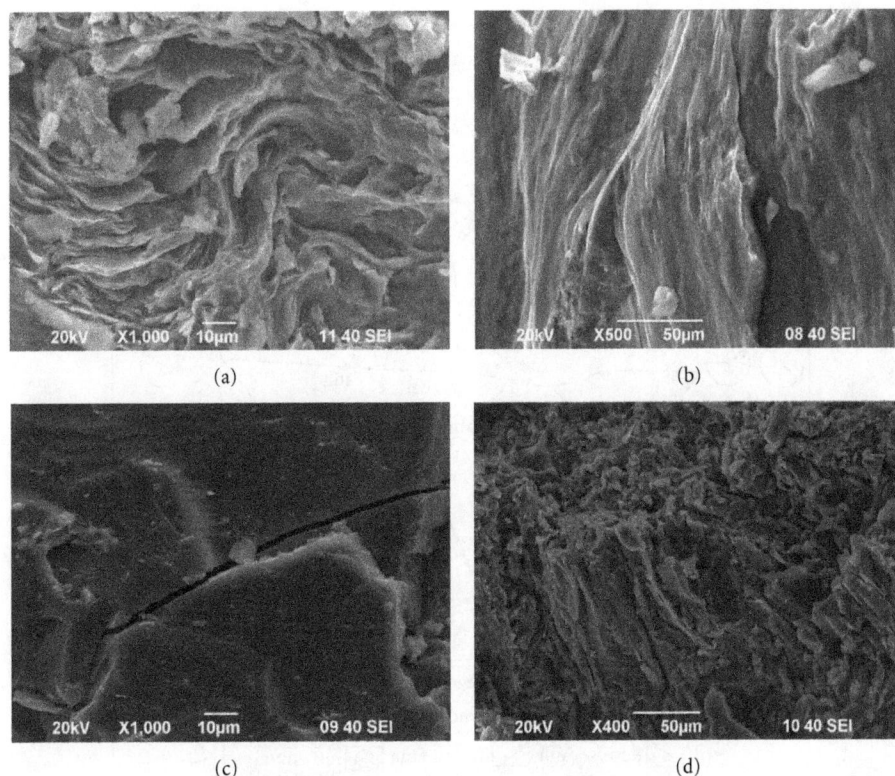

FIGURE 8: SEM photographs showing different types of deformation. (a) YZM16, strongly folded deformation; (b) LHM05, ductile shear deformation; (c) HZM03, slight deformation, brittle fractures; (d) HZM10, strong deformation, locally folded.

As the suggested route given above, the style of uniaxial negative might take place either in the third stage or in the transition between the first and the second stages. More evidence is needed to constrain the actual process of physical deformation and metamorphism. During coalification, the coal samples collected from three regions likely belong to different evolution stages.

(1) Samples of Region C experienced more effects of tectonic deformation and shearing heat. With significant anisotropy magnitude (higher $B_w/R_{max}$ and $R_{am}$ value) and biaxial negative style, they have more likely reached into the third stage.

(2) Samples of Region A experienced more effect of burial metamorphism and less influenced by magmatism and tectonic agency. With unimpressive anisotropy magnitude (lower $B_w/R_{max}$ and $R_{am}$ value) and biaxial negative style, they more likely belong to the transition between the first to the second stage.

(3) Samples of Region B influenced by the Mesozoic magmatism, representing polarized anisotropy magnitude (diversified value of $R_o$, $B_w/R_{max}$ and $R_{am}$). With complicated and various RIS style, they probably belong to the transition between the second and the third stage. Scanning electron microscopic (SEM) photos of coals in the study area indicated that some coal seams have strong deformation (Figure 8). The local stress field was presumably influenced by the

complicated volcanic activity such as granite pegmatite and diorite porphyrite during the Yanshanian orogeny [50].

5.3. Implications of RIS-Orientations. As mentioned in Section 1, previous works suggest that minimum reflectance develops incrementally parallel to $\sigma_1$ and maximum reflectance develops incrementally parallel to $\sigma_3$ during coalification [1, 14]. These viewpoints were verified by simple shear experiments under high temperature-pressure environment [28, 33, 34]. On this basis, the vitrinite reflectance may provide important information about tectonic stress in each period during the coalification, and the $\sigma_1$-direction could be indicated by the minimum reflectance axis.

Eastern North China Craton experienced an important tectonic inversion during Mesozoic. The EW-trending tectonic grain was transformed to NE-NNE-trending and the contractional regime to an extensional regime during Jurassic-Middle Cretaceous [51]. The Huaibei Coalfield lies in the southeastern margin of North China Craton, and influenced by multistage and complicated tectonic events, which could be represented by the triaxial rotation of RIS.

The orientations of RIS axes are respectively projected on the horizontal plane, (Figure 9) and vertical section (Figure 10), and analyzed as follows:

5.3.1. Horizontal Projection. The horizontal projections are based on the stereographic polar method. The radius of base

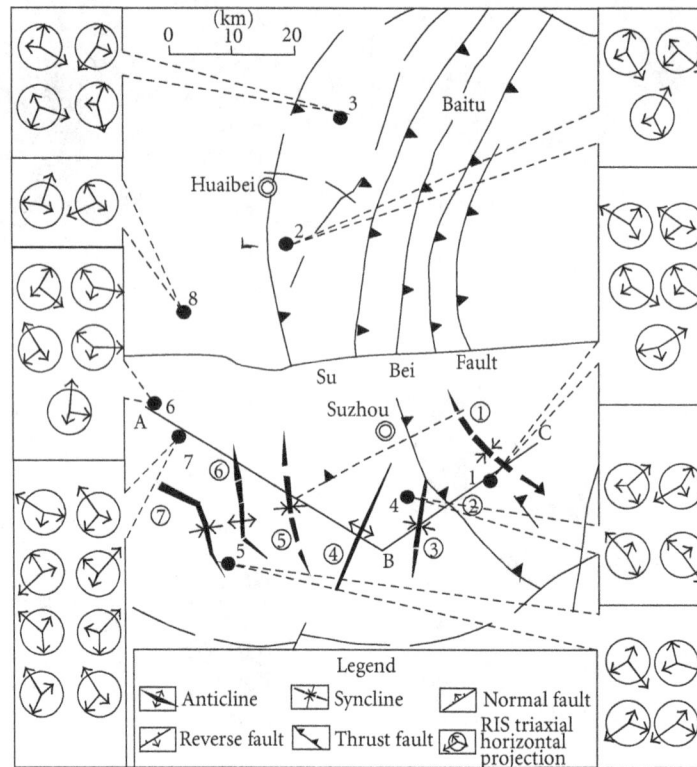

FIGURE 9: RIS-triaxial horizontal Projection. Coal mine label: 1-Luling mine, 2-Yangzhuang mine, 3-Shitai mine, 4-Taoyuan mine, 5-Xutuan mine, 6-Haizi mine, 7-Linhuan mine, 8-Baishan mine. Structure label: (1) Luling syncline; (2) Xisipo fault; (3) Sunan syncline; (4) Sunan anticline; (5) Nanping syncline; (6) Tongting anticline; (7) Wugou syncline.The radius of base circle represents the value of Rint; long (middle and short) arrows represent $R_{max}$($R_{int}$ and $R_{min}$).

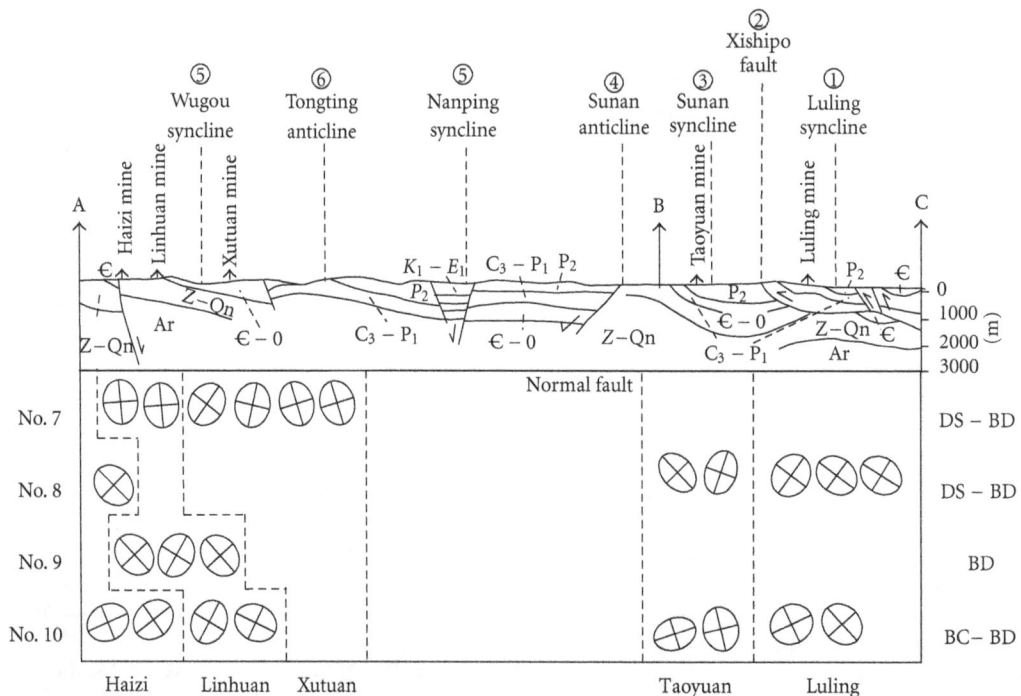

FIGURE 10: RIS-vertical projection. Generally speaking, stronger tectonic agency played the primary role and brought about coal seams to be RIS-orientation basically alike.

circle represents the value of Rint. Based on the feature that minimum reflectance develops incrementally parallel to $\sigma_1$ [1, 14], the cumulative effect of multistage tectonism on each sample is indicated by the orientation of $R_{min}$ axis.

The samples collected from different working faces have different RIS orientation, even though they are in the same coal mine. The samples of Region C are fan-shaped distributing on the thrust arc, and their Rmin axial directions are mainly pointing to the center of the circle such as, from north to south in order, NW-SW or NNW-SSE direction in the Luling mine), NEE-SWW direction in Yangzhuang mine, SSE-NNW or NNE-SSW direction in the Shitai mine. In other words, the $R_{min}$ axial direction was parallel to the thrust- direction.

The samples of Region A and B are distributing in the front of the thrust arc and a series of NNW-NE-trending folds. Their $R_{min}$ axial directions are mainly perpendicular to the corresponding fold-hinge as shown in Figure 9. These features suggest that the RIS-orientation in these areas was primarily controlled by the thrust-fold system.

*5.3.2. Vertical Projections.* The cross-section (A-B-C) is about perpendicular to the hinges of the major folds and its position is shown in Figure 9. The vertical projections on the cross-section are based on RIS-analyses of the samples from five coal mines. Since most of the $R_{min}$ axial directions are located in the cross-section, the projection provides information on RIS-dip angles.

Samples of the Luling mine belong to Region C, and their RIS orientations were following the direction of thrust. Number 8 coal seam has similar RIS dips as Number 10. However, samples of the Taoyuan mine, Xutuan mine, Linhuan mine and Haizi mine seem to be more influenced by the regional tectonic stress, the distinctions between samples of different coal seam are obvious.

Combining with the deformation type of coal samples (as listed in Table 3), No. 7, No. 8, and No. 9 coal seams are mainly characterized as ductile-type or brittle-ductile-type, with lower $R_{min}$ axial dip angles. In contrast, N0. 10 coal seam is mainly characterized as brittle-type or brittle-ductile type, and their RIS $R_{min}$ axial dip angles are higher. The contrastive fact indicates that the coal samples in different deformation types have remarkably different RIS orientation, although they belong to the same structural unit.

## 6. Conclusions

Our results indicate that:

(1) there are obvious relationships between the anisotropic parameters and tectonic stress action. Based on the tectonic situation, the Huaibei Coalfield is divided into three regions. Samples of each region have special RIS-shape and magnitude. Samples of Region C have significant anisotropy magnitude (higher $B_w/R_{max}$ and $R_{am}$ value) and biaxial negative style. By contrast, samples of the Region A have unimpressive anisotropy magnitude and biaxial negative style. Samples of Region B represent polarized anisotropy magnitude (diversified value of $R_o$, $B_w/R_{max}$ and $R_{am}$) with complicated and various RIS styles,

(2) orientation of RIS is mainly controlled by regional tectonic stress: either parallel to the direction of thrusting or perpendicular to the hinge of folds. It is likely influenced by deformation mechanisms of coal.

## Acknowledgments

These research results are part of a key project carried out in 2006–2011 and financially supported by The national Natural Science Foundation of China (Grant no. 41030422; 40940014; 40772135) and National Basic Research Program of China (Grant no. 2009CB219601). The authors are grateful to Dai Jiming and Ai Tianjie at China University of Mining and Technology (Beijing) for their assistance in the experiments. Many thanks to Zheng Yadong, Wang Guiliang, Jin Weijun and Liu Qinfu, Liu Dameng, and three reviewers for their help with improving the paper.

## References

[1] I. Stone and A. Cook, "The influence of some tectonic structures upon vitrinite reflectance," *The Journal of Geology*, vol. 87, pp. 497–508, 1979.

[2] E. Stach, M.-T. Mackowsky, M. Techmuller, G. H. Taylor, D. Chandra, and R. Techmuller, *Stach's Textbook of Coal Petrology*, 3rd edition, 1982.

[3] J. V. Ross, R. M. Bustin, and J. N. Rouzaud, "Graphitization of high rank coals—the role of shear strain: experimental considerations," *Organic Geochemistry*, vol. 17, no. 5, pp. 585–596, 1991.

[4] R. M. Bustin, J. V. Ross, and J. N. Rouzaud, "Mechanisms of graphite formation from kerogen: experimental evidence," *International Journal of Coal Geology*, vol. 28, no. 1, pp. 1–36, 1995.

[5] J. Komorek and R. Morga, "Relationship between the maximum and the random reflectance of vitrinite for coal from the Upper Silesian Coal Basin (Poland)," *Fuel*, vol. 81, no. 7, pp. 969–971, 2002.

[6] S. Pusz, S. Duber, and B. K. Kwiecińska, "The study of textural and structural transformations of carbonized anthracites," *Fuel Processing Technology*, vol. 77-78, pp. 173–180, 2002.

[7] J. Komorek and R. Morga, "Vitrinite reflectance property change during heating under inert conditions," *International Journal of Coal Geology*, vol. 54, no. 1-2, pp. 125–136, 2003.

[8] S. Pusz, B. K. Kwiecińska, and S. Duber, "Textural transformation of thermally treated anthracites," *International Journal of Coal Geology*, vol. 54, no. 1-2, pp. 115–123, 2003.

[9] M. Krzesińska, S. Pusz, and A. Koszorek, "Elastic and optical anisotropy of the single-coal monolithic high-temperature (HT) carbonization products obtained on a laboratory scale," *Energy and Fuels*, vol. 19, no. 5, pp. 1962–1970, 2005.

[10] D. Y. Cao, X. M. Li, and S. R. Zhang, "Influence of tectonic stress on coalification: stress degradation mechanism and stress polycondensation mechanism," *Science in China, Series D*, vol. 50, no. 1, pp. 43–54, 2007.

[11] J. P. Jiang, G. Y. Gao, and J. W. Kang, "Tests on vitrinite reflectance of coal and analysis of tectonic stress field," *Chinese*

*Journal of Geophysics (Acta Geophysica Sinica)*, vol. 50, no. 1, pp. 138–145, 2007.

[12] J. Komorek and R. Morga, "Evolution of optical properties of vitrinite, sporinite and semifusinite in response to heating under inert conditions," *International Journal of Coal Geology*, vol. 71, no. 4, pp. 389–404, 2007.

[13] I. Suárez-Ruiz and A. B. García, "Optical parameters as a tool to study the microstructural evolution of carbonized anthracites during high-temperature treatment," *Energy and Fuels*, vol. 21, no. 5, pp. 2935–2941, 2007.

[14] J. R. Levine and A. Davis, "Optical anisotropy of coals as an indicator of tectonic deformation, Broad Top Coal Field, Pennsylvania ( USA)," *Geological Society of America Bulletin*, vol. 95, no. 1, pp. 100–108, 1984.

[15] J. R. Levine and A. Davis, "Reflectance anisotropy of Upper Carboniferous coals in the Appalachian foreland basin, Pennsylvania, U.S.A," *International Journal of Coal Geology*, vol. 13, no. 1-4, pp. 341–373, 1989.

[16] B. Jiang and Y. Qin, "Experimental research on deformation of optical fabric of coal vitrinite reflectance experimental research on deformation of optical fabric of coal vitrinite reflectance," *Coal Geology & Exploration*, vol. 25, pp. 11–15, 1997 (Chinese).

[17] J. C. Hower and A. Davis, "Vitrinite reflectance anisotropy as a tectonic fabric element," *Geology*, vol. 9, pp. 165–168, 1981.

[18] D. Y. Cao, "The vitrinite reflectance anisotropy in the nappe structure in the Huaibei coalfield, Anhni province," *Geology Review*, vol. 36, pp. 333–340, 1990 (Chinese).

[19] W. Langenberg and W. Kalkreuth, "Reflectance anisotropy and syn-deformational coalification of the Jewel seam in the Cadomin area, Alberta, Canada," *International Journal of Coal Geology*, vol. 19, no. 1-4, pp. 303–317, 1991.

[20] B. Jiang, F. Y. Xu, Y. Liu, and F. L. Jin, "Vitrinite optical fabric and stress-strain analysis of northern margin of Chaidamu Basin," *Journal of China University of Mining & Technology*, vol. 31, pp. 561–564, 2002.

[21] J. Jones and S. Creaney, "Optical character of thermally metamorphosed coals of northern England," *Journal of Microscopy*, vol. 109, pp. 105–118, 1977.

[22] G. K. Khorasani, D. G. Murchison, and A. C. Raymond, "Molecular disordering in natural cokes approaching dyke and sill contacts," *Fuel*, vol. 69, no. 8, pp. 1037–1046, 1990.

[23] W. E. Kilby, "Recognition of vitrinite with non-uniaxial negative reflectance characteristics," *International Journal of Coal Geology*, vol. 9, no. 3, pp. 267–285, 1988.

[24] W. E. Kilby, "Vitrinite reflectance measurement—some technique enhancements and relationships," *International Journal of Coal Geology*, vol. 19, no. 1-4, pp. 201–218, 1991.

[25] J. C. Hower, R. F. Rathbone, G. D. Wild, and A. Davis, "Observations on the use of vitrinite maximum reflectance versus vitrinite random reflectance for high volatile bituminous coals," *Journal of Coal Quality*, vol. 13, pp. 71–76, 1994.

[26] K. R. Wilks, M. Mastalerz, R. M. Bustin, and J. V. Ross, "The role of shear strain in the graphitization of a high-volatile bituminous and an anthracitic coal," *International Journal of Coal Geology*, vol. 22, no. 3-4, pp. 247–277, 1993.

[27] D. W. Houseknecht and C. M. B. Weesner, "Rotational reflectance of dispersed vitrinite from the Arkoma basin," *Organic Geochemistry*, vol. 26, no. 3-4, pp. 191–206, 1997.

[28] J. V. Ross and R. M. Bustin, "Vitrinite anisotropy resulting from simple shear experiments at high temperature and high confining pressure," *International Journal of Coal Geology*, vol. 33, no. 2, pp. 153–168, 1997.

[29] S. Duber and J. N. Rouzaud, "Calculation of relectance values for two models of texture of carbon materials," *International Journal of Coal Geology*, vol. 38, no. 3-4, pp. 333–348, 1999.

[30] S. Duber, S. Pusz, B. K. Kwiecińska, and J. N. Rouzaud, "On the optically biaxial character and heterogeneity of anthracites," *International Journal of Coal Geology*, vol. 44, no. 3-4, pp. 227–250, 2000.

[31] A. C. Cook, D. G. Murchison, and E. Scott, "Optically biaxial anthracitic vitrinites," *Fuel*, vol. 51, no. 3, pp. 180–184, 1972.

[32] F. Ting, "Uniaxial and biaxial vitrinite reflectance models and their relationship to paleotectonics," in *Organic Maturation Studies and Fossil Fuel Exploration*, J. Brooks, Ed., pp. 379–392, Academic Press, London, UK, 1981.

[33] R. M. Bustin, J. V. Ross, and I. Moffat, "Vitrinite anisotropy under differential stress and high confining pressure and temperature: preliminary observations," *International Journal of Coal Geology*, vol. 6, no. 4, pp. 343–351, 1986.

[34] B. Jiang, F. Jin, Q. Zhou, and W. Wang, "Experimental research on deformation of optical fabric of coal vitrinite reflectance," *Coal Geology & Exploration*, vol. 25, pp. 11–15, 1997 (Chinese).

[35] J. G. Ramsay, *Folding and Fracturing of Rocks*, Mc Graw-Hill, New York, NY, USA, 1967.

[36] Y. D. Wu, Y. W. Ju, Q. L. Hou, S. Hu, S. Q. Ni, and J. J. Fan, "Characteristics of tectono-thermal modeling and restriction on coalbed-gas generation in Sulin mining area, Huaibei coalfield," *Progress in Natural Science*, vol. 19, pp. 1134–1141, 2009 (Chinese).

[37] W. Guiliang, J. Bo, C. Daiyong, Z. Hai, and J. Weijun, "On the Xuzhou-Suzhou arcuate duplex-imbricate fan thrust system," *Acta Geologica Sinica*, vol. 72, no. 3, pp. 235–236, 1998.

[38] Y. W. Ju, Q. L. Hou, B. Jiang, G. L. Wang, and A. M. Fang, "Tectonic coals: structure and physical properties of reservoirs," in *Proceedings of the Proceedings of the 6th International Workshop on CBM/CMM in China*, Beijing, China, 2006.

[39] G. Zhu, C. Song, D. Wang, G. Liu, and J. Xu, "Studies on $^{40}$Ar/$^{39}$Ar thermochronology of strike-slip time of the Tan-Lu fault zone and their tectonic implications," *Science in China, Series D*, vol. 44, no. 11, pp. 1002–1009, 2001.

[40] G. Zhu, Y. Wang, G. Liu, M. Niu, C. Xie, and C. Li, "$^{40}$Ar/$^{39}$Ar dating of strike-slip motion on the Tan-Lu fault zone, East China," *Journal of Structural Geology*, vol. 27, no. 8, pp. 1379–1398, 2005.

[41] G. Zhu, C. L. Xie, Y. S. Wang, M. L. Niu, and G. S. Liu, "Characteristics of the Tan-Lu high-pressure strike-slip ductile shear zone and its $^{40}$Ar/$^{39}$Ar dating," *Acta Petrologica Sinica*, vol. 21, no. 6, pp. 1687–1702, 2005.

[42] A. Yin and S. Nie, "An indentation model for the north and south China collision and development of the Tan-Lu and Honam fault systems, Eastern Asia," *Tectonics*, vol. 12, pp. 810–813, 1993.

[43] G. Zhu, Y. S. Wang, N. M. Lan, G. S. Liu, and C. L. Xie, "Synorogenic movement of the Tan-Lu fault zone," *Earth Science Frontiers*, vol. 11, pp. 169–182, 2004 (Chinese).

[44] D. Liu, Y. Yao, D. Tang, S. Tang, Y. Che, and W. Huang, "Coal reservoir characteristics and coalbed methane resource assessment in Huainan and Huaibei coalfields, Southern North China," *International Journal of Coal Geology*, vol. 79, no. 3, pp. 97–112, 2009.

[45] D. Flinn, "On folding during three-dimensional progressive deformation," *Quarterly Journal of the Geological Society*, vol. 118, pp. 385–428, 1962.

[46] C. Barker and M. Pawlewicz, "The correlation of vitrinite reflectance with maximum temperature in humic organic matter," *Lecture Notes in Earth Science*, vol. 5, pp. 79–81, 1986.

[47] F. Goodarzi and D. G. Murchison, "Optical properties of carbonized vitrinites," *Fuel*, vol. 51, no. 4, pp. 322–328, 1972.

[48] D. G. Murchison, "Petrographic aspects of coal structure: reactivity of macerals in laboratory and natural environments," *Fuel*, vol. 70, no. 3, pp. 296–315, 1991.

[49] S. Duber, J. N. Rouzaud, C. Clinard, and S. Pusz, "Microporosity and optical properties of some activated chars," *Fuel Processing Technology*, vol. 77-78, pp. 221–227, 2002.

[50] G. L. Wang, B. Jiang, D. Y. Cao, H. Zou, and W. J. Jin, "On the Xuzhou-Suzhou arcuate duplex-imbricate fan thrust system.," *Acta Geologica Sinica (in Chinese with English summary)*, vol. 72, pp. 228–236, 1998.

[51] M. Zhai, R. Zhu, J. Liu et al., "Time range of Mesozoic tectonic regime inversion in eastern North China Block," *Science in China, Series D*, vol. 47, no. 2, pp. 151–159, 2004.

# 12

# Magmatism in the Asunción-Sapucai-Villarrica Graben (Eastern Paraguay) Revisited: Petrological, Geophysical, Geochemical, and Geodynamic Inferences

**Piero Comin-Chiaramonti,[1] Angelo De Min,[1] Aldo Cundari,[2] Vicente A. V. Girardi,[3] Marcia Ernesto,[4] Celso B. Gomes,[3] and Claudio Riccomini[3]**

[1] Mathematics and Geosciences Department, Trieste University, Via Weiss 8, 34127 Trieste, Italy
[2] Geotrack International, 37 Melville Road, Brunswick West, Melbourne, VIC 3055, Australia
[3] Geosciences Institute, University of São Paulo, Cidade Universitária, Rua do Lago 562, 05508-900 São Paulo, SP, Brazil
[4] Astronomical and Geophysical Institute (IAG) of the São Paulo University, Rua do Matão 1226, 05508-090 São Paulo, SP, Brazil

Correspondence should be addressed to Piero Comin-Chiaramonti; comin@units.it

Academic Editor: David T. A. Symons

The Asunción-Sapucai-Villarrica graben (ASV) in Eastern Paraguay at the westernmost part of the Paraná Basin was the site of intense magmatic activity in Mesozoic and Tertiary times. Geological, petrological, mineralogical, and geochemical results indicate that the following magmatic events are dominant in the area: (1) tholeiitic basalt and basaltic andesites, flows and sills of low- and high-titanium types; (2) K-alkaline magmatism, where two suites are distinguished, that is, basanite to phonolite and alkali basalt to trachyte and their intrusive analogues; (3) ankaratrite to phonolite with strong Na-alkaline affinity, where mantle xenoliths in ultramafic rocks are high- and low-potassium suites, respectively. The structural and geophysical data show extensional characteristics for ASV. On the whole, the geochemical features imply different mantle sources, consistently with Sr-Nd isotopes that are Rb-Nd enriched and depleted for the potassic and sodic rocks, respectively. Nd model ages suggest that some notional distinct "metasomatic events" may have occurred during Paleoproterozoic to Neoproterozoic times as precursor to the alkaline and tholeiitic magmas. It seems, therefore, that the genesis of the ASV magmatism is dominated by a lithospheric mantle, characterized by small-scale heterogeneity.

## 1. Introduction

Velázquez et al. [1] presented a structural analysis of the central segment of the "Asunción Rift," mainly based on the previous papers related to the Eastern Paraguay magmatism in general, and to the Asunción-Sapucai-Villarrica graben (ASV) in particular, and also based on extensive field data collected earlier on the dyke swarms cropping out in the area. However, some aspects as, for example, the close association in space of potassic and sodic alkaline rock-types with tholeiitic dykes and flows (both of high- and low-Ti types; cf. [2]), have not been discussed in detail by the above authors.

This paper focuses on general aspects of the magmatism from the western margin of the Paraná-Angola-Etendeka

system (PAE), where tholeiitic flows and dykes (Early Cretaceous, both of high-Ti and low-Ti types; cf. Figure 1(a) and [3]) are associated in time and space with a wide variety of alkaline rock-types (both potassic and sodic) and carbonatites.

The investigated alkaline and tholeiitic rocks from Eastern Paraguay span in age mainly from Early Cretaceous to Paleogene times. Therefore, they are germane to the magmatic and tectonic evolution of the PAE and of the Atlantic Ocean (cf. Figure 1(b) and [4, 5]).

This review aims at discussing (1) the relationships between tectonics and magmatic activity on the basis of field and geophysical data; (2) a detailed description of the geological characteristics of the Asunción-Sapucai-Villarrica

Magmatism in the Asunción-Sapucai-Villarrica Graben (Eastern Paraguay) Revisited: Petrological, Geophysical, Geochemical, and Geodynamic Inferences

143

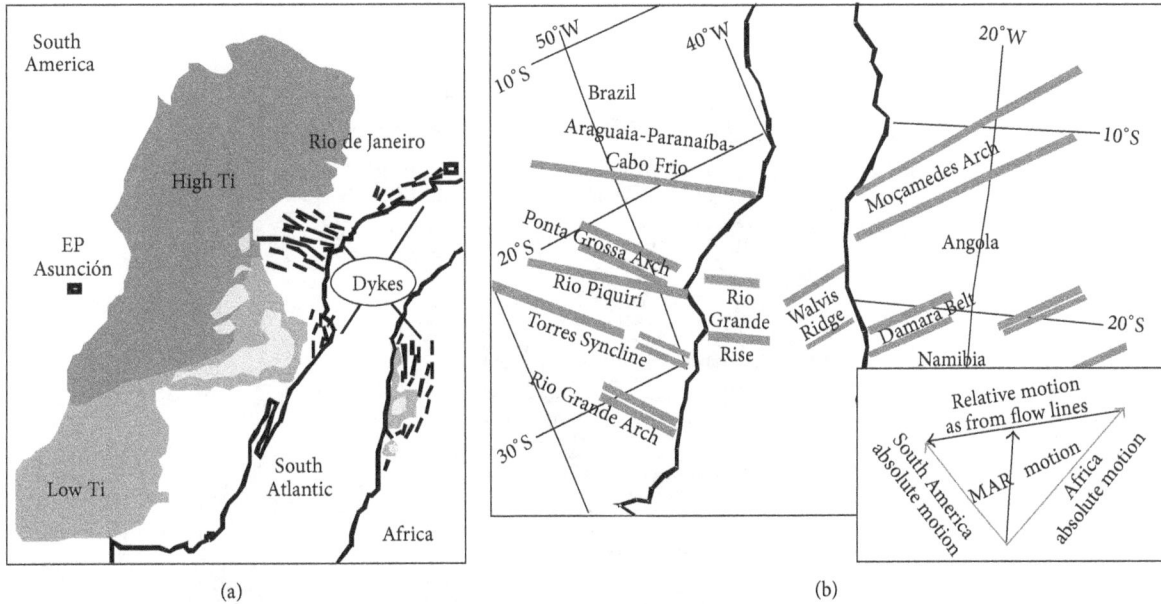

FIGURE 1: (a) Sketch map of the Paraná-Angola-Etendeka system [3] where the arrows indicate the occurrences of the main dyke swarms. The basaltic lavas are subdivided into broad high- and low-Ti groups and late-stage rhyolites (yellow fields). EP: Eastern Paraguay. (b) Main lineaments in the Paraná-Angola-Etendeka system (PAE), western Gondwana at ca. 110 Ma (modified after [6]), corresponding to the main lineaments of the alkaline and alkaline-carbonatitic complex. Inset: vector diagram showing relationships among absolute plate motions, relative motions, and motion of the Mid-Atlantic Ridge (MAR, after [7]).

(ASV) graben; (3) petrography and petrochemistry of the magmatic rock-types (and the associated mantle xenoliths) from ASV; (4) the most important geochemical and Sr-Nd isotopic features of the magmatism; (5) the petrogenesis and geodynamic implications.

## 2. Eastern Paraguay: Geological Outlines and Geophysical Aspects

Eastern Paraguay shows a complex block-faulted structure between the southern Precambrian tip of the Amazonian Craton (Apa block) and the northern one of the Rio de la Plata Craton (Caacupú high; Figure 2). This intercratonic area, including the westernmost fringe of the Paraná Basin (PB), represents an undeformed basin along the western Gondwana (Figure 1), where the sedimentation, started in Cambrian times, was topped by Early Cretaceous tholeiitic basalts of the Serra Geral Formation [8, 9] and followed by younger sedimentation (Figure 2).

Moreover, it should be also noted that Eastern Paraguay lies along the former western margin of the Gondwana, bounded by an anticlinal structure established since Early Paleozoic, the Asunción Arch, which separates the Paraná Basin (East) from the Gran Chaco Basin (West) [10]. PB shows a high-velocity upper mantle "lid" with a maximum S-wave velocity of 4.7 km/s (Moho 37 km depth), with an unresolved low-velocity zone to a depth of at least 200 km [11]. Between the two blocks, that is, Apa and Caacupú of Figure 2, Eastern Paraguay was subjected to NE-SW-trending crustal extension during Late Jurassic-Early Cretaceous, probably related to the western Gondwana breakup (Figure 1; [10, 12]).

and therein references). NW-SE trending faults, paralleling the dominant orientation of Mesozoic alkaline and tholeiitic dykes, reflect this type of structure [13, 14].

From the aeromagnetic survey [15, 16], the linear magnetic anomalies trend mainly N40-45W (Figure 3). These anomalies have been interpreted as resulting from Early Cretaceous tholeiitic dyke swarms [16], but field evidence does not support this hypothesis. Most magnetic anomalies correspond to older Precambrian tectonic lineament [17]. The Landsat lineaments, trending mainly NE and EW, may therefore reflect tectonic lineaments from the basement [18].

The Bouguer gravity map (Figure 3) consists of prevailing NW-trending gravity "highs" and "lows" that represent shallow to exposed basement and sedimentary basins, respectively [19]. The boundaries between gravity highs and lows are generally marked by step gradients that may reflect abrupt basement offsets along faults, or basement dip changes by crustal warping. Notably, the gravity lows and highs parallel the dominant NW attitude of the magnetic lineaments, with post-Palaeozoic magmatism (both tholeiitic and alkaline) associated with gravity lows. Therefore, the ASV represents the most important geological structure of the region (see later).

Notably, the present seismic activity (earthquakes with hypocenters <70 km) indicates that the NW-trending fault systems continue up to present day, that is, the Pilcomayo lineament (inset of Figure 3), also aligned with the Piquirí lineament and the Torres syncline (Figure 1(b)).

In conclusion, the resulting structural pattern controlled the development of grabens or semigrabens as a response to NE-SW-directed extension and continued evolving into Cenozoic times [12, 24]. According to Tommasi and

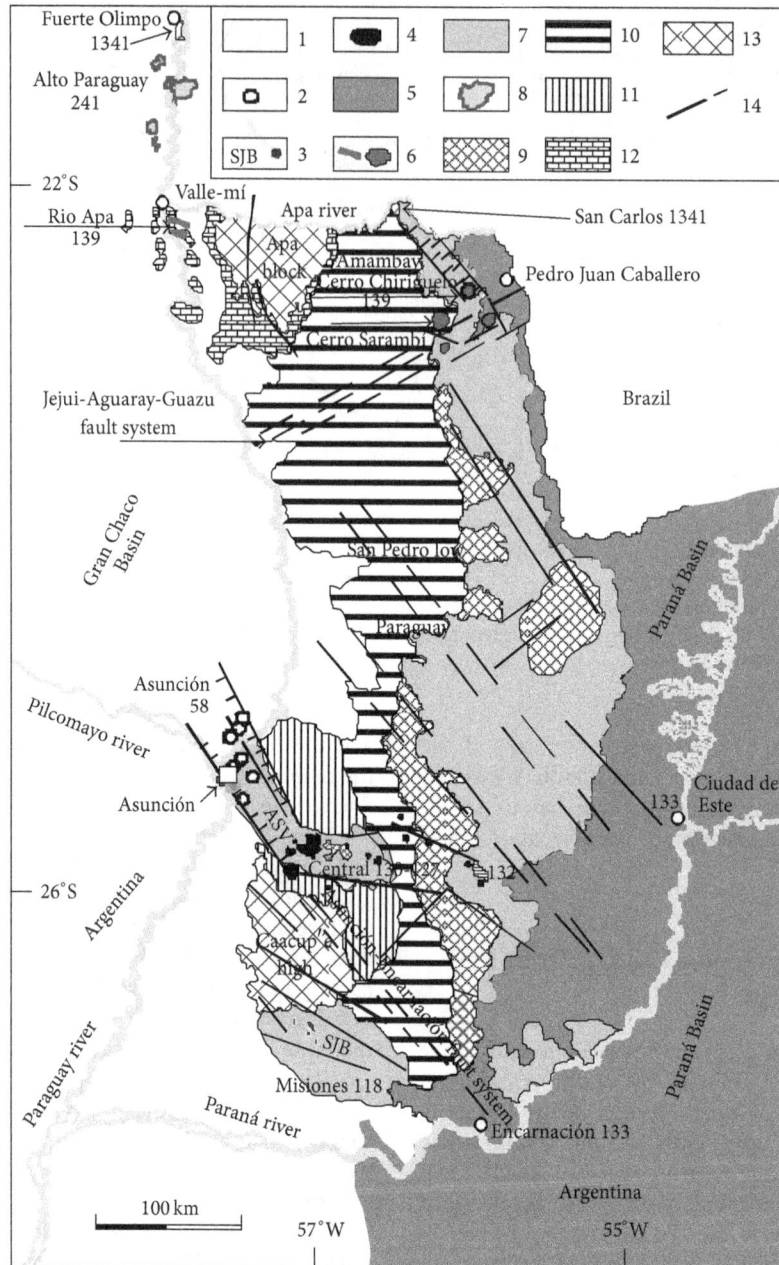

FIGURE 2: Geological sketch map of Eastern Paraguay (modified after [10, 13, 20–24]). ASV: Asunción-Sapucai-Villarrica graben. (1): Neogene and Paleogene sedimentary cover (e.g., Gran Chaco); (2): Paleogene sodic alkaline rocks, Asunción Province; (3): Late-Early Cretaceous sodic alkaline rocks (Misiones Province, San Juan Bautista, SJB); (4): Early Cretaceous potassic alkaline rocks (posttholeiites; ASU: Asunción-Sapucai-Villarica graben, Central Province); (5): Early Cretaceous tholeiites of the Paraná Basin (Serra Geral Formation in Brazil and Alto Paraná Formation in Paraguay; cf. [4]); (6): Early Cretaceous potassic alkaline rocks (pretholeiites, Apa and Amambay Provinces); (7): Jurassic-Cretaceous sedimentary rocks (Misiones Formation); (8): Permo-Triassic alkaline rocks (Alto Paraguay Province); (9): Permian sedimentary rocks (Independencia Group); (10): Permo-Carboniferous sedimentary rocks (Coronel Oviedo Group); (11): Ordovician-Silurian sedimentary rocks (Caacupé and Itacurubí Groups); (12): Cambro-Ordovician platform carbonates (Itacupumí Group); (13): Archean to Early Paleozoic crystalline basement: high- to low-grade metasedimentary rocks, metarhyolites, and granitic intrusions; (14): major tectonic lineaments and faults. The radiometric values, for example, 133 Ma, are referred to $^{40}Ar/^{39}Ar$ plateau ages for the main magmatic types [25–27] and to Rb/Sr ages for the Precambrian rhyolitic rocks of Fuerte Olimpo and Fuerte San Carlos [28].

Magmatism in the Asunción-Sapucai-Villarrica Graben (Eastern Paraguay) Revisited: Petrological, Geophysical, Geochemical, and Geodynamic Inferences

145

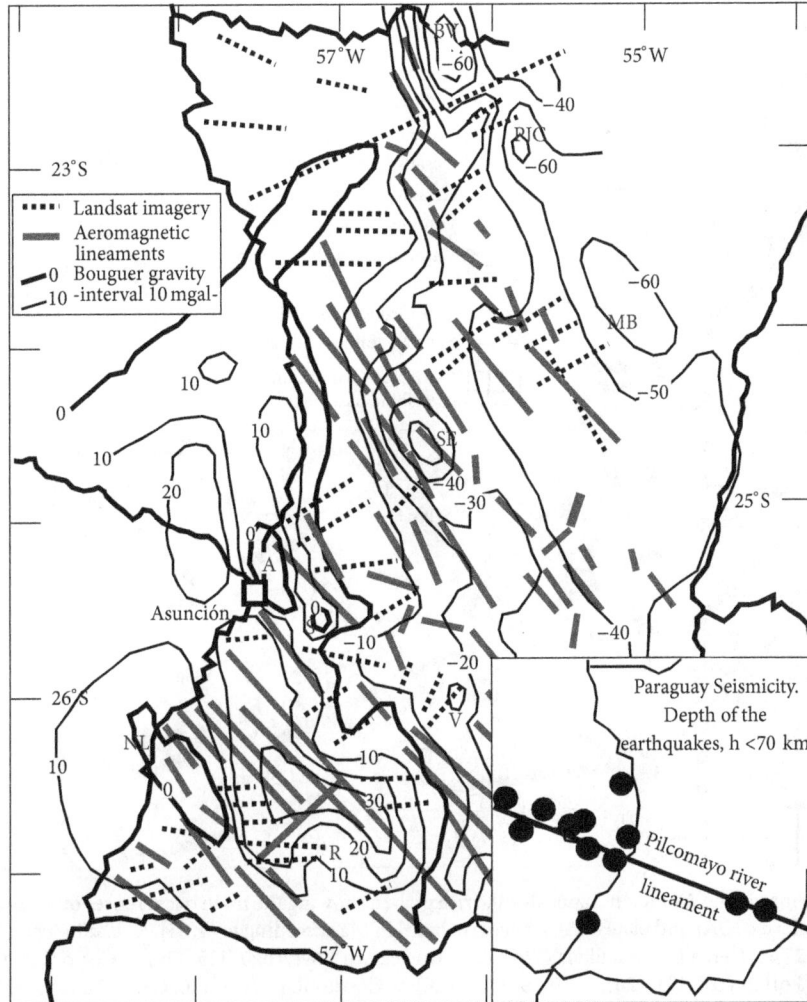

FIGURE 3: Geological sketch map showing aeromagnetic and Landsat lineaments in Eastern Paraguay and Bouguer gravity data [16–18]. Gravity lows: BV, Bella Vista; PJC, Pedro Juan Caballero; MB, Mbacarayú; SE, San Estanislao (formerly San Pedro); A, Asunción, S, Sapucai, V, Villarrica (ASV graben); R, Santa Rosa. Inset: distribution of earthquakes with depth <70 km [29].

Vauchez [30], rift orientations seem to have been controlled by the preexisting lithospheric mantle fabric, as indicated by deep geophysical data.

The thermal history, using apatite fission track analyses (AFTA), reveals that at least two main episodes have been identified in sedimentary and igneous/metamorphic samples ranging in age from Late Ordovician to Early Cretaceous [31, 32]. AFTA data from ASV show evidence for rapid cooling beginning sometime between 90 and 80 Ma, similar to the results found along Brazilian and Uruguayan coasts. A Tertiary heating/cooling episode is also suggested by AFTA data (60–10 Ma), supporting early work in the area. The time of the first event is significantly younger than any rifting activity related to the Paraná flood basalts and to the opening of the South Atlantic. Late Cretaceous cooling may have involved several kilometers of differential uplift and erosion and would have played an important role in the control of the geomorphology and drainage patterns of the region, especially in the ASV system [31, 32].

Beginning from Mesozoic times (Figure 2), almost five main alkaline magmatic events have occurred in Eastern Paraguay, other than the Early Cretaceous tholeiitic magmatism (133-134 Ma; [4]). Three of these include rocks of sodic affinity, corresponding geographically to the provinces of Alto Paraguay (241.5 ± 1.3 Ma), Misiones (118.3 ± 1.6 Ma), and Asunción (58.7 ± 2.4 Ma), whereas two involve rocks of potassic affinity associated with the Apa and Amambay provinces, both of similar age (138.9 ± 0.7), and with the Central Province (ASV, 126.4 ± 0.4 Ma).

## 3. Asunción-Sapucai-Villarrica Graben (ASV)

The ASV graben (Figure 4) is the dominant rifting structure in Eastern Paraguay, linked to NE-SW extensional vectors [13], and it is characterized by a gravity anomaly trending N30-45W near Asunción townships (Figure 3). The graben, nearly symmetrical and defined by major faults along each margin [1, 12, 18, 37], may be subdivided into three main

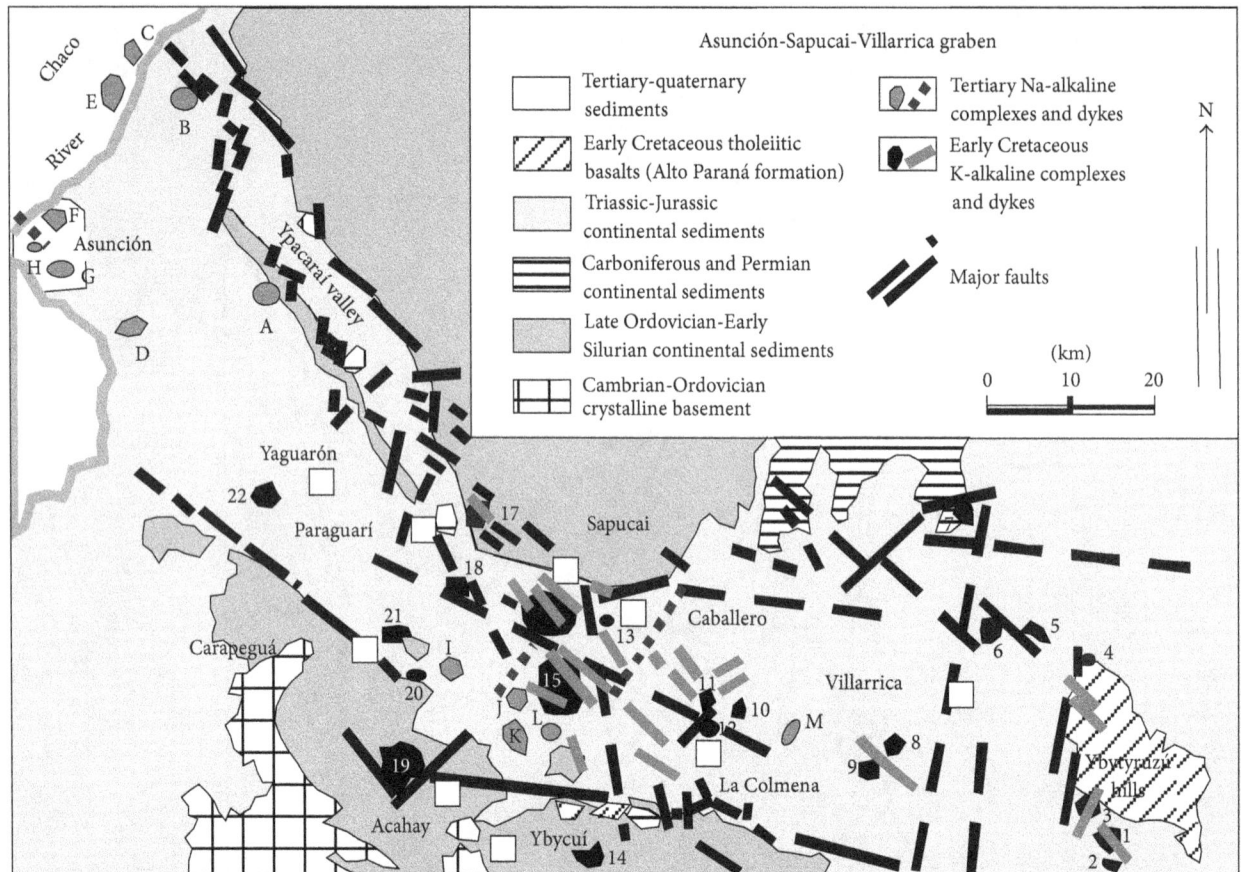

FIGURE 4: Geological sketch map of the Asunción-Sapucai-Villarrica graben showing the main occurrences of magmatic rock-types, modified after [33]. In parentheses preferred K/Ar and plateau Ar/Ar ages (in bold) in Ma according to [4, 33]. *K-alkaline complexes*: 1, Cerro Km 23 (132, **128**); 2, Cerro San Benito (**127**); 3, Cerro E. Santa Elena (**126**); 4, Northwestern Ybytyruzú (125–129); 5, Cerro Capiitindy (n.a.); 6, Mbocayaty (126–130, **126**); 7, Aguapety Portón (128–133, **126**); 8, Cerro Itapé (n.a.); 9, Cerrito Itapé (n.a.); 10, Cerro Cañada (**127**); 11, Cerro Chobí (n.a.); 12, Potrero Garay (n.a.); 13, Catalán (n.a.); 14, Cerro San José (**127**); 15, Potrero Ybaté (126–128, **128**); 16, Sapucai (119–131, **126**); 17, Cerro Santo Tomás (126–130, **127**); 18, Cerro Porteño (n.a.); 19, Cerro Acahay (118, **127**); 20, Cerro Pinto (n.a.); 21, Cerro Ybypyté (124); 22, Cerro Arrúa-í (126–132, **129**). *Na-alkaline complexes*: A, Cerro Patiño (39); B, Limpio (50); C, Cerro Verde (57, **61**); D, Ñemby (46, **61**); E, Cerro Confuso (55–61); F, Nueva Teblada (46–57); G, Lambaré (49); H, Tacumbú (41–46, **58**); I, Cerro Yariguaá (n.a.); J, Cerrito (**56**); K, Cerro Gimenez (66); L, Cerro Medina (n.a.); M, Colonia Vega (n.a.). Detailed information on the single occurrences, that is, geological map, location, country rocks, forms, and main rock-types, is provided in [33–36]; n.a.: not available.

segments (Figure 4): (1) the northwestern part, between the Asunción and Paraguarí townships, indicated also by the intragraben Ypacaray rifting and by the presence of Na-alkaline mafic-ultramafic plugs carrying mantle xenoliths (Asunción Province; age between 66 and 39 Ma; cf. Figure 4); (2) a central area, defined between Paraguarí and La Colmena townships, characterized by K-alkaline complexes and dykes (average age 127 Ma; cf. Figures 3 and 4) and by subordinate Na-alkaline complexes and dykes (age around 66–60 Ma); (3) the eastern part, between La Colmena and the Ybytyruzú hills, characterized by K-alkaline complexes and dykes (age around 130–125 Ma) and by tholeiitic lavas and flows (age around 133–130 Ma; cf. [4]).

Geology and gravity results indicate that the ASV graben extends up to 100 km into the Chaco Basin (Figures 2 and 3) and turns to an N80W trend between Paraguarí and Villarrica townships, in the area marked by a gravity low anomaly that increases eastwards (Figures 3 and 4). Paleomagnetic

data [38] revealed that the potassic Cretaceous rocks (75 samples) have reversal polarity remnants corresponding to a paleomagnetic pole at 62.3°E and 85.4°S, similar to that calculated for the Serra Geral Formation in the Paraná Basin. It is therefore inferred that ASV K-alkaline rocks were emplaced, while the tholeiitic magmatic activity was still taking place (Figure 1).

Tertiary sodic rocks, carrying mantle xenoliths, mainly occur in an area characterized by relatively gravity highs (Figure 3), whereas the potassic complexes and dykes occur in a gravity low belt characterized by block faulting (Figures 3 and 4). Finally, an NS-trending fault system bounds the western side of the Ybytyruzú hills, close to several alkaline intrusions [24] East of Villarrica.

A total of 527 samples from the ASV area, including 220 dykes, have been investigated in detail, including field relationships, age, orientation (dykes), petrography, and mineral chemistry of the rocks reported in [10, 13, 33–36].

Magmatism in the Asunción-Sapucai-Villarrica Graben (Eastern Paraguay) Revisited: Petrological, Geophysical, Geochemical, and Geodynamic Inferences

147

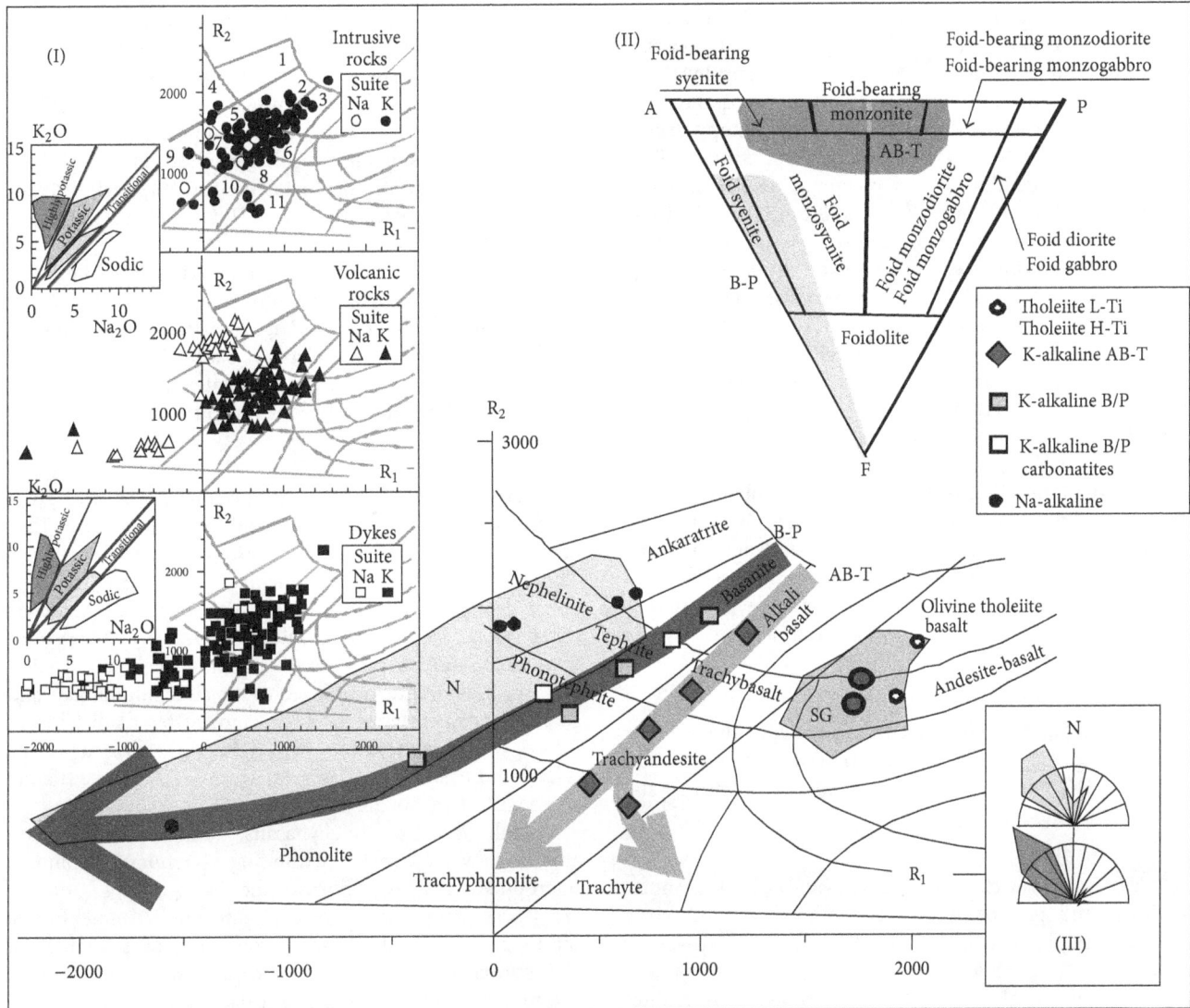

FIGURE 5: Compositional variation of the rock-types from ASV in terms of $R_1 = 4Si - 11(Na + K) - 2(Fe + Ti)$ and $R_2 = 6Ca + Mg + Al$ [39] (also Figure 2 of [13]). Data source: [34]. SG, field of Serra Geral tholeiites; ASV, field of Na-alkaline rocks; B-P and AB-T, lineages of the K-alkaline rocks (basanite to phonolitic suite and alkaline basalt to trachyphonolite/trachyte suite, resp.; white squares represent a lava flow and a theralite with carbonatitic affinity, resp. (see text)). Insets: (I), the same with subdivision into intrusive and volcanic rocks and dykes, plus Le Maitre [40] classification, as in Figure 2 of [10]. (II), QAPF diagram [41] for intrusive potassic suites from ASV, as from Figure 3 of [10]. (III), Azimutal frequence of the B-P and AB-T dykes [42]. The symbols represent selected compositions as in Tables 1 to 4.

55% of the dykes are preferentially oriented N20-60 W (Figure 5, (III)).

## 4. Classification, Petrography, and Petrochemistry of the Magmatic Rocks

The fine-grained texture of the ASV rocks, including the intrusive variants, is plotted on the De La Roche chemical classification [39] diagram (Figure 5), with an additional subdivision based on wt% $K_2O$ and $Na_2O$ on Le Maitre's [40] diagram (inset (I)).

4% of the dykes have tholeiitic affinity, sills, lava flows, and dykes, both high- and low-Ti variants, according to [2, 43] and straddle the fields of the olivine tholeiitic basalts and

andesite-basalts, whereas 17% are sodic and 79% potassic in composition. The distribution of the potassic rocks shows that the variation of the dykes is consistent with that of the associated intrusive and volcanic rocks (inset (I)). Two main lineages are apparent for the potassic rock-types (Figure 5 and inset (II)): (1) a silica undersatured lineage (B-P) ranging from basanite to phonolite and peralkaline phonolite and (2) a silica-saturated lineage (AB-T) ranging from alkali basalt to trachyphonolite and trachyte [13, 42, 44].

In summary, tholeiitic rock-types (both high-titanium, H-Ti, and low-titanium, L-Ti, variants), K-alkaline, and Na-alkaline rocks are widespread in a relatively narrow area represented by ASV graben (ca. $35 \times 200$ km; cf. Figure 3). In Tables 1 to 4, representative and average analyses of these rocks are reported.

*4.1. Tholeiites.* The tholeiitic rocks are mainly lava flows (Ybytyruzú hills) and gabbro sills usually occurring near the Ybycuí township (Figure 4). The textures are porphyritic to aphanitic (lava flows and dykes) and equi- to subequigranular (sills), characterized by the presence of two clinopyroxenes (augite and pigeonite) and by pronounced variations of $TiO_2$ (1.4 to 3.2 wt%; cf. Table 1) and incompatible elements (IE: e.g., Ba ca. 150 to 400 ppm, Zr ca. 100 to 200 ppm). These variations are similar to those common to tholeiitic basalts of the Paraná Basin that are dominated by low and high contents of $TiO_2$ (L-Ti and H-Ti variants: southern and northern Paraná Basin with $TiO_2$ <2 and >2.5 wt%, resp.; cf. Figure 1) and IE [2, 45, 46]. The mineral assemblages (augite, pigeonite, olivine, plagioclase, magnetite, and ilmenite) show crystallization temperatures around 1200–1000°C and $fO_2$ conditions between NNO and QMF buffers [3].

Mass balance calculations and the remarkable differences in the ratios of the IE between L-Ti and H-Ti tholeiites suggest that such differences cannot be explained by fractional crystallization, or by means of melting and zone refining processes of a homogeneous mantle source. A basalt genesis from chemically different lithospheric mantle materials is therefore indicated. Mixing processes seem to have played a minor role [2, 47].

*4.2. Potassic Suites.* The different lithotypes of the potassic rocks (cf. Figure 5 and Tables 2 and 3) maybe subdivided into B-P and AB-T lineages, a distinction consistent with crystal fractionation models [13].

*4.2.1. B-P Suite.* The effusive rock-types and dykes (basanites, tephrites, and phonotephrites) typically show porphyritic textures with phenocrysts of clinopyroxene ($Wo_{40-50}Fs_{10-19}$), olivine ($Fo_{60-85}$), and leucite pseudomorphed by sanidine + nepheline in a glassy groundmass showing microlites of clinopyroxene ± olivine, Ti-magnetite ± ilmenite, Ti-phlogopite-biotite, alkali feldspar ($Or_{15-88}$), and nepheline-analcime ($Ne_{44-59}Ks_{17-26}$). Phenocrystal plagioclase (up to $An_{74}$) is also present. Accessory phases are amphibole (pargasite-kaersutite) and apatite ± zircon. Phonolites are characterized by phenocrysts of leucite-pseudomorphs, alkali feldspar ($Or_{47-75}$), clinopyroxene ($Wo_{48-50}Fs_{11-34}$), Fe-pargasite, and nepheline ± biotite ± titanite ± melanite (Ti-andradite up to 68 wt%) ± magnetite or haematite. Glassy groundmass contains microlites of alkali feldspar, nepheline, and clinopyroxene ± Ti-andradite ± magnetite. Haematite is found replacing magnetite or as a common groundmass constituent in peralkaline phonolites.

The intrusive analogues (theralites, essexitic gabbros, ijolites and essexites) show a holocrystalline inequigranular texture, with diopsidic pyroxene ($Wo_{44-51}Fs_{8-17}$), olivine ($Fo_{75-82}$ to $Fo_{44-66}$), mica (Ti-phlogopite to Ti-biotite), Ti-magnetite, alkali feldspar, and nepheline ($Ne_{64-80}Ks_{20-36}$) ± leucite ± amphibole. "Intrusive" leucite and leucite pseudomorphed by analcime and plagioclase are common, testifying to subvolcanic conditions [24]. Cumulitic textures are represented by clinopyroxene with alkali feldspar + nepheline + carbonate as intercumulus phases.

On the whole, the temperatures calculated for the mineral assemblages span from 1250 to 1200°C and from 1130 to 1080°C, pheno- and microphenocrysts, respectively, under about 1 kb $H_2O$ pressure and around 1000°C under atmospheric pressure; lower temperatures, for example, 900–700°C, suggest possible subsolidus exchange reactions (e.g., albitization). Magnetite-ilmenite pairs define temperatures between 1000 and 1100°C, along the NNO buffer [12].

Notably, the B-P suite is also characterized by carbonatitic magmatism, although it is very limited in volume. A silicobeforsitic flow is present in the Sapucai lava sequences and as "ocelli" at the Canãda and Cerro E. Santa Helena K-alkaline complexes [48]. These primary carbonates are relevant as they reveal a $CO_2$ imprinting after the tholeiitic magmatism.

*4.2.2. AB-T Suite.* Alkali gabbros, syenogabbros, and syenodiorites are usually porphyritic and seriate in texture. They contain clinopyroxene ($Wo_{43-50}Fs_{6-14}$), olivine ($Fo_{43-82}$), Ti-biotite, Ti-magnetite ± ilmenite, plagioclase ($An_{31-78}$), alkali feldspar ($Or_{60-84}$), interstitial nepheline-analcime ($Ne_{37-82}Ks_{15-23}$), and alkali feldspar ($Or_{80-97}$). Accessories are apatite ± amphibole ± titanite ± zircon. Nepheline syenites and syenites are texturally equi- to subequigranular and seriate. The rock-types are characterized by alkali feldspar ($Or_{32-63}$), clinopyroxene ($Wo_{43-48}Fs_{10-32}$), and nepheline ($Ne_{85}$) and hastingsite. Common accessories include titanite and apatite ± carbonate ± zircon.

Alkali basalts, trachybasalts, and trachyandesites are porphyritic rocks exhibiting phenocrysts and/or microphenocrysts of clinopyroxene ($Wo_{44-49}Fs_{7-15}$), olivine ($Fo_{65-83}$), plagioclase ($An_{28-76}$), magnetite, and biotite set in a glassy groundmasses consisting of microlites of clinopyroxene ($Wo_{46-49}Fs_{13-18}$), magnetite, ilmenite, biotite, plagioclase ($An_{20-45}$), alkali feldspar ($Or_{52-65}$), nepheline-analcime ($Ne_{37-73}Ks_{22-38}$), amphibole, and apatite ± titanite ± zircon.

Trachyphonolites and trachytes are porphyritic to aphyric. The phenocrysts are alkali feldspar ($Or_{60-65}$) ± clinopyroxene ($Wo_{46-49}Fs_{14-20}$) ± plagioclase ($An_{14-16}$), pseudomorphosed leucite, amphibole, and biotite in a hypocrystalline to glassy groundmass containing microlites of alkali feldspar, biotite, and clinopyroxene ± biotite ± amphibole ± magnetite ± Ti-andradite ± haematite.

Notably, the equilibration temperatures of the mineral assemblages are very similar to those suggested for the B-P suite [10, 13, 49].

The differentiation history of the potassic ASV magmas remained largely into the stability fields of the ferromagnesian phases [12], that is, olivine + clinopyroxene, in the system $Mg_2SiO_4$-$KAlSiO_4$ [10, 13, 51], suggesting that relatively high temperature subvolcanic conditions were prevalent. Moreover, the extreme differentiates from both B-P and AB-T suites approaches to the composition of peralkaline residua, pointing to an extensive subvolcanic crystallization of K-rich, aluminous phases, likely phlogopite and leucite, at least for the B-P suite [52].

Magmatism in the Asunción-Sapucai-Villarrica Graben (Eastern Paraguay) Revisited: Petrological, Geophysical, Geochemical, and Geodynamic Inferences

149

TABLE 1: Averages of representative chemical analyses (recalculated to 100 wt% on anhydrous basis; in parentheses, standard deviation) of tholeiites from ASV [33, 34, 44]. L-Ti and H-Ti tholeiites according to [2, 45]. The $\varepsilon$ time-integrated notations are calculated using the present day Bulk Earth parameters [50], that is, UR = 0.7047 ($^{87}Rb/^{86}Sr$ = 0.0816) and CHUR = 0.512638 ($^{147}Sm/^{144}Nd$ = 0.1967. Mg#: molar ratio MgO/(MgO + FeO)).

| Suite: tholeiite | L-Ti basalt | L-Ti andesite-basalt | H-Ti basalt | H-Ti Andesite-basalt |
|---|---|---|---|---|
| No. of samples | (3) | (3) | (6) | (5) |
| Wt% | | | | |
| $SiO_2$ | 50.20 (0.79) | 53.06 (0.40) | 50.83 (0.84) | 50.52 (0.60) |
| $TiO_2$ | 1.40 (0.06) | 1.36 (0.11) | 2.58 (0.36) | 3.19 (0.60) |
| $Al_2O_3$ | 14.64 (0.21) | 13.78 (0.09) | 13.98 (0.45) | 14.25 (0.63) |
| $FeO_{tot}$ | 12.73 (0.39) | 13.42 (0.49) | 13.54 (0.56) | 14.55 (1.13) |
| MnO | 0.19 (0.05) | 0.25 (0.05) | 0.21 (0.02) | 0.21 (0.02) |
| MgO | 7.36 (0.22) | 5.53 (0.08) | 5.13 (0.44) | 4.19 (0.65) |
| CaO | 10.63 (0.66) | 8.55 (0.05) | 9.65 (0.62) | 8.56 (0.44) |
| $Na_2O$ | 2.36 (0.11) | 2.66 (0.20) | 2.61 (0.15) | 2.49 (0.29) |
| $K_2O$ | 0.33 (0.05) | 1.07 (0.18) | 1.10 (0.17) | 1.14 (0.10) |
| $P_2O_5$ | 0.15 (0.02) | 0.19 (0.01) | 0.37 (0.09) | 0.36 (0.04) |
| Mg# | 0.66 | 0.55 | 0.54 | 0.48 |
| ppm | | | | |
| La | 6.2 (0.8) | 9.0 (1.0) | 25.3 (6.3) | 26.0 (6.0) |
| Ce | 16.0 (1.7) | 21.0 (2.9) | 60.4 (12.2) | 50.0 (2.0) |
| Nd | 11.1 (1.3) | 14.1 (2.2) | 28.3 (3.6) | 25.0 (3.0) |
| Sm | 3.4 (0.3) | 4.3 (0.4) | 7.7 (1.5) | 6.8 (1.1) |
| Eu | 1.2 (0.1) | 1.5 (0.1) | 2.2 (0.2) | 1.9 (0.2) |
| Gd | 4.3 (0.6) | 5.4 (0.8) | 7.7 (1.2) | 6.9 (1.1) |
| Dy | 4.8 (0.5) | 6.0 (0.9) | 6.3 (1.5) | 5.7 (1.3) |
| Er | 2.9 (0.4) | 3.6 (0.6) | 3.9 (1.3) | 3.5 (0.9) |
| Yb | 2.6 (0.4) | 3.2 (0.4) | 3.4 (1.2) | 3.2 (1.0) |
| Lu | 0.34 (0.07) | 0.42 (0.09) | 0.49 (0.15) | 0.50 (0.19) |
| Cr | 325 (27) | 62 (1) | 119 (15) | 48 (2) |
| Ni | 103 (18) | 55 (7) | 60 (15) | 54 (12) |
| Rb | 13 (4) | 13 (1) | 24 (3) | 33 (11) |
| Sr | 188 (42) | 187 (2) | 386 (90) | 334 (15) |
| Ba | 154 (19) | 190 (4) | 409 (78) | 347 (37) |
| Th | 1.1 (0.1) | 1.7 (0.2) | 2.0 (0.1) | 3.0 (0.2) |
| U | 0.24 (0.05) | 0.37 (0.08) | 0.43 (0.08) | 0.65 (0.04) |
| Ta | 0.28 (0.04) | 0.42 (0.07) | 1.2 (0.3) | 1.8 (0.4) |
| Nb | 4.0 (1.0) | 6.1 (1.0) | 13 (2) | 20 (1) |
| Zr | 83 (10) | 113 (2) | 190 (36) | 197 (6) |
| Y | 28 (2) | 42 (1) | 33 (4) | 36 (3) |
| Measured | | | | |
| ($^{87}Sr/^{86}Sr$) | 0.70548 (20) | 0.70535 (20) | 0.70617 (21) | 0.70620 (18) |
| ($^{143}Nd/^{144}Nd$) | 0.51271 (9) | 0.51272 (10) | 0.51239 (2) | 0.51240 (1) |
| Age (Ma): 131.6 | | | | |
| Initial ratio | | | | |
| ($^{87}Sr/^{86}Sr$)$_i$ | 0.70502 | 0.70456 | 0.70576 | 0.70554 |
| ($^{143}Nd/^{144}Nd$)$_i$ | 0.51256 | 0.51256 | 0.51225 | 0.51226 |
| $T_{DM}$ | 1991 | 1976 | 2127 | 2037 |
| $R_1$ | 2037 | 1929 | 1758 | 1722 |
| $R_2$ | 1789 | 1459 | 1561 | 1403 |
| $\varepsilon Sr$ | 6.70 | 0.12 | 17.16 | 14.12 |
| $\varepsilon Nd$ | 1.60 | 1.79 | −4.30 | −4.10 |

*4.3. The Sodic Rocks.* These consist mainly of ankaratrites + (mela)nephelinites (45%) and phonolites (42%) (Comin-Chiaramonti et al. [10, 13, 44]). Representative chemical analyses are reported in Table 4.

Phenocrysts-microphenocrysts in ankaratrites and nephelinites are characterized by olivine (phenocrysts 1–7 vol%, $Fo_{89-85}$ mole%; 1–6 vol% microphenocrysts, $Fo_{82-77}$ mole%), clinopyroxene (phenocrysts 1–6 vol%, Mg# ~0.8), titanomagnetite (0.3–0.7 vol%, up to 38 ulv. mole%), and occasionally phlogopite microphenocrysts. The hypocrystalline groundmass contains clinopyroxene (39–46 vol%, Mg# ~0.75), olivine (3–6 vol%, $Fo_{74-76}$ mole%), titanomagnetite (4–7 vol%, up to 43% ulv. mole%), nepheline (16–21 vol%), and glass (11–25 vol%) [53].

Phonolites are typically microphyric to hypocrystalline with alkali feldspar phenocrysts or microphenocrysts ($Or_{43-83}$), nepheline ($Ne_{67-79}$), occasionally altered to cancrinite, acmitic clinopyroxene (acmite up to 63 wt%), and ferroedenite-ferropargasite amphibole. Haüyne, mica, haematite, and zircon are typical accessory minerals. Analcime occurs as a nepheline pseudomorph. Ti-andradite and titanite phenomicrophenocrysts are present in some dykes [10].

Lherzolite, harzburgite, dunite mantle xenoliths, and xenocrystic debris are common and abundant in the ankaratrites and nephelinites [54] from eastern ASV (Figure 3).

The generation of the ASV nephelinites may be modelled for 4–6% degrees of partial melting, with crystallization temperatures between 1200 and 1000°C from garnet peridotite sources [10, 55].

*4.4. Mantle Xenoliths.* The xenoliths are mainly spinel-lherzolites, harzburgites, and subordinate dunites. The dominant texture is protogranular, rarely tabular or porphyroclastic [54]. The Paraguay mantle xenoliths contain variable amounts of glassy patches (blebs) and glassy drops in clinopyroxenes [56]. The latter show an overprinted spongy texture [54]. The blebs (whole composition: Mg# 0.88–0.91) consist mainly of a glassy matrix having microlites of olivine (Mg# 0.91-0.92), clinopyroxene (Mg# 0.91–0.93), Cr-spinel, and (rarely) phlogopite (Mg# 0.86–0.92). According to [56, 57], they were formed by decompression melting of amphibole and phlogopite.

The Paraguayan xenoliths are characterized by a large range of $K_2O$ (0.02 to 0.51 wt%). Some xenoliths have $K_2O$ abundances comparable or even higher than those reported for metasomatized mantle peridotites [58], resembling in some cases to amphibole-mica-apatite-bearing mantle-xenolith suites [59]. $K_2O$ contents and the abundance of blebs and glassy drops allow to group the Paraguay xenoliths into two main suites, that is, a low-K suite (LK, $K_2O$ <0.15 wt%) and a high-K suite, the latter with abundant glassy drops and/or variable amounts of blebs and spongy clinopyroxenes (HK, $K_2O$ ≥0.2 wt%). Representative analyses are listed in Table 5, and complete sets of chemical analyses are in [54].

Based on Spera's [60] approach, Comin-Chiaramonti et al. [55] suggested that the ascent of the xenoliths to the surface took place in a very short time, for example, less than nine days (assuming a diameter of 45 cm, corresponding to the largest xenolith size, a density of 3.3 g/cm$^3$, and an origin of the hosting liquids from depth ~70–75 km, i.e., ~boundary between garnet and spinel peridotite). These mantle xenoliths provide a direct sampling of the subcontinental mantle in ASV along with the metasomatic processes [61, 62].

Coherent variations of major element in both suites follow a dunite-lherzolite sequence trending to the mantle composition [54]. The population is represented mainly by lherzolitic-harzburgitic (dunitic) compositions, mostly with a 0.55–0.63 range of $(SiO_2 + Al_2O_3)/(MgO + FeO_t)$ molar ratio. The residual character of the harzburgitic mantle xenoliths (believed to be consistent with melting and basalt-component removal) is also indicated by the decrease in the cpx/opx modal ratio with decreasing modal cpx, which fits the model variation trend induced by partial melting of lherzolite [56, 63].

Equilibration temperatures for orthopyroxene-clinopyroxene pairs [64] and for olivine-spinel pairs [65] vary between 862° and 1075°C and between 748° and 968°C, respectively. Intracrystalline temperatures [66–68] for clinopyroxene and orthopyroxene pairs vary between 936° and 1033°C and between 920 and 1120°C (LK and HK suites, resp., with pressures in the range of 1.1–2.3 GPa for both suites, with the higher values related to the more depleted xenoliths from the LK suite [53, 56]).

The oxygen isotope compositions on separates of clinopyroxene and coexisting olivine ($\delta^{18}O‰$) vary from 5.5 to 6.0‰and from 5.0 to 6.1‰, respectively [53]. These isotopic ratios are in the range for worldwide mantle phases: olivine, 4.4 to 7.5‰, and clinopyroxene, 4.8 to 6.7‰ [69, 70], and for South America mantle xenoliths: olivine, 4.9 to 6.4‰, and clinopyroxene, 5.0 to 6.0‰ [71]. The calculated Cpx-Ol isotopic temperatures [53, 71] are around 970–1070°C, 1030–1130°C (LK), and 1100–1180°C (HK), suggesting equilibration temperatures higher in the HK suite than in LK suite.

The highest temperature of the HK suite is substantially due to higher $\delta^{18}O‰$ in olivine than $\delta^{18}O‰$ in clinopyroxene. According to [72], the relatively low $\delta^{18}O$ in the pyroxenes reflects metasomatism by a silicate melt from subducted altered oceanic crust. Therefore, the Paraguayan HK suite would be for some geologists the best candidate for a subduction-related environment.

Notably, the ASV clinopyroxenes [73] have $V_{Cell}$ and $V_{M1}$ sites intermediate between those of plagioclase- and garnet-bearing mantle peridotites, that is, in a pressure range between 1.2 and 2.2 GPa. Thus, isotopic and crystallographic results are internally consistent.

On the whole, the ASV xenoliths define a geotherm [53] which, starting at about 830°C and 1.0 GPa, at the transition mantle-crust, intersects the hydrous peridotite solidus at about 1140°C and 2.1-2.2 GPa in the spinel peridotite facies, near to the transition with the garnet peridotite facies, believed to be the source of the sodic alkaline magmatism.

## 5. Geochemistry

Incompatible elements (i.e., large ion lithophile elements, IE, LILE, and high field strength elements (HFSE)),

Magmatism in the Asunción-Sapucai-Villarrica Graben (Eastern Paraguay) Revisited: Petrological, Geophysical, Geochemical, and Geodynamic Inferences

151

TABLE 2: Representative chemical analyses (recalculated to 100 wt% on anhydrous basis; in parentheses, standard deviation) of K-alkaline rocks (B-P suite, effusive and intrusive equivalents and dykes) from ASV [10, 13, 24]. Analysis of a silicobeforsitic flow, on anhydrous basis, is also shown [48, 74]. Ages as in [4].

| Suite B-P |  |  |  |  |  |  |
|---|---|---|---|---|---|---|
| K-alkaline | Basanite | Tephrite | Phonotephrite | Phonolite | Silicobeforsite 22% dolomite | Silicobeforsite carb. fraction |
| No. of samples | (4) | (6) | (4) | (1) | (1) |  |
| Wt% |  |  |  |  |  |  |
| $SiO_2$ | 46.95 (2.14) | 49.08 (1.31) | 52.36 (1.64) | 48.69 | 49.05 | — |
| $TiO_2$ | 1.78 (0.19) | 1.94 (0.26) | 1.42 (0.19) | 1.97 | 1.94 | — |
| $Al_2O_3$ | 12.84 (1.63) | 12.94 (1.17) | 17.27 (1.24) | 16.42 | 15.32 | — |
| $FeO_{tot}$ | 10.20 (1.80) | 10.02 (0.97) | 7.39 (0.75) | 8.45 | 9.60 | — |
| MnO | 0.18 (0.03) | 0.17 (0.02) | 0.15 (0.02) | 0.15 | 0.17 | — |
| MgO | 9.95 (2.02) | 7.88 (1.64) | 3.66 (0.58) | 3.86 | 5.28 | — |
| CaO | 11.31 (1.41) | 9.36 (0.85) | 7.00 (1.02) | 6.96 | 8.81 | — |
| $Na_2O$ | 2.42 (0.61) | 2.99 (0.92) | 4.34 (0.58) | 3.79 | 3.50 | — |
| $K_2O$ | 3.78 (1.18) | 4.99 (1.24) | 5.73 (0.17) | 8.60 | 6.14 | — |
| $P_2O_5$ | 0.59 (0.21) | 0.66 (0.18) | 0.58 (0.07) | 0.69 | 0.34 | — |
| Mg# | 0.67 | 0.62 | 0.51 | 0.45 | — |  |
| ppm |  |  |  |  |  |  |
| La | 82.1 (19.0) | 91.2 (13.0) | 93.3 (13.6) | 98 | 80 | 239.06 |
| Ce | 162.6 (30.4) | 164.1 (18.1) | 170.0 (25.0) | 186 | 145 | 420.34 |
| Nd | 64.4 (10.9) | 69.3 (4.9) | 66.6 (12.1) | 73.2 | 62 | 159.89 |
| Sm | 12.8 (2.4) | 12.8 (3.6) | 11.4 (1.6) | 15.1 | 11.3 | 19.06 |
| Eu | 3.0 (0.6) | 3.0 (0.4) | 2.8 (0.3) | 2.02 | 2.3 | 3.90 |
| Gd | 7.4 (1.9) | 6.8 (1.1) | 7.1 (0.6) | 5.93 | 8.8 | 14.90 |
| Dy | 4.4 (0.6) | 3.6 (0.2) | 3.9 (0.4) | 2.86 | 6.2 | 10.57 |
| Er | 1.9 (0.4) | 1.7 (0.3) | 1.9 (0.3) | 1.59 | 3.3 | 5.40 |
| Yb | 1.6 (0.1) | 1.4 (0.4) | 1.5 (0.3) | 1.59 | 1.9 | 3.16 |
| Lu | 0.25 (0.01) | 0.21 (0.06) | 0.26 (0.06) | 0.23 | 0.28 | 0.32 |
| Cr | 275 (106) | 378 (260) | 36 (23) | 53 | 51 | — |
| Ni | 81 (29) | 111 (42) | 28 (18) | 22 | 18 | — |
| Rb | 88 (18) | 117 (51) | 111 (35) | 260 | 148 | 0.21 |
| Sr | 1134 (279) | 1661 (227) | 1569 (10) | 1626 | 1126 | 268 |
| Ba | 1179 (252) | 1582 (147) | 1462 (215) | 1935 | 1350 | — |
| Th | 14.4 (9.9) | 11.4 (3.5) | 12.1 (1.6) | 12.7 | 15.2 | — |
| U | 2.9 (1.0) | 2.4 (0.7) | 2.5 (0.1) | 3.5 | 4.2 | — |
| Ta | 3.2 (0.2) | 3.0 (0.7) | 3.2 (0.3) | 5.1 | 2.7 | — |
| Nb | 38.0 (7.0) | 44 (10) | 47 (1) | 62 | 35 | — |
| Zr | 246 (42) | 282 (91) | 259 (35) | 365 | 280 | — |
| Y | 14.0 (2.8) | 17.3 (6.7) | 18.9 (3.0) | 13 | 18 | 23.75 |
| Measured |  |  |  |  |  |  |
| $(^{87}Sr/^{86}Sr)$ | 0.70761 (4) | 0.70774 (3) | 0.70727 (4) | 0.70752 (2) | 0.70807 (2) | 0.70762 (3) |
| $(^{143}Nd/^{144}Nd)$ | 0.51182 (4) | 0.51171 (1) | 0.51160 (4) | 0.51187 (2) | 0.511804 (6) | 0.511371 (8) |
| Age | 130 | 129 | 124 | 125 | 128 | — |
| Initial ratio |  |  |  |  |  |  |
| $(^{87}Sr/^{86}Sr)_i$ | 0.70697 | 0.70716 | 0.70671 | 0.70645 | 0.70700 | 0.70717 |
| $(^{143}Nd/^{144}Nd)_i$ | 0.51172 | 0.51162 | 0.51152 | 0.51177 | 0.51171 | 0.51131 |
| $T_{DM}$ | 2049 | 2042 | 2043 | 2067 | 1884 | 1844 |
| $R_1$ | 1054 | 713 | 365 | −397 | — | — |
| $R_2$ | 1955 | 1646 | 1269 | 1258 | — | — |
| $\varepsilon Sr$ | 34.31 | 37.04 | 30.58 | 26.88 | 34.7 | 34.17 |
| $\varepsilon Nd$ | −14.69 | −16.71 | −18.78 | −13.83 | −14.8 | −22.7 |

TABLE 3: Representative chemical analyses (recalculated to 100 wt% on anhydrous basis; in parentheses, standard deviation) of K-alkaline rocks (AB-T suite, effusive and intrusive equivalents and dykes) from ASV [13, 74]. Ages as in [4].

| Suite AB-T | | | | | |
|---|---|---|---|---|---|
| K-alkaline | Alkaline basalt | Trachybasalt | Trachyandesite | Trachyphonolite | Trachyte |
| No. of samples | (3) | (3) | (7) | (5) | (1) |
| Wt% | | | | | |
| $SiO_2$ | 48.69 (1.50) | 50.90 (1.56) | 53.16 (1.18) | 58.25 (1.44) | 58.95 |
| $TiO_2$ | 1.69 (0.48) | 1.49 (0.31) | 1.42 (0.21) | 0.91 (0.23) | 1.94 |
| $Al_2O_3$ | 15.17 (1.86) | 17.07 (1.06) | 16.99 (1.45) | 18.39 (1.01) | 19.03 |
| $FeO_{tot}$ | 10.31 (0.76) | 9.04 (0.57) | 7.96 (0.75) | 4.87 (1.51) | 4.49 |
| MnO | 0.19 (0.01) | 0.16 (0.02) | 0.15 (0.02) | 0.16 (0.01) | 0.12 |
| MgO | 6.71 (0.89) | 5.01 (0.24) | 4.03 (0.58) | 0.89 (0.90) | 1.69 |
| CaO | 11.09 (0.63) | 8.58 (1.11) | 6.95 (0.66) | 4.59 (1.09) | 3.37 |
| $Na_2O$ | 3.16 (0.42) | 3.64 (0.44) | 4.10 (0.38) | 5.89 (1.38) | 4.21 |
| $K_2O$ | 2.46 (0.42) | 3.65 (0.72) | 4.61 (1.06) | 5.76 (1.13) | 7.00 |
| $P_2O_5$ | 0.55 (0.36) | 0.50 (0.17) | 0.55 (0.09) | 0.29 (0.10) | 0.25 |
| Mg# | 0.58 | 0.53 | 0.51 | 0.32 | 0.40 |
| ppm | | | | | |
| La | 51.2 (16.0) | 54.0 (18.6) | 82.8 (29.7) | 154.0 (33.9) | 136 |
| Ce | 106.7 (24.4) | 109.0 (21.0) | 154.2 (55.7) | 261.1 (44.2) | 206 |
| Nd | 52.7 (19.7) | 47.9 (12.9) | 64.7 (15.7) | 84.0 (17.7) | 78 |
| Sm | 10.5 (1.6) | 9.4 (30.5) | 10.8 (2.1) | 12.7 (3.5) | 15.1 |
| Eu | 2.8 (1.0) | 2.5 (0.3) | 2.8 (0.5) | 3.28 (0.8) | 3.9 |
| Gd | 6.8 (2.4) | 6.0 (1.2) | 6.4 (1.0) | 5.70 (1.6) | 9.4 |
| Dy | 4.1 (0.7) | 4.2 (0.6) | 4.2 (0.7) | 5.44 (1.1) | 6.3 |
| Er | 1.7 (0.1) | 1.8 (0.3) | 1.6 (0.4) | 2.59 (0.58) | 2.3 |
| Yb | 1.5 (0.1) | 1.6 (0.4) | 1.6 (0.4) | 2.25 (0.46) | 2.0 |
| Lu | 0.22 (0.05) | 0.24 (0.03) | 0.28 (0.08) | 0.35 (0.08) | 0.31 |
| Cr | 143 (24) | 92 (59) | 38 (34) | 10 (6) | 8 |
| Ni | 49 (6) | 31 (18) | 17 (12) | 6 (2) | 2 |
| Rb | 53 (13) | 98 (16) | 87 (20) | 7 (3) | 133 |
| Sr | 1573 (263) | 1389 (407) | 1474 (194) | 77 (34) | 600 |
| Ba | 1033 (321) | 1280 (170) | 1211 (168) | 1377 (215) | 1726 |
| Th | 4.3 (2.1) | 7.9 (1.5) | 14.6 (8.5) | 31.7 (8.5) | — |
| U | 1.3 (0.5) | 1.7 (0.6) | 3.4 (3.0) | 9.5 (3.3) | — |
| Ta | 2.3 (1.1) | 1.6 (0.4) | 2.8 (0.7) | 2.6 (0.6) | 2.5 |
| Nb | 18.1 (9.0) | 25.0 (6.0) | 40 (12) | 62 (18) | 61 |
| Zr | 260 (31) | 220 (33) | 268 (14) | 411 (128) | 392 |
| Y | 14.5 (9.2) | 14.6 (6.5) | 19.5 (3.2) | 25 (4) | 23 |
| Measured | | | | | |
| $(^{87}Sr/^{86}Sr)$ | 0.70717 (4) | 0.70721 (5) | 0.70753 (4) | 0.70752 (2) | 0.70807 (2) |
| $(^{143}Nd/^{144}Nd)$ | 0.51185 (4) | 0.51170 (3) | 0.51160 (4) | 0.51187 (2) | 0.51181 (6) |
| Age | 125 | 119 | 129 | 125 | 128 |
| Initial ratio | | | | | |
| $(^{87}Sr/^{86}Sr)_i$ | 0.70693 | 0.70673 | 0.70709 | 0.70708 | 0.70639 |
| $(^{143}Nd/^{144}Nd)_i$ | 0.51175 | 0.51161 | 0.51151 | 0.51179 | 0.51171 |
| $T_{DM}$ | 2009 | 2199 | 1999 | 1530 | 2001 |
| $R_1$ | 1215 | 954 | 749 | 545 | 649 |
| $R_2$ | 1817 | 1501 | 1277 | 1014 | 818 |
| $\varepsilon Sr$ | 33.68 | 30.74 | 36.09 | 35.85 | 26.0 |
| $\varepsilon Nd$ | −14.16 | −17.12 | −18.68 | −13.31 | −14.9 |

Magmatism in the Asunción-Sapucai-Villarrica Graben (Eastern Paraguay) Revisited: Petrological, Geophysical, Geochemical, and Geodynamic Inferences

153

TABLE 4: Representative chemical analyses (recalculated to 100 wt% on anhydrous basis; in parentheses, standard deviation) of Na-alkaline rocks from ASV ASV [13, 74]. Ages as in [4].

| Suite: sodic Na-Alkaline | Ankaratrite | Ankaratrite | Ankaratrite | Melanephelinite | Peralkaline Phonolite |
|---|---|---|---|---|---|
| No. of samples | (1) | (1) | (5) | (1) | (1) |
| Wt% | | | | | |
| $SiO_2$ | 43.71 | 43.97 | 43.29 (0.65) | 45.67 | 55.96 |
| $TiO_2$ | 2.39 | 2.12 | 2.50 (0.48) | 2.21 | 0.43 |
| $Al_2O_3$ | 14.84 | 13.87 | 13.42 (0.95) | 14.83 | 21.19 |
| $FeO_{tot}$ | 10.01 | 10.44 | 10.71 (0.40) | 7.83 | 4.04 |
| MnO | 0.18 | 0.21 | 0.19 (0.01) | 0.20 | 0.22 |
| MgO | 10.07 | 9.87 | 11.47 (1.23) | 8.92 | 0.21 |
| CaO | 12.06 | 10.81 | 11.82 (0.28) | 10.78 | 1.34 |
| $Na_2O$ | 5.15 | 6.00 | 4.12 (0.21) | 6.73 | 10.77 |
| $K_2O$ | 0.89 | 1.51 | 1.61 (0.32) | 1.54 | 5.76 |
| $P_2O_5$ | 0.90 | 1.22 | 0.88 (0.14) | 1.28 | 0.08 |
| Mg# | 0.74 | 0.72 | 0.69 | 0.58 | 0.22 |
| ppm | | | | | |
| La | 81 | 119 | 84.2 (12.4) | 118 | 120 |
| Ce | 145 | 186 | 162.8 (24.3) | 190 | 191 |
| Nd | 43.4 | 63.7 | 55.2 (8.5) | 65.1 | 65 |
| Sm | 8.15 | 11.23 | 9.4 (1.2) | 11.45 | 11 |
| Eu | 1.89 | 2.15 | 2.7 (0.3) | 2.18 | — |
| Gd | 5.77 | 5.16 | 6.7 (0.7) | 5.23 | — |
| Dy | 4.81 | 4.81 | 5.5 (0.7) | 5.00 | — |
| Er | 2.13 | 2.75 | 2.4 (0.3) | 2.86 | — |
| Yb | 1.65 | 1.79 | 1.8 (0.3) | 1.87 | — |
| Lu | 0.23 | 0.27 | 0.27 (0.04) | 0.28 | — |
| Cr | 542 | 648 | 490 (74) | 470 | 2 |
| Ni | 207 | 273 | 249 (48) | 239 | 6 |
| Rb | 22 | 59 | 53 (13) | 63 | 60 |
| Sr | 1016 | 1013 | 1109 (63) | 1095 | 654 |
| Ba | 1090 | 980 | 1090 (95) | 1094 | 367 |
| Th | 10.5 | 10.5 | 11.0 (2.4) | 11.6 | — |
| U | 2.3 | 2.3 | 2.4 (0.5) | 2.5 | — |
| Ta | 5.9 | 5.9 | 8.4 (1.0) | 6.5 | — |
| Nb | 86 | 101 | 105 (13) | 113 | — |
| Zr | 152 | 234 | 250 (43) | 228 | 955 |
| Y | 26 | 33 | 29 (4) | 32 | 39 |
| Measured | | | | | |
| $(^{87}Sr/^{86}Sr)$ | 0.70395 (1) | 0.70374 (1) | 0.70381 (8) | 0.70392 (1) | 0.70405 (2) |
| $(^{143}Nd/^{144}Nd)$ | 0.51275 (1) | 0.51276 (1) | 0.512724 (61) | 0.51274 (2) | 0.51268 (2) |
| Age | 46 | 46 | 46 | 50 | 60 |
| Initial ratio | | | | | |
| $(^{87}Sr/^{86}Sr)_i$ | 0.70391 | 0.70364 | 0.70373 | 0.70381 | 0.70384 |
| $(^{143}Nd/^{144}Nd)_i$ | 0.51272 | 0.51273 | 0.51269 | 0.51271 | 0.51264 |
| $T_{DM}$ | 581 | 532 | 562 | 558 | 617 |
| $R_1$ | 582 | 100 | 682 | 18 | −1567 |
| $R_2$ | 2080 | 1918 | 2096 | 1886 | 570 |
| $\varepsilon Sr$ | −10.42 | −14.30 | −13.05 | −11.79 | −11.19 |
| $\varepsilon Nd$ | 2.67 | 2.91 | 2.23 | 2.57 | 1.54 |

TABLE 5: Average chemical analyses (recalculated to 100 wt% on anhydrous basis; in parentheses, standard deviation) of mantle xenoliths (LK and HK suites, $K_2O$ <0.15 wt%, and >0.2 wt%, resp.from ASV [53, 54, 56]. The age is the average of ages of Table 4. The bleb is a representative glassy drop in an H-K harzburgite [54, 56].

| Whol-rock | L-K lherzolite | L-K harzburgite | H-K lherzolite | H-K harzburgite | H-K bleb |
|---|---|---|---|---|---|
| No. of samples | (5) | (6) | (3) | (2) | (1) |
| Wt% | | | | | |
| $SiO_2$ | 44.02 (1.15) | 43.95 (0.35) | 44.65 (0.11) | 44.63 (0.20) | 46.15 |
| $TiO_2$ | 0.03 (0.02) | 0.01 (0.00) | 0.07 (0.03) | 0.02 (0.00) | 0.70 |
| $Al_2O_3$ | 1.76 (0.86) | 1.14 (0.50) | 2.37 (0.04) | 1.36 (0.03) | 10.10 |
| $FeO_{tot}$ | 7.97 (0.13) | 7.85 (0.22) | 7.92 (0.24) | 7.52 (0.01) | 6.74 |
| MnO | 0.12 (0.01) | 0.11 (0.00) | 0.11 (0.01) | 0.11 (0.00) | 0.14 |
| MgO | 44.17 (1.87) | 45.58 (1.60) | 41.55 (0.46) | 44.76 (0.21) | 23.67 |
| CaO | 1.75 (0.86) | 1.18 (0.69) | 2.64 (0.23) | 1.12 (0.02) | 6.56 |
| $Na_2O$ | 0.12 (0.01) | 0.10 (0.05) | 0.28 (0.01) | 0.16 (0.01) | 2.00 |
| $K_2O$ | 0.09 (0.04) | 0.06 (0.04) | 0.39 (0.10) | 0.32 (0.00) | 3.10 |
| $P_2O_5$ | 0.21 (0.10) | 0.01 (0.01) | 0.01 (0.00) | 0.02 (0.00) | 0.83 |
| ppm | | | | | |
| La | 1.69 (0.12) | 2.87 (0.51) | 2.66 (0.52) | 6.71 (0.69) | 77.39 |
| Ce | 1.53 (0.09) | 4.09 (0.75) | 3.35 (0.67) | 8.45 (0.81) | 106.25 |
| Nd | 0.20 (0.01) | 0.28 (0.16) | 0.96 (0.55) | 2.42 (0.25) | 27.74 |
| Sm | 0.012 (0.004) | 0.038 (0.015) | 0.260 (0.014) | 0.51 (0.20) | 4.16 |
| Eu | 0.001 (0.001) | 0.006 (0.002) | 0.010 (0.01) | 0.020 (0.006) | 1.00 |
| Gd | 0.07 (0.008) | 0.04 (0.01) | 0.37 (0.02) | 0.74 (0.03) | 2.92 |
| Dy | 0.10 (0.01) | 0.05 (0.01) | 0.43 (0.02) | 0.91 (0.04) | 2.61 |
| Er | 0.07 (0.001) | 0.05 (0.01) | 0.30 (0.01) | 0.65 (0.06) | 1.68 |
| Yb | 0.07 (0.001) | 0.08 (0.01) | 0.29 (0.01) | 0.63 (0.03) | 1.23 |
| Lu | — | — | — | — | |
| Cr | 2601 (350) | 2421 (357) | 2360 (88) | 2635 (30) | 38114 |
| Ni | 2307 (160) | 2309 (102) | 2120 (4) | 2307 (73) | 1124 |
| Rb | 3.33 (9.51) | 2.67 (1.63) | 7.50 (2.12) | 6.00 (1.41) | 120 |
| Sr | 22.41 (10.47) | 11.15 (4.58) | 28.89 (10.29) | 64.67 (7.62) | 1414 |
| Ba | 20.0 (10.0) | 4.67 (1.18) | 40 (8) | 34.5 (14.8) | 1781 |
| Th | — | — | — | — | — |
| U | — | — | — | — | — |
| Ta | — | | — | — | — |
| Nb | 5.91 (0.58) | 1.78 (0.20) | 6.39 (1.77) | 6.98 (1.01) | 155 |
| Zr | 5.95 (0.41) | 6.08 (0.45) | 6.90 (0.13) | 7.86 (0.93) | 189 |
| Y | 0.51 (0.10) | 0.72 (0.09) | 2.62 (0.09) | 5.70 (1.08) | 14.70 |
| Measured | | | | | |
| $(^{87}Sr/^{86}Sr)$ | 0.70426 (0.00005) | 0.70421 (0.00038) | 0.70416 (0.00006) | 0.70395 (0.00002) | — |
| $(^{143}Nd/^{144}Nd)$ | 0.51264 (0.00007) | 0.51275 (0.00039) | 0.51299 (0.00033) | 0.51288 (0.00005) | — |
| Age | (50 ± 6) | (50 ± 6) | (50 ± 6) | (50 ± 6) | — |
| Initial ratio | | | | | |
| $(^{87}Sr/^{86}Sr)_i$ | 0.70398 | 0.70376 | 0.70367 | 0.70377 | |
| $(^{143}Nd/^{144}Nd)_i$ | 0.51263 | 0.51272 | 0.51294 | 0.51285 | |
| $T_{DM}$ | 427 | 448 | 446 | 453 | — |
| $\varepsilon Sr$ | −9.40 | −12.55 | −13.80 | −12.31 | — |
| $\varepsilon Nd$ | 1.06 | 2.92 | 7.08 | 5.16 | — |

Magmatism in the Asunción-Sapucai-Villarrica Graben (Eastern Paraguay) Revisited: Petrological, Geophysical, Geochemical, and Geodynamic Inferences

155

considered with the Sr-Nd isotopic composition, indicate that the ASV magmatic was generated from geochemically distinct (enriched versus depleted) mantle sources (cf. [75]).

### 5.1. Incompatible Elements.

*5.1. Incompatible Elements.* Mantle-normalized IE patterns for the various and different magmatic rock groups are represented in Figure 6.

The L-Ti and H-Ti ASV tholeiites are distinct in terms of their relatively low elemental abundances and high LILE/HFSE ratios. In particular, their marked Ta-Nb negative spikes are similar to those of the potassic alkaline magmas from ASV, but clearly different in comparison to the Cenozoic sodic alkaline rocks from the same area.

The suites are similarly enriched in REE and exhibit steep, subparallel LREE trends ($(La/Lu)_{CN}$ = 26–161, 17–62 and 11–46 for B-P, AB-T and Na rocks, resp.), which tend to flatten out for HREE ($(Dy/Lu)_{CN}$ = 1.24–1.96, 1.09–2.00 and 0.56–2.05 for B-P, AB-T and Na rocks, resp.); HK-dykes are excepted. REE profiles with LREE enrichment and flat HREE suggest mantle sources previously depleted by melt extraction and subsequently enriched [76].

Multielemental diagrams, normalized to a primordial mantle composition (Figure 6) show a substantial overlap of B-P and AB-T compositions, negative Nb, Ta, P, Ti, and Y spikes, and positive U, K, and Sr anomalies. In general, B-P compositions are higher in Rb, K, Zr, Hf, Ti, Y, LREE, and MREE than AB-T compositions. On the contrary, the Na rocks yielded La/Nb and La/Ta ratios close to unity, respectively, and Nb/K and Ta/K ratios greater than 1.0, respectively. Notably, the variations of incompatible elements of the AB-T compositions mimic to some extent those of the ASV H-Ti (and L-Ti) tholeiitic basalts, which approach the lower (Rb to Nd) and higher (Sm to Lu) elemental concentrations of that suite (Figure 6).

In summary, substantial overlap in bulk-rock chemistry exists between the investigated B-P and AB-T rocks, both characterized by variable K/Na ratio, the K types being dominant. REE and other incompatible elements show similar concentration levels and variation trends in the two suites. The mantle normalized incompatible element patterns of both ASV suites show strong affinities, including negative "Ta-Nb-Ti anomalies," with the Paraná tholeiites [10].

On the whole, the geochemical features suggest that the enrichment processes were related to small-volume melts in a lithospheric mantle [4, 10, 74].

*5.2. Sr-Nd Isotopes.* The investigated rocks from ASV show a large distribution of Sr-Nd isotopic compositions (Figure 7), delineating a trend similar to the "low Nd" array of [77] (cf. also "Paraguay array" of [4]). Due to the high Sr and Nd of the most "primitive" alkaline rocks and associated carbonatites from Eastern Paraguay, Comin-Chiaramonti et al. [12] suggested that initial Sr-Nd isotopic ratios of these rocks can be considered uncontaminated by the crust and, as a result, representative of the isotopic composition of their mantle source(s). The ASV potassic rocks have the highest (time integrated) $Sr_i$ and the lowest $Nd_i$. Including the carbonatites of the BP-suite [24, 78], the $Sr_i$ and $Nd_i$ range from

0.70645 to 0.70716 and from 0.51151 to 0.51179, respectively (Tables 2 and 3). These values are quite distinct from those of the ASV Paleocene sodic rocks (ca. 60 Ma), which plot within the depleted quadrant ($Sr_i$ = 0.70364–0.70391, $Nd_i$ = 0.51264–0.51273; cf. Table 4), towards the HIMU-DMM depleted mantle components (Figure 7). Notably, $Sr_i$ and $Nd_i$ of the tholeiites, believed to be uncontaminated [10], both H- and L-Ti, are intermediate between the potassic and sodic rocks: 0.70456–0.70576 and 0.51225–0.51256, respectively (cf. Table 4). These values approach to the range of the Early Cretaceous uncontaminated tholeiites from the Paraná Basin [47], that is, $Sr_i$ = 0.70527 ± 0.00051 and $Nd_i$ = 0.51264 ± 0.00011. To be stressed that the genesis of these tholeiites requires lithospheric mantle components, as indicated by K-alkaline and carbonatitic rocks from ASV [4, 5].

Considering the whole Paraná-Agola-Etendeka system (PAE), and that the ASV is located at the central westernmost side of the PAE [5, 6], the different geochemical behaviour in the different PAE sectors implies also different sources. Utilizing the $T_{DM}$ (Nd) model ages [78, 79], it should be noted that (1) the H-Ti flood tholeiites and dyke swarms from the Paraná Basin and the Early Cretaceous potassic rocks and carbonatites from Eastern Paraguay range mainly from 0.9 to 2.1 Ga, whereas in Angola and Namibia the Early Cretaceous K-alkaline rocks vary from 0.4 to 0.9 Ga; (2) the L-Ti tholeiites display a major $T_{DM}$ variation, from 0.7 to 2.4 Ga (mean 1.6 ± 0.3) with an increase of the model ages from North to South; (3) Late Cretaceous alkaline rocks show model ages ranging from 0.6 to 1 Ga, similar to the age span shown by the Triassic to Paleocene sodic alkaline rock-types lying along the Paraguay river [4].

Thus, model ages suggest that some distinct "metasomatic events" may have occurred during Paleoproterozoic to Neoproterozoic times as precursor to the alkaline and tholeiitic magmas in the PAE [78].

The significance of model ages is supported by (1) the isotopic overlapping of different igneous rocks (e.g., H-Ti and L-Ti tholeiites or K-alkaline rocks and carbonatites), which cannot be accidental and points to sampling of ancient reservoirs formed at different times from the same subcontinental upper mantle [79]; (2) whatever the implication, that is, heterogeneity induced by recycled crust in the mantle [80, 81], or occurrence of variably veined material in the subcontinental upper mantle, or both [82, 83], Pb isotope data indicate a mantle source of ca. 1.8 Ga for the Paraná H-Ti tholeiites. Since much of the crust in southern Brazil appears to have been formed at ca. 2 Ga ago [84], it follows that magma genesis involved ancient lithospheric mantle reset at well-defined isotopic ranges. A veined lithospheric mantle (amphibole/phlogopite-carbonate-lherzolite and amphibole-lherzolite + $CO_2$-fluid type III and IV veins of Meen et al. [85] of Proterozoic age) may well account for the magmatism both of the PAE and ASV (Figure 8).

*5.3. Mantle Sources.* The origin of the ASV magmas is closely related to, and probably constrained by, the geodynamic processes which promoted the generation of the adjacent and coeval magmatism in Brazil [10]. The origin and

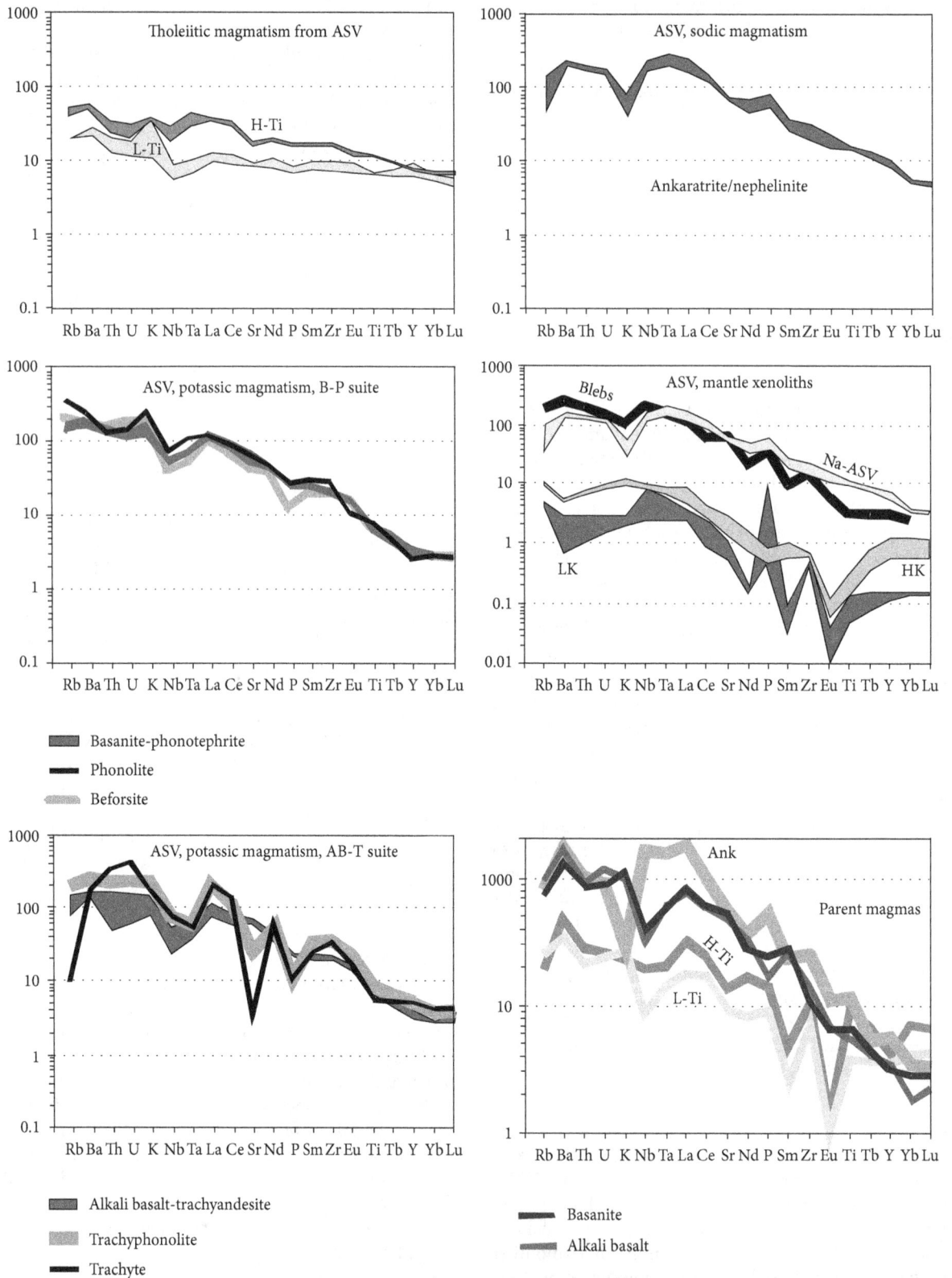

FIGURE 6: ASV representative compositions (data source as in Tables 1 to 5): incompatible elements normalized to the primitive mantle [86]. Parent magmas: normalizations relative to the calculated parental magmas as proposed by [2, 10, 43, 46]; L-Ti (Mg# 0.64, Ni 250 ppm); H-Ti (Mg# 0.60, Ni 250 ppm); basanite (Mg# 0.74, Ni 710 ppm); alkaline basalt (Mg# 0.74, Ni 323 ppm); ankaratrite (Mg# 0.70, Ni 993 ppm).

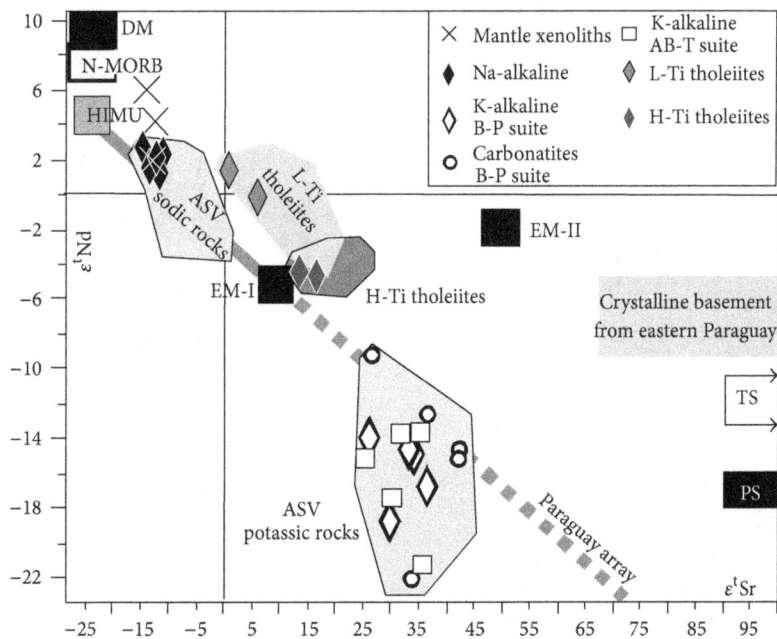

FIGURE 7: Sr and Nd isotopic plot in $\varepsilon$-$\varepsilon$ notation: $\varepsilon_t$Sr versus $\varepsilon_t$Nd correlation diagram for igneous rocks from ASV. Data source: Tables 1 to 5 and [4, 56, 78, 87–89]. DM, HIMU, EMI, and EMII mantle components, terrigenous and pelagic sediments, and TS and PS, respectively [77]; crystalline basement as in [10]. The $\varepsilon_t$Sr and $\varepsilon_t$Nd time-integrated values were calculated using the following values for Bulk Earth: $^{87}Sr/^{86}Sr = 0.7047$; $^{87}Rb/^{86}Rb = 0.0827$; $^{143}Nd/^{144}Nd = 0.512638$; $^{147}Sm/^{144}Nd = 0.1967$ [50]. Paraguay array is according to [4, 5].

emplacement of the ASV magmatism occurred before, during, and after the opening of the South Atlantic and appears to be related to early stages of lithospheric extension, associated with an anomalously hot mantle [3, 90, 91]. The K/Na and trace element variations and Sr-Nd isotope characteristics of the ASV suites support the view that the lithospheric mantle played an important role in their genesis as well as that of the Paraná flood basalts in Brazil [47, 92, 93].

It should be noted that the most mafic tholeiitic and potassic ASV rock-types are relatively evolved compared with expected primary compositions (e.g., Mg# >0.60–0.65, Ni >235 ppm), ankaratrite is excepted. Possible ASV primary melts (e.g., Mg# ~0.74) in equilibrium with Fo$_{90-91}$ are expected to have fractionated olivine (Ol) and clinopyroxene (Cpx) at or near the mantle source and certainly during the ascent to the surface [13]. Neglecting the effects of polythermal-polybaric fractionation on the chemistry of the fractionates, crystal fractions have been calculated to restore the selected ASV parental compositions to possible near-primary melts in equilibrium with their mantle sources (cf. Table 3 of [10]).

Mass balance calculations, starting from different garnet-phlogopite peridotites (cf. [94]; whole-rock and mineral compositions in Table 3 of [10] and Table 10 of [43]), indicate that the ASV composition of primary melts can be derived from high and relatively high melting degrees of anhydrous garnet or phlogopite-bearing peridotite, that is, 12 and 30% for tholeiitic basalts (H-Ti and L-Ti, respectively) and 4–11% for the alkaline rock-types (Table 3 of [10]). However, the presence of a relative K enrichment in the ASV potassic

and tholeiitic rocks (Figure 6) suggests that a K-bearing phase (e.g., phlogopite) did not represent a residual phase during partial melting. Phlogopite, instead, was probably a residual phase in the mantle source(s) of the ASV sodic rocks. Notably, the parent melts (Figure 6) show high abundances of IE and high LREE/HREE ratios, which require mantle sources enriched in IE before or subcontemporaneously with melting process [94]. Nd model ages approximately indicate that the IE enrichment in the source mantle of the ASV tholeiitic and potassic rocks probably occurred during Paleoproterozoic times (mean 2.03 ± 0.08 Ga), whereas those relative to the sodic rocks would be related to Late Neoproterozoic events (0.57 ± 0.03 Ga). On the other hand, the mantle xenoliths display Nd model ages of 444 ± 11 Ma (cf. Tables 1 to 5).

Concluding, the patterns of the mantle sources of the ASV potassic rocks (along with H-Ti and L-Ti tholeiites) are characterized by negative "Ta-Nb-Ti" and positive Ba and Sm spikes. On the contrary, the patterns of the mantle sources of the ASV sodic rocks (ankaratrites-melanephelinites) show positive Ta-Nb and Zr and negative K and Sm spikes. It seems, therefore, that the genesis of the ASV alkaline magmatism is dominated by a lithospheric mantle, characterized by small-scale heterogeneity, and documented by the occurrence of bleb-like glass in spinel peridotite nodules from the sodic ultramafic rocks [54, 55].

Finally, the isotopic signature of the tholeiitic and K-Na-alkaline-magmatism from the PAE in general, and from ASV in particular, may reflect ancient heterogeneities preserved in the subcontinental lithospheric mantle. As matter of fact, all the geochemical data indicate that the PAE magmatism

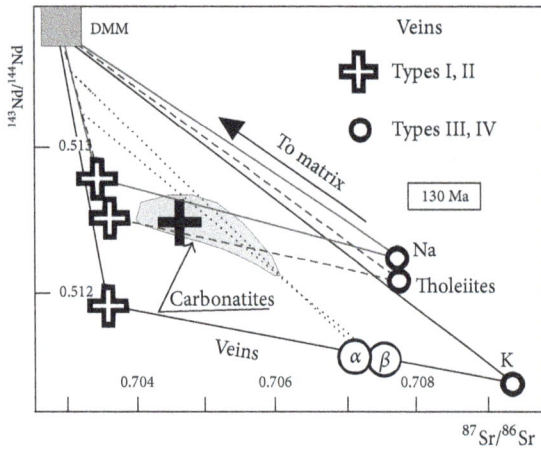

FIGURE 8: Calculated subcontinental upper mantle (SCUM) isotopic composition at 2.0 Ga ago, projected to 130 Ma [85], modified after Figure 10 of [6]. Parental melts with various Rb/Sr and Sm/Nd ratios are assumed for K, Na (potassic and sodic rocks from ASV [10]), and PAE tholeiites [3]. It should be noted that the compositions of metasomites formed from a single metasomatizing melt vary with the evolution of the melt. Consequently, the veins will define a trend of shallow slope, and mixing curves between vein and matrix will define an array towards the matrix (cf. $\alpha$ and $\beta$ regression lines). Model depleted mantle (DMM): Rb = 0, Sr = 0.133, Sm = 0.314, and Nd = 0.628; present day Bulk Earth: $^{87}Sr/^{86}Sr$ = 0.70475, $^{87}Rb/^{86}Sr$ = 0.0816, $^{143}Nd/^{144}Nd$ = 0.512638, and $^{147}Sm/^{143}Nd$ = 0.1967; $(Rb/Sr)_{diopside} : (Rb/Sr)_{melt} \approx 0.125$, $(Sm/Nd)_{diopside} : (Sm/Nd)_{melt} \approx 1.5$; K : Rb/Sr = 0.0957, Sm/Nd = 0.1344; Na : Rb/Sr = 0.0732, Sm/Nd = 0.2295; Th : Rb/Sr = 0.0733, Sm/Nd = 0.2082.

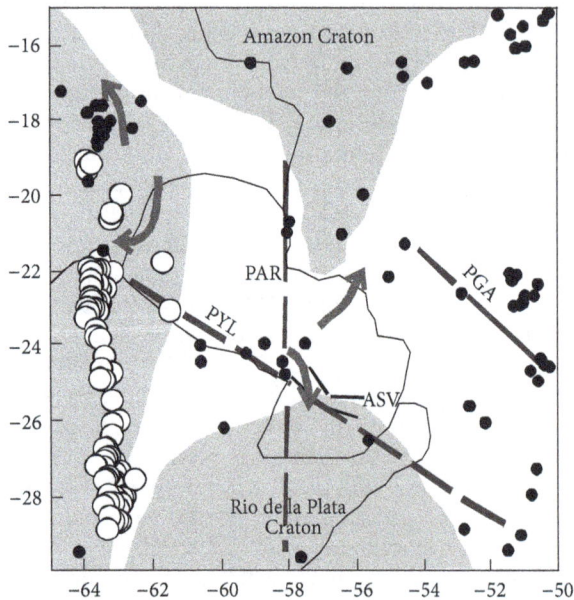

(a)

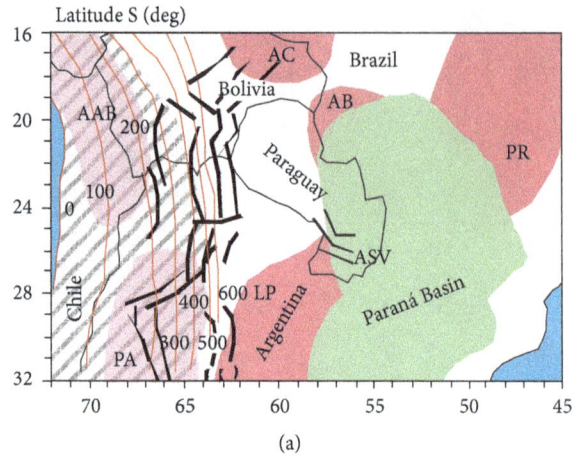

$\delta$Vp/Vp (%)

−1.5　　−0.9 −0.6 −0.3　0.0　0.3　0.6 0.9　　1.5

(b)

FIGURE 10: (a) General map, including ASV, showing contours (red lines) of the depth (km) of the subducting Nazca slab based on seismic data of [96]. Heavy lines (black) outline the Cretaceous rift systems. The hatched area roughly marks the extension of intense early Paleozoic reworking of Proterozoic material, but the exact border to the Brazilian Craton remains unknown [97]. Pink fields delineate inferred positions of major cratonic fragments below Phanerozoic cover [98]: AAB, Arequipa-Antofalla; AC, Amazon Craton; AB, Apa Block; PR, Paranapanema; LP, Rio de la Plata; PA, Pampia. (b) Seismic tomography image [99] across a profile approximately at 24° Lat. S. Note that the low-velocity feature in the mantle to the East has been interpreted as a fossil mantle plume [100].

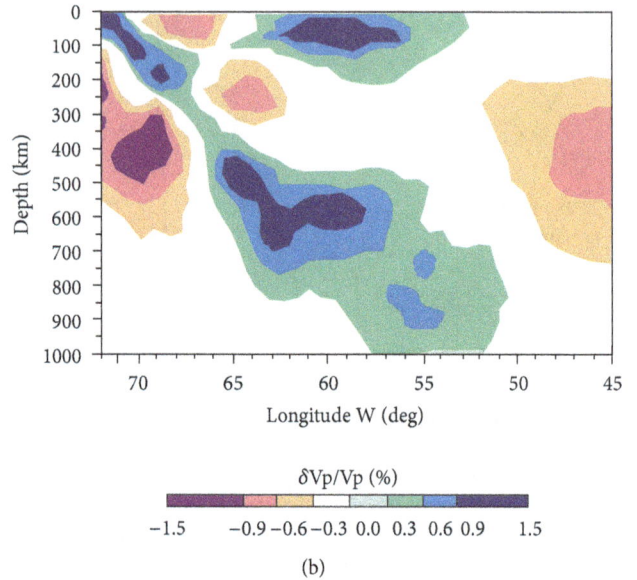

FIGURE 9: Earthquakes distribution at the Paraguay and neighbouring regions: open and full circles, earthquake hypocenters >500 and <70 km, respectively [29]. Heavy black lines indicate extensional lineaments: PGA, Ponta Grossa Arch; PAR and PYL, Paraguays and Pylcomaio lineaments, respectively (cf. also inset of Figure 3). The dark grey delineates the rotational subplate trends [24, 95]. ASV: Asunción-Sapucai-Villarrica graben.

requires heterogeneous mantle sources, also in terms of radiogenic isotopes [6, 10]: such heterogeneities probably represent metasomatic processes which have occurred mainly at 0.5–1.0 Ga and at 1.5–2.0 Ga (cf. $T_{DM}$ of Tables 1–4).

## 6. Geodynamic Implications

Some general considerations on the geodynamics are discussed in [4–6, 87–89]. The ASV geodynamic evolution is related to that of the Western Gondwana (Figure 1), which reflects the regional amalgamation processes and requires an overall understanding of the pre-Gondwanic setting, at least at the time of the Brasiliano cycle, both at the Atlantic and Pacific systems. The Brasiliano cycle was developed diachronically between about 890 and 480 Ma, until the final basement of the South American Platform was attained [101]. This occurred during the Lower Ordovician, where Unrug [102] suggested a mosaic of lithospheric fragments linked by several (accretionary, collisional) Neoproterozoic mobile belts. After amalgamation, the Gondwana supercontinent accumulated Paleozoic and Triassic-Jurassic sediments. Concomitantly, it was continuously and laterally accreted to the western borders by successive orogenic belts, until the Pangea Formation [103, 104]. The main cratonic fragments descending from ancestors of the Pangea, for example, the Amazonic and Rio de La Plata cratons, were revoked, and smaller ancient crustal blocks formed at the Paraguay boundaries [105]. The magmatism was likely driven by the relative extensional regimes derived from the relative movements of the ancient blocks (Figure 9), probably induced by counterclockwise and clockwise movements (North and South, resp.) hinged at about 20° and 24°S [106].

The earthquake distribution and mechanisms [29] with hypocenters >500 km and <70 km, respectively (Figure 9), coincide with the inferred Nazca plate subduction at the Paraguay latitude (Figure 10), where the depth of the lithospheric earthquakes, together with the paleomagnetic results [38, 95, 107], delineates different rotational paths at about 18–24°S (i.e., Chaco-Pantanal Basin). These data indicate extensional subplate tectonics along the Andean system [95], coupled with the lineaments of lithospheric earthquakes versus the Atlantic system that parallels the Ponta Grossa Arch and the lineaments of Rio Paraguay and Rio Pilcomayo, also clearly related to extensional environments [24, 29, 78].

Crucial to the genesis of the ASV magma types is the link with the whole PAE system and with the geodynamic processes responsible for the opening of the South Atlantic. According to Nürberg and Müller [108], the sea-floor spreading in the South Atlantic at ASV latitude started at about 125–127 Ma (Chron M4), but North of the Walvis-Rio Grande ridge (Figure 1(b)) the onset of the oceanic crust would be younger, for example, 113 Ma [109]. Therefore, the Early Cretaceous alkaline and alkaline-carbonatite complexes from ASV (and PAE) are subcoeval with respect to the main flood tholeiites of the Paraná Basin and occurred during the early stages of rifting, before the main continental separation.

The origin of alkaline-carbonatitic magmatism in terms of plate tectonics is currently debated [77, 110] and often considered in terms of models only. In the present context, the different westward angular velocity of the lithospheric fragments in the South American plate, as defined by the "second order plate boundaries" (e.g., the Pylcomaio lineament [18]) as well as the different rotational trends at S19–24° Latitude, would favour decompression and melting at different times of variously metasomatized (wet spot) portion of the lithospheric mantle with variable isotopic signatures [24]. Small-scale metasomatism is indicated by the different suites of mantle xenoliths and significant H-O-C and F

variations, expected in the mantle source from the occurrence of the related carbonatites. The latter imply lowering of the solidus and, along with an extensional tectonic, favour the melting of the mantle and magma ascent. This scenario also accounts for the presence of Late Cretaceous to Tertiary sodic magmatism in the PAE and in the Eastern Paraguay, where rifting structures are evident (cf. Figures 2 and 4).

## 7. Concluding Remarks

Geological and geophysical results show that the Mesozoic-Tertiary block-faulting tectonics in Eastern Paraguay is responsible for NW-trending grabens (ASV and Amambay), fault systems, and fault-controlled basins, for example, Jejui-Aguaray-Guazu, Asunción-Encarnación, San Pedro (cf. Figure 2).

Distinct magmatic events are widespread in the Asunción-Sapucai-Villarrica graben, the major extensional structure in Eastern Paraguay, that is, Early Cretaceous tholeiites and K-alkaline-carbonatitic complexes and dykes, along with Tertiary sodic magmatism. It is expected that beneath a rifted continental area, as ASV, the lithospheric mantle may have been thinned and subjected to high heat flow. This allowed the geotherm to intersect the peridotite solidus in the presence of $H_2O$-$CO_2$-rich fluids and favour the melt formations. Under these conditions, the derived melts may form, dissolving most of the "volatile" phases and, percolating through a porous and deformed rock matrix, invaded the base of the lithosphere and favoured metasomatic processes at different levels in the peridotitic lithosphere with the formation of amphibole and/or phlogopite ("veined mantle" of [82, 83]). Notably, the different tholeiitic (high- and low-Ti), K-alkaline suites (B-P and Ab-T), and sodic magmatism with HK and LK xenoliths are consistent with variously depleted lithospheric mantle at different times, pervasively and locally invaded by metasomatizing fluids and/or melts. Thereafter, the newly formed veins ("enriched component") and peridotitic matrix ("depleted component") undervent different isotopic compositions with time, depending on their parent/daughter ratio (cf. Figure 8), testifying heterogeneous mantle sources beneath ASV.

## Acknowledgments

This paper is dedicated to the memory of Enzo Michele Piccirillo who was an inspiration to the authors over many years of geological work in the region. P. Comin-Chiaramonti gratefully acknowledges CNR and MURST (Italian Agencies) for the field financial support since 1981 and the Department of Geosciences of the Trieste University for the use of chemical laboratories where all the analyses were performed. Thanks are due to Exxon Co., USDMA, ITAIPU, and YACIRETA (Paraguayan Agencies) for their collaborations and to engineer E. Debernardi (deceased) and geologists D. Orué, Luis Lúcia, and L. A. Martinez. This research was also supported by grants (Procs. 08/3807-4 and 10/50887-3) to C. B. Gomes from the Brazilian agency Fapesp.

The Ms benefit of the accurate revision of Keith Bell and Lalou Gwalani.

# References

[1] V. C. Velázquez, C. B. Gomes, C. Riccomini, and J. Kirk, "The cretaceous alkaline dyke swarm in the central segment of the Asunción rift, eastern Paraguay: its regional distribution, mechanism of emplacement, and tectonic significance," *Journal of Geological Research*, vol. 2011, Article ID 946701, 18 pages, 2011.

[2] G. Bellieni, P. C. Chiaramonti, L. S. Marques et al., "Continental flood basalts from the central-western regions of the Parana plateau (Paraguay and Argentina): petrology and petrogenetic aspects," *Neues Jahrbuch für Mineralogie, Abhandlungen*, vol. 154, no. 2, pp. 111–139, 1986.

[3] E. M. Piccirillo and A. J. Melfi, Eds., *The Mesozoic Flood Volcanism of the Paranà Basin: Petrogenetic and Geophysical Aspects*, IAG, São Paulo, Brazil, 1988.

[4] P. Comin-Chiaramonti, A. Marzoli, C. De Barros Gomes et al., "The origin of post-Paleozoic magmatism in eastern Paraguay," *Special Paper of the Geological Society of America*, no. 430, pp. 603–633, 2007.

[5] P. Comin-Chiaramonti, C. B. Gomes, M. Ernesto, A. Marzoli, and C. Riccomini, *Eastern Paraguay: Post-Paleozoic Magmatism*, Large Igneous Province of the Month, 2007, http://www.largeigneousprovinces.org/07jan .

[6] P. Comin-Chiaramonti, A. De Min, V. A. V. Girardi, and E. Ruberti, "Post-Paleozoic magmatism in Angola and Namibia: a review," in *Volcanism and Evolution of the African Lithosphere, the Geological Society of America*, L. Beccaluva, G. Bianchini, and M. Wilson M, Eds., Special Paper 478, pp. 223–247, 2011.

[7] J. D. Fairhead and M. Wilson, "Plate tectonic processes in the South Atlantic Ocean: do we need deep mantle plumes?" *Special Paper of the Geological Society of America*, no. 388, pp. 537–553, 2005.

[8] P. V. Zalan, S. Wolff, M. A. Astolfi et al., "The Paraná basin, Brazil," in *Interior Cratonic Basins*, M. W. Leighton, D. R. Kolata, D. F. Oltz, and J. J. Eidel, Eds., vol. 51 of *Memoir*, pp. 601–708, American Association of Petroleum Geology, Tulsa, Okla, USA, 1990.

[9] J. J. W. Rogers, R. Unrug, and M. Sultan, "Tectonic assembly of Gondwana," *Journal of Geodynamics*, vol. 19, no. 1, pp. 1–34, 1995.

[10] P. Comin-Chiaramonti, A. Cundari, E. M. Piccirillo et al., "Potassic and sodic igneous rocks from Eastern Paraguay: their origin from the lithospheric mantle and genetic relationships with the associated Paraná flood tholeiites," *Journal of Petrology*, vol. 38, no. 4, pp. 495–528, 1997.

[11] M. Feng, V. D. S. Lee, and M. Assumpção, "Upper mantle srructure of South America from joint inversions of waveforms and fundamental mode group velocities of Rayleigh waves," *Journal of Geophysical Research*, vol. 112, pp. 1–16, 2007.

[12] P. Comin-Chiaramonti, A. Cundari, J. M. DeGraff, C. B. Gomes, and E. M. Piccirillo, "Early Cretaceous-Tertiary magmatism in Eastern Paraguay (western Parana basin): geological, geophysical and geochemical relationships," *Journal of Geodynamics*, vol. 28, no. 4-5, pp. 375–391, 1999.

[13] P. Comin-Chiaramonti, A. Cundari, C. B. Gomes et al., "Potassic dyke swarm in the Sapucai Graben, eastern Paraguay: petrographical, mineralogical and geochemical outlines," *LITHOS*, vol. 28, no. 3-6, pp. 283–301, 1992.

[14] C. Riccomini, V. F. Velázquez, and C. De Barros Gomes, "Cenozoic lithospheric faulting in the Asunción Rift, eastern Paraguay," *Journal of South American Earth Sciences*, vol. 14, no. 6, pp. 625–630, 2001.

[15] Anschutz Co., *Geologic Map of Eastern Paraguay (1:500,000)*, F. Wiens Compiler, Denver, Colo, USA, 1981.

[16] M. D. Druecker and S. P. Gay Jr., "Mafic dyke swarms associated with Mesozoic rifting in Eastern Paraguay, South America," in *Mafic Dyke Swarms, Geological Association of Canada*, H. C. Halls and A. R. Fahrig, Eds., pp. 187–193, 1987.

[17] N. Ussami, A. Kolisnyk, M. I. B. Raposo, F. J. F. Ferreira, E. C. Molina, and M. Ernesto, "Detectabilitade magnetica de diques do Arco de Ponta Grossa: um estudo integrado de magnetometria terrestre/aerea e magnetismo de rocha," *Revista Brasileira de Geociências*, vol. 21, pp. 317–327, 1994.

[18] J. M. de Graff, "Late Mesozoic crustal extension and rifting on the western edge of the Paraná Basin, Paraguay," *Geological Society of America, Abstracts with Programs*, vol. 17, p. 560, 1985.

[19] PHOTO GRAVITY Co., *Regional Bouguer Gravity Data and Station Location Map of the Paraguay (Scale 1:2,000,000)*, Archivo DRM-MOPC, Asunción, Paraguay, 1991.

[20] D. S. Hutchinson, "Geology of the Apa High," Internal Report, TAC, Asunción, Paraguay, 1979.

[21] F. Wiens, "Mapa geologica de la region oriental, Republica del Paraguay, escala 1:500,0002," *Simposio Recursos Naturales, Paraguay, Asunción*, p. 9, 1982.

[22] R. A. Livieres and H. Quade, "Distribución regional y asentamiento tectónico de los complejos alcalinos del Paraguay," *Zentralblatt für Geologie und Paläontologie*, vol. 7, pp. 791–805, 1987.

[23] A. Kanzler, "The southern Precambrian in Paraguay. Geological inventory and age relations," *Zentralblatt für Geologie und Paläontologie*, vol. 7, pp. 753–765, 1987.

[24] P. Comin-Chiaramonti and C. B. Gomes, Eds., *Alkaline Magmatism in Central-Eastern Paraguay. Relationships with Coeval Magmatism in Brazil*, Edusp/Fapesp, São Paulo, Brazil, 1996.

[25] P. R. Renne, M. Ernesto, I. G. Pacca et al., "The age of Paraná flood volcanism, rifting of gondwanaland, and the Jurassic-Cretaceous boundary," *Science*, vol. 258, no. 5084, pp. 975–979, 1992.

[26] P. R. Renne, D. F. Mertz, W. Teixeira, H. Ens, and M. Richards, "Geochronologic constraints on magmatic and tectonic evolution of the Paraná Province," *The American Geophysical Union*, vol. 74, abstract, p. 553, 1993.

[27] P. R. Renne, K. Deckart, M. Ernesto, G. Féraud, and E. M. Piccirillo, "Age of the Ponta Grossa dike swarm (Brazil), and implications to Paraná flood volcanism," *Earth and Planetary Science Letters*, vol. 144, no. 1-2, pp. 199–211, 1996.

[28] C. B. Gomes, P. Comin-Chiaramonti, and V. F. Velázquez, "The Mesoproterozoic rhyolite occurrences of Fuerte Olimpo and Fuerte San Carlos, Northern Paraguay," *Revista Brasileira De Geociências*, vol. 30, pp. 785–788, 2000.

[29] J. Berrocal and C. Fernandes, "Seismicity in Paraguay and neighbouring regions," in *Alkaline Magmatism in Central-Eastern Paraguay, Relationships with Coevalmagmatism in Brazil*, P. Comin-Chiaramonti and C. B. Gomes, Eds., pp. 57–66, Edusp/Fapesp, São Paulo, Brazil, 1996.

[30] A. Tommasi and A. Vauchez, "Continental rifting parallel to ancient collisional belts: an effect of the mechanical anisotropy of the lithospheric mantle," *Earth and Planetary Science Letters*, vol. 185, no. 1-2, pp. 199–210, 2001.

Magmatism in the Asunción-Sapucai-Villarrica Graben (Eastern Paraguay) Revisited: Petrological, Geophysical, Geochemical, and Geodynamic Inferences

161

[31] P. F. Green, I. R. Duddy, P. O'Sullivan, K. A. Hegarty, P. Comin-Chiaramonti, and C. B. Gomes, "Mesozoic potassic magmatism from the Asunción-Sapucai graben: apatite fission track analysis of the Acahai suite and implication for hydrocarbon exploration," *Geochimica Brasiliensis*, vol. 5, pp. 79–88, 1991.

[32] K. A. Hegarty, I. R. Duddy, and P. F. Green, "The thermal history in around the Paraná basin using apatite fission track analysis-implications for hydrocarbon occurrences and basin formation," in *Alkaline magmatism in Central-Eastern Paraguay, Relationships with Coevalmagmatism in Brazil*, P. Comin-Chiaramonti and C. B. Gomes, Eds., pp. 67–84, Edusp/Fapesp, São Paulo, Brazil, 1996.

[33] P. Comin-Chiaramonti, A. De Min, and C. B. Gomes, "Magmatic rock-types from the Asunción-Sapucai graben: description of the occurrences and petrographical notes," in *Alkaline Magmatism in Central-Eastern Paraguay, Relationships with coevalmagmatism in Brazil*, P. Comin-Chiaramonti and C. B. Gomes, Eds., pp. 275–330, Edusp/Fapesp, São Paulo, Brazil, 1996.

[34] P. Comin-Chiaramonti, A. De Min, and A. Marzoli, "Magmatic rock-types from the Asunción-Sapucai graben: chemical analyses," in *Alkaline Magmatism in Central-Eastern Paraguay, Relationships with coevalmagmatism in Brazil*, P. Comin-Chiaramonti and C. B. Gomes, Eds., pp. 331–387, Edusp/Fapesp, São Paulo, Brazil, 1996.

[35] P. Comin-Chiaramonti, A. Cundari, and G. Bellieni, "Mineral analyses of alkaline rock-types from the Asunción-Sapucai graben," in *Alkaline Magmatism in Central-Eastern Paraguay, Relationships with coevalmagmatism In Brazil*, P. Comin-Chiaramonti and C. B. Gomes, Eds., pp. 389–458, Edusp/Fapesp, São Paulo, Brazil, 1996.

[36] P. Comin-Chiaramonti, A. Cundari, A. De Min, C. B. Gomes, and V. F. Velázquez, "Magmatism in Eastern Paraguay: occurrence and petrography," in *Alkaline Magmatism in Central-Eastern Paraguay, Relationships with coevalmagmatism in Brazil*, P. Comin-Chiaramonti and C. B. Gomes, Eds., pp. 103–122, Edusp/Fapesp, São Paulo, Brazil, 1996.

[37] J. M. de Graff, S. P. Gay Jr., and D. Orué, "Interpretación geofísica y geológica del Valle de Ypacaraí (Paraguay) y su formación," *Revista de la Asociación Geológica Argentina*, vol. 36, no. 3, pp. 240–256, 1981.

[38] M. Ernesto, P. Comin-Chiaramonti, C. B. Gomes, A. M. C. Castillo, and V. F. Velazquez, "Palaeomagnetic data from the central alkaline province, Eastern Paraguay," in *Alkaline Magmatism in Central-Eastern Paraguay, Relationships with coevalmagmatism in Brazil*, P. Comin-Chiaramonti and C. B. Gomes, Eds., pp. 85–102, Edusp/Fapesp, São Paulo, Brazil, 1996.

[39] H. de La Roche, "Classification and nomenclature des roches ignées: un essai de restauration de la convergence entre systématique quantitative, typologie d'usage et modélisation génétique," *Bulletin de la Societé Géologique de France*, vol. 8, pp. 337–353, 1986.

[40] R. P. Le Maitre, *A Classification of Igneous Rocks and Glossary of Terms*, Blackwell, Oxford, UK, 1989.

[41] A. Streckeisen, "To each plutonic rock its proper name," *Earth Science Reviews*, vol. 12, no. 1, pp. 1–33, 1976.

[42] C. B. Gomes, P. Comin-Chiaramonti, A. DeMin et al., "Atividade filoniana asociada ao complexo alcalino de Sapukai, Paraguay Oriental," *Geochimica Brasiliensis*, vol. 3, pp. 93–114, 1989.

[43] G. Bellieni, P. Comin-Chiaramonti, L. S. Marques et al., "Petrogenetic aspects of acid and basaltic lavas from the Paraná plateau (Brazil): geological, mineralogical and petrochemical relationships," *Journal of Petrology*, vol. 27, no. 4, pp. 915–944, 1986.

[44] P. Comin-Chiaramonti, C. B. Gomes, P. Censi, A. DeMin, S. Rotolo, and V. F. Velazquez, "Geoquimica do magmatismo Post-Paleozoico no Paraguai Centro-Oriental," *Geochimica Brasiliensis*, vol. 7, pp. 19–34, 1993.

[45] G. Bellieni, P. Comin-Chiaramonti, L. S. Marques et al., "High- and low-TiO$_2$ flood basalts of Paraná plateau (Brazil): petrology, petrogenetic aspects and mantle source relationships," *Neues Jahrbuch für Mineralogie Abhandlungen*, vol. 150, pp. 273–306, 1984.

[46] E. M. Piccirillo, G. Bellieni, P. Comin-Chiaramonti et al., "Continental flood volcanism from the Parana' Basin (Brazil)," in *Continental Flood Volcanism*, J. D. McDougal, Ed., pp. 195–238, Kluver Academic, London, UK, 1988.

[47] E. M. Piccirillo, L. Civetta, R. Petrini et al., "Regional variations within the Paraná flood basalts (southern Brazil): evidence for subcontinental mantle heterogeneity and crustal contamination," *Chemical Geology*, vol. 75, no. 1-2, pp. 103–122, 1989.

[48] P. Comin-Chiaramonti, P. Censi, A. Cundari, and C. B. Gomes, "A silico-beforsitic flow from the Sapucai Complex (central-eastern Paraguay)," *Geochimica Brasiliensis*, vol. 6, no. 1, pp. 87–91, 1992.

[49] P. Comin-Chiaramonti, A. Cundari, C. B. Gomes et al., "Mineral chemistry and its genetic significance of major and accessory minerals from a potassic dyke swarm in the Sapucai graben, Central-Eastern Paraguay," *Brasileira Geoquímica*, vol. 4, pp. 175–206, 1990.

[50] G. Faure, *Origin of Igneous Rocks: The Isotopic Evidence*, Springer, Berlin, Germany, 2001.

[51] A. D. Edgar, "Role of subduction in the genesis of leucite-bearing rocks: discussion," *Contributions to Mineralogy and Petrology*, vol. 73, no. 4, pp. 429–431, 1980.

[52] A. Cundari and A. K. Ferguson, "Significance of the pyroxene chemistry from leucite-bearing and related assemblages," *TMPM Tschermaks Mineralogische und Petrographische Mitteilungen*, vol. 30, no. 3, pp. 189–204, 1982.

[53] P. Comin-Chiaramonti, F. Lucassen, V. A. V. Girardi, A. De Min, and C. B. Gomes, "Lavas and their mantle xenoliths from intracratonic eastern Paraguay (South America Platform) and andean domain, NW-Argentina: a comparative review," *Mineralogy and Petrology*, vol. 98, no. 1-2, pp. 143–165, 2010.

[54] G. DeMarchi, P. Comin-Chiaramonti, P. DeVito, S. Sinigoi, and A. M. C. Castillo, "Lherzolite-dunite xenoliths from Eastern Paraguay: petrological constraints to mantle metasomatism," in *The Mesozoic Flood Volcanism of the Paran' Basin: Petrogenetic and Geophysical Aspects*, E. M. Piccirillo and A. J. Melfi, Eds., pp. 207–228, IAG, São Paulo, Brazil, 1988.

[55] P. Comin-Chiaramonti, L. Civetta, E. M. Piccirillo et al., "Tertiary nephelinitic magmatism in eastern Paraguay: petrology, Sr- Nd isotopes and genetic relationships with associated spinel-peridotite xenoliths," *European Journal of Mineralogy*, vol. 3, no. 3, pp. 507–525, 1991.

[56] P. Comin-Chiaramonti, F. Princivalle, V. A. V. Girardi, C. B. Gomes, A. Laurora, and R. Zanetti, "Mantle xenoliths from Ñemby, Eastern Paraguay: O-Sr-Nd isotopes, trace elements and crystal chemistry of hosted clinopyroxenes," *Periodico di Mineralogia*, vol. 70, pp. 205–230, 2001.

[57] J. Wang, K. H. Hattori, R. Kilian, and C. R. Stern, "Metasomatism of sub-arc mantle peridotites below southernmost

South America: reduction of $fO_2$ by slab-melt," *Contributions to Mineralogy and Petrology*, vol. 153, no. 5, pp. 607–624, 2007.

[58] M. F. Roden, F. A. Frey, and D. M. Francis, "An example of consequent mantle metasomatism in peridotite inclusions from Nunivak Island, Alaska," *Journal of Petrology*, vol. 25, no. 2, pp. 546–577, 1984.

[59] S. Y. O'Reilly and W. L. Griffin, "Mantle metasomatism beneath western Victoria, Australia: I. Metasomatic processes in Cr-diopside lherzolites," *Geochimica et Cosmochimica Acta*, vol. 52, no. 2, pp. 433–447, 1988.

[60] F. J. Spera, "Carbon dioxide in petrogenesis III: role of volatiles in the ascent of alkaline magma with special reference to xenolith-bearing mafic lavas," *Contributions to Mineralogy and Petrology*, vol. 88, no. 3, pp. 217–232, 1984.

[61] F. A. Frey and D. H. Green, "The mineralogy, geochemistry and origin of lherzolite inclusions in Victorian basanite," *Geochimica et Cosmochimica Acta*, vol. 38, pp. 1023–1050, 1979.

[62] D. K. Koustpoulos, "Melting of the shallow upper mantle: a new perspective," *Journal of Perology*, vol. 32, pp. 671–699, 1991.

[63] G. Rivalenti, R. Vannucci, E. Rampone et al., "Peridotite clinopyroxene chemistry reflects mantle processes rather than continental versus oceanic settings," *Earth and Planetary Science Letters*, vol. 139, no. 3-4, pp. 423–437, 1996.

[64] P. R. A. Wells, "Pyroxene thermometry in simple and complex systems," *Contributions to Mineralogy and Petrology*, vol. 42, pp. 109–121, 1977.

[65] J. Fabriès, "Spinel-olivine geothermometry in peridotites from ultramafic complexes," *Contributions to Mineralogy and Petrology*, vol. 69, pp. 329–336, 1979.

[66] J. C. Mercier, "Single-pyroxene geothermometry and geo-barometry," *The American Journal of Sciences*, vol. 61, no. 7-8, pp. 603–615, 1980.

[67] J. C. C. Mercier, V. Benoit, and J. Girardeau, "Equilibrium state of diopside-bearing harzburgites from ophiolites: geobarometric and geodynamic implications," *Contributions to Mineralogy and Petrology*, vol. 85, no. 4, pp. 391–403, 1984.

[68] G. Sen, F. A. Frey, N. Shimizu, and W. P. Leeman, "Evolution of the lithosphere beneath Oahu, Hawaii: rare earth element abundances in mantle xenoliths," *Earth and Planetary Science Letters*, vol. 119, no. 1-2, pp. 53–69, 1993.

[69] H. Chiba, T. Chack, R. N. Clayton, and J. Goldsmith, "Oxygen isotope fractionations involving diopside, forsterite, magnetite, and calcite: application to geothermometry," *Geochimica et Cosmochimica Acta*, vol. 53, pp. 2985–2989, 1989.

[70] D. Mattey, D. Lowry, and C. Macpherson, "Oxygen isotope composition of mantle peridotite," *Earth and Planetary Science Letters*, vol. 128, no. 3-4, pp. 231–241, 1994.

[71] T. K. Kyser, "1990. Stable isotopes in the continental lithospheric mantle," in *Continental Mantle*, M. A. Menzies, Ed., pp. 127–156, Oxford Clarendon Press, Oxford, UK, 1990.

[72] T. K. Kyser, J. R. O'Neil, and I. S. E. Carmichael, "Oxygen isotope thermometry of basic lavas and mantle nodules," *Contributions to Mineralogy and Petrology*, vol. 77, no. 1, pp. 11–23, 1981.

[73] F. Princivalle, M. Tirone, and P. Comin-Chiaramonti, "Clinopyroxenes from metasomatized spinel-peridotite mantle xenoliths from Nemby (Paraguay): crystal chemistry and petrological implications," *Mineralogy and Petrology*, vol. 70, no. 1-2, pp. 25–35, 2000.

[74] F. Castorina, P. Censi, P. Comin-Chiaramonti et al., "Carbonatites from Eastern Paraguay and genetic relationships with potassic magmatism: C, O, Sr and Nd isotopes," *Mineralogy and Petrology*, vol. 61, no. 1-4, pp. 237–260, 1998.

[75] P. Antonini, M. Gasparon, P. Comin-Chiaramonti, and C. B. Gomes, "Post-palaeozoic magmatism in Eastern paraguay: Sr-Nd-Pb isotope compositions," in *Mesozoic to Cenozoic Alkaline Magmatism in the Brazilian Platform*, P. Comin-Chiaramonti and C. B. Gomes, Eds., pp. 57–70, EdUSP, Fapesp, São Paulo, Brazil, 2005.

[76] D. Mckenzie and R. K. O'nions, "The source regions of ocean island basalts," *Journal of Petrology*, vol. 36, no. 1, pp. 133–159, 1995.

[77] S. R. Hart and A. Zindler, "Constraints on the nature and the development of chemical heterogeneities in the mantle," in *Mantle Convection Plate Tectonics and Global Dynamics*, W. R. Peltier, Ed., pp. 261–388, Gordon and Breach Sciences, New York, NY, USA, 1989.

[78] P. Comin-Chiaramonti, C. B. Gomes, and Eds, *Mesozoic to Cenozoic Alkaline Magmatism in the Brazilian Platform*, EdUSP, Fapesp, São Paulo, Brazil, 2005.

[79] R. W. Carlson, S. Esperança, and D. P. Svisero, "Chemical and Os isotopic study of Cretaceous potassic rocks from Southern Brazil," *Contributions to Mineralogy and Petrology*, vol. 125, no. 4, pp. 393–405, 1996.

[80] M. A. Menzies, Ed., *Continental Mantle*, Oxford Clarendon Press, 1990.

[81] B. L. Weaver, "The origin of ocean island basalt end-member compositions: trace element and isotopic constraints," *Earth and Planetary Science Letters*, vol. 104, no. 2–4, pp. 381–397, 1991.

[82] S. Foley, "Petrological characterization of the source components of potassic magmas: geochemical and experimental constraints," *LITHOS*, vol. 28, no. 3–6, pp. 187–204, 1992.

[83] S. Foley, "Vein-plus-wall-rock melting mechanisms in the lithosphere and the origin of potassic alkaline magmas," *LITHOS*, vol. 28, no. 3–6, pp. 435–453, 1992.

[84] C. J. Hawkesworth, M. S. M. Mantovani, P. N. Taylor, and Z. Palacz, "Evidence from the Parana of south Brazil for a continental contribution to Dupal basalts," *Nature*, vol. 322, no. 6077, pp. 356–359, 1986.

[85] J. K. Meen, J. C. Ayers, and E. J. Fregeau, "A model of mantle metasomatism by carbonated alkaline melts: trace element and isotopic compositions of mantle source regions of carbonatite and other continental igneous roks," in *Carbonatites, Genesis and Evolution*, K. Bell, Ed., pp. 464–499, Unwin Hyman, London, UK, 1989.

[86] S.-S. Sun and W. F. McDonough, "Chemical and isotopic systematics of oceanic basalts: implications for mantle composition and processes," in *Magmatism in the Ocean Basins*, A. D. Saunders and M. J. Norry, Eds., Special Paper 42, pp. 313–345, Geological Society of London, 1989.

[87] P. Comin-Chiaramonti, F. Castorina, A. Cundari, R. Petrini, and C. B. Gomes, "Dykes and sills from Eastern paraguay: Sr and Nd isotope systematic," in *IDC-3, Physics and Chemistry of Dykes*, G. Baer and A. Heiman, Eds., pp. 267–278, Balkema, Rotterdam, The Netherlands, 1995.

[88] P. Comin-Chiaramonti, C. B. Gomes, A. Cundari, F. Castorina, and P. Censi, "A review of carbonatitic magmatism in the Paraná-Angola-Etendeka (Pan) system," *Periodico di Mineralogia*, vol. 76, pp. 25–78, 2007.

[89] P. Comin-Chiaramonti, C. B. Gomes, P. Censi et al., "Alkaline complexes from the alto paraguay province at the border of Brazil (Mato Grosso do Sul State) and Paraguay," in *Mesozoic To Cenozoic Alkaline Magmatism in the Brazilian Platform*, P. Comin-Chiaramonti and C. B. Gomes, Eds., pp. 71–148, Edusp-Fapesp, SãoPaulo, Brazil, 2005.

Magmatism in the Asunción-Sapucai-Villarrica Graben (Eastern Paraguay) Revisited: Petrological, Geophysical, Geochemical, and Geodynamic Inferences

163

[90] C. J. Hawkesworth, K. Gallagher, S. Kelley et al., "Paraná magmatism and the opening of the South Atlantic," in *Magmatism and the Causes of Continental Break-Up*, B. C. Storey, Ed., vol. 68, pp. 221–240, Geological Society of London, 1992.

[91] S. Turner, M. Regelous, S. Kelley, C. Hawkesworth, and M. Mantovani, "Magmatism and continental break-up in the South Atlantic: high precision $^{40}Ar^{39}Ar$ geochronology," *Earth and Planetary Science Letters*, vol. 121, no. 3-4, pp. 333–348, 1994.

[92] D. W. Peate, "The parana-etendeka province," in *Large Igneuos Province: Continetal Oceanic, Planetary Flood Volcanism*, J. Mahoney and M. F. Coffin, Eds., pp. 217–245, American Geophyscial Union, Washington, DC, USA, 1992.

[93] D. W. Peate and C. J. Hawkesworth, "Lithospheric to asthenospheric transition in low-Ti flood basalts from southern Paraná, Brazil," *Chemical Geology*, vol. 127, no. 1–3, pp. 1–24, 1996.

[94] A. J. Erlank, F. G. Waters, C. J. Hawkesworth et al., "Evidence for mantle metasomatism in peridotite nodules from the Kimberley pipes, South Africa," in *Mantle Metasomatism*, M. A. Menzies and C. J. Hawkesworth, Eds., pp. 221–309, Academic Press, London, UK, 1987.

[95] D. E. Randall, "A new Jurassic-Recent apparent polar wander path for South America and a review of central Andean tectonic models," *Tectonophysics*, vol. 299, no. 1-3, pp. 49–74, 1998.

[96] O. Gudmundsson and M. Sambridge, "A Regionalized Upper Mantle (RUM) seismic model," *Journal of Geophysical Research B*, vol. 103, no. 4, pp. 7121–7136, 1998.

[97] F. Lucassen, R. Becchio, H. G. Wilke et al., "Proterozoic-Paleozoic development of the basement of the Central Andes (18-26°S) - A mobile belt of the South American craton," *Journal of South American Earth Sciences*, vol. 13, no. 8, pp. 697–715, 2000.

[98] J. H. Laux, M. M. Pimentel, E. L. Dantas, R. Armstrong, and S. L. Junges, "Two neoproterozoic crustal accretion events in the Brasília belt, central Brazil," *Journal of South American Earth Sciences*, vol. 18, no. 2, pp. 183–198, 2005.

[99] H. K. Liu, S. S. Gao, P. G. Silver, and Y. Zhang, "Mantle layering across central South America," *Journal of Geophysical Research*, vol. 108, no. 1, p. 2510, 2003.

[100] J. C. Van Decar, D. E. James, and M. Assumpção, "Seismic evidence for a fossil mantle plume beneath South America and implications for plate driving forces," *Nature*, vol. 378, no. 6552, pp. 25–31, 1995.

[101] M. S. M. Mantovani, M. C. L. Quintas, W. Shukowsky, and B. B. de Brito Neves, "Delimitation of the Paranapanema proterozoic block: a geophysical contribution," *Episodes*, vol. 28, no. 1, pp. 18–22, 2005.

[102] R. Unrug, "The Assembly of Gondwanaland," *Episodes*, vol. 19, pp. 11–20, 1996.

[103] U. G. :Cordani, K. Sato, W. Teixeira, C. C. G. Tassinari, and M. A. S. Basei, "Crustal evolution of the South American platform," in *Tectonic Evolution of South America, 31st International Geological Congress, Rio De Janeiro*, U. G. :Cordani, E. J. Milani, A. A. Thomaz Filho, and D. A. Campos, Eds., pp. 19–40, 2000.

[104] U. G. :Cordani, C. C. G. Tassinari, and D. R. Rolim, "The basement of the Rio Apa Craton in Mato Grosso do Sul (Brazil) and northern Paraguay: a geochronological correlation with the tectonic provinces of the south-western Amazonian Craton," in *Gondwana Conference, Abstracts Volume, Mendoza, Argentina*, pp. 1–12, 2005.

[105] A. Kröner and U. Cordani, "African, southern Indian and South American cratons were not part of the Rodinia supercontinent: evidence from field relationships and geochronology," *Tectonophysics*, vol. 375, no. 1–4, pp. 325–352, 2003.

[106] C. B. Prezzi and R. N. Alonso, "New paleomagnetic data from the northern Argentine Puna: central Andes rotation pattern reanalyzed," *Journal of Geophysical Research B*, vol. 107, no. 2, pp. 1–18, 2002.

[107] A. E. Rapalini, "The accretionary history of southern South America from the latest Proterozoic to the Late Palaeozoic: some palaeomagnetic constraints," *Geological Society*, vol. 246, pp. 305–328, 2005.

[108] D. Nürberg and R. D. Müller, "The tectonic evolution of South Atlantic from Late Jurassic to present," *Tectonophysics*, vol. 191, pp. 27–43, 1991.

[109] H. K. Chang, R. O. Kowsmann, and A. M. F. de Figueiredo, "New concepts on the development of east Brazilian marginal basins," *Episodes*, vol. 11, no. 3, pp. 194–202, 1988.

[110] W. S. Holbrook and P. B. Kelemen, "Large igneous province on the US atlantic margin and implications for magmatism during continental breakup," *Nature*, vol. 364, no. 6436, pp. 433–436, 1993.

# 13

# FTIR and Raman Spectral Research on Metamorphism and Deformation of Coal

## Xiaoshi Li,[1] Yiwen Ju,[1] Quanlin Hou,[1] Zhuo Li,[2] and Junjia Fan[3]

[1] Key Laboratory of Computational Geodynamics, College of Earth Science, Graduate University of Chinese Academy of Sciences, Beijing 100049, China
[2] State Key Laboratory of Petroleum Resource and Prospecting, China University of Petroleum, Beijing 102249, China
[3] Key Lab of Basin Structure and Petroleum Accumulation, PetroChina Research Institute of Petroleum Exploration and Development, Beijing 100083, China

Correspondence should be addressed to Yiwen Ju, juyw03@163.com

Academic Editor: Hongyuan Zhang

Under different metamorphic environments, coal will form different types of tectonically deformed coal (TDC) by tectonic stress and even the macromolecular structure can be changed. The structure and composition evolution of TDC have been investigated in details using Fourier transform infrared spectroscopy and Raman spectroscopy. The ductile deformation can generate strain energy via increase of dislocation in molecular structure of TDC, and it can exert an obvious influence on degradation and polycondensation. The brittle deformation can generate frictional heat energy and promote the metamorphism and degradation, but less effect on polycondensation. Furthermore, degradation affects the structural evolution of coal in lower metamorphic stage primarily, whereas polycondensation is the most important controlling factor in higher metamorphic stage. Tectonic deformation can produce secondary structural defects in macromolecular structure of TDC. Under the control of metamorphism and deformation, the small molecules which break and fall off from the macromolecular structure of TDC are replenished and embedded into the secondary structural defects preferentially and form aromatic rings by polycondensation. These processes improved the stability of macromolecular structure greatly. It is easier for ductile deformation to induce secondary structural defects than in brittle deformation.

## 1. Introduction

The study of macromolecular structure and complicated composition of coal is the most difficult and important topic in coal chemistry [1–4]. Effective analytical methods for coal structure and chemical composition research, such as Fourier transform infrared (FTIR) spectroscopy and Raman spectroscopy are, becoming more and more important for coal chemical researchers [5–15]. Previous studies discussed that with the upgrade of metamorphism, the aromatic structure of primary structure coal increased and expanded, whereas the side chain of aliphatic compound and functional group decreased. The hydrogen and oxygen deplete in coals; as a result, the condensation degree gradually improved. Aromatic structure is mainly composed of anthracite, and the condensation degree is further improved [2, 3, 5,

16, 17]. The coal basins in China experienced multiple tectonic movements and developed widely distribution of tectonically deformed coal (TDC) under the strong tectonic deformation environments [18]. Compared with primary structure coal, the evolution characteristics and ways of macromolecular structure of TDC are more complicated [8, 17–21]. The formation of lignite due to diagenesis could experience different degrees of metamorphism under various temperature-pressure conditions during geological history. If the coal seams are apparently affected by tectonic stress, they can produce deformation in certain degrees. For all the deformational coal seams, the process and mechanism of deformation (brittle deformation and ductile deformation) are actually different [17–21]. Many studies have been conducted in order to investigate the relationship between the group absorption frequency of FTIR and the metamorphic grade of coals, as well as Raman spectral analysis on carbon materials [3–15, 19]; however, products about the studies of different deformation mechanisms and grades of TDC using FTIR and Raman spectrum analysis, and the analysis of evolution process and mechanisms in macromolecular structure of TDC, were rarely reported. Different deformation mechanisms will exert different influences on the macromolecular structure and composition of TDC [6, 7, 18, 19]. Based on the FTIR and Raman spectrum analysis of TDC samples from Huaibei coalfield, the evolution characteristics of macromolecular structure affected by the change of metamorphism and deformation grade are discussed. The aim of this paper is to explain the mechanism of tectonic deformational influence on the evolution of structural defects and chemical composition.

## 2. Samples and Experimental Methods

The TDC samples with different deformation and metamorphism ($0.7\% < R_{o,max} < 3.1\%$) were collected from Permo-Carboniferous coal bed in Huaibei coal field, which was strongly affected by the Mesozoic tectonic deformation. The coal seams were mainly distributed in graben, especially in syncline. The tectonic deformation altered the structure of coal seams remarkably and formed various types of TDC.

All samples are pretreated through demineralization and vitrinite centrifugation processes in order to better delineate the characteristics of the deformation and metamorphism of TDC samples. The demineralization process utilized HCl and HF to reduce the proportion of mineral matter in each sample (<2%). The vitrinite centrifugation process used benzene and $CCl_4$ to increase the composition percentage of vitrinite to 80–90%. The group maceral and vitrinite reflectance ($R_{o,max}$) were tested as well.

The FTIR analysis on 32 samples and Raman analysis on 19 samples are conducted to further understand the evolution of macromolecular structure affected by deformation and metamorphism. It can be determined that the type of functional group and its change correlated with deformation and metamorphism in coal by analysis of absorption band, shown on infrared spectrum [7, 9]. The structural analysis and qualitative identification of molecular can be studied by Raman spectrum because the formation of Raman spectra is closely connected with the structure and the ordering degree of coal macromolecule [22]. FTIR was performed on Nicolet 750 microscopic infrared spectrometer in the Analytical Instrumentation Center in Peking University. The scan ranges from $4000\,cm^{-1}$ to $650\,cm^{-1}$. Spectra were recorded by coadding 128 scans at a resolution of $8\,cm^{-1}$ using an MCT/A detector and a KBr/Ge beam splitter. Raman was performed on Renishaw RM-1000 microscopic Raman spectrometer at the State Key Laboratory of Geological Processes and Mineral Resources in China University of Geosciences. The spectral resolution was $2.4\,cm^{-1}$, spit width was $2.5\,cm^{-1}$, and laser power on the sample surface was 0.8 mw; measurements were carried out using the 514.5 nm excitation wavelength with $Ar^+$ laser; acquisition time was 30 s. The spectra were measured under $25°C$ temperature.

## 3. Results and Analysis

Part of experiment results are listed in Table 1.

*3.1. Parameter of FTIR Analysis.* The types of functional group in TDC samples collected from Huaibei coal field are basically the same with other researchers [3, 7, 23]. But the peak positions and the changes of absorption peaks with the increase of metamorphic grade are different from others (Figure 1). The biggest difference between TDC and the primary structure coal showed on FTIR spectrum is reflected in the change of relative absorption strength but not the absorption frequency.

*3.1.1. Evolution of Characteristic Frequency of Aromatic Structure.* The characteristic frequency of aromatic structure includes the absorption strength of (1) $3049\,cm^{-1}$ related to stretching vibration of CH in aromatic ring, (2) $1600\,cm^{-1}$ related to vibration of C=C in aromatic ring, and (3) $749\,cm^{-1}$, $810\,cm^{-1}$, and $871\,cm^{-1}$ related to the plane deformation vibration of CH in aromatic rings. With the increase of metamorphic grade, the absorption strength of $1600\,cm^{-1}$ has little change in brittle deformational coal but decreases first and increases later in ductile deformational coal, and with the range from 0.994 to 1 (Figure 2(a)).

Generally, the change of absorption strength of $1600\,cm^{-1}$ was not so obvious compared with the other frequencies. The absorption strength of $749\,cm^{-1}$, $810\,cm^{-1}$, and $871\,cm^{-1}$ related to the plane deformation vibration of CH in aromatic ring is correlated with independent, two and more adjacent hydrogen atoms state, respectively. The strongest absorption strength of these frequencies is in the middle metamorphic grade and then in the higher and lower metamorphic grades it is the weakest (Figure 1). Figure 2(b) shows that this change is influenced by brittle and ductile deformation. The strongest absorption strength of $3049\,cm^{-1}$ related to stretching vibration of CH in aromatic ring is also in the middle metamorphic grade and gradually decreases towards lower and higher metamorphic grades.

With the increase of deformational intensity (brittle and ductile deformation), the absorption strength of $1600\,cm^{-1}$ increases at first and then decreases, which is inversed with the variation of absorption strength of $749\,cm^{-1}$ (Figures 2(c) and 2(d)).

*3.1.2. Evolution of Characteristic Frequency of Aliphatic Structure.* The frequency of aliphatic structure includes absorption strength of (1) $2923\,cm^{-1}$ and $2862\,cm^{-1}$ related to the asymmetric stretching vibration of $CH_2$ and symmetrical stretching vibration of $CH_3$, shown as shoulder absorption of $2923\,cm^{-1}$ in aliphatic structure. These frequencies are the weakest in the middle metamorphic grade and increase in the lower and higher metamorphic grades and (2) $1442\,cm^{-1}$ related to the asymmetric deformation vibration of $CH_2$ and $CH_3$ in alkane structure. With the increase of metamorphic grade, the change of absorption strength of $1442\,cm^{-1}$ range from 0.8 to 0.99.

TABLE 1: Part of experiment results of TDC samples.

| Series of deformation | Sample ID | $R_{o,max}/(\%)$ | $R_{o,min}/(\%)$ | $\Delta R_o/R_{o,max}$ [1] | FTIR | | | | | Raman[2] | |
| | | | | | CH$_3$CH$_2$ | | C=C | CH$_3$CH$_2$ | C–H | $A_G$ | $A_D$ |
| | | | | | 2923 | 2826 | 1600 | 1442 | 749 | | |
| Brittle deformation coal | LHM06 | 0.98 | 0.83 | 0.15 | 0.557 | 0.326 | 0.991 | 0.815 | 0.317 | 113967 | 55594 |
| | HZM03 | 1.93 | 1.67 | 0.13 | 0.344 | 0.220 | 0.994 | 0.822 | 0.348 | 515720 | 286732 |
| | SK04 | 1.00 | 0.91 | 0.09 | 0.522 | 0.372 | 0.998 | 0.920 | 0.369 | 115158 | 58378 |
| | HZM02 | 1.93 | 1.67 | 0.13 | 0.323 | 0.199 | 0.998 | 0.935 | 0.420 | — | — |
| | LHM12 | 1.37 | 1.13 | 0.18 | 0.579 | 0.383 | 0.998 | 0.933 | 0.385 | 215980 | 131925 |
| | STM02 | 1.41 | 1.12 | 0.21 | 0.647 | 0.403 | 0.994 | 0.806 | 0.380 | 272570 | 153720 |
| | TYM04 | 0.95 | 0.8 | 0.16 | 0.504 | 0.354 | 1.000 | 0.895 | 0.176 | 137498 | 64296 |
| | SK03 | 0.98 | 0.88 | 0.10 | 0.363 | 0.264 | 0.996 | 0.886 | 0.548 | 174251 | 78056 |
| Ductile deformation coal | LHM04 | 1.40 | 1.18 | 0.16 | 0.657 | 0.414 | 0.988 | 0.893 | 0.450 | 394537 | 202143 |
| | LHM09 | 1.39 | 1.12 | 0.19 | 0.599 | 0.406 | 0.996 | 0.951 | 0.362 | — | — |
| | LLM04 | 0.83 | 0.60 | 0.28 | 0.547 | 0.378 | 0.999 | 0.918 | 0.282 | 324725 | 164211 |
| | HZM10 | 2.62 | 2.02 | 0.23 | 0.671 | 0.406 | 0.996 | 0.913 | 0.301 | 461139 | 157064 |
| | LHM02 | 1.38 | 1.08 | 0.22 | 0.582 | 0.38 | 0.998 | 0.946 | 0.385 | — | — |
| | LHM03 | 1.58 | 1.18 | 0.25 | 0.555 | 0.351 | 0.988 | 0.840 | 0.406 | 420961 | 229994 |
| | STM05 | 1.66 | 1.12 | 0.33 | 0.491 | 0.312 | 0.997 | 0.992 | 0.455 | 511926 | 306591 |
| | XTM08 | 1.92 | 1.63 | 0.15 | 0.260 | 0.180 | 1.000 | 0.910 | 0.500 | 499840 | 260745 |

[1] $\Delta R_o = R_{o,max} - R_{o,min}$, [2] The data of Raman was cited in [22].

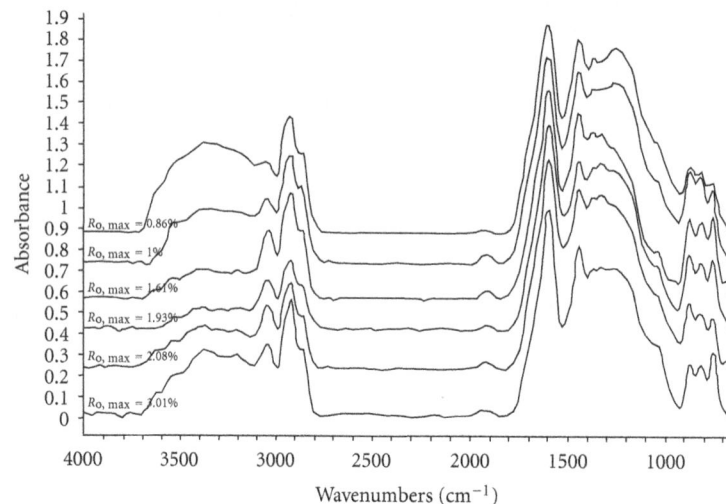

FIGURE 1: The infrared spectrum of TDC with different metamorphic stage.

With the increase of metamorphic grade, the absorption strength of aliphatic structure is much more complicated (Figures 2 and 3). The absorption strength of 2923 cm$^{-1}$ increases first and then decreases in brittle and ductile deformational coal, but absorption strength change of 1442 cm$^{-1}$ is not obvious. It is indicated that the aliphatic structure is gradually degraded with the increase of metamorphic grade in these two deformational mechanisms and this results in the decrease of aliphatic structure.

With the increase of deformational intensity, under the lower metamorphism grade, the absorption strength of 2923 cm$^{-1}$ increases first and decreases later, which is contrary to the absorption strength variation of 1442 cm$^{-1}$ in brittle and ductile deformational coal (Figures 3(c) and 3(d)). Under the lower deformational intensity, the aliphatic functional groups, alkane branched chains, and a few aromatic rings break off at first, which promotes the metamorphism because of the brittle deformation. In the meantime, the dropped small moleculars have not got enough time to form aromatic structure because of the rapid strain rate [19–22]. Part of dropped alkane branched chains are transformed into aliphatic functional groups, and others are turned to free macromolecules. Under the slow strain rate of ductile deformational coal [22], the dropped small

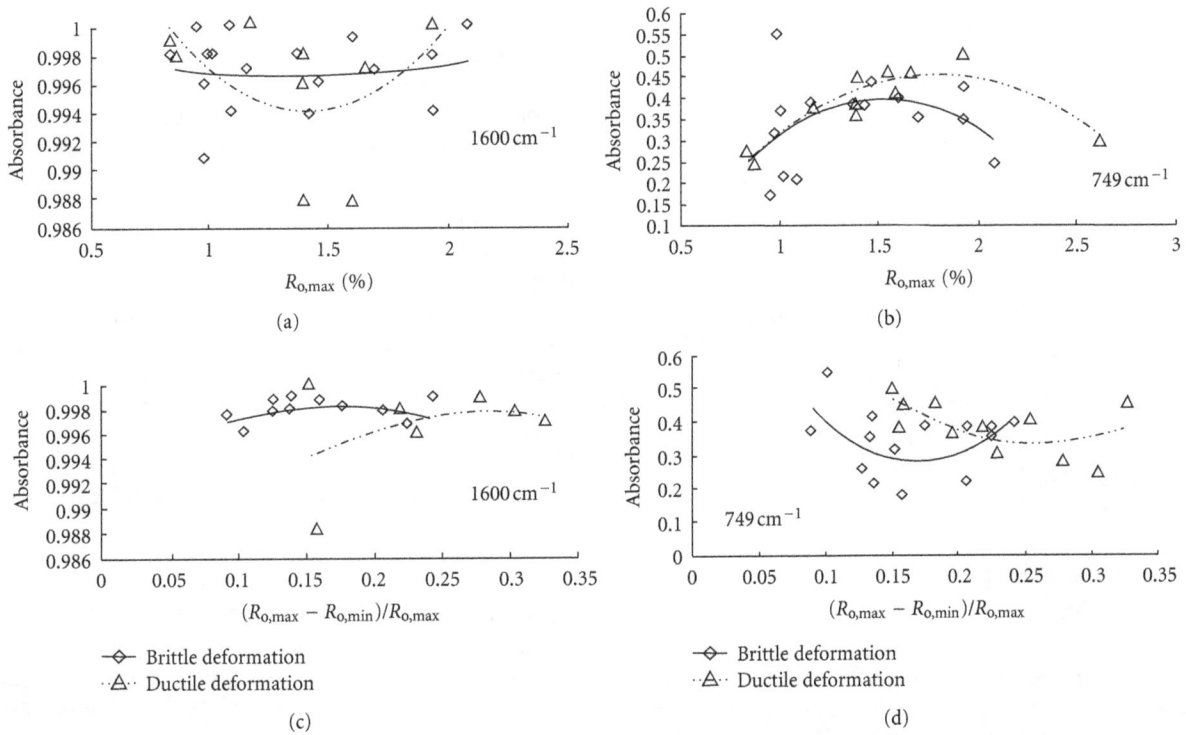

FIGURE 2: Relationship of aromatic absorbance peaks of TDC with different metamorphic and deformation stages. (a) and (b) relationship between aromatic absorbance peaks and metamorphic stages. (c) and (d) relationship between aromatic absorbance peaks and deformation stages.

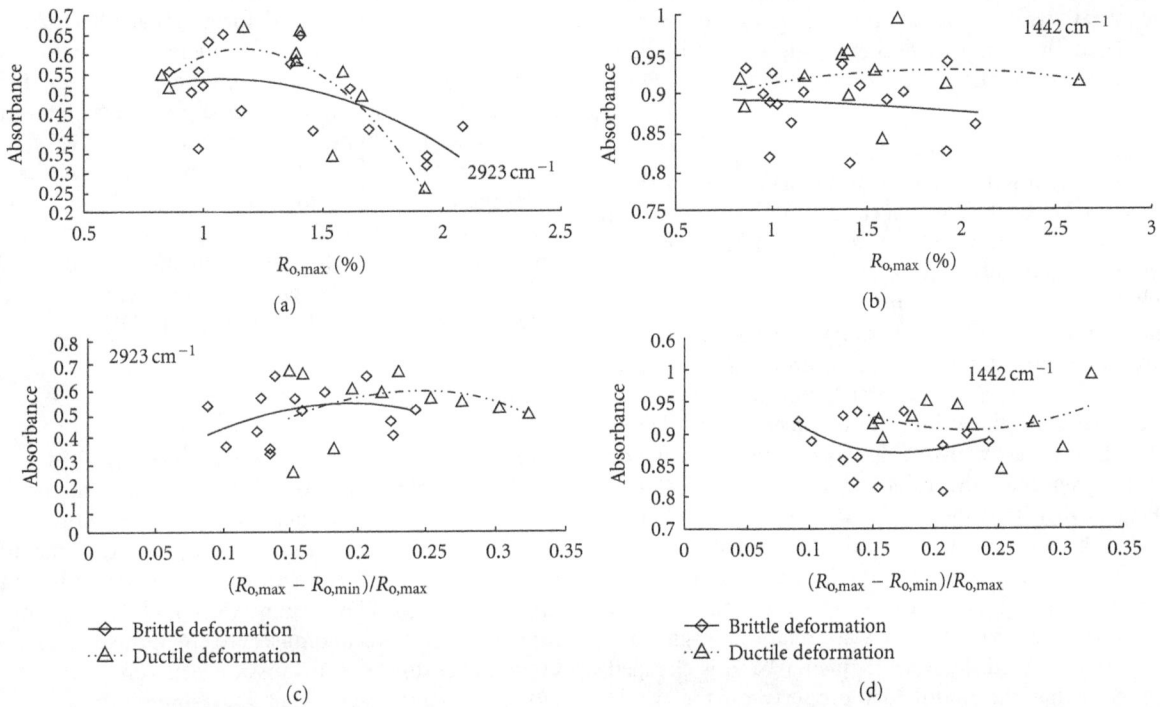

FIGURE 3: Relationship of aliphatic absorbance peaks of TDC with metamorphic and deformation stages. (a) and (b) relationship between aliphatic absorbance peaks and metamorphic stages. (c) and (d) relationship between aliphatic absorbance peaks and deformation stages.

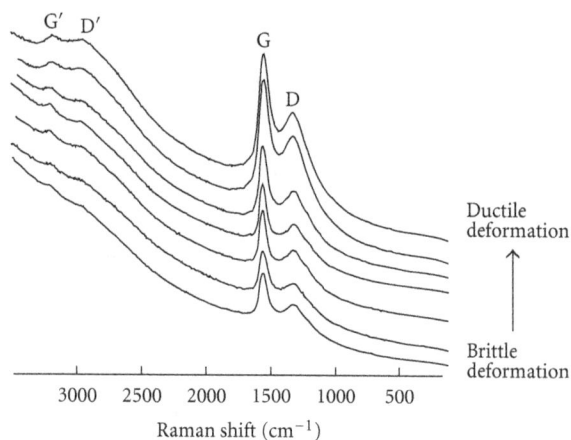

FIGURE 4: The Raman spectrum of TDC with different deformation mechanisms (from [22]).

molecules are combined to aromatic rings and increase the absorption strength of aromatic structure.

*3.2. The Raman Spectral Analysis.* Nakamizo studied the grinded graphite and graphitized coke applying Raman spectrum and found that the peak G in Raman spectrum is related to the deformation vibration of C=C on aromatic plane; the peak area of G ($A_G$) stands for the total number of aromatic rings [24, 25]. The grinding process could form structure defects. It is also suggested that secondary structural defects are produced by exogenic action like grinding process. Previous researchers suggested that the tectonic shear stress process and the grinding process were similar under the geological environment; both could generate secondary structural defects in coal graphite [26].

*3.2.1. Evolution of the Peak Area of G ($A_G$).* Two peaks (G and D) are observed in Raman spectrum in TDC samples from Huainan-Huaibei coal field, which are in the range from $1590 \, cm^{-1}$ to $1600 \, cm^{-1}$ and the range from $1340 \, cm^{-1}$ to $1356 \, cm^{-1}$, respectively (Figure 4).

Peak G in Raman spectrum is related to the deformational vibration of C=C on aromatic plane; $A_G$ means peak area of G, the total number of aromatic rings [22, 26]. With the increase of metamorphic grade, forbrittle deformational coal, the values of $A_G$ increase but change gently like a convex curve in ductile deformational coal which is always higher than the values in brittle deformational coal (Figure 5(a)). It is indicated that the total number of aromatic rings increase in brittle deformational coal, but increase first and decrease later in ductile deformational coal when metamorphic grade increases. The FTIR data show that the change of total aromatic rings are relatively slight, but the absorption strength of every frequency band is changed which means that the undulatory property on the whole aromatic structure is caused by fracture, abscission, and cyclization of aliphatic structure in macromolecular structure of coal. With the increase of deformational intensity, the value of $A_G$ increases in brittle deformational coal, but

first decreases and then increases in ductile deformational coal (Figure 5). The total number of aromatic rings formed in brittle deformational coal are relatively less than in ductile deformational coal. It is indicated that the ductile deformation played a more important role in the process of polycondensation in macromolecular structure of coal.

*3.2.2. Evolution of the Peak Area of D ($A_D$).* The Peak D is related to the lattice vibration of irregular hexagon in disordered $sp^2$ carbonaceous material connected with secondary structural defects between molecular structures [22, 25, 26]. $A_D$ refers to peak area of D and reflects the change of secondary structural defects in the macromolecular structure and the degree of structure order. With the increase of metamorphic grade, the change of $A_D$ is the same as $A_G$. The values of $A_D$ increase in brittle deformational coal, but increase first and then decrease in ductile deformational coal (Figure 6(a)).

Previous research discussed that there were two types in D peak of Raman spectrum; the first one is induced by defects of primary structures ($1370 \, cm^{-1}$), and the other is induced by secondary structural defects ($1360 \, cm^{-1}$) which are also related to tectonic stress [26]. The peak D in Huaibei TDC samples is all distributed at $1360 \, cm^{-1}$, indicating that the secondary structural defects generated by tectonic stress exist in TDC. With the increase of metamorphic grade, the secondary structural defects increase rapidly in brittle deformational coal, but increase prior to decrease in ductile deformational coal.

With the increase of deformational intensity, the value of $A_D$ increases in brittle deformational coal, but decreases prior to increases in ductile deformational coal (Figure 6(b)). It is suggested that with the increase of deformational intensity, the secondary structural defects increase in brittle deformational coal, but decrease first and then increase in ductile deformational coal. However, the secondary structural defects in brittle deformational coal are always fewer than in ductile deformational coal. The increasing and accumulating of unit dislocation may transform the stress into strain energy in ductile deformational coal [19, 20, 22, 27], which is easier to generate the secondary structural defects in macromolecular structure of TDC.

## 4. Discussion

Compared with primary structure coal [2–7, 16, 17, 23, 25, 26], the difference of macromolecular structure shown by FTIR and Raman data of TDC is obvious with the increase of metamorphic grade. Ju et al. studied the different metamorphism-deformation environments and ultrastructure of various TDC using XRD and Nuclear magnetic resonance methods and observed the ultrastructure directly by High-resolution transmission electron microscope [28, 29]. The results reveal that the temperature and tectonic deformation could affect the metamorphism-deformation environment as shown by the change of the stacking of the basal structural units (BSU) layer $L_c$ and the ratio of extension and stacking of the BSU layer $L_a/L_c$. Under the effect

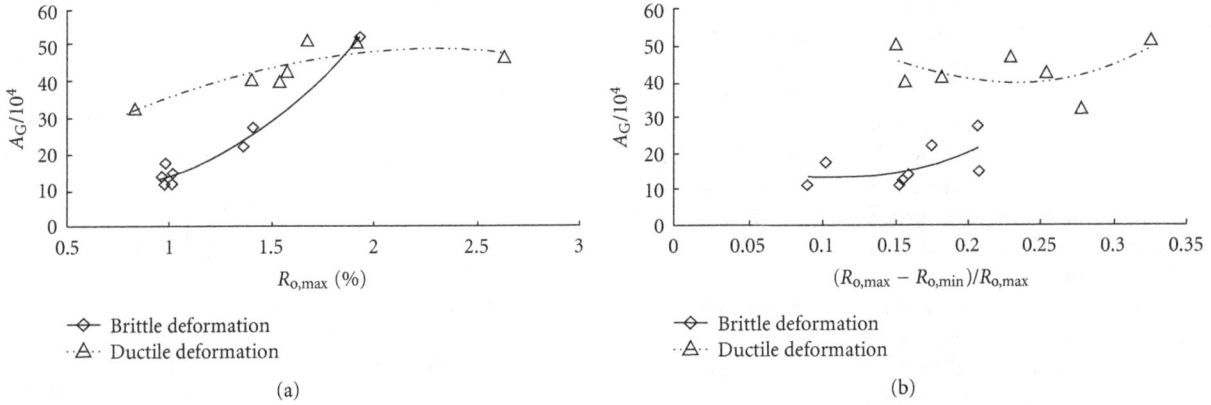

FIGURE 5: Relationship of $A_G$ of TDC with their metamorphic and deformation stages: (a) relationship between $A_G$ and metamorphic stages, (b) relationship between $A_G$ and deformation stages.

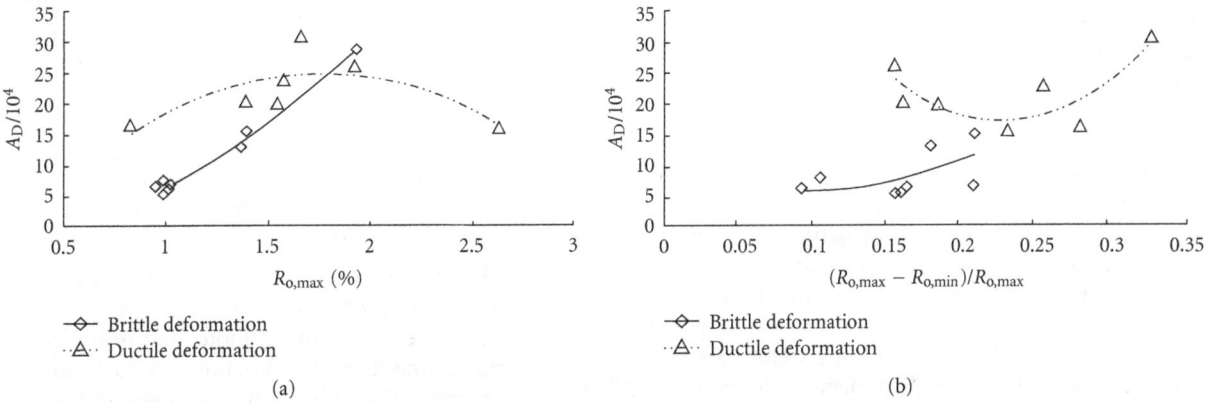

FIGURE 6: Relationship of $A_D$ of TDC with their metamorphic and deformation stages: (a) relationship between $A_D$ and metamorphic stages, (b) relationship between $A_D$ and deformation stages.

of oriented stress, the orientation of the macromolecular structure becomes locally stronger, and the ordering degree of C-nets as well as the arrangement of the BSU is obviously enhanced. It is suggested that the temperature is not the only reason that results in carbonification process of coal, but also the tectonic deformation plays a very important role in this process [17, 28].

With the increase of metamorphic grade, the aliphatic structure and functional groups break off, and the aromatic structure is enriched in ductile deformational coal. However, for brittle deformational coal, the increase of aromatic structure is not obvious. It is indicated that the ductile deformation could produce apparent effect on degradation and polycondensation, and the increased metamorphic grade could promote the ductile deformation and the polycondensation process. The brittle deformation can only produce apparent effect on degradation and has little influence on polycondensation. The absorption strength of aromatic structure is developed as complementary to the aliphatic structure bands compared to the absorption peaks of aromatic and aliphatic structure. In lower metamorphic grade, there is much more aliphatic structures in TDC. Because of the fracture, abscission and cyclization of aliphatic

functional groups and alkane branched chains made the aromatic structure increase with the metamorphic grade rise.

With the increase of deformational intensity, the brittle deformation transforms the stress into frictional heat energy [19, 20, 22, 27], increases the kinetic energy, and promotes the degradation. The aliphatic functional groups and alkane branched chains break off first because of the high strain rate, and the residual of them will be transformed into aromatic structure so that the rangeability of aromatic structure in brittle deformation is relatively low. The ductile deformation transforms the stress into strain energy by increasing and accumulating of unit dislocation [19, 20, 22, 27], which promotes the polycondensation. In lower metamorphism stage with degradation as the main reaction, the aliphatic functional groups and alkane branched chains break off by ductile deformation. The dropped small molecules have enough time to form aromatic rings because of the slow strain rate, which results in more and more aromatic structure enrichment in ductile deformational coal.

Raman data show that the $D$ peak of Raman spectrum is observed around $1360\,\mathrm{cm}^{-1}$, indicating that secondary structural defects in TDC samples from Huaibei coal field are induced by tectonic deformation. In lower metamorphic grade, the increase of the total number of aromatic rings and secondary structural defects are induced by brittle and ductile deformation, whereas by ductile deformation in higher metamorphic grade. This feature suggests that lower metamorphic grade is a benefit to brittle deformation, while the higher metamorphic grade can promote the ductile deformation. The FTIR spectrum shows that the ductile deformation can produce apparent effect on degradation and polycondensation, but brittle deformation exerts little influence on polycondensation. It is observed that degradation is the main reaction for macromolecular structure formation in lower metamorphic grade, instead of polycondensation in higher metamorphic grade. With the increase of deformational intensity, fewer secondary structural defects exist in brittle deformational coal than in ductile deformational coal, indicating that the ductile deformation can produce the secondary structural defects easier than brittle deformation.

To summarize, different deformational mechanisms change the chemical structure and produce the secondary structural defects, which are the main reason for various structure evolution of TDC compared with primary structure coals. Based on the XRD test about those TDC samples we studied early [29], with the increase of deformational intensity, the degradation is the main effect under brittle deformational coal. The brittle deformation can transform stress into frictional heat energy, increase the kinetic energy, and accelerate the movement of molecule (functional group) [19, 27, 29, 30]. The aliphatic functional groups and alkane branched chains break off by brittle deformation and promote the degradation as well, shown by decrease of $L_c$ and $L_a$. With increasing deformation and accumulating thermal energy, the cyclization and aromatization of aliphatic functional groups increase and form aromatic rings and make the $L_c$ and $L_a$ increase. Meanwhile, for the ductile deformational coal, the aliphatic functional groups, alkane branched chains, and a fraction of aromatic rings break off by ductile deformational in the lower metamorphism stage shown by the decrease of $L_c$. The accumulating of unit dislocation may transform the stress into strain energy and the slow strain rate [19, 27, 29, 30]; parts of dropped small molecules have enough time to form aromatic rings and make $L_a$ increase. In the higher-middle metamorphism stage, with the polycondensation as the main effect, this promotes more dropped small molecules to form aromatic rings and the polycondensation on the other side. With the increase of ductile deformation and accumulating of unit dislocation, the secondary structural defects are generated in this stage which can reduce the structural stability of TDC. The secondary structural defects make the aliphatic functional groups, alkane branched chains, and a fraction of aromatic rings break off selectivity shown by decrease of $L_a$. With more secondary structural defects and small molecules dropped, these small molecules started to fill into secondary structural defects to stabilize the macromolecular structure. So the dropped small molecules splice and embed preferentially in secondary structural defects or residual aromatic structures and form aromatic rings to make the macromolecular structure of much stability.

## 5. Conclusions

(1) With the increase of deformation and metamorphism, the change of FTIR and Raman spectrum shows different ways. The tectonic deformation made a very important role which affects the macromolecular structure of TDC. Different deformational mechanism induced different evolution process of macromolecular structure of TDC. The ductile deformation can produce apparent effect on degradation and polycondensation, but brittle deformation has little influence on polycondensation in lower metamorphic grade. In higher metamorphic grade, polycondensation is the main reaction in macromolecular structure of coal. It means that the degradation is the main effect under brittle deformation and the polycondensation under ductile deformation.

(2) Tectonic dmaineformation can produce the secondary structural defects in macromolecular structure of TDC. The increase of total number of aromatic rings and secondary structural defects is mainly caused by brittle deformation in lower metamorphic grade, but ductile deformation in higher metamorphic grade. Furthermore, the ductile deformation can produce the secondary structural defects easier than brittle deformation.

(3) The existence of secondary structural defects reduces the stability of macromolecular structure in TDC. Brittle deformation promotes the degradation and makes the aliphatic functional groups and alkane branched chains break off selectively in lower metamorphic grade. With the increase of deformation and metamorphism, more secondary structural defects are produced and small molecules are dropped; the ductile deformation promotes the polycondensation, so the dropped small molecules splice and embed preferentially in secondary structural defects or residual aromatic structures and form aromatic rings to make the macromolecular structure of much stability.

## Acknowledgments

This work is supported by the National Natural Science Foundation of China (Grants nos. 40772135, 40972131, and 41030422), the National Basic Research Program of China (Grants nos. 2009CB219601 and 2006CB202201), and the Strategic Priority Research Program of the Chinese Academy of Sciences (XDA05030100).

## References

[1] J. R. Levine and A. Davis, "The relationship of coal optical fabrics to alleghanian tectonic deformation in the central Appalachian fold-and-thrust belt," *Pennsylvania Geological Society of America Bulletin*, vol. 101, no. 10, pp. 1333–1347, 1989.

[2] D. Y. Cao, S. R. Zhang, and D. Y. Ren, "The influence of structural deformation on coalification: a case study of

carboniferous coal measures in the Northern foothills of the dabie orogenic belt," *Geological Review*, vol. 48, no. 3, pp. 313–317, 2002.

[3] D. J. Zhang and X. F. Xian, "I.R. spectroscopy analysis of the groups in coal macromolecule," *Journal of Chongqing University*, vol. 13, no. 5, pp. 6–7, 1990.

[4] X. D. Zhu, Z. B. Zhu, and C. J. Han, "Quantitative determination of oxygen-containing functional groups in coal by FTIR spectroscopy," *Journal of Fuel Chemistry and Technology*, vol. 27, no. 4, pp. 335–339, 1999.

[5] J. Ibarra, E. Muñoz, and R. Moliner, "FTIR study of the evolution of coal structure during the coalification process," *Organic Geochemistry*, vol. 24, no. 6-7, pp. 725–735, 1996.

[6] Y. W. Ju, B. Jiang, G. L. Wang et al., *Tectonically Deformed Coals: Structure and Physical Properties of Reservoirs*, China University of Mining and Technology Press, Xuzhou, China, 2005.

[7] Y. W. Ju, B. Jiang, Q. L. Hou, and G. L. Wang, "FTIR spectroscopic study on the stress effect of compositions of macromolecular structure in tectonically deformed coals," *Spectroscopy and Spectral Analysis*, vol. 25, no. 8, pp. 1216–1220, 2005.

[8] J. Jehlička, O. Urban, and J. Pokorný, "Raman spectroscopy of carbon and solid bitumens in sedimentary and metamorphic rocks," *Spectrochimica Acta A*, vol. 59, no. 10, pp. 2341–2352, 2003.

[9] C. P. Marshall, E. J. Javaux, A. H. Knoll, and M. R. Walter, "Combined micro-Fourier transform infrared (FTIR) spectroscopy and micro-Raman spectroscopy of proterozoic acritarchs: a new approach to Palaeobiology," *Precambrian Research*, vol. 138, no. 3-4, pp. 208–224, 2005.

[10] S. Bernard, O. Beyssac, K. Benzerara, N. Findling, G. Tzvetkov, and G. E. Brown, "XANES, Raman and XRD study of anthracene-based cokes and saccharose-based chars submitted to high-temperature pyrolysis," *Carbon*, vol. 48, no. 9, pp. 2506–2516, 2010.

[11] A. Cuesta, P. Dhamelincourt, J. Laureyns, and J. M. D. Tascón, "Comparative performance of X-ray diffraction and Raman microprobe techniques for the study of carbon materials," *Journal of Materials Chemistry*, vol. 8, pp. 2875–2879, 1998.

[12] O. Beyssac, L. Bollinger, J. P. Avouac, and B. Goffé, "Thermal metamorphism in the lesser Himalaya of Nepal determined from Raman spectroscopy of carbonaceous material," *Earth and Planetary Science Letters*, vol. 225, no. 1-2, pp. 233–241, 2004.

[13] J. Jehlička, O. Urban, and J. Pokorný, "Raman spectroscopy of carbon and solid bitumens in sedimentary and metamorphic rocks," *Spectrochimica Acta A*, vol. 59, no. 10, pp. 2341–2352, 2003.

[14] J. Jehlička and C. Beny, "First and second order Raman spectra of natural highly carbonified organic compounds from metamorphic rocks," *Journal of Molecular Structure*, vol. 480-481, pp. 541–545, 1999.

[15] O. Urban, J. Jehlička, J. Pokorný, and J. N. Rouzaud, "Influence of laminar flow on preorientation of coal tar pitch structural units: Raman microspectroscopic study," *Spectrochimica Acta A*, vol. 59, no. 10, pp. 2331–2340, 2003.

[16] Y. Qin, *Micropetrology and Structural Evolution of High-Rank Coals in P. R. China*, China University of Mining and Technology Press, Xuzhou, China, 1994.

[17] Y. W. Ju and X. S. Li, "New research progress on the ultrastructure of tectonically deformed coals," *Progress in Natural Science*, vol. 19, no. 11, pp. 1455–1466, 2009.

[18] Y. W. Ju, B. Jiang, Q. L. Hou, and G. L. Wang, "The new structure-genetic classification system in tectonically deformed coals and its geological significance," *Journal of China Coal Society*, vol. 29, no. 5, pp. 513–517, 2004.

[19] Y. W. Ju, G. L. Wang, B. Jiang, and Q. Hou, "Microcosmic analysis of ductile shearing zones of coal seams of brittle deformation domain in superficial lithosphere," *Science in China D*, vol. 47, no. 5, pp. 393–404, 2004.

[20] Y. W. Ju, H. Lin, X. S. Li et al., "Tectonic deformation and dynamic metamorphism of coal," *Earth Science Frontiers*, vol. 16, no. 1, pp. 158–166, 2009.

[21] D. Y. Cao, X. M. Li, and S. R. Zhang, "Influence of tectonic stress on coalification: stress degradation mechanism and stress polycondensation mechanism," *Science in China D*, vol. 50, no. 1, pp. 43–54, 2007.

[22] H. Lin, Y. W. Ju, Q. L. Hou et al., "Raman spectra of tectonically deformed coals in brittle and ductile deformation mechanisms and its response to structural components," *Progress in Natural Science*, vol. 19, no. 10, pp. 1117–1125, 2009.

[23] D. J. Zhang and X. F. Xian, "The study of the macromolecular structure of coal by FTIR spectroscopy," *Spectroscopy and Spectral Analysis*, vol. 9, no. 3, pp. 17–19, 1989.

[24] M. Nakamizo, R. Kammereck, and P. L. Walker Jr., "Laser Raman studies on carbons," *Carbon*, vol. 12, no. 3, pp. 259–267, 1974.

[25] M. F. Li, F. G. Zeng, F. H. Qi, and B. L. Sun, "Raman spectroscopic characteristics of different rank coals and the relation with XRD structural parameters," *Spectroscopy and Spectral Analysis*, vol. 29, no. 9, pp. 2446–2449, 2009.

[26] Z. Zheng and X. H. Chen, "Raman spectra of coal-based graphite," *Science in China B*, vol. 38, no. 1, pp. 97–106, 1995.

[27] Q. L. Hou, J. L. Li, S. Sun et al., "Discovery and mechanism discussion of supergene micro-ductile shear zone," *Chinese Science Bulletin*, vol. 40, no. 10, pp. 824–827, 1995.

[28] Y. W. Ju, B. Jiang, Q. L. Hou, G. Wang, and S. Ni, "$^{13}$C NMR spectra of tectonic coals and the effects of stress on structural components," *Science in China D*, vol. 5, no. 9, pp. 847–861, 2005.

[29] X. S. Li, Y. W. Ju, Q. L. Hou et al., "Structural response to deformation of the tectonically deformed coal macromolecular," *Acta Geologica Sinica (English Eition)*, In press, 2012.

[30] Q. L. Hou and D. L. Zhong, "The deformation and metamorphism in the wuliangshan ductile shear zone in Western Yunnan, China," in *Memoir of Lithospheric Tectonic Evolution Research*, pp. 24–29, Seismology Press, Beijing, China, 1st edition, 1993.

# 14

# Geochemistry of the Neoarchaean Volcanic Rocks of the Kilimafedha Greenstone Belt, Northeastern Tanzania

**Charles W. Messo, Shukrani Manya, and Makenya A. H. Maboko**

*Department of Geology, University of Dar es Salaam, P.O. Box 35052, Dar es Salaam, Tanzania*

Correspondence should be addressed to Shukrani Manya, shukrani73@udsm.ac.tz

Academic Editor: Michael O. Garcia

The Neoarchaean volcanic rocks of the Kilimafedha greenstone belt consist of three petrological types that are closely associated in space and time: the predominant intermediate volcanic rocks with intermediate calc-alkaline to tholeiitic affinities, the volumetrically minor tholeiitic basalts, and rhyolites. The tholeiitic basalts are characterized by slightly depleted LREE to nearly flat REE patterns with no Eu anomalies but have negative anomalies of Nb. The intermediate volcanic rocks exhibit very coherent, fractionated REE patterns, slightly negative to absent Eu anomalies, depletion in Nb, Ta, and Ti in multielement spidergrams, and enrichment of HFSE relative to MORB. Compared to the other two suites, the rhyolites are characterized by low concentrations of $TiO_2$ and overall low abundances of total REE, as well as large negative Ti, Sr, and Eu anomalies. The three suites have a $\varepsilon$Nd (2.7 Ga) values in the range of $-0.51$ to $+5.17$. The geochemical features of the tholeiitic basalts are interpreted in terms of derivation from higher degrees of partial melting of a peridotite mantle wedge that has been variably metasomatized by aqueous fluids derived from dehydration of the subducting slab. The rocks showing intermediate affinities are interpreted to have been formed as differentiates of a primary magma formed later by lower degrees of partial melting of a garnet free mantle wedge that was strongly metasomatized by both fluid and melt derived from the subducting oceanic slab. The rhyolites are best interpreted as having been formed by shallow level fractional crystallization of the intermediate volcanic rocks involving plagioclase and Ti-rich phases like ilmenite and magnetite as well as REE-rich phases like apatite, zircon, monazite, and allanite. The close spatial association of the three petrological types in the Kilimafedha greenstone belt is interpreted as reflecting their formation in an evolving late Archaean island arc.

## 1. Introduction

The Kilimafedha greenstone belt of northeast Tanzania is one of the six greenstone belts of the Tanzania Craton occurring in the northern part of the country in the area south and east of the Lake Victoria. Other greenstone belts include the Sukumaland, Shinyanga-Malita, Nzega, Musoma-Mara, and Iramba-Sekenke [1, Figure 1]. All of these greenstone belts are prospective for gold mineralization with several large-scale mines now in operation including the Bulyanhulu, Tulawaka, Geita, Buzwagi, North Mara, and Golden Pride (Figure 1). Because of their economic significance, the greenstone belts of the Tanzania Craton have recently been the focus of research on the processes that control gold mineralization (e.g., [2, 3]), lithostratigraphical relationships

(e.g., [4, 5]), geochemistry, and geochronology (e.g., [6–12]). These studies have helped us to better understand, among other things, the timing of and the processes responsible for the formation of the earliest continental crust in Tanzania as well as the ancient tectonic settings in which the greenstone belts formed.

Previous geological work in the Kilimafedha greenstone belt is limited to the geological mapping done by Macfarlane [14] and more recently to geochronological investigation by Wirth et al. [9] who reported zircon Pb-Pb ages from rhyolites and granitic intrusions in the area. In this paper, we present whole-rock major and trace element as well as the Nd-isotopic compositions for the volcanic rocks of the Kilimafedha greenstone belt around the Ikoma area with the aim of unraveling their petrogenesis and tectonic

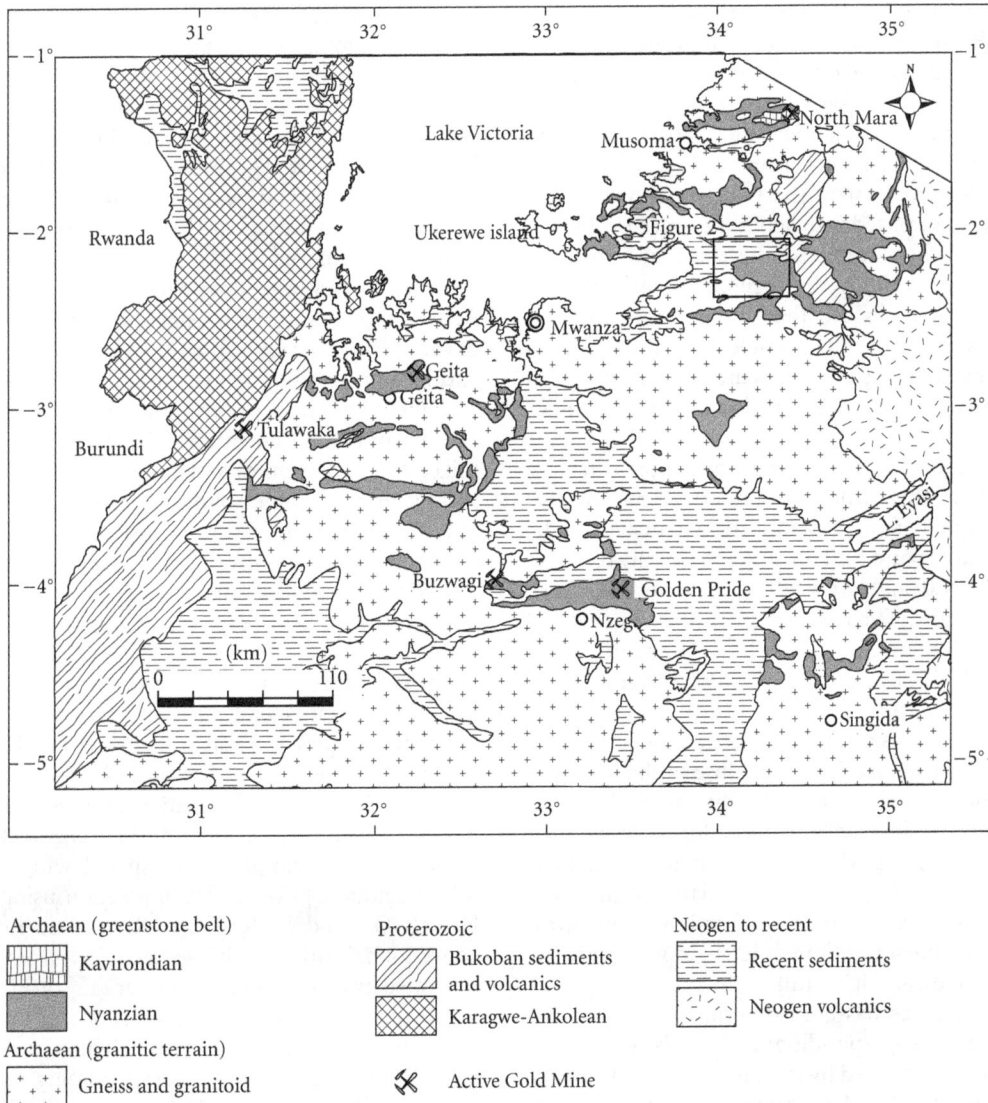

FIGURE 1: Geological map of the northern part of the Tanzania Craton showing the greenstone belts in the area around Lake Victoria (modified after [13]). The inset frame indicates the area of study in Figure 2.

setting of eruption. The results of this study complement the information available from other greenstone belts of the Tanzania Craton on the processes that led to the growth of the continental crust during the late Archaean.

## 2. Geological Setting

The Tanzania Craton forms the central nucleus of Tanzania and extends northwards into southwestern Kenya and southeastern Uganda. The Craton is divided into two main lithological units: the Dodoman belt which is comprised of high-grade metamorphic rocks, granite, granitic gneisses, and migmatites of central Tanzania and the low-grade granite-greenstone terrane of northern Tanzania, southwestern Kenya, and southwestern Uganda [15]. The low-grade granite-greenstone terrane comprises mafic to felsic

volcanic rocks and metasedimentary rocks including shales, sandstones, siltstones, chert, and banded iron formation (BIF) which are in turn intruded by granites.

Robust geochronological data shows that the oldest greenstones in the Tanzania Craton are the mafic volcanic rocks of the Rwamagana area in the Sukumaland greenstone belt (Figure 1). These yielded a Sm-Nd isochron age of 2823 ± 44 Ma reported by Manya and Maboko [7]. The youngest volcanism in the greenstone belts of the Tanzania Craton is from the Musoma-Mara greenstone belt reported by Manya et al. [10] as indicated by a zircon U-Pb age of 2667 ± 8 Ma obtained from dacites collected near Tarime (Figure 1). A more thorough review of the geochronology of the greenstones and intruding granites of the Tanzania Craton can be found in Borg and Krogh [6] and Manya et al. [10].

The Kilimafedha greenstone belt forms an asymmetrical horseshoe-shaped exposure of metavolcanic and minor metasedimentary rocks in the area east and southeast of Lake Victoria (Figure 2). Most of the greenstone exposures including the old Kilimafedha mining district lie within the Serengeti National park. The samples collected for this study, however, were sampled outside the National park boundary in the Ikoma area (Figure 2). According to Macfarlane [14], the greenstone sequences start with a poorly preserved mafic volcanic unit now converted into actinolite and hornblende schist (Figure 3) in the extreme southeastern and northern margins of the belt (within the Serengeti National Park boundary), and the rocks are thus metamorphosed into greenschist facies except for the hornblende schists that are proximal to granitic intrusions. This unit has locally been found to be pillowed suggesting extrusion of the lavas under water.

The mafic volcanic rocks are overlain by a more extensive and better preserved thick sequence of intermediate volcanic rocks with infrequent felsic volcanic rocks patched in the intermediate rocks (Figure 3). This sequence locally contains thin horizons of tuff and metasediments including chert, jaspilite, and quartzite. The felsic volcanic rocks were dated by Wirth et al. [9] who reported $^{207}Pb/^{206}Pb$ zircon ages of $2712 \pm 5$ Ma (MSWD = 0.35) and $2720 \pm 5$ Ma (MSWD = 1.9).

More fresh exposures of the greenstone sequence occur in the area near Fort Ikoma (Figure 2) close to the northern boundary of the greenstone belt. In this area, the most common rock types include amygdaloidal andesite with streams of vesicles filled up with quartz, epidote, and chlorite (Figure 4). Other less vesicular types have large phenocrysts of albite-oligoclase. Sediments intercalated with metavolcanics are largely metamorphosed ferruginous quartzite, siltstones, mudstones, and felsic tuffs.

The whole sequence has been deformed resulting into the development of a steeply dipping N to NW trending foliation. Folding is indicated by isoclinal contortions in the ferruginous quartzite [14]. Late orogenic granites outcrop along the eastern, northern, and southwestern margins of the greenstone belt. The granite-greenstone contact along the northern margin has been decorated with minor metagabbroic intrusions that have been correlated with the larger Nyamongo gabbros of the Musoma-Mara greenstone belt [16]. Neoproterozoic arenaceous to argillaceous sedimentary rocks of the Ikorongo Group [17] unconformably overlie all the Archaean rocks in the area.

## 3. Sampling and Analytical Methodology

Samples were obtained from surface outcrops, and sampling was dictated by the degree of accessibility, exposure, and freshness of the outcrops. Fifty volcanic rocks samples were collected in the field and were subsequently prepared for laboratory chemical analyses. All 50 samples were analyzed for major element compositions, but only 18 representative samples from distinguished suites were analyzed for trace element compositions (see Figure 2 for sample locations). For chemical analyses, the samples were pulverized in an agate mill at the Southern and Eastern Africa Mineral Centre (SEAMIC) Laboratories, Dar es Salaam. Samples were

first dried in an oven at 110°C, and 1 g of the powdered sample was mixed with 7 g of lithium metaborate and fused in a furnace at 1000°C for about 10 minutes to make glass beads. The glass beads were analyzed using an SRS 3000 Siemens X-ray Fluorescence Spectrometer at the same laboratories following procedures reported in Messo [18]. Loss on ignition (LOI) was determined by repeatedly heating the samples in a furnace at 1010°C and cooling until constant weight was achieved.

The samples were also analyzed for trace elements at the Activation Laboratories of Ancaster, Ontario, Canada. 0.25 g of each sample was mixed with a flux of lithium metaborate and lithium tetraborate and fused in an induction furnace. The melt was immediately poured into a solution of 5% $HNO_3$ containing an internal In standard and thoroughly mixed for ~30 minutes to achieve complete dissolution. An aliquot of the sample solution was spiked with internal In and Rh standards to cover the entire mass range and diluted 6000 times prior to introduction into a Perkin Elmer SCIEX ELAN 6000 ICP-MS for trace elements analysis. Precision and accuracy as deduced from replicate analyses of the BIR-1 and W2 standards are 5–10%. The analytical reproducibility deduced from replicate analyses of the samples is better than 8% for most trace elements.

Nine samples were also analysed for Sm-Nd isotopic compositions as well as Sm and Nd concentrations using a Triton-MC Thermal Ionization Mass Spectrometer at the Activation Laboratories of Ontario, Canada. Aliquots of the powdered rock samples were spiked with a $^{149}Sm$-$^{146}Nd$ mixed solution prior to decomposition using a mixture of HF, $HNO_3$, and $HClO_4$. The REEs were separated using conventional cation-exchange techniques. Sm and Nd were separated by extraction chromatography on HDEHP covered teflon powder. Total blanks are 0.1-0.2 ng for Sm and 0.1–0.5 ng for Nd and are negligible. The accuracy of the Sm and Nd analyses is ±0.5% corresponding to errors in the $^{147}Sm/^{144}Nd$ ratios of ±0.5% (2$\sigma$). The $^{143}Nd/^{144}Nd$ ratios are calculated relative to the value of 0.511860 for the La Jolla standard. During the period of analysis, the weighted average of 10 La Jolla Nd-standard runs yielded $0.511872 \pm 15$ (2$\sigma$) for $^{143}Nd/^{144}Nd$, using a $^{146}Nd/^{144}Nd$ value of 0.7219 for normalization.

## 4. Geochemistry

*4.1. Alteration and Element Mobility.* Alteration of metavolcanic rocks is a common phenomenon, in particular for Archaean greenstones, and is typically characterized by high loss on ignition (LOI) values and increased scatter of major and large ion lithophile elements. In this regard, the volcanic rocks, collected for this study, are variably affected by greenschist metamorphism, so it is expected that major and LILE are also affected by alteration. However, numerous studies have demonstrated that rare earth elements (REEs) and high field strength elements (HFSEs) remain relatively undisturbed at greenschist facies and even higher grades of metamorphism [19, 20]. So in this study major and LILE are used with great care, and emphasize is placed on the REE and HFSE.

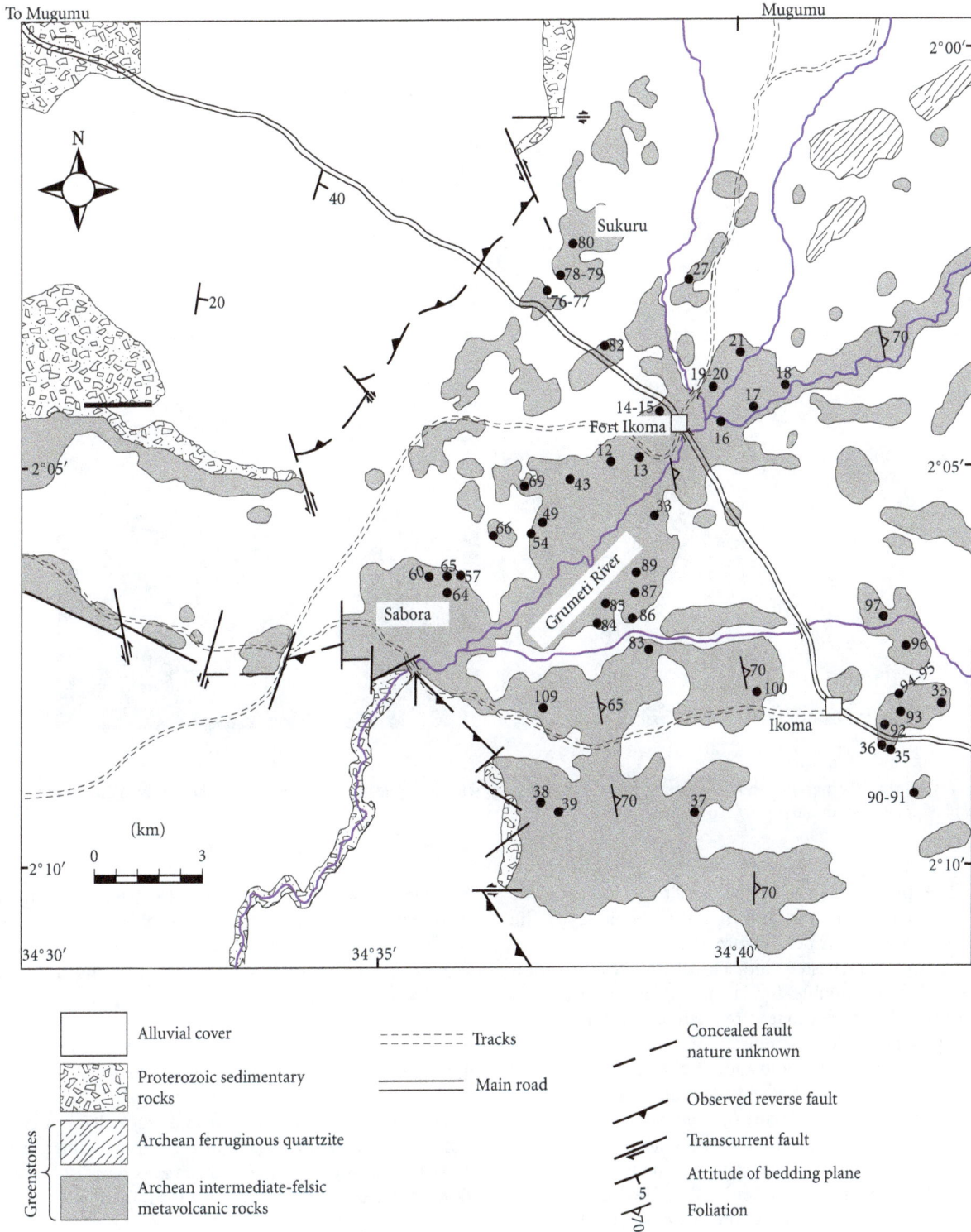

FIGURE 2: Geological map of the Kilimafedha greenstone belt showing sample locations (modified after [14]).

4.2. *Classification and Petrography.* The volcanic rocks of the Kilimafedha greenstone belt represent a mafic, intermediate to felsic compositional continuum as indicated by their wide range of $SiO_2$ contents (48.48–76.02 wt%). Out of a total of 50 samples that were analyzed for major elements and as shown in Figure 4 and 7 samples are basaltic in

composition ($SiO_2$ = 48.48–51.57 wt%), 40 samples show intermediate compositions ranging from basaltic andesites to basaltic trachyandesites, which are predominant; andesites to dacites ($SiO_2$ = 52.51–66.80 wt%); only 3 samples are rhyolites ($SiO_2$ = 75.52–76.02 wt%, values quoted on water free basis), revealing the sporadic nature of the more felsic

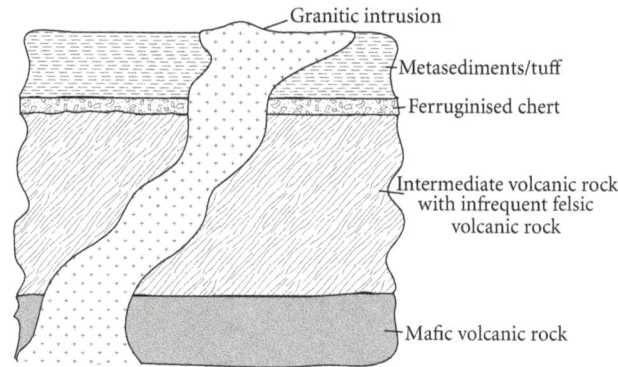

FIGURE 3: Diagrammatic representation of the stratigraphic sequence of Kilimafedha greenstone belt, not to scale (modified after [14]).

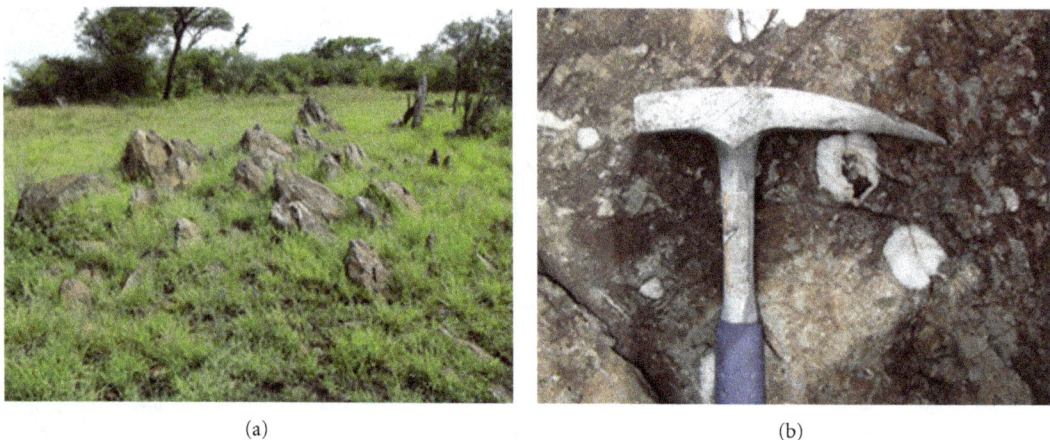

(a)                                                                      (b)

FIGURE 4: Outcrop photographs showing Ikoma foliated metabasalts with an appearance of a schist (a) and amygdaloidal andesite with vesicles filled in with secondary quartz (b).

rocks. The predominance of the basaltic trachyandesitic rocks and the nature of these rocks are unique to the Kilimafedha greenstone belt (Figure 5).

Using the Winchester and Floyd [22] classification scheme which is suitable for classifying metamorphosed rocks, of the 18 samples that were analyzed for both major and trace elements, four samples plot along the boundary between subalkaline basalt and andesite/basalt, eleven others plot in the fields of andesite/basalt and andesites, and three samples plot as rhyolites (Figure 6). The basalts to andesitic basalts exhibit tholeiitic affinity, the intermediate rocks intermediate geochemical characteristics between tholeiitic and calc-alkaline affinities, whereas the rhyolites are calc-alkaline. Accordingly, the rocks are henceforth subdivided into three suites: the tholeiitic basalts, the intermediate volcanic rocks, and the rhyolites.

The primary petrographic features of the tholeiitic basalts are strongly obliterated by alteration which has resulted in the formation of chlorite, epidote, and hornblendic amphibole which appear to form after olivines and pyroxenes. The predominant intermediate rocks are often amygdaloidal, with streams of vesicles filled up with quartz, epidote, and chlorite. Other less vesicular types have large phenocrysts of albite-oligoclase. Most felsic rocks have

fine matrix of quartz and sericitized feldspars with sparse phenocrysts of quartz and altered feldspars. The presence of chlorite and epidote in the Kilimafedha volcanic rocks suggests that these rocks have mainly been metamorphosed into greenschist facies.

### 4.3. Major and Trace Element Geochemistry

4.3.1. Tholeiitic Basalts. Major and trace element composition of the tholeiitic basalts of the Kilimafedha greenstone belt are presented in Table 1. The rocks have $SiO_2$ compositions that are in the range of 48.48–51.57 wt%, $TiO_2$ = 0.61–1.80 wt%, $Fe_2O_3$ = 6.32–13.92 wt%, MgO = 5.56–8.49 wt%, and Mg numbers, calculated as $100 \times Mg^{2+}/(Mg^{2+} + Fe_{total}^{2+})$, range from 47 to 73 (the major elements presented on a water free basis). Cr and Ni contents are 160–240 ppm and 90–190 ppm, respectively. La contents vary considerably (1.7–9.2 ppm), whereas Yb contents range from 1.7 to 3.1 ppm resulting in La/Yb ratios of 1.00–4.00 (Table 1). On major and trace element bivariate plots (Figure 7), $Fe_2O_3$, MgO, CaO, and Ni correlate negatively with $SiO_2$ pointing to fractionation signature or cogenetic relationship. The tholeiitic basalts alone do not show any major trends most

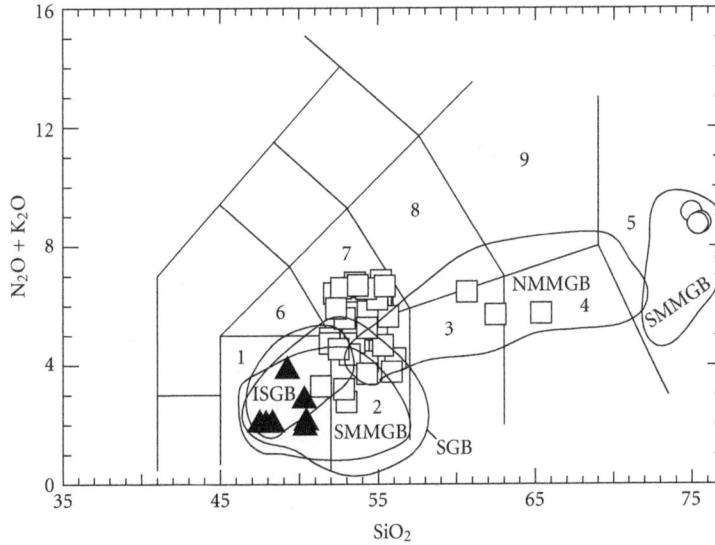

FIGURE 5: TAS classification diagram of Le Maitre et al. [21] for the Kilimafedha volcanic rocks. Also shown in the diagram are fields for volcanic rocks from other greenstone belts of the Tanzania Craton: ISGB: Iramba—Sekenke greenstone belt, SGB: Sukumaland greenstone belt, SMMGB: Southern Musoma—Mara greenstone belt, and NMMGB: Northern Musoma—Mara greenstone belt. Numbers in the diagram indicate fields as follows: 1: basalt, 2: basaltic andesite, 3: andesite, 4: dacite, 5: rhyolite, 6: trachybasalt, 7: basaltic trachyandesite, 8: trachyandesite, and 9: trachydacite.

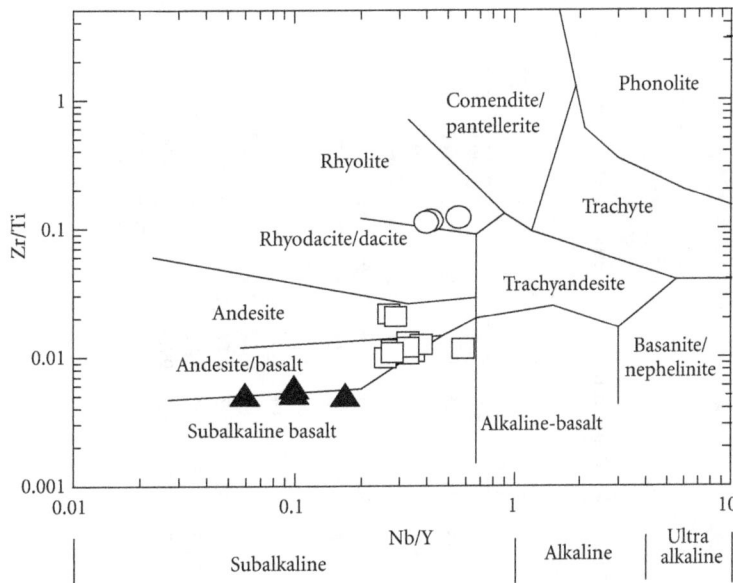

FIGURE 6: Classification of the Kilimafedha volcanic rocks using the Nb/Y-Zr/Ti diagram of Winchester and Floyd [22]. Symbols are filled triangles: tholeiitic basalts, open squares: intermediate volcanic rocks, and open cycles: rhyolites.

likely due to the small number of samples and their restricted range in $SiO_2$ content.

The rocks display slightly depleted LREE to nearly flat REE patterns (Figure 8(a)) that are characterized by $La/Sm_{CN}$ and $La/Yb_{CN}$ ratios of 0.8-0.9 and 0.72–1.06, respectively, except for sample MU 69 which is relatively enriched in LREE ($La/Sm_{CN} = 1.7$ and $La/Yb_{CN} = 2.89$) compared to the other three samples (where CN refers to chondrite normalized values). The $La/Sm_{CN}$ and $La/Yb_{CN}$ values of the three samples (MU 69 excluded) are slightly higher than those of NMORB

($La/Sm_{CN} = 0.6$, $La/Yb_{CN} = 0.59$; [23]) indicating relative enrichment of the LREE in the Kilimafedha tholeiites. The samples do not show any Eu anomalies ($Eu/Eu^* = 0.90$–1.10). The samples were also plotted on primitive mantle spidergrams (Figure 9(a)), where they display enrichment in Rb, Ba,Th, K, and Pb relative to the more compatible elements with flatter multielement patterns characterized by negative Nb anomalies ($Nb/La_{pm} = 0.42$–0.64) and minor negative Ti anomalies (where pm refers to primitive mantle normalized values).

TABLE 1: Major (wt%) and trace (ppm) element composition for the Kilimafedha greenstone belt volcanic rocks.

| | Tholeiitic basalts | | | | | | | | | | | | Transitional intermediate volcanic rocks | | | | | | | | | | | | | |
|---|---|---|---|---|---|---|---|---|---|---|---|---|---|---|---|---|---|---|---|---|---|---|---|---|---|---|
| | MU 14 | MU 15 | MU 19 | MU 20 | MU 43 | MU 69 | MU 109 | MU 12 | MU 13 | MU 21 | MU 27 | MU 33 | MU 35 | MU 36 | MU 37 | MU 38 | MU 39 | MU 49 | MU 54 | MU 57 | MU 60 | MU 64 | MU 65 | MU 66 | MU 68 |
| $SiO_2$ | 50.3 | 47.4 | 50.4 | 50.3 | 47.8 | 48.2 | 49.2 | 55.2 | 52.9 | 52.8 | 65.3 | 60.6 | 53.7 | 55.1 | 54.3 | 54.1 | 53.5 | 53.7 | 53.5 | 54.1 | 54.9 | 56.0 | 54.9 | 55.3 | 62.4 |
| $TiO_2$ | 1.36 | 0.92 | 0.60 | 0.66 | 1.73 | 1.75 | 0.62 | 1.53 | 1.44 | 0.84 | 0.58 | 0.67 | 1.45 | 1.46 | 1.36 | 1.54 | 1.60 | 1.46 | 1.26 | 1.35 | 1.38 | 0.91 | 1.24 | 1.41 | 0.84 |
| $Al_2O_3$ | 14.7 | 16.0 | 14.4 | 13.5 | 14.6 | 14.1 | 17.3 | 13.1 | 15.1 | 15.0 | 15.5 | 16.5 | 14.0 | 13.4 | 14.1 | 14.3 | 14.9 | 14.2 | 13.7 | 13.2 | 13.5 | 14.7 | 14.3 | 13.6 | 14.2 |
| $Fe_2O_3$ | 12.3 | 12.4 | 11.9 | 12.9 | 13.3 | 13.5 | 6.2 | 9.3 | 10.8 | 11.2 | 4.2 | 6.7 | 11.3 | 12.0 | 11.2 | 11.9 | 12.1 | 12.1 | 11.8 | 11.3 | 11.7 | 9.1 | 10.8 | 12.0 | 6.6 |
| $MnO$ | 0.27 | 0.17 | 0.17 | 0.18 | 0.19 | 0.18 | 0.08 | 0.25 | 0.24 | 0.16 | 0.06 | 0.10 | 0.11 | 0.14 | 0.11 | 0.14 | 0.14 | 0.15 | 0.12 | 0.12 | 0.12 | 0.19 | 0.13 | 0.13 | 0.09 |
| $MgO$ | 5.45 | 8.09 | 7.53 | 7.32 | 6.62 | 6.76 | 8.30 | 6.00 | 4.66 | 5.58 | 2.68 | 3.57 | 3.42 | 3.71 | 3.43 | 3.46 | 3.62 | 3.85 | 4.24 | 3.53 | 3.77 | 5.27 | 3.80 | 3.50 | 2.60 |
| $CaO$ | 10.9 | 10.9 | 10.7 | 10.7 | 10.3 | 10.5 | 12.3 | 7.54 | 10.94 | 8.54 | 3.65 | 2.73 | 9.62 | 6.25 | 9.49 | 6.41 | 5.30 | 5.69 | 8.30 | 6.40 | 6.94 | 6.28 | 7.16 | 7.43 | 4.53 |
| $Na_2O$ | 2.29 | 1.84 | 1.92 | 1.79 | 1.96 | 1.91 | 2.99 | 5.14 | 2.49 | 1.74 | 4.91 | 6.15 | 3.12 | 4.96 | 2.70 | 4.02 | 5.00 | 5.34 | 2.67 | 4.69 | 3.10 | 3.85 | 3.01 | 3.94 | 4.20 |
| $K_2O$ | 0.52 | 0.17 | 0.15 | 0.11 | 0.03 | 0.10 | 0.81 | 0.79 | 0.27 | 1.45 | 0.83 | 0.31 | 1.44 | 1.09 | 1.43 | 1.88 | 1.77 | 1.34 | 2.09 | 1.44 | 2.75 | 0.37 | 2.20 | 0.73 | 1.50 |
| $P_2O_5$ | 0.04 | 0.01 | 0.01 | 0.01 | 0.19 | 0.09 | 0.01 | 0.07 | 0.06 | 0.01 | 0.06 | 0.08 | 0.12 | 0.12 | 0.12 | 0.15 | 0.15 | 0.13 | 0.12 | 0.13 | 0.03 | 0.01 | 0.11 | 0.13 | 0.16 |
| $LOI$ | 1.46 | 2.14 | 2.17 | 2.45 | 3.19 | 2.55 | 2.15 | 0.92 | 1.03 | 2.29 | 2.19 | 2.53 | 1.39 | 1.34 | 1.72 | 1.81 | 1.83 | 1.75 | 1.97 | 3.30 | 1.66 | 2.93 | 2.00 | 1.69 | 2.80 |
| Total | 99.5 | 100 | 99.9 | 100 | 99.9 | 99.8 | 99.9 | 99.9 | 99.9 | 99.6 | 100 | 100 | 99.7 | 99.6 | 99.9 | 99.7 | 99.9 | 99.6 | 99.8 | 99.5 | 99.8 | 99.7 | 99.6 | 99.8 | 100 |
| $Mg^{\#}$ | 46.8 | 56.5 | 55.6 | 52.8 | 49.6 | 49.7 | 72.7 | 56.1 | 46.1 | 49.8 | 56.0 | 51.2 | 37.4 | 38.0 | 37.7 | 36.5 | 37.3 | 38.7 | 41.5 | 38.2 | 39.1 | 53.3 | 41.0 | 36.7 | 43.9 |

TABLE 1: Continued.

| | Tholeiitic basalts | | | | | | | | | | | | | | Transitional intermediate volcanic rocks | | | | | | | | | | |
|---|---|---|---|---|---|---|---|---|---|---|---|---|---|---|---|---|---|---|---|---|---|---|---|---|---|
| | MU 14 | MU 15 | MU 19 | MU 20 | MU 43 | MU 69 | MU 109 | MU 12 | MU 13 | MU 21 | MU 27 | MU 33 | MU 35 | MU 36 | MU 37 | MU 38 | MU 39 | MU 49 | MU 54 | MU 57 | MU 60 | MU 64 | MU 65 | MU 66 | MU 68 |
| Ba | 204 | 27 | 66 | | | 140 | | | | | 1230 | | | 418 | | | | 329 | | 97 | 763 | | | | 755 |
| Rb | 26 | 9 | 5 | | | 5 | | | | | 16 | | | 35 | | | | 40 | | 2 | 97 | | | | 18 |
| Sr | 147 | 106 | 101 | | | 214 | | | | | 428 | | | 236 | | | | 239 | | 293 | 320 | | | | 423 |
| Ni | 100 | 190 | 90 | | | 120 | | | | | 30 | | | 80 | | | | 90 | | 60 | 90 | | | | 30 |
| Cr | 160 | 200 | 240 | | | 170 | | | | | 30 | | | 40 | | | | 70 | | 40 | 90 | | | | 20 |
| V | 342 | 228 | 232 | | | 249 | | | | | 76 | | | 171 | | | | 154 | | 162 | 153 | | | | 108 |
| Th | 0.5 | 0.3 | 0.2 | | | 1 | | | | | 3.9 | | | 8 | | | | 8.7 | | 7 | 8.9 | | | | 5.8 |
| Pb | 5 | 5 | 5 | | | 5 | | | | | 7 | | | 5 | | | | 9 | | 42 | 6 | | | | 10 |
| U | 0.2 | 0.1 | 0.1 | | | 0.4 | | | | | 1.3 | | | 2.1 | | | | 2.9 | | 1.9 | 2.5 | | | | 1.4 |
| Nb | 3 | 2 | 1 | | | 4 | | | | | 4 | | | 8 | | | | 8 | | 7 | 8 | | | | 7 |
| Ta | 0.3 | 0.2 | 0.1 | | | 0.3 | | | | | 0.4 | | | 0.6 | | | | 0.8 | | 0.6 | 0.7 | | | | 0.5 |
| Zr | 80 | 53 | 31 | | | 88 | | | | | 126 | | | 163 | | | | 177 | | 147 | 184 | | | | 176 |
| Hf | 2.2 | 1.4 | 1 | | | 2.4 | | | | | 3.2 | | | 4.3 | | | | 4.9 | | 3.9 | 4.7 | | | | 4.6 |
| Y | 29 | 20 | 16 | | | 23 | | | | | 15 | | | 23 | | | | 24 | | 21 | 24 | | | | 24 |
| La | 4.60 | 3.00 | 1.70 | | | 9.20 | | | | | 22.1 | | | 26.2 | | | | 28.1 | | 23.8 | 28.2 | | | | 33.7 |
| Ce | 12.4 | 8.2 | 4.6 | | | 21.6 | | | | | 42.2 | | | 53.5 | | | | 60 | | 49.4 | 59.9 | | | | 69.3 |
| Pr | 1.96 | 1.27 | 0.75 | | | 3.05 | | | | | 4.99 | | | 6.46 | | | | 7.74 | | 6.21 | 7.49 | | | | 9 |
| Nd | 6.58 | 6.70 | 4.10 | | | 17.82 | | | | | 17.9 | | | 24.4 | | | | 28.1 | | 22.9 | 28.5 | | | | 31.6 |
| Sm | 1.63 | 2.1 | 1.3 | | | 6.1 | | | | | 3.5 | | | 5.7 | | | | 6.4 | | 5.4 | 6.2 | | | | 6.5 |
| Eu | 1.11 | 0.84 | 0.57 | | | 1.36 | | | | | 1.06 | | | 1.37 | | | | 1.66 | | 1.68 | 1.65 | | | | 1.62 |
| Gd | 4.3 | 3.0 | 2.0 | | | 4.1 | | | | | 3.3 | | | 5.7 | | | | 5.8 | | 4.8 | 6.1 | | | | 5.3 |
| Tb | 0.8 | 0.5 | 0.4 | | | 0.7 | | | | | 0.5 | | | 0.8 | | | | 0.9 | | 0.8 | 0.9 | | | | 0.8 |
| Dy | 5.0 | 3.4 | 2.5 | | | 4.0 | | | | | 2.6 | | | 4.7 | | | | 4.7 | | 4.0 | 4.8 | | | | 4.2 |
| Ho | 1.1 | 0.7 | 0.6 | | | 0.8 | | | | | 0.5 | | | 0.9 | | | | 0.9 | | 0.8 | 0.9 | | | | 0.8 |
| Er | 3.3 | 2.1 | 1.7 | | | 2.4 | | | | | 1.5 | | | 2.4 | | | | 2.5 | | 2.1 | 2.5 | | | | 2.5 |
| Tm | 0.48 | 0.32 | 0.26 | | | 0.36 | | | | | 0.22 | | | 0.34 | | | | 0.35 | | 0.29 | 0.34 | | | | 0.37 |
| Yb | 3.10 | 2.10 | 1.70 | | | 2.30 | | | | | 1.40 | | | 2.10 | | | | 2.20 | | 1.80 | 2.10 | | | | 2.30 |
| Lu | 0.46 | 0.31 | 0.26 | | | 0.35 | | | | | 0.22 | | | 0.31 | | | | 0.31 | | 0.26 | 0.3 | | | | 0.34 |

TABLE 1: Continued.

| | Tholeiitic basalts | | | | | | Transitional intermediate volcanic rocks | | | | | | | | | | | | | | | | | Rhyolites | | |
|---|---|---|---|---|---|---|---|---|---|---|---|---|---|---|---|---|---|---|---|---|---|---|---|---|---|---|
| | MU 14 | MU 15 | MU 19 | MU 20 | MU 43 | MU 69 | MU 109 | MU 12 | MU 13 | MU 21 | MU 27 | MU 33 | MU 35 | MU 36 | MU 37 | MU 38 | MU 39 | MU 49 | MU 54 | MU 57 | MU 60 | MU 64 | MU 65 | MU 66 | MU 68 |
| La/Yb | 1.5 | 1.4 | 1.0 | | | 4.0 | | | | | 15.8 | | | 12.5 | | | | 12.8 | | 13.2 | 13.4 | | | | 14.7 |
| La/Nb | 1.53 | 1.5 | 1.7 | | | 2.3 | | | | | 5.53 | | | 3.28 | | | | 3.51 | | 3.40 | 3.53 | | | | 4.81 |
| Eu/Eu* | 0.90 | 1.02 | 1.08 | | | 1.10 | | | | | 0.95 | | | 0.74 | | | | 0.83 | | 1.01 | 0.82 | | | | 0.84 |
| La/Sm$_{CN}$ | 0.90 | 0.92 | 0.84 | | | 1.70 | | | | | 4.08 | | | 2.97 | | | | 2.83 | | 2.85 | 2.94 | | | | 3.35 |
| La/Yb$_{CN}$ | 1.06 | 1.02 | 0.72 | | | 2.87 | | | | | 11.3 | | | 8.95 | | | | 9.16 | | 9.48 | 9.63 | | | | 10.5 |
| Nb/La$_{pm}$ | 0.63 | 0.64 | 0.57 | | | 0.42 | | | | | 0.17 | | | 0.29 | | | | 0.27 | | 0.28 | 0.27 | | | | 0.20 |

| | Transitional intermediate volcanic rocks | | | | | | | | | | | | | | | | | | | | | | Rhyolites | | |
|---|---|---|---|---|---|---|---|---|---|---|---|---|---|---|---|---|---|---|---|---|---|---|---|---|---|---|
| | MU 76 | MU 77 | MU 78 | MU 79 | MU 80 | MU 81 | MU 82 | MU 83 | MU 84 | MU 85 | MU 86 | MU 87 | MU 89 | MU 90 | MU 91 | MU 92 | MU 93 | MU 94 | MU 95 | MU 96 | MU 97 | MU 100 | MU 16 | MU 17 | MU 18 |
| SiO$_2$ | 51.9 | 51.9 | 53.2 | 52.4 | 53.1 | 53.3 | 54.3 | 53.5 | 52.2 | 54.3 | 55.1 | 54.3 | 55.6 | 55.8 | 52.8 | 51.3 | 52.6 | 54.9 | 54.2 | 53.7 | 55.4 | 52.3 | 74.9 | 75.5 | 75.3 |
| TiO$_2$ | 1.42 | 1.38 | 1.34 | 1.37 | 1.49 | 1.49 | 0.70 | 1.51 | 1.62 | 1.25 | 1.35 | 1.41 | 1.41 | 1.43 | 1.52 | 0.84 | 1.20 | 1.37 | 1.41 | 1.09 | 1.07 | 1.42 | 0.03 | 0.03 | 0.03 |
| Al$_2$O$_3$ | 14.7 | 14.2 | 14.1 | 14.4 | 14.2 | 13.9 | 15.1 | 14.0 | 14.7 | 13.5 | 14.4 | 13.8 | 13.6 | 13.4 | 14.4 | 14.1 | 14.2 | 14.6 | 14.2 | 15.1 | 14.2 | 15.6 | 14.28 | 14.38 | 14.46 |
| Fe$_2$O$_3$ | 12.4 | 12.6 | 12.2 | 12.4 | 12.5 | 12.6 | 8.2 | 12.4 | 13.0 | 11.3 | 11.3 | 12.2 | 11.8 | 11.4 | 13.5 | 11.6 | 12.5 | 11.4 | 11.6 | 12.5 | 11.3 | 12.9 | 0.34 | 0.28 | 0.34 |
| MnO | 0.14 | 0.14 | 0.14 | 0.14 | 0.14 | 0.14 | 0.13 | 0.14 | 0.14 | 0.13 | 0.14 | 0.13 | 0.13 | 0.12 | 0.16 | 0.15 | 0.19 | 0.13 | 0.14 | 0.15 | 0.17 | 0.15 | 0.02 | 0.01 | 0.01 |
| MgO | 4.40 | 4.20 | 4.04 | 4.10 | 3.88 | 3.92 | 6.47 | 3.86 | 3.89 | 4.44 | 3.60 | 3.68 | 3.72 | 3.55 | 3.82 | 7.08 | 4.89 | 3.66 | 4.10 | 4.56 | 4.26 | 3.64 | 0.23 | 0.22 | 0.25 |
| CaO | 7.36 | 7.97 | 8.36 | 8.38 | 6.74 | 7.00 | 8.87 | 6.82 | 6.01 | 7.40 | 6.36 | 7.03 | 6.39 | 7.13 | 6.32 | 9.37 | 5.66 | 5.84 | 5.56 | 5.28 | 5.40 | 5.97 | 0.26 | 0.10 | 0.15 |
| Na$_2$O | 3.66 | 2.83 | 2.88 | 2.87 | 4.32 | 3.92 | 3.14 | 4.20 | 4.63 | 2.91 | 5.57 | 3.79 | 3.73 | 3.62 | 4.63 | 2.41 | 4.53 | 4.55 | 4.00 | 4.91 | 5.04 | 4.75 | 5.14 | 5.17 | 5.32 |
| K$_2$O | 1.25 | 1.91 | 1.48 | 1.67 | 1.60 | 1.52 | 0.57 | 1.57 | 1.79 | 2.89 | 1.27 | 1.47 | 1.91 | 0.15 | 0.95 | 0.88 | 2.06 | 1.66 | 2.57 | 1.77 | 1.62 | 1.16 | 3.98 | 3.61 | 3.41 |
| P$_2$O$_5$ | 0.11 | 0.02 | 0.11 | 0.11 | 0.12 | 0.13 | 0.11 | 0.13 | 0.14 | 0.11 | 0.12 | 0.13 | 0.14 | 0.13 | 0.13 | 0.03 | 0.09 | 0.10 | 0.13 | 0.10 | 0.09 | 0.12 | 0.01 | 0.01 | 0.01 |
| LOI | 2.48 | 2.35 | 1.78 | 1.79 | 1.59 | 1.57 | 2.46 | 1.52 | 1.49 | 1.70 | 1.40 | 1.97 | 1.65 | 2.94 | 1.70 | 2.27 | 1.87 | 1.40 | 1.63 | 0.94 | 1.50 | 1.99 | 0.75 | 0.69 | 0.69 |
| Total | 99.8 | 99.5 | 99.6 | 99.7 | 99.7 | 99.6 | 100 | 99.6 | 99.6 | 99.9 | 101 | 99.9 | 100 | 99.6 | 100 | 100 | 99.8 | 99.6 | 99.5 | 100 | 100 | 100 | 100 | 100 | 100 |
| Mg# | 41.3 | 39.7 | 39.6 | 39.5 | 38.1 | 38.1 | 60.9 | 38.2 | 37.2 | 43.9 | 38.8 | 37.4 | 38.4 | 38.3 | 35.9 | 54.8 | 43.6 | 38.9 | 41.2 | 42.0 | 42.8 | 35.8 | 57.3 | 60.9 | 59.3 |

TABLE 1: Continued.

| | Transitional intermediate volcanic rocks | | | | | | | | | | | | | | | | | | | | | Rhyolites | | |
|---|---|---|---|---|---|---|---|---|---|---|---|---|---|---|---|---|---|---|---|---|---|---|---|---|---|
| | MU 76 | MU 77 | MU 78 | MU 79 | MU 80 | MU 81 | MU 82 | MU 83 | MU 84 | MU 85 | MU 86 | MU 87 | MU 89 | MU 90 | MU 91 | MU 92 | MU 93 | MU 94 | MU 95 | MU 96 | MU 97 | MU 100 | MU 16 | MU 17 | MU 18 |
| Ba | | | | 504 | | | | | | | | 415 | 643 | | | | | 626 | | | | 485 | 301 | 220 | 217 |
| Rb | | | | 56 | | | | | | | | 48 | 61 | | | | | 54 | | | | 40 | 91 | 83 | 77 |
| Sr | | | | 349 | | | | | | | | 398 | 395 | | | | | 296 | | | | 312 | 37 | 32 | 31 |
| Ni | | | | 160 | | | | | | | | 90 | 80 | | | | | 80 | | | | 43 | 20 | 20 | 20 |
| Cr | | | | 40 | | | | | | | | 70 | 70 | | | | | 30 | | | | 40 | 20 | 20 | 20 |
| V | | | | 180 | | | | | | | | 155 | 149 | | | | | 163 | | | | 175 | 5 | 5 | 5 |
| Th | | | | 7.8 | | | | | | | | 8.9 | 8.8 | | | | | 7.2 | | | | 7.5 | 3.1 | 2.9 | 2.7 |
| Pb | | | | 5 | | | | | | | | 14 | 18 | | | | | 5 | | | | 5 | 9 | 5 | 5 |
| U | | | | 2.2 | | | | | | | | 2.5 | 3.1 | | | | | 2 | | | | 2.2 | 2.5 | 2.1 | 2 |
| Nb | | | | 7 | | | | | | | | 9 | 8 | | | | | 6 | | | | 7 | 5 | 5 | 4 |
| Ta | | | | 0.6 | | | | | | | | 0.7 | 0.8 | | | | | 0.6 | | | | 0.6 | 0.6 | 0.6 | 0.6 |
| Zr | | | | 158 | | | | | | | | 179 | 173 | | | | | 137 | | | | 155 | 34 | 36 | 33 |
| Hf | | | | 4.3 | | | | | | | | 4.6 | 4.9 | | | | | 3.7 | | | | 4.1 | 1.9 | 2 | 1.9 |
| Y | | | | 25 | | | | | | | | 24 | 24 | | | | | 23 | | | | 25 | 12 | 9 | 10 |
| La | | | | 23.6 | | | | | | | | 27.4 | 28.0 | | | | | 22.8 | | | | 20.1 | 6.60 | 3.90 | 3.70 |
| Ce | | | | 50.7 | | | | | | | | 58.3 | 59.5 | | | | | 47.4 | | | | 46.6 | 13.2 | 9.3 | 12.3 |
| Pr | | | | 6.35 | | | | | | | | 7.2 | 7.68 | | | | | 5.82 | | | | 6.24 | 1.90 | 1.31 | 1.12 |
| Nd | | | | 24.4 | | | | | | | | 27.3 | 27.9 | | | | | 22.7 | | | | 23.6 | 7.80 | 5.40 | 4.60 |
| Sm | | | | 5.8 | | | | | | | | 6.2 | 6.6 | | | | | 5.2 | | | | 5.7 | 2.1 | 1.4 | 1.4 |
| Eu | | | | 1.78 | | | | | | | | 1.6 | 1.81 | | | | | 1.63 | | | | 1.62 | 0.12 | 0.09 | 0.1 |
| Gd | | | | 6.0 | | | | | | | | 6.2 | 5.7 | | | | | 5.3 | | | | 5.5 | 2.1 | 1.4 | 1.5 |
| Tb | | | | 0.9 | | | | | | | | 0.9 | 0.9 | | | | | 0.8 | | | | 0.9 | 0.4 | 0.3 | 0.3 |
| Dy | | | | 4.9 | | | | | | | | 4.7 | 4.7 | | | | | 4.4 | | | | 4.7 | 2.0 | 1.5 | 1.5 |
| Ho | | | | 0.9 | | | | | | | | 0.9 | 0.9 | | | | | 0.8 | | | | 0.9 | 0.4 | 0.3 | 0.3 |
| Er | | | | 2.6 | | | | | | | | 2.4 | 2.5 | | | | | 2.3 | | | | 2.6 | 1 | 0.8 | 0.9 |
| Tm | | | | 0.36 | | | | | | | | 0.34 | 0.36 | | | | | 0.32 | | | | 0.36 | 0.15 | 0.13 | 0.14 |
| Yb | | | | 2.20 | | | | | | | | 2.00 | 2.20 | | | | | 2.00 | | | | 2.20 | 0.90 | 0.80 | 0.90 |
| Lu | | | | 0.32 | | | | | | | | 0.3 | 0.31 | | | | | 0.3 | | | | 0.33 | 0.13 | 0.12 | 0.12 |

TABLE 1: Continued.

| | MU 76 | MU 77 | MU 78 | MU 79 | MU 80 | MU 81 | MU 82 | MU 83 | MU 84 | MU 85 | MU 86 | MU 87 | MU 89 | MU 90 | MU 91 | MU 92 | MU 93 | MU 94 | MU 95 | MU 96 | MU 97 | MU 100 | MU 16 | MU 17 | MU 18 |
|---|---|---|---|---|---|---|---|---|---|---|---|---|---|---|---|---|---|---|---|---|---|---|---|---|---|
| | | | | | | | | | | Transitional intermediate volcanic rocks | | | | | | | | | | | | | Rhyolites | | |
| La/Yb | | | | 10.7 | | | | | | | | 13.7 | 12.7 | | | | | 11.4 | | | | 9.1 | 7.3 | 4.9 | 4.1 |
| La/Nb | | | | 3.37 | | | | | | | | 3.04 | 3.50 | | | | | 3.80 | | | | 2.87 | 1.32 | 0.78 | 0.925 |
| Eu/Eu* | | | | 0.92 | | | | | | | | 0.79 | 0.90 | | | | | 0.95 | | | | 0.89 | 0.17 | 0.20 | 0.21 |
| La/Sm$_{CN}$ | | | | 2.63 | | | | | | | | 2.85 | 2.74 | | | | | 2.83 | | | | 2.28 | 2.03 | 1.80 | 1.71 |
| La/Yb$_{CN}$ | | | | 7.69 | | | | | | | | 9.83 | 9.13 | | | | | 8.18 | | | | 6.55 | 5.26 | 3.50 | 2.95 |
| Nb/La$_{pm}$ | | | | 0.29 | | | | | | | | 0.32 | 0.28 | | | | | 0.25 | | | | 0.34 | 0.73 | 1.24 | 1.04 |

Mg$^{\#}$ (Magnesium number) = 100 ×Mg$^{2+}$/(Mg$^{2+}$ + Fe$_{total}^{2+}$) as in Section 4.3.1, and Eu/Eu* = Eu/((SmN × GdN)$^{1/2}$).

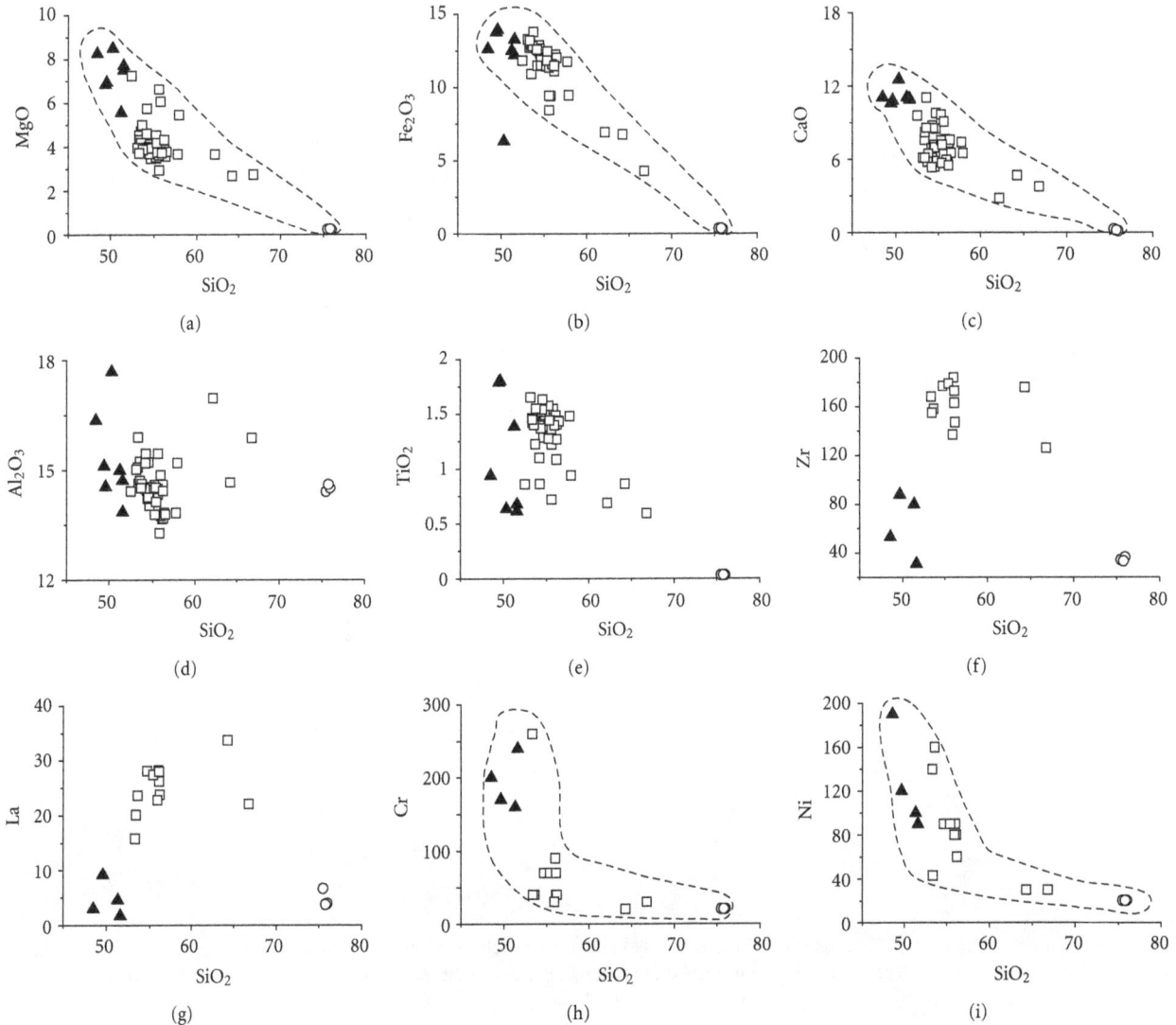

FIGURE 7: Major and trace elements variation diagram for the Kilimafedha volcanic rocks. Symbols as in Figure 6.

*4.3.2. Intermediate Volcanic Rocks.* The intermediate volcanic rocks have a wide range of $SiO_2$ contents (52.51–66.80 wt%) spanning from basaltic andesites through basaltic trachyandesites and andesites to dacites. The compositions, however, are skewed to the basaltic andesitic compositions as indicated by the averages ($SiO_2$ = 52.51–66.80 wt%, average = 55.67 wt%, $n$ = 40, Table 1, Figures 5 and 6), and only 5 samples have $SiO_2$ > 57 wt%. $TiO_2$ contents are 0.59–1.65 wt%, $Fe_2O_3$ = 4.26–13.78 wt%, MgO = 2.68–7.25 wt%, and Mg numbers range from 36 to 61. Cr and Ni contents are 20–90 ppm and 30–160 ppm, respectively. La varies from 20.1–33.7 ppm and Yb from 1.4–2.3 ppm which results in La/Yb ratios of 9.14–15.8.

The intermediate volcanic rocks display fractionated REE patterns (Figure 8(b)) and in comparison with the tholeiitic basalts are characterized by an enrichment of the LREE relative to the MREE and HREE (La/Sm$_{CN}$ = 2.3–4.1, La/Yb$_{CN}$ = 6.5–11.3). The degree of REE fractionation,

however, is less than that of adakites which have La/Sm$_{CN}$ = 5.8, La/Yb$_{CN}$ = 43.5 [25, Figure 8(b)]. The rocks show slightly negative to nonexistent Eu anomalies (Eu/Eu* = 0.74–1.01). On the primitive mantle-normalized diagram (Figure 9(b)), the samples display fractionated patterns with enrichment of the incompatible elements (Rb, Ba, Th, K, and Pb) relative to the compatibles ones and are associated with negative anomalies of Nb, Ta, and Ti relative to adjacent elements (Nb/La$_{pm}$ = 0.17–0.34).

*4.3.3. Rhyolites.* The rhyolite samples have a restricted range in $SiO_2$ contents (75.52–76.02 wt%, $n$ = 3). $TiO_2$ contents are 0.03 wt% for all the 3 samples, whereas $Fe_2O_3$ and MgO vary from 0.28 to 0.34 wt% and from 0.22 to 0.25 wt%, respectively. The samples are depleted in Cr and Ni (≤20 ppm) as well as in Zr (33–36 ppm). Compared with the tholeiitic basalts and the intermediate volcanic rocks, the rhyolites have lower total REE contents. La varies from 3.7 to

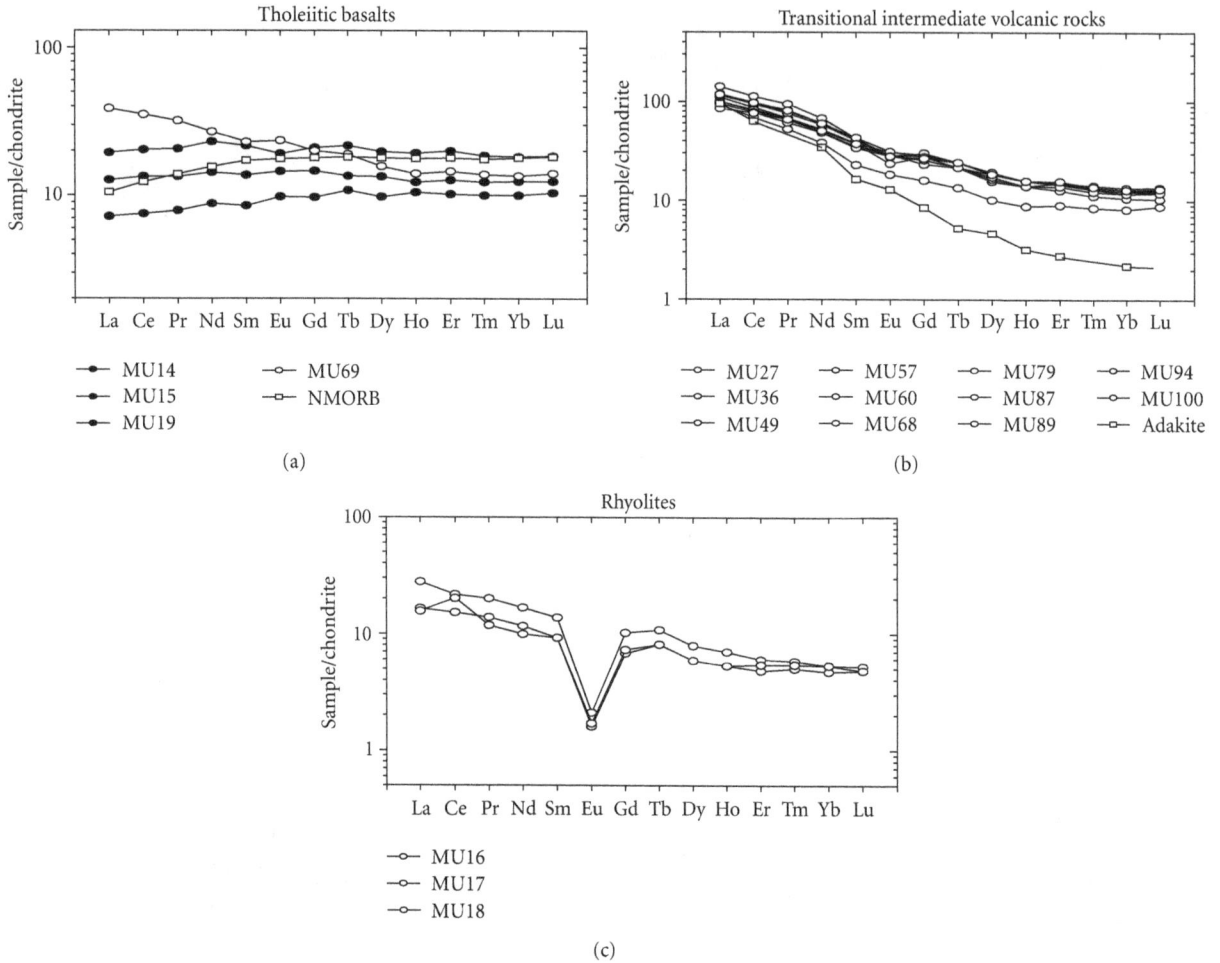

FIGURE 8: Chondrite normalized REE diagrams for the Kilimafedha volcanic rocks normalizing values from Sun and McDonough [23]. (a) Tholeiitic basalts superimposed with NMORB, (b) intermediate volcanic rocks superimposed with adakites, and (c) rhyolites.

6.6 ppm, whereas Yb is restricted in the range of 0.8-0.9 ppm resulting into La/Yb ratios of 4.11–7.13.

The rhyolites are characterized by slight enrichment of the LREE relative to MREE and HREE (La/Sm$_{CN}$ = 1.7–2.0, La/Yb$_{CN}$ = 2.95–5.26) with characteristic strong negative Eu anomalies (Eu/Eu* = 0.17–0.21, Figure 8(c)). On primitive mantle-normalized plots (Figure 9(c)), the samples show enrichment in incompatible elements (Rb, Ba, Th, K, and Pb), negative anomalies of Nb, Ta, Eu, Sr, and Ti anomalies relative to adjacent elements.

*4.4. Sm-Nd Isotopic Composition.* Sm-Nd isotopic compositions for the Kilimafedha greenstone belt rhyolites are reported in Table 2. Also shown in the Table are the εNd values calculated assuming a crystallization age of 2712 ± 5 Ma reported by Wirth et al. [9]. The εNd (2.7 Ga) values range from +1.87 to +2.18 for the tholeiitic basalts, +1.57 to +2.46 for the intermediate volcanic rocks, and −0.51 to +5.16 for the rhyolites (Figure 10), and these values are comparable with those from the volcanic rocks from the northern Musoma-Mara greenstone belt reported by Manya et al. [11, 12], some few hundreds of km north of

the Kilimafedha greenstone belt. Their respective depleted mantle (T$_{DM}$) ages are 2980–3763 Ma, 2846–2970 Ma, and 2557–3914 Ma (Table 2).

## 5. Discussion

### 5.1. Petrogenesis

*5.1.1. Tholeiitic Basalts.* The slight depletion in LREE to nearly flat REE patterns shown by the tholeiitic basalts coupled with their close compositional similarity to N-MORB suggests that these rocks were generated in a source similar to that generating modern N-MORB. Unlike NMORB, however, these patterns display negative anomalies in Nb and Ti, features which together with tectonic setting discrimination diagrams (see next section) are suggestive of derivation in a subduction setting. The nature of the mantle source rocks can further be constrained by the trace element ratios Nb/Yb, Zr/Yb, and Th/Yb [26]. When plotted on the Nb/Yb versus Zr/Yb diagram (Figure 11), the tholeiitic basalts plot around NMORB with a general trend towards increasing mantle enrichment to E-MORB within the MORB

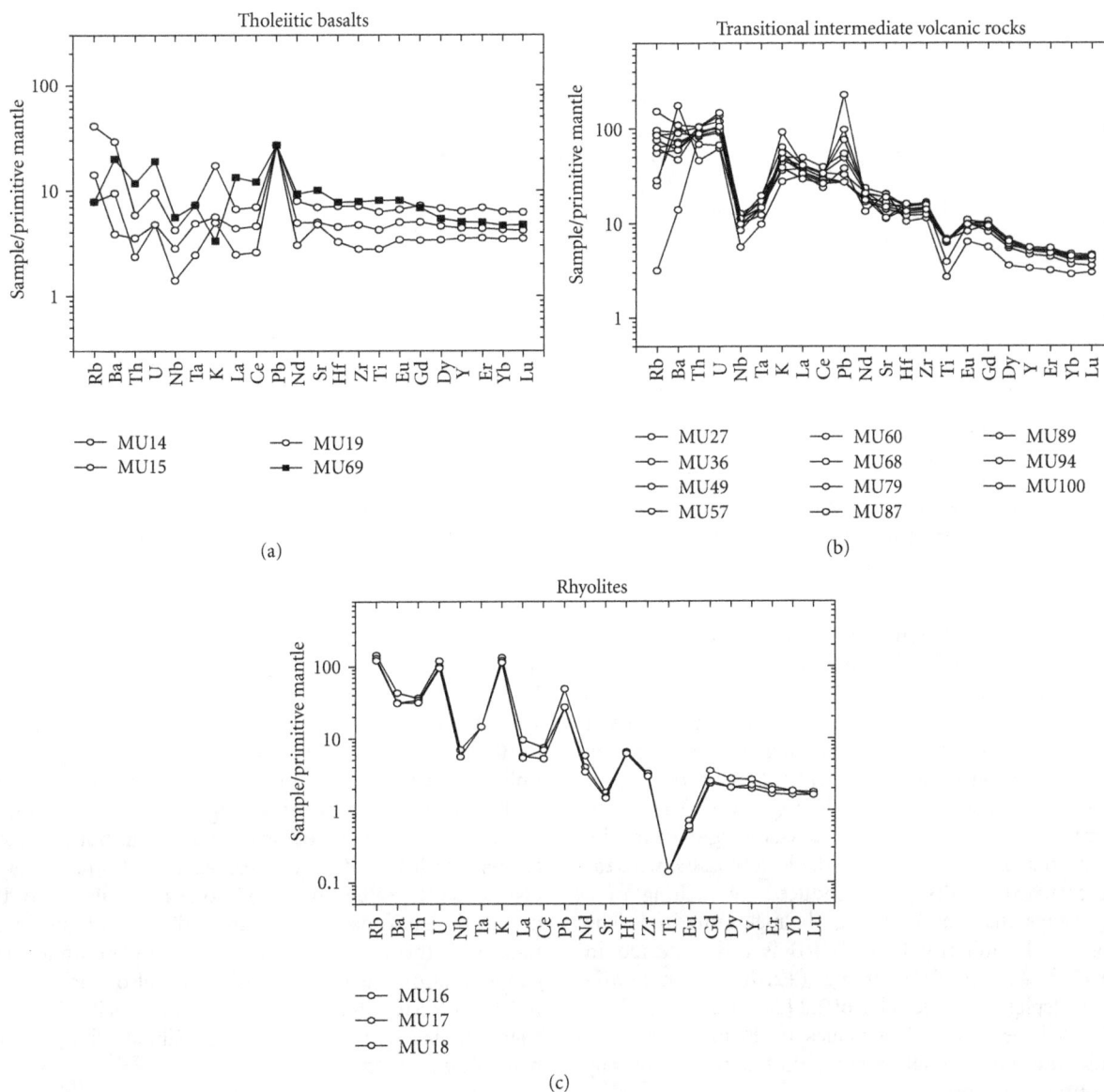

FIGURE 9: Primitive mantle normalized diagrams for the Kilimafedha volcanic, normalizing values from Sun and McDonough [23]. (a) Tholeiitic basalts, (b) intermediate volcanic rocks, and (c) rhyolites.

TABLE 2: Sm-Nd isotopic data for the Kilimafedha greenstone belt volcanic rocks.

| Sample | Rock suite | Sm (ppm) | Nd (ppm) | $^{147}Sm/^{144}Nd$ | $^{143}Nd/^{144}Nd$ | $\varepsilon Nd$ (at 2.7 Ga) | $T_{DM}$ (Ma) |
|---|---|---|---|---|---|---|---|
| MU 14 | Tholeiite basalts | 1.68 | 6.57 | 0.1545 | 0.511995 | 2.18 | 2980 |
| MU 69 | | 6.12 | 18.82 | 0.1965 | 0.512731 | 1.87 | 3763 |
| MU 49 | | 6.09 | 29.4 | 0.1252 | 0.511451 | 1.82 | 2923 |
| MU 57 | | 5.06 | 24.2 | 0.1263 | 0.511481 | 1.99 | 2910 |
| MU 68 | Transitional intermediate volcanic rocks | 6.08 | 33.3 | 0.1103 | 0.511218 | 2.46 | 2846 |
| MU 89 | | 6.02 | 28.7 | 0.1267 | 0.51148 | 1.83 | 2926 |
| MU 100 | | 5.45 | 24.6 | 0.1339 | 0.511594 | 1.57 | 2970 |
| MU 16 | Rhyolites | 1.94 | 6.81 | 0.1721 | 0.512459 | 5.10 | 2557 |
| MU 17 | | 1.37 | 4.51 | 0.1835 | 0.512378 | −0.51 | 3914 |

Calculations are based on a decay constant of $6.54 \times 10^{-12}$ per year for $^{147}Sm$ and DM values for Nd are $(^{143}Nd/^{144}Nd)_{today} = 0.51316$, $(^{147}Sm/^{144}Nd)_{today} = 0.2137$.

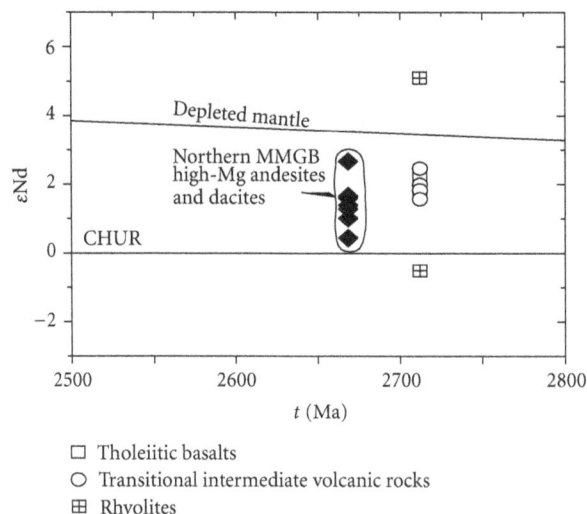

FIGURE 10: Plot of εNd versus t (Ma) for the Kilimafedha greenstone belt volcanic rocks. The depleted mantle model is from DePaolo [24] and the northern Musoma-Mara greenstone belt high-Mg andesites and dacites data is from Manya et al. [11].

array. This suggests that the enrichment observed in the Kilimafedha tholeiites can be explained by their derivation from an initially homogeneous MORB-like source that was differentially metasomatized by an aqueous fluid derived from the subducting slab [26]. Fluxing of the source by the metasomatizing fluid most likely enhanced melting of the mantle wedge at a relatively low pressure. Melting of the mantle wedge which was not affected significantly by the metasomatism yielded the La-depleted basalts, whereas the La-enriched basalts were produced by melting of a mantle wedge that has been slightly metasomatized. The compositional similarity to NMORB is also reflected in similar εNd (2.7 Ga) of the samples (+2.18 for sample MU 14) to the depleted mantle value of 2.2 [24] at the same time. The slightly lower εNd (2.7Ga) values of 1.87 for sample MU 69 would indicate minimal contamination of the magmas by older continental crust.

### 5.1.2. Intermediate Volcanic Rocks.

Unlike the flat REE patterns that characterize the tholeiitic basalts, the intermediate rocks show fractionated patterns characterized by La/Yb ratios of 9.14–15.8. These ratios are, however, lower than those reported in adakites (La/Yb = 40; [23]). In adakites, such high ratios are indicative of the presence of garnet ± amphibole in the source during partial melting. Thus, the lower La/Yb ratios of the rocks preclude the involvement of garnet ± amphibole in their magma genesis. In Figure 11, the intermediate volcanic rocks cluster just above E-MORB out of the MORB array towards increasing Zr content, which according to Pearce and Peate [26] signifies the involvement of both slab melt and hydrous fluid in metasomatising the source rocks. The metasomatism resulted in the enrichment of the source mantle wedge in the HFSE that has been scavenged from the subducting oceanic slab thereby explaining the observed enrichment in the HFSE relative to MORB (Figure 11). The involvement of slab partial melts in the petrogenesis of the intermediate volcanic rocks suggests

that temperatures in the subduction zone were sufficiently high to initiate partial melting of the slab. According to Pearce and Peate [26], the onset of melting of the subducting slab in the Phanerozoic occurs at relatively greater depth (>100 km) beneath the subduction zone. This suggests that, unlike the tholeiitic basalts that were formed at relatively shallow depths, the primary magmas for the intermediate volcanic rocks originated at greater depth, but outside the garnet stability field. The intermediate volcanic rocks have εNd (2.7 Ga) values of +1.57 to +2.46 similar to those of the tholeiitic basalts and are indicative of the juvenile nature of the magmas accompanied by minimal crustal contamination. Such a conclusion was also reached for the northern MMGB high-Mg andesites and dacites [11] which share similar εNd values with the Kilimafedha greenstobe belt volcanic rocks.

### 5.1.3. Rhyolites.

Rhyolites differ from the other two suites in having lower contents of $TiO_2$, $P_2O_5$, Zr, and overall lower abundances of the REE (Table 1). In chondrite normalized REE diagrams (Figure 8) and extended trace element diagram (Figure 9), the rhyolites are characterized by large negative Eu (Eu/Eu* = 0.17–0.21, Table 1) accompanied by negative Sr anomalies as well as Nb and Ti anomalies. The close spatial association of the rhyolites and other suites of the Kilimafedha greenstone belt coupled with their trends towards lower $Fe_2O_3$, MgO, CaO, $TiO_2$, Cr, and Ni with increasing $SiO_2$ (Figure 7) may suggest that the rhyolites may be products of extensive fractional crystallization of the same magmas that generated the more basic members. Such a model is also supported by experimental studies which showed that low-pressure fractional crystallization of olivine, pyroxene, plagioclase, and Fe-Ti oxides can produce rhyolites [27] with relatively flat HREE patterns. Thus, the generally lower REE abundances, $TiO_2$, $P_2O_5$, and Zr contents can be explained by shallow level fractionation of Ti-rich phases (e.g., titanomagnetite) and REE-rich phases such as apatite,

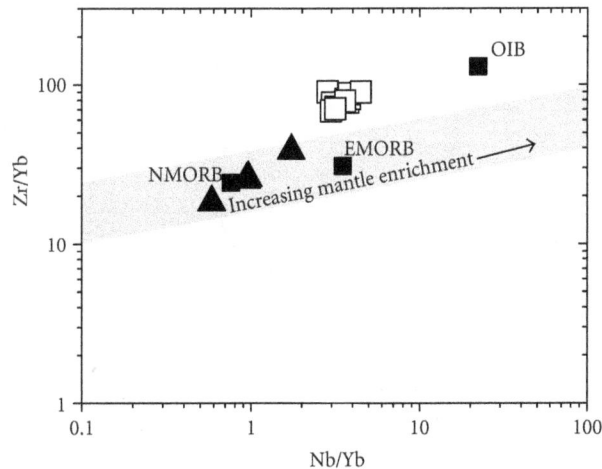

FIGURE 11: Nb/Yb-Zr/Yb diagram (after Pearce and Peate, [26]) for the Kilimafedha tholeiitic basalts (filled triangles) and intermediate volcanic rocks (open squares). The tholeiitic basalt samples plot around the NMORB field tending towards the mantle enrichment direction within the MORB array suggestive of metasomatism by aqueous fluid. The intermediate rocks plot just above the EMORB tending to higher Zr values illustrating input of the HFSE from the subducted slab. NMORB, EMORB, and OIB data are from Sun and McDonough [23].

monazite, zircon, and allanites, whereas the large negative Eu and Sr anomalies could be due to plagioclase fractionation. Compared to the other two suites, the rhyolites show variable $\varepsilon$Nd (2.7 Ga) values of $-0.51$ to $+5.17$ and suggest that older crustal involvement in the genesis of the rhyolites either by partial melting of older crust or contamination of evolving magmatic liquid cannot be ruled out as an important process in the genesis of these rocks. The wide variation in $\varepsilon$Nd towards more depleted signatures demands reexamination of the rocks.

*5.2. Tectonic Setting.* Trace element discrimination diagrams developed for Phanerozoic rocks have been used together with their ratios to infer the tectonic setting for Archaean rocks (e.g., [19]). Using this approach, the Kilimafedha tholeiitic basalts and basaltic andesites were plotted on the Th-Hf-Nb triangular diagram of Wood [28] which is suitable for mafic as well as intermediate volcanic rocks. On this diagram, three of the four tholeiitic basalt samples plot along the boundary between the N-MORB and E-MORB fields, while the other samples together with all intermediate volcanic samples plot in the field of volcanic arc basalts (Figure 12). The similarity of the three tholeiitic basalt samples with N-MORB on the Th-Hf-Nb diagram is also reflected in Figure 8(a). The La/Nb ratio of basaltic samples is particularly important in discriminating basalts that erupted in ocean ridges and ocean plateaus from those that erupted in arcs [29, 30]. According to Rudnick [29] and Condie [30], ocean ridge and ocean plateau basalts have La/Nb < 1.4, whereas arc basalts have La/Nb > 1.4. Both the tholeiitic basalts and intermediate rocks of the Kilimafedha greenstone belt show La/Nb > 1.4 (1.5–2.30 and 2.87–5.53, resp.) suggestive of arc affinities. Thus, the results obtained from the discrimination diagrams combined with trace element ratios data are suggestive of an arc tectonic setting for the Kilimafedha greenstone belts rocks.

This conclusion is supported by the fact that the rocks exhibit negative anomalies of Nb, Ti, and/or Ta anomalies in extended trace element spidergrams (Figure 9), features attributed to magmas generated at subduction zones [26].

## 6. Comparison with Other Greenstone Belts of the Tanzania Craton

A closer review of geochemistry and geochronology by Manya et al. [10] and Manya and Maboko [8] showed that the individual greenstone belts of the Tanzania Craton exhibit different formation ages, and their formation occurred in different tectonic settings. This suggestion is also evident in different volcanic rock packages found in these belts. Kilimafedha greenstone belt (KGB) differs from the Sukumaland (SGB) to the west and Iramba-Sekenke (ISGB) to the far south in having predominantly intermediate volcanic rocks with tholeiitic to calc-alkaline intermediate affinities, rare mafic, and felsic volcanic package, which are in contrast to the later that are dominated by tholeiitic basalts and rare intermediate volcanic rocks. The volcanic package in KGB also differs from those of the southern Musoma-Mara greenstone belt (MMGB) to the near north as the later is comprised of bimodal volcanic assemblage [31]. Although the northern part of the MMGB is predominantly comprised of intermediate rocks similar to KGB, the former lacks mafic members.

The foregoing discussion corroborates the findings by Manya et al. [10] that the individual greenstone belts evolved as separate entities at different time intervals having different volcanic rocks assemblages. Although volcanism in greenstone belts of the Tanzania Craton seems to have erupted at different time intervals (2823–2780 Ma for SGB, 2755–2712 Ma for MMGB, ISGB, and KGB, 2676–2667 Ma for northern MMGB, [8] and references therein); one thing is common to all of them: they formed exclusively

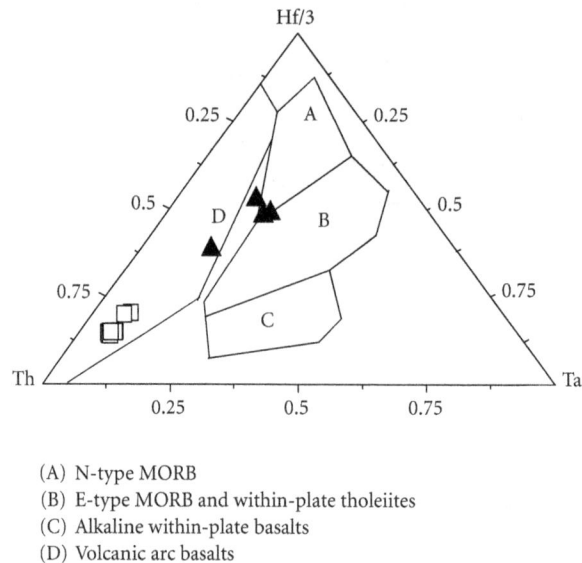

(A) N-type MORB
(B) E-type MORB and within-plate tholeiites
(C) Alkaline within-plate basalts
(D) Volcanic arc basalts

FIGURE 12: Ta-Hf-Th tectonic setting discrimination diagram [28] for the tholeiitic basalts (filled triangles) and intermediate volcanic rocks (open squares) for the Kilimafedha greenstone belts.

at convergent margins. This signifies the importance of formation and growth of late Archaean continental crust at convergent margins.

## 7. Conclusion

The Neoarchaean Kilimafedha greenstone belt of northeastern Tanzania consists of three closely associated suites of volcanic rocks: the predominant intermediate basaltic andesite and dacites, and the volumetrically minor tholeiitic basalts and rhyolites. The tholeiitic basalts have nearly flat REE patterns and show close compositional similarity to NMORB. Trace element systematics of the tholeiites suggest that they were formed by shallow partial melting of a mantle wedge that has been variably metasomatized by an aqueous fluid in a convergent tectonic setting. The intermediate rocks are characterized by fractionated REE patterns, enrichment of the HFSE relative to NMORB, and negative anomalies of Nb and Ta. Such geochemical features are consistent with derivation of these rocks by partial melting of a mantle wedge that has been metasomatized by both fluid and slab melt at a greater depth than the tholeiitic basalts source but outside the garnet stability field. The geochemical features defining the Kilimafedha greenstone belt rhyolites include low $TiO_2$, $P_2O_5$, Zr, and overall lower abundance of total REE compared with the other two suites and large negative Eu, Sr, and Ti anomalies in extended trace element spidergrams. These features can be explained by shallow level fractional crystallization of the parent magma of the intermediate volcanic rocks involving plagioclase, Ti-rich phases like ilmenite and magnetite as well as REE-rich phases like apatite, zircon, monazite, and allanite. The close spatial association of the three petrological types in the Kilimafedha greenstone belt is interpreted as reflecting their formation in an evolving late Achaean island arc.

## Acknowledgments

This research was financially supported by Sida/SAREC through the project Research Capacity Building at the Faculty of Science, now College of Natural and Applied Sciences (CoNAS), University of Dar es Salaam, to which the authors are greatly indebted. The authors are also thankful to Michael O. Garcia, the Journal Editor and two anonymous reviewers for their insightful comments that helped shape the paper.

## References

[1] G. Borg and R. M. Shackleton, "The Tanzania and NE Zaire cratons," in *Greenstone Belts*, M. J. de Wit and L. D. Ashwal, Eds., pp. 608–619, Clarendon Press, Oxford, UK, 1997.

[2] G. Borg, "The Geita gold deposit in NW Tanzania—geology, ore petrology, geochemistry and timing of events," *Geologisches Jahrbuch D*, vol. 100, pp. 545–595, 1994.

[3] E. Kazimoto, *Study of integrated geochemical techniques in the exploration for gold in North Mara mines, Tanzania [M.S. thesis]*, University of Dar es Salaam, 2008.

[4] G. Borg, "New aspects of the lithostratigraphy and evolution of the Siga Hills, an Archaean granite-greenstone terrain in NW Tanzania," *Zeitschrift fur Angewandte Geologie*, vol. 38, no. 2, pp. 89–93, 1992.

[5] S. Manya and M. A. H. Maboko, "Geochemistry of the Neoarchaean mafic volcanic rocks of the Geita area, NW Tanzania: implications for stratigraphical relationships in the Sukumaland greenstone belt," *Journal of African Earth Sciences*, vol. 52, no. 4-5, pp. 152–160, 2008.

[6] G. Borg and T. Krogh, "Isotopic age data of single zircons from the Archaean Sukumaland Greenstone Belt, Tanzania," *Journal of African Earth Sciences*, vol. 29, no. 2, pp. 301–312, 1999.

[7] S. Manya and M. A. H. Maboko, "Dating basaltic volcanism in the Neoarchaean Sukumaland Greenstone Belt of the Tanzania Craton using the Sm-Nd method: implications for

the geological evolution of the Tanzania Craton," *Precambrian Research*, vol. 121, no. 1-2, pp. 35–45, 2003.

[8] S. Manya and M. A. H. Maboko, "Geochemistry and geochronology of Neoarchaean volcanic rocks of the Iramba-Sekenke greenstone belt, central Tanzania," *Precambrian Research*, vol. 163, no. 3-4, pp. 265–278, 2008.

[9] K. R. Wirth, J. D. Vervoot, and B. Weisberger, "Origin and evolution of the Kilimafedha greenstone belt, eastern Tanzania Craton: evidence from Pb isotopes," *Geological Society of America Abstracts with Programs*, vol. 36, p. 244, 2004.

[10] S. Manya, K. Kobayashi, M. A. H. Maboko, and E. Nakamura, "Ion microprobe zircon U-Pb dating of the late Archaean metavolcanics and associated granites of the Musoma-Mara Greenstone Belt, Northeast Tanzania: implications for the geological evolution of the Tanzania Craton," *Journal of African Earth Sciences*, vol. 45, no. 3, pp. 355–366, 2006.

[11] S. Manya, M. A. H. Maboko, and E. Nakamura, "The geochemistry of high-Mg andesite and associated adakitic rocks in the Musoma-Mara Greenstone Belt, northern Tanzania: possible evidence for Neoarchaean ridge subduction?" *Precambrian Research*, vol. 159, no. 3-4, pp. 241–259, 2007.

[12] S. Manya, M. A. H. Maboko, and E. Nakamura, "Geochemistry and Nd-isotopic composition of potassic magmatism in the Neoarchaean Musoma-Mara Greenstone Belt, northern Tanzania," *Precambrian Research*, vol. 159, no. 3-4, pp. 231–240, 2007.

[13] P. Pinna, S. Muhongo, B. A. Mcharo et al., "Geology and Mineral Map of Tanzania. Scale: 1:2.000.000," BRGM-UDSM-GST team, 2008.

[14] M. Macfarlane, "Brief explanation of the geology of quarter degree sheet 25, East Mara," Mineral Resource Division, Dodoma, Tanzania, 1965.

[15] T. N. Clifford, "The structural framework of Africa," in *African Magmatism and Tectonics, Oliver and Boyd*, T. N. Clifford and I. G. Gass, Eds., pp. 1–26, Edinburgh, UK, 1970.

[16] I. M. Gray and A. S. Macdonald, "Brief explanation of the geology of quarter degree sheet 6 and 14, Seronera," Mineral Resource Division, Dodoma, Tanzania, 1965.

[17] C. Kasanzu, M. A. H. Maboko, and S. Manya, "Geochemistry of fine-grained clastic sedimentary rocks of the Neoproterozoic Ikorongo Group, NE Tanzania: implications for provenance and source rock weathering," *Precambrian Research*, vol. 164, no. 3-4, pp. 201–213, 2008.

[18] C. W. A. Messo, *Geochemistry of Neoarchaean volcanic rocks of the Ikoma area in the Kilimafedha greenstone belt, Northwestern Tanzania [M.S. thesis]*, University of Dar es Salaam, 2004.

[19] H. R. Rollinson, *Using Geochemical Data: Evaluation, Presentation, Interpretation*, Longman, Essex, UK, 1993.

[20] K. P. Jochum and S. P. Verma, "Extreme enrichment of Sb, Tl and other trace elements in altered MORB," *Chemical Geology*, vol. 130, no. 3-4, pp. 289–299, 1996.

[21] R. W. Le Maitre, P. Bateman, A. Dudek et al., *A Classification of Igneous Rocks and Glossary of Terms*, Blackwell, Oxford, UK, 1989.

[22] J. A. Winchester and P. A. Floyd, "Geochemical discrimination of different magma series and their differentiation products using immobile elements," *Chemical Geology*, vol. 20, no. C, pp. 325–343, 1977.

[23] S. S. Sun and W. F. McDonough, "Chemical and isotopic systematics of oceanic basalts: implications for mantle composition and processes," *Magmatism in the ocean basins*, pp. 313–345, 1989.

[24] D. J. DePaolo, "Neodymium isotopes in the Colorado Front Range and crust-mantle evolution in the Proterozoic," *Nature*, vol. 291, no. 5812, pp. 193–196, 1981.

[25] H. Martin, "Adakitic magmas: modern analogues of Archaean granitoids," *Lithos*, vol. 46, no. 3, pp. 411–429, 1999.

[26] J. A. Pearce and D. W. Peate, "Tectonic implications of the composition of volcanic arc magmas," *Annual Review of Earth & Planetary Sciences*, vol. 23, pp. 251–285, 1995.

[27] S. D. Spulber and M. J. Rutherford, "The origin of rhyolite and plagiogranite in oceanic crust: an experimental study." *Journal of Petrology*, vol. 24, no. 1, pp. 1–25, 1983.

[28] D. A. Wood, "The application of a Th-Hf-Ta diagram to problems of tectonomagmatic classification and to establishing the nature of crustal contamination of basaltic lavas of the British Tertiary Volcanic Province," *Earth and Planetary Science Letters*, vol. 50, no. 1, pp. 11–30, 1980.

[29] R. L. Rudnick, "Making continental crust," *Nature*, vol. 378, no. 6557, pp. 571–578, 1995.

[30] K. C. Condie, "Mafic crustal xenoliths and the origin of the lower continental crust," *Lithos*, vol. 46, no. 1, pp. 95–101, 1999.

[31] M. Mtoro, M. A. H. Maboko, and S. Manya, "Geochemistry and geochronology of the bimodal volcanic rocks of the Suguti area in the southern part of the Musoma-Mara Greenstone Belt, Northern Tanzania," *Precambrian Research*, vol. 174, no. 3-4, pp. 241–257, 2009.

# 15

# Different Origins of the Fractionation of Platinum-Group Elements in Raobazhai and Bixiling Mafic-Ultramafic Rocks from the Dabie Orogen, Central China

**Qing Liu,[1] Quanlin Hou,[1] Liewen Xie,[2] Hui Li,[3] Shanqin Ni,[4] and Yudong Wu[4]**

[1] Key Laboratory of Computational Geodynamics, Graduate University of Chinese Academy of Sciences,
Chinese Academy of Sciences, 19 Yuquan Road, Beijing 100049, China

[2] State Key Laboratory of Lithospheric Evolution, Institute of Geology and Geophysics, Chinese Academy of Sciences,
P.O. Box 9825, Beijing 10029, China

[3] 416 Geological Prospecting Party, Bureau of Geology, Mineral Exploration and Development, Hunan Province,
Zhuzhou 412007, China

[4] Institute of Mineral Resources, Chinese Academy of Geological Sciences, 26 Baiwanzhuang Road, Beijing 100037, China

Correspondence should be addressed to Qing Liu, 790908619@qq.com

Academic Editor: Yi-Wen Ju

Concentrations of the platinum group elements (PGEs), including Ir, Ru, Rh, Pt, and Pd, have been determined for both Raobazhai and Bixiling mafic-ultramafic rocks from the Dabie Orogen by fire assay method. Geochemical compositions suggest that the Raobazhai mafic-ultramafic rocks represent mantle residues after variable degrees of partial melting. They show consistent PGE patterns, in which the IPGEs (i.e., Ir and Ru) are strongly enriched over the PPGEs (i.e., Pt and Pd). Both REE and PGE data of the Raobazhai mafic-ultramafic rocks suggest that they have interacted with slab-derived melts during subduction and/or exhumation. The Bixiling ultramafic rocks were produced through fractional crystallization and cumulation from magmas, which led to the fractionated PGE patterns. During fractional crystallization, Pd is in nonsulfide phases, whereas both Ir and Ru must be compatible in some mantle phases. We suggest that the PGE budgets of the ultramafic rocks could be fractionated by interaction with slab-derived melts and fractional crystallization processes.

## 1. Introduction

The platinum group elements (PGEs), including Os, Ir, Ru, Rh, Pt, and Pd, are strongly siderophile and chalcophile elements. They have similar geochemical behaviors during magmatic processes. Traditionally, the PGEs are subdivided into two groups, the compatible IPGEs (Os, Ir, and Ru) and the incompatible PPGEs (Rh, Pd, and Pt) [1]. It has been suggested that the IPGEs are refractory and tend to be retained in the mantle peridotites during partial melting [2]. In contrast, the PPGEs are concentrated in the base metal sulphides (e.g., pentlandite, chalcopyrite), which are released to the melts along with the molten sulfide melts [2]. Because of their unique geochemical characteristics, the PGEs can be used to identify the magma sources and unravel the complex petrogenetic processes, such as partial melting, melt percolation, and metasomatism in the mantle [1]. Mafic-ultramafic rocks have lower REE contents but higher PGE contents than other rocks, so the PGEs have advantages in studying their petrogenetic processes [1, 3–7].

In this study, we present the PGE data of both Raobazhai and Bixiling mafic-ultramafic rocks from Dabie Orogen, central China, to discuss their fractionation behaviours during magma evolution. The mechanisms of differentiation between these elements will be examined below, taking into account the geochemical affinities of the PGE and their partition in the mineral phases. The results also demonstrate that the PGEs can provide important information on the genesis of magmas.

Different Origins of the Fractionation of Platinum-Group Elements in Raobazhai and Bixiling Mafic-Ultramafic Rocks from the Dabie Orogen, Central China

191

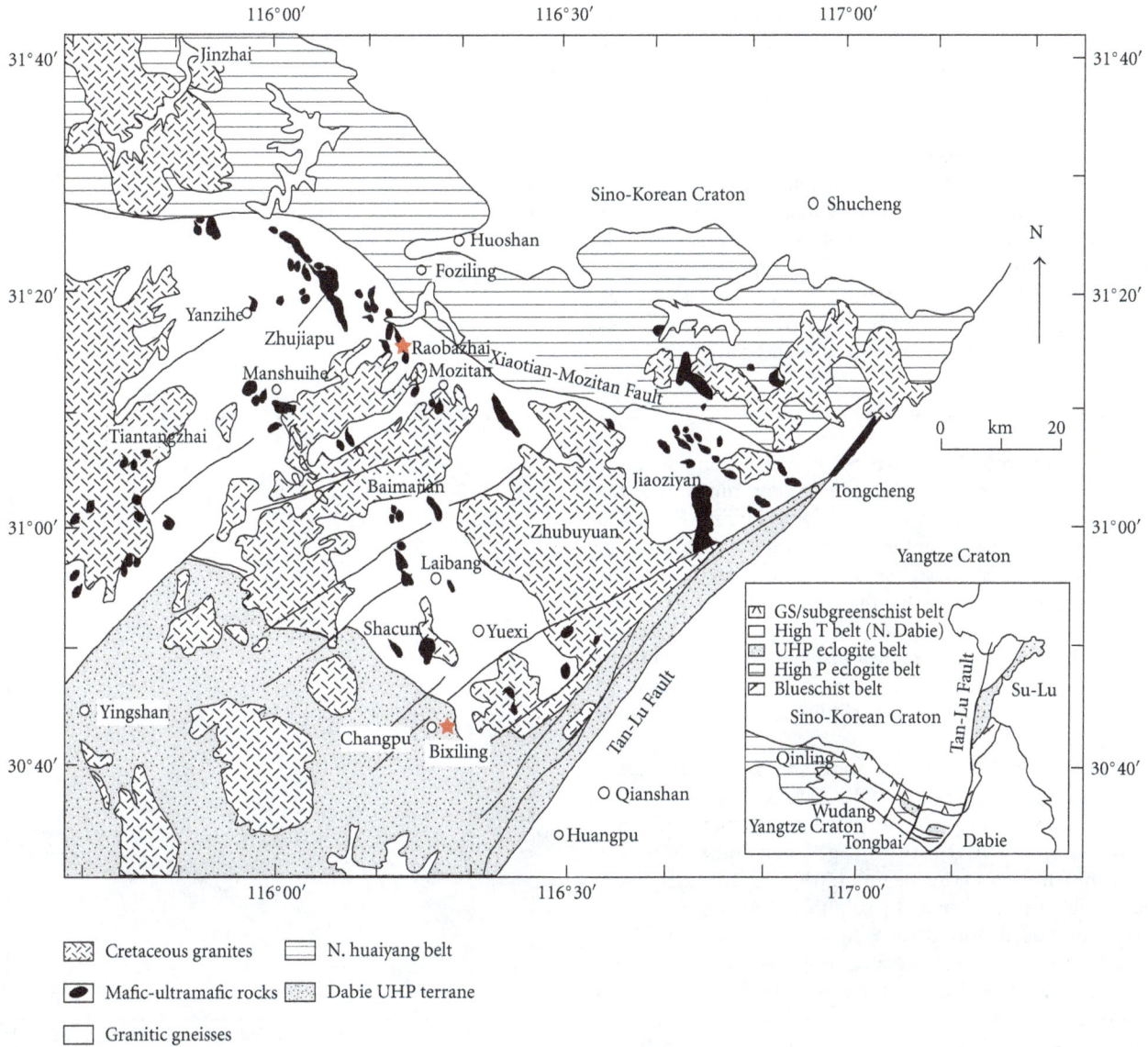

FIGURE 1: Simplified geological map of the Dabie Complex [8]. Sampling localities are roughly indicated by the names of Raobazhai and Bixiling.

## 2. Geological Background and Occurrence

The Dabie Orogen is the eastern segment of the Qinling-Dabie Orogen, which was formed by the continental collision between the Yangtze Craton and North China Craton (Figure 1). It has been subdivided, from north to south, into five main tectonic zones by several large-scale EW-trending faults [8, 10, 11].

The Raobazhai ultramafic massif is outcropped in the North Dabie high-temperature and ultra-high-pressure (HT/UHP) granulite-facies zone. It is located at ca 5 km south of the Xiaotian-Mozitan Fault (Figure 1), which is a major strike-slip fault along the eastern part of the Qinling-Tongbai-Dabie orogenic belt that might have witnessed the early evolutions of this orogenic belt [13]. Previous studies have suggested that the Raobazhai massif is a sheet-like peridotitic slice, which is in fault contact with the surrounding

amphibolite facies orthogneisses [14–16]. Migmatization can be locally observed in field. The long dimension of the massif is subparallel to the strike of the Xiaotian-Mozitan Fault and the regional foliation. The Raobazhai ultramafic body mainly consists of spinel harzburgites, with minor dunites and lherzolites [16]. They are all highly deformed and metamorphosed. Previous petrographic, geochemical, and thermobarometric studies have suggested that they represent a tectonic slice of the subcontinental lithospheric mantle [16–18]. Five representative samples have been selected in this study.

The Bixiling Complex is the largest (∼1.5 km²) coesite-bearing mafic-ultramafic body in the Dabie Orogen, which occurs as a tectonic block that is enclosed within the foliated quartzofeldspathic gneisses in the eastern part of the Dabie UHP terrane (Figure 1). It consists predominantly of banded eclogites and about 20 elongated lenses of garnet-bearing ultramafic rocks, for example, garnet peridotites, garnet pyroxenites, and wehrlites, which range from 50 to 300 m in length and from 5 to 50 m in width [19]. The contact between eclogite and ultramafic rocks is gradational. Field relationships and petrological evidence indicate a cumulate origin of the mafic-ultramafic rocks [19]. Therefore, the diverse rock types are considered, at least to a first approximation, as magmatically cogenetic [20]. The selected samples include three nattier blue eclogites (garnet, omphacite, kyanite, phengite, and rutile), two greenish black eclogites (garnet, omphacite, rutile, and quartz), and two garnet peridotites (olivine, orthopyroxene, clinopyroxene and garnet).

## 3. Materials and Methods

Samples were ground to 200 mesh powders using an agate mill. Whole-rock major elements were determined by X-ray fluorescence spectrometry (XRF) using a Phillips PW 2400 sequential XRF instrument at the Institute of Geology and Geophysics, Chinese Academy of Sciences (IGGCAS). The analytical precision is better than ±2% for major oxides. Bulk-rock rare earth elements (REE) were analyzed on the Plasma PQ2 inductively coupled plasma mass spectrometry (ICP-MS) at IGGCAS. Replicate analyses of a monitor sample suggest that the reproducibility for the REE analysis is better than 3.5%.

Whole-rock PGE contents were analyzed by fire assay (FA) method and measured on a Plasma PQ2 ICP-MS at IGGCAS. About 15 g sample powder, together with 20 g $Na_2B_4O_7$, 10 g $Na_2CO_3$, 2 g Ni, 2 g S, and some $SiO_2$, was fused in a fire-clay crucible at 1150°C for 2 hours. Then, the crucible was broken and a sulphide bead was recovered. The bead was dissolved in a Teflon beaker using 15 mL HCl. After the bead disintegrated into powder, 2 mL Te and 4 mL $SnCl_2$ were added into the solution. The solution was heated to become clear and then was filtered to collect the insoluble residue. The residue was cleaned and transferred into a Teflon beaker containing 2.5 mL aqua regia. Once the solution became clear, appropriate amounts of Re and Cd spike solutions were then added to the mixture, which was diluted with 50 mL $H_2O$ for TJA Pro Excel inductively coupled plasma mass spectrometry (ICP-MS) determination. The detection limits, which are defined as average blank plus three standard deviations, for Ir, Ru, Rh, Pt and Pd were 0.002, 0.0086, 0.0048, 0.082, and 0.043 ppb, respectively. The PGE contents of most samples are higher than the detection limits, whereas both Ir and Ru contents of some samples are close to their detection limits. Replicate analyses of standard WPR-1 have respectively given values of 13.6 ppb Ir, 9.7 ppb Ru, 13.7 ppb Rh, 257 ppb Pt and 248 ppb Pd. The average element concentrations of replicate analyses of WPR-1 are within 10% of the certified value except Ir which is 13% lower.

## 4. Results

*4.1. Raobazhai Mafic-Ultramafic Rocks.* The major-, trace-elements and PGE concentration data of the Raobazhai

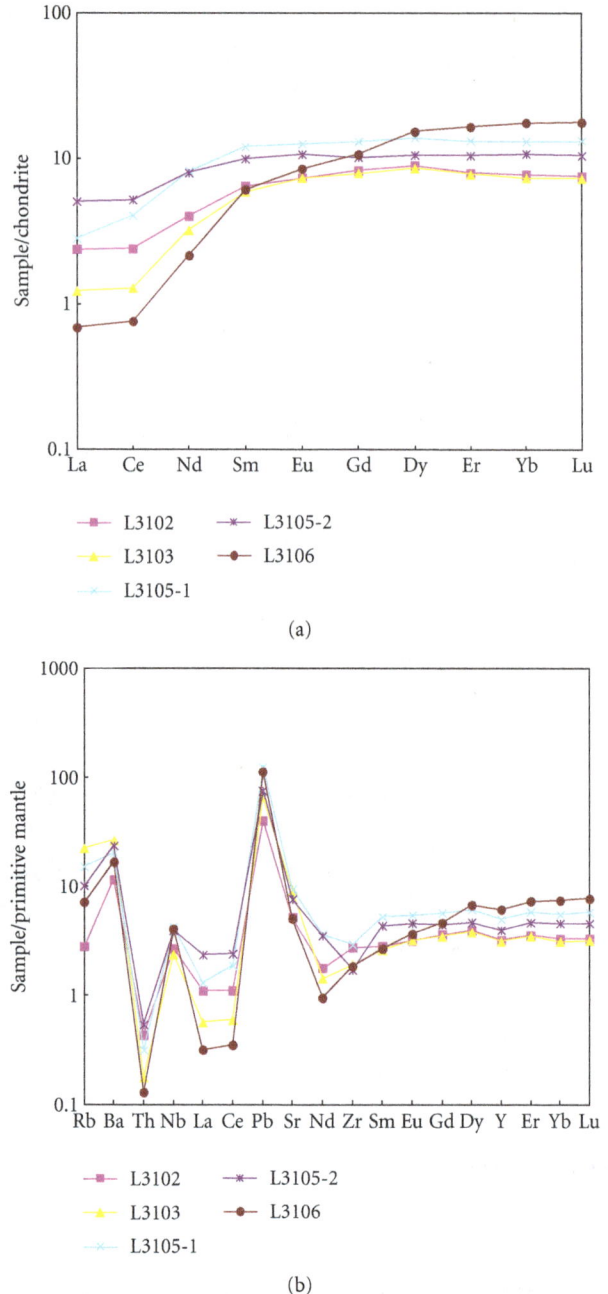

(a)

(b)

FIGURE 2: Chondrite-normalized REE patterns (a) and spider patterns (b) of Raobazhai mafic-ultramafic rocks (chondrite and primitive mantle values are from [9]).

mafic-ultramafic rocks are given in Table 1. The Raobazhai samples show consistent REE patterns (Figure 2(a)), that is, flat HREE patterns but variable depletion in LREE. Similar results have been reported in a previous study on the Raobazhai peridotites [9]. The REE are incompatible elements during partial melting of mantle peridotites; removal of basaltic components tend to decrease the REE contents of the mantle peridotites. In comparison, the LREE are more incompatible within mantle minerals than the HREE; therefore, the residual peridotites are depleted in LREE

TABLE 1: Major elements (wt%), trace elements (ppm), and PGE (ppb) concentration of Raobazhai and Bixiling mafic-ultramafic rocks.

| Location | Raobazhai mafic-ultramafic rocks | | | | | Bixiling mafic-ultramafic rocks | | | | | | | |
|---|---|---|---|---|---|---|---|---|---|---|---|---|---|
| | | | | | | Nattier blue eclogite | | | Greenish black eclogite | | Garnet peridotite | | |
| sample | L3102 | L3103 | L3105-1 | L3105-2 | L3106 | L2906 | L2907 | L2912 | L2908 | L2909 | L2910 | L2911 | L2911P |
| SiO2 | 46.81 | 47.99 | 46.23 | 45.64 | 44.72 | 50.65 | 50.66 | 46.88 | 42.96 | 42.97 | 42.11 | 38.34 | 38.42 |
| TiO2 | 0.60 | 0.77 | 1.39 | 0.79 | 1.19 | 0.77 | 0.77 | 2.69 | 2.05 | 2.08 | 0.25 | 0.35 | 0.35 |
| Al2O3 | 13.94 | 11.94 | 13.33 | 14.11 | 15.02 | 16.64 | 16.09 | 13.95 | 13.57 | 13.84 | 8.85 | 6.59 | 6.61 |
| Fe2O3 | 12.99 | 11.21 | 12.35 | 12.87 | 14.82 | 9.31 | 9.34 | 18.83 | 20.89 | 20.63 | 14.3 | 15.76 | 15.91 |
| MnO | 0.23 | 0.18 | 0.20 | 0.21 | 0.28 | 0.15 | 0.15 | 0.24 | 0.14 | 0.14 | 0.2 | 0.21 | 0.2 |
| MgO | 11.00 | 11.53 | 10.45 | 10.83 | 9.46 | 7.47 | 7.76 | 5.49 | 7.08 | 7 | 24.65 | 28.78 | 28.71 |
| CaO | 11.72 | 12.20 | 12.04 | 11.48 | 11.50 | 11.92 | 12.01 | 9.26 | 12.02 | 11.99 | 5.33 | 3.08 | 3.07 |
| Na2O | 2.11 | 2.11 | 2.13 | 2.28 | 1.82 | 2.1 | 2.21 | 2.37 | 1.23 | 1.22 | 0.21 | 0.32 | 0.31 |
| K2O | 0.06 | 0.40 | 0.24 | 0.20 | 0.04 | 0.24 | 0.19 | 0.03 | 0.01 | 0.02 | 0.01 | 0.05 | 0.05 |
| P2O5 | 0.05 | 0.01 | 0.09 | 0.08 | 0.08 | 0.05 | 0.08 | 0.15 | 0.02 | 0.01 | 0.04 | 0.13 | 0.13 |
| LOI | 0.48 | 1.55 | 1.05 | 1.16 | 0.38 | 0.39 | 0.17 | | | | 3.56 | 4.79 | 4.89 |
| Total | 99.99 | 99.90 | 99.49 | 99.64 | 99.32 | 99.7 | 99.42 | 99.24 | 99.39 | 99.48 | 99.51 | 98.4 | 98.66 |
| Sc | 42 | 44 | 49 | 46 | 56 | 40 | 39 | 50 | 55 | 57 | 15 | 12 | 11 |
| V | 257 | 326 | 347 | 272 | 318 | 225 | 232 | 389 | 911 | 918 | 99 | 116 | 102 |
| Cr | 1744 | 1258 | 1112 | 995 | 851 | 400 | 661 | 1 | | | 2673 | 7448 | 7313 |
| Co | 45 | 46 | 55 | 56 | 46 | 41 | 42 | 45 | 93 | 82 | 118 | 146 | 142 |
| Ni | 185 | 224 | 223 | 240 | 99 | 72 | 79 | 6 | 12 | 10 | 802 | 1062 | 1012 |
| Cu | 14 | 145 | 53 | 96 | 82 | 97 | 97 | 177 | 63 | 62 | 80 | 29 | 27 |
| Zn | 51 | 60 | 75 | 75 | 61 | 53 | 55 | 109 | 88 | 94 | 80 | 101 | 97 |
| Ga | 11 | 12 | 14 | 12 | 12 | 14 | 13 | 18 | 18 | 18 | 6 | 7 | 6 |
| Rb | 1.8 | 14.1 | 9.6 | 6.4 | 4.5 | 6.1 | 4.3 | 1.5 | 0.4 | 0.6 | 0.4 | 1.4 | 1.4 |
| Sr | 110 | 195 | 201 | 161 | 107 | 345 | 230 | 70 | 75 | 66 | 89 | 122 | 123 |
| Y | 15 | 14 | 23 | 18 | 28 | 10 | 10 | 19 | 7 | 7 | 5 | 5 | 5 |
| Zr | 31 | 22 | 33 | 19 | 21 | 7 | 10 | 17 | 11 | 12 | 11 | 14 | 13 |
| Nb | 1.9 | 1.7 | 3.0 | 2.8 | 2.9 | 0.5 | 0.5 | | | | 0.3 | 0.6 | 0.6 |
| Cs | 0.6 | 0.5 | 0.6 | 1.0 | 0.9 | 0.4 | 0.3 | 0.1 | | 0.1 | 0.3 | 0.2 | 0.2 |
| Ba | 81 | 184 | 143 | 164 | 117 | 87 | 74 | 27 | 11 | 17 | 14 | 36 | 48 |
| La | 0.8 | 0.4 | 0.9 | 1.6 | 0.2 | 3.1 | 2.1 | 3.2 | 0.3 | 0.3 | 1.7 | 4.6 | 4.2 |
| Ce | 2.0 | 1.1 | 3.3 | 4.3 | 0.6 | 7.1 | 3.8 | 10.2 | 0.8 | 0.8 | 4.3 | 9.7 | 8.6 |
| Pr | 0.4 | 0.3 | 0.8 | 0.8 | 0.2 | 1.2 | 0.6 | 1.7 | 0.2 | 0.2 | 0.8 | 1.4 | 1.2 |
| Nd | 2.4 | 1.9 | 4.9 | 4.8 | 1.3 | 5.8 | 2.7 | 9.2 | 1.4 | 1.4 | 3.6 | 6 | 5.5 |
| Sm | 1.26 | 1.15 | 2.34 | 1.93 | 1.19 | 1.54 | 0.91 | 2.55 | 0.93 | 0.98 | 0.95 | 1.23 | 1.16 |
| Eu | 0.54 | 0.54 | 0.91 | 0.78 | 0.62 | 0.66 | 0.46 | 1.09 | 0.43 | 0.45 | 0.32 | 0.37 | 0.37 |
| Gd | 2.16 | 2.07 | 3.41 | 2.65 | 2.77 | 1.76 | 1.42 | 3.34 | 1.24 | 1.28 | 0.9 | 1.15 | 1.09 |
| Tb | 0.43 | 0.41 | 0.66 | 0.52 | 0.65 | 0.32 | 0.29 | 0.59 | 0.22 | 0.24 | 0.15 | 0.18 | 0.17 |
| Dy | 2.9 | 2.8 | 4.5 | 3.5 | 5.0 | 2.1 | 2.1 | 3.9 | 1.5 | 1.6 | 1 | 1.1 | 1 |
| Ho | 0.6 | 0.6 | 0.9 | 0.7 | 1.1 | 0.4 | 0.4 | 0.8 | 0.3 | 0.3 | 0.2 | 0.2 | 0.2 |
| Er | 1.7 | 1.7 | 2.8 | 2.2 | 3.5 | 1.3 | 1.3 | 2.2 | 0.9 | 0.9 | 0.6 | 0.6 | 0.6 |
| Tm | 0.2 | 0.2 | 0.4 | 0.3 | 0.6 | 0.2 | 0.2 | 0.3 | 0.1 | 0.1 | 0.1 | 0.1 | 0.1 |
| Yb | 1.6 | 1.5 | 2.7 | 2.3 | 3.7 | 1.2 | 1.2 | 2.1 | 0.8 | 0.8 | 0.6 | 0.6 | 0.6 |
| Lu | 0.25 | 0.24 | 0.43 | 0.34 | 0.57 | 0.18 | 0.19 | 0.31 | 0.12 | 0.12 | 0.1 | 0.09 | 0.1 |
| Hf | 0.92 | 0.83 | 1.19 | 0.64 | 0.78 | 0.24 | 0.31 | 0.54 | 0.43 | 0.47 | 0.29 | 0.41 | 0.39 |

Table 1: Continued.

| Location | Raobazhai mafic-ultramafic rocks | | | | | Nattier blue eclogite | | | Greenish black eclogite | | Garnet peridotite | | |
|---|---|---|---|---|---|---|---|---|---|---|---|---|---|
| Ta | 0.209 | 0.08 | 0.14 | 0.17 | 0.147 | 0.031 | 0.029 | 0.002 | 0.001 | 0.003 | 0.021 | 0.037 | 0.086 |
| Pb | 7.4 | 12.2 | 23.0 | 13.9 | 20.8 | 2.7 | 0.9 | 9.7 | 2 | 1.3 | 2.5 | 7.7 | 7.8 |
| Th | 0.0 | 0.0 | 0.0 | 0.0 | 0.0 | 0.2 | 0.1 | 0.6 | 0.2 | 0.2 | 0.2 | 0.3 | 0.3 |
| U | 0.02 | 0.01 | 0.05 | 0.06 | 0.02 | 0.06 | 0.04 | 0.17 | 0.08 | 0.08 | 0.05 | 0.1 | 0.1 |
| Ir | 0.153 | 0.060 | 0.257 | 0.180 | 0.360 | — | 0.003 | — | — | — | 0.837 | 1.18 | 0.52 |
| Ru | 0.487 | 0.297 | 0.367 | 0.520 | 0.513 | 0.003 | — | 0.007 | — | 0.03 | 0.723 | 1.76 | 1.357 |
| Rh | 0.237 | 0.327 | 0.447 | 0.343 | 0.633 | 0.108 | 0.083 | 0.002 | 0.109 | 0.142 | 0.665 | 0.648 | 0.422 |
| Pt | 3.86 | 8.77 | 9.71 | 7.65 | 14.33 | 0.34 | 0.20 | 0.08 | 0.16 | 0.29 | 12.00 | 3.40 | 3.75 |
| Pd | 4.86 | 31.98 | 4.44 | 7.69 | 23.38 | 0.47 | 0.03 | 0.16 | 0.09 | 0.50 | 7.42 | 8.67 | 8.34 |

(—): lower than the limits of detection.

Bixiling mafic-ultramafic rocks

Different Origins of the Fractionation of Platinum-Group Elements in Raobazhai and Bixiling Mafic-Ultramafic Rocks from the Dabie Orogen, Central China

195

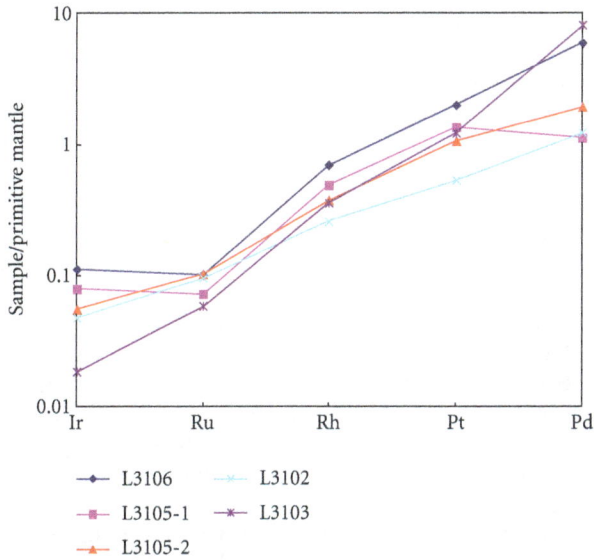

FIGURE 3: Primitive mantle-normalized PGE abundances for the Raobazhai mafic-ultramafic rocks (Normalizing values are after [1]).

relative to the HREE [21]. The Raobazhai peridotites are ubiquitously depleted in LREE, which suggest that they represent mantle residues after variable degrees of partial melting. In the (primitive mantle)-PM-normalized trace element diagram (Figure 2(b)), most Raobazhai samples show variable enrichment in LILE (e.g., Sr, Ba, and Rb). In particular, both Rb and Ba have concentrations ten times higher than those of the primitive mantle. The enrichment of LILE is an important feature of the Raobazhai mafic-ultramafic rocks. It has been suggested that enrichment of Rb and Ba in mantle peridotites could result from subduction-related metasomatism [21]. Zhi et al. [18] also concluded that the enrichment of LILE shown by the Raobazhai ultramafic rocks might be probably related to the slab-released fluids.

The total PGE contents of the Raobazhai mafic-ultramafic rocks range from 9.6 to 41.4 ppb, with an average of 24.4 ppb, which are higher than the estimated values of the primitive mantle (20.1 ppb) but similar to the Alpine-type orogenic peridotites, for example, Ronda (17.5–39.5 ppb) and the Beni Bousera (17.2–32.5 ppb) [22]. All Raobazhai mafic-ultramafic rocks display consistent and pronounced positive PGE patterns (Figure 3); they are strongly enriched in PPGEs (e.g., Pt and Pd) over IPGEs (e.g., Ir and Ru). Their Pd/Ir ratios vary from 17 to 65. Both Ir and Ru show good positive correlations with Ni, which suggest they behave as compatible elements during partial melting [23, 24]. In contrast, both Pt and Pd behave as incompatible elements. Therefore, the concentrations of Ir and Ru increase in the residual peridotites along with the melt extraction, whereas melts are enriched in both Pd and Pt relative to Ir and Ru. This suggests that the residual mantle peridotites should enrich in IPGEs over PPGEs, which is in contrast to the PGEs patterns shown by the Raobazhai mafic-ultramafic rocks. Therefore, the PGE budgets of the Raobazhai peridotites have been affected by processes other than partial melting.

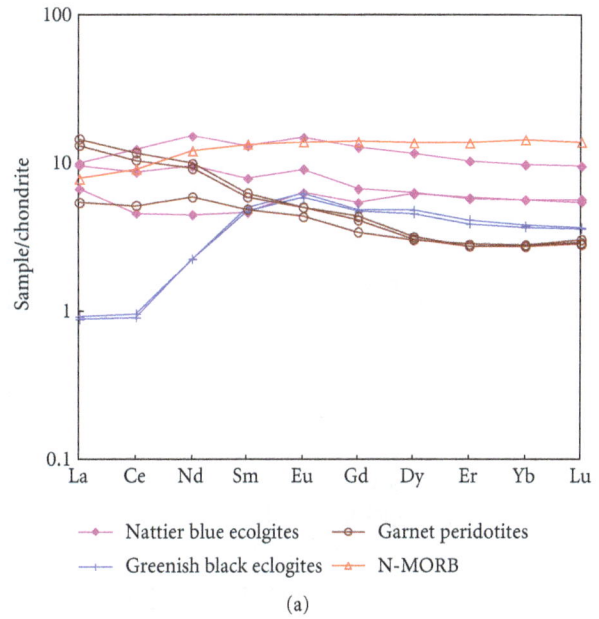

(a)

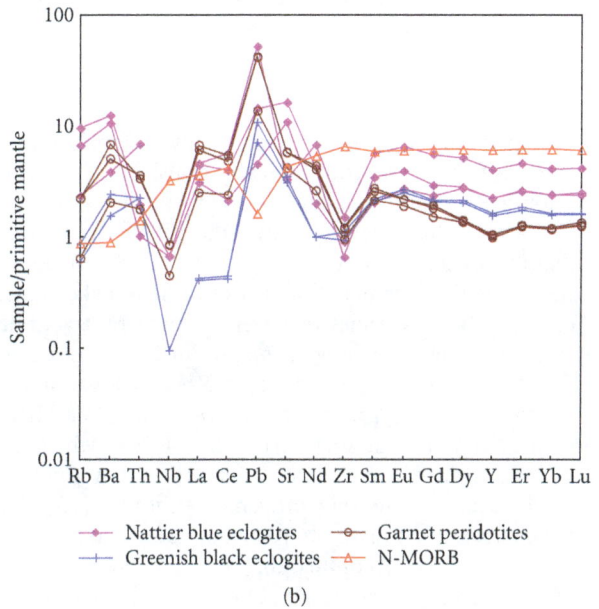

(b)

FIGURE 4: Chondrite-normalized REE patterns (a) and spider patterns (b) of Bixiling mafic-ultramafic rocks (Chondrite, Primitive mantle, and N-MORB values are from [9]).

4.2. Bixiling Eclogites and Peridotites. Both REE and PGE data for five eclogites and two garnet peridotites from Bixiling are given in Table 1, and their distribution patterns are shown in Figures 4 and 5. All eclogites show remarkable positive Eu anomalies, suggesting that they were originally transformed from rocks with cumulated plagioclase. Three nattier blue eclogites selected in this study have quiet similar REE patterns with variable enrichment in LREE (Figure 4(a)), which are consistent with results reported in a previous study [20]. Compared to the normal mid-ocean ridge basalts (N-MORB), the Bixiling nattierblue eclogites have relatively low HREE contents and are slightly enriched in LREE.

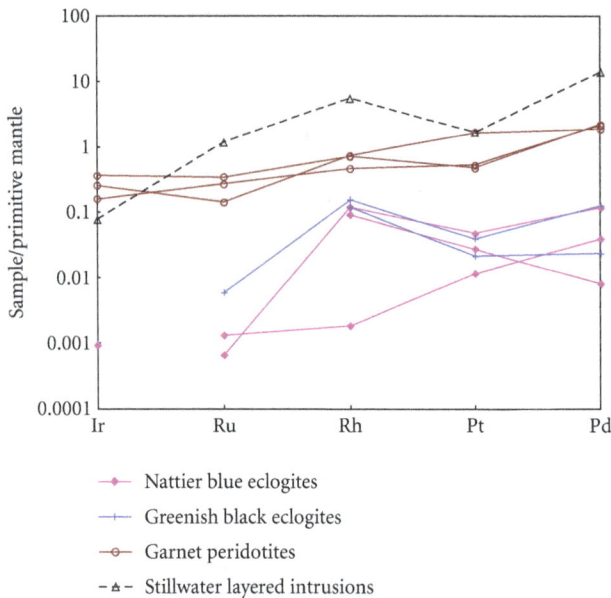

FIGURE 5: Primitive mantle-normalized PGE abundances for the Bixiling mafic-ultramafic rocks (Normalizing values are after [1]).

This suggests that they were not originally metamorphosed from N-MORB [20]. In contrast, the greenish black eclogites have lower REE contents and are strongly depleted in LREE (Figure 4(a)). This may be because the LREEs are incompatible in olivine, pyroxene, and garnet, but rich in liquid phase during the differentiation process, which make cumulation conglomerate facies that were formed by fractional crystallization have a relative depletion in LREE, and their HREE abundance also different from N-MORB. Compared to the eclogites, the garnet peridotites have lower HREE contents. They are variably enriched in LREE, which are also differed from both garnet peridotites entrained in kimberlites and Alpine-type orogenic peridotites [25]. The patterns suggest that these two garnet peridotites represent a more advanced crystal cumulation [20]. In the trace-element diagram (Figure 4(b)), both garnet peridotites and eclogites show pronounced negative Zr and Nb anomalies. Depletion of Zr has been suggested to be an indigenous feature of the upper mantle origin. The negative Nb anomaly is an important fingerprint of subduction-related magmas and continental crust. Based on Sr-Nd isotopes of the Bixiling eclogites and garnet peridotites, Chavagnac and Jahn [20] suggested that the Bixiling magmas were contaminated, probably by lower-crustal granulites that are depleted in LILE. The contamination did not significantly modify the Sr isotops, but could result in lowering in εNd values and negative Nb anomaly as observed [20]. A previous study on oxygen isotopic compositions of silicate minerals in the Bixiling eclogites and garnet peridotites has suggested that the original magmas were derived from the upper mantle but probably contaminated by small amounts of crustal rocks during their differentiation processes [26].

The Bixiling mafic-ultramafic rocks display PGE patterns increasing from IPGEs to PPGEs (Figure 5), which

are different from that the mantle wedge xenoliths from Kamchatka [27] but very similar to to the PGE patterns shown by the Stillerwater layered intrusions [3]. The total PGE contents of garnet peridotites range from 14.39 ppb to 21.65 ppb. They have high Pd/Ir ratios up to 10 that are remarkably higher than that of the primitive mantle. It has been suggested that the typical mantle xenoliths and Alpine-type orogenic peridotites have flat PGE patterns, in which the PGEs are not fractionated [3]. Furthermore, partial melting would lead to depletion of PPGEs over IPGEs in the mantle residues, which should display flat to negative PGE patterns [3, 22, 28, 29]. Therefore, we suggest that the Bixiling garnet peridotites are refractory mantle residues after melt extraction but represent fractional crystallization products of mantle-derived melts, which is also supported by both trace-element and isotope compositions as discussed above.

The PGE patterns of both nattier blue eclogites and greenish black eclogites are distinguished from each other due to their very low contents of Ir and Ru, which are even lower than their detection limits. The Bixiling eclogites have low total contents of PGEs, which range from 0.25 ppb to 0.96 ppb. They display fractionated PGE patterns, which increase from IPGEs to PPGEs. This implies that they represent the late-stage products of magmatic differentiation. The PGE patterns of the Bixiling mafic-ultramafic rocks vary with the lithologies. The PGE content systematically decreases from the garnet peridotites to the eclogites, which show positive correlations with the MgO contents. Along with the magmatic differentiation, variations in PPGEs (i.e., Pt and Pd) are more limited than those observed for IPGEs (i.e., Ir and Ru). This suggests that the IPGEs are compatible during fractional crystallization and controlled by phases (e.g., metal alloys, chromite, olivine, or clinopyroxene) other than low-temperature sulphides [7].

## 5. Discussion

Although it has been suggested that the PGEs can be mobilized and fractionated by secondary post-magmatic events, such as hydrothermal alteration or weathering [30], it has been widely accepted that the PGEs are immobile under near-surface conditions. For example, Rehkämper et al. [31] suggested that the PGE budgets of abyssal peridotites have not been significantly disturbed by low-temperature alteration (<150°C). Furthermore, it has also been suggested that the PGEs are immobile during serpentinization (<600°C) processes, for which is commonly taken place under very reducing conditions [31, 32]. Büchl et al. [33] demonstrated that the PGEs are also significantly fractionated by hydrothermal fluids. Therefore, the fractionated PGE patterns observed in the Raobazhai peridotites cannot be ascribed to any secondary process; their PGE patterns reflect the magmatic processes occurred in the mantle, such as partial melting and melt percolation. The Raobazhai peridotites display enrichment in IPGEs over PPGEs, which is in stark contrast to the predicated PGE patterns of residual mantle peridotites. Hence, we believe that their PGE budgets have been affected by other processes than partial melting.

Recently, various studies have shown that the PGE budgets of mantle peridotites could be significantly disturbed by metasomatic processes, including melt/fluid infiltration and percolation [34–40]. Unlike hydrous fluids, slab-derived melts are capable of carrying HFSE (e.g., Zr, Hf, Nb and Ta) at some instances [41–43]. The mantle wedge could achieve such distinct geochemical signatures through extensive interaction with slab-derived melts [27]. It has also been suggested that slab-derived melts could fractionate the IPGEs from the PPGEs [27]. The positive relationship between Pt/Pd ratio and Hf concentration shown by the Raobazhai mafic-ultramafic rocks indicates that they have been metasomatized by slab-derived melts. Occurrence of hydrous mineral in Raobazhai mafic-ultramafic rocks also supports that they have been interacted with hydrous melts during subduction and/or exhumation [44]. Segregation of secondary sulfides from the volatile-rich melts into the mantle peridotites could increase their Pd abundances [34]. Enrichment of PPGEs over IPGEs in the Raobazhai peridotites could be interpreted by addition of secondary sulfides. In conclusion, both PGE and trace-elements data suggest that the Raobazhai mafic-ultramafic rocks have been metasomatized by slab-derived melts.

The partition coefficient of Cu between sulfide and silicate melts ($D^{sulfide/silicate}$) has been experimentally determined to be 900–1400, which is 3000–90000 for Pd [45]. The covariation between Cu and Pd is a useful indicator for sulphide fractionation. Because Cu is much less chalcophile than Pd, the Cu/Pd ratio should increase if sulphide is fractionated from a magma. In the Bixiling garnet peridotites and eclogites, The Cu/Pd ratio of the Bixiling garnet peridotites varies from $10^3$ to $10^4$, whereas it ranges from $10^5$ to $10^7$ for the Bixiling eclogites. As shown in the Cu/Pd versus Pd diagram (Figure 6), the Cu/Pd ratios of the garnet peridotites are close to the mantle values, which suggests that they have not experienced sulphide segregation prior to their emplacement [12]. If sulfide segregation had occurred in the Bixiling garnet peridotites, then their Cu/Pd ratios should be greater than the normal mantle values because Pd is preferentially partitioning into sulfide liquid relative to Cu. The PGE patterns of the Bixiling eclogites are consistent with sulfide segregation from these samples. That is, the Cu/Pd ratio becomes elevated along with the increase of fractional crystallization.

Sulfide segregation is an inevitable process during fractional crystallization [7]. Fractional crystallization tend to decrease the FeO content in the residual magma, which might result in S saturation and thus formation of immiscible sulfide liquids [46]. The Bixiling garnet peridotites and eclogites were formed along with fractional crystallization, during which the S contents of the magmas became saturated to segregate sulfides. Removal of sulfides would lead to depletion of Pd in the residual magmas and increase in Cu/Pd ratio (Figure 6).

The Bixiling mafic-ultramafic rocks display a positive correlation between Ru and Pd, indicating that both elements are partitioning into the same kind of sulfides. The PGE data of the Bixiling mafic-ultramafic rocks suggest that

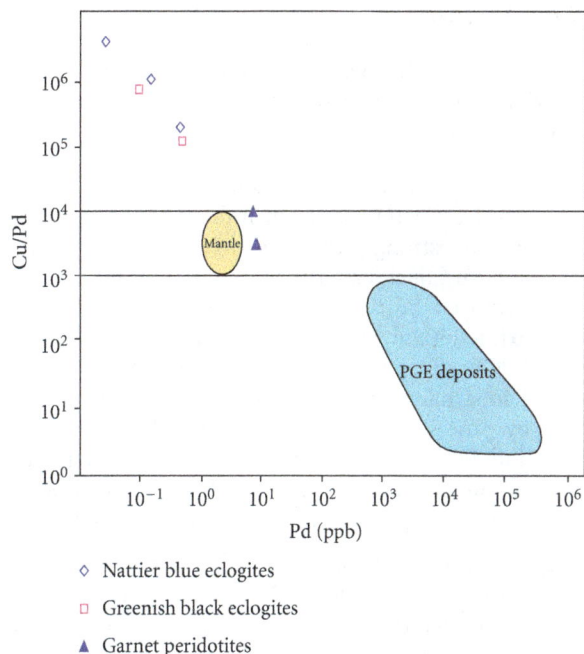

FIGURE 6: Plots of Cu/Pd ratios versus Pd for Bixiling mafic-ultramafic rocks (filled fields are after [12]).

the PGEs can be fractionated during magma differentiation. Both Ir and Ru are compatible in a nonsulfide phase, in which Pd is incompatible. Both petrological and geochemical studies have suggested that olivine [47, 48], spinel or chromite [49, 50], and refractory alloys [51, 52] are the most likely candidates [7]. However, modeling calculations have shown that neither olivine nor clinopyroxene can significantly fractionate the PGEs [7]. It has been suggested that the PGEs are not incorporated into the lattice of chromites but concentrated within tiny inclusions, such as sulfides or alloys. Therefore, chromites themselves are not able to fractionate the PGEs [52]. Keays [2] has suggested that both Ir and Os exist as Os-Ir alloys in the upper mantle. These tiny alloys can be physically segregated into the magmas during partial melting. They can be trapped within the crystallized silicate (e.g., olivine) and oxide (e.g., chromite) minerals, which results in the fractionation of the IPGEs from the PPGEs. Tredoux et al. [53] has also suggested that the PGEs and other nonlithophile elements could be aggregated together as clusters in silicate magmas. Theoretically, the IPGEs (i.e., Os and Ir) are more likely to form clusters than the PPGE (i.e., Rh and Pd) [53]. Therefore, we explain the enrichment of IPGEs rather than PPGEs in the Bixiling mafic-ultramafic rocks as entrainment of such clusters in these samples. The PGE alloys have been rarely reported in mantle peridotites, which could be due to their extremely small sizes. Microinclusions of Os, Ir, and Pt have been identified in Merensky sulfides, which support the occurrence of PGEs as polymetallic clusters in silicates [54]. However, future studies are still needed to investigate the PGE-rich phases in the mantle.

## 6. Conclusions

On the basis of the geochemical compositions of both Raobazhai and Bixiling mafic-ultramafic rocks, in particular PGE data, we can draw the following primary conclusions:

(1) The Raobazhai mafic-ultramafic rocks show consistent PGE patterns, in which the PPGEs (e.g., Pt and Pd) are strongly enriched over the IPGEs (e.g., Ir and Ru). Such patterns are in stark contrast to those displayed by refractory mantle residues after melt extraction, in which the IPGEs are enriched over the PPGEs. This indicates that the PGE budgets of the Raobazhai mafic-ultramafic rocks have been affected by processes other than partial melting. Both REE and PGE data support that the Raobazhai mafic-ultramafic rocks have interacted with slab-derived melts during subduction and/or exhumation.

(2) The Bixiling mafic-ultramafic rocks were formed through fractional crystallization and cumulation from magmas. The PGE patterns shown by the Bixiling mafic-ultramafic rocks were produced by the magmatic differentiation processes, during which fractional crystallization of silicate minerals and segregation of immiscible sulfide liquids are involved. The fractionated PGE patterns of the Bixiling mafic-ultramafic rocks reflect that Pd is incompatible in the nonsulfide phases, whereas both Ir and Ru are compatible in some mantle phases.

## Acknowledgments

This research has been financially supported by the Nature Science Foundation of China (Grant no. 40702009; 41030422). The authors thank Tianshan Gao and Hongyuan Zhang for help in the field work, He Li and Xindi Jin for major and trace element analyses, and Caifen Nu and Hongliao He for their help in PGE analyses. Comments from two reviewers greatly improved the quality of this paper.

## References

[1] S. J. Barnes, R. Boyd, and A. Korneliussen, "The use of mantle normalization and metal ratios in discriminating between the effects of partial melting, crystal fractionation and sulphide segregation on platinum-group elements, gold, nickel and copper: examples from Norway," in *Proceedings of the Geo-Platinum Conference*, H. M. Prichard, J. W. Bowles, and P. Potts, Eds., pp. 113–143, Elsevier, Amsterdam, The Netherlands, 1988.

[2] R. R. Keays, "The role of komatiitic and picritic magmatism and S-saturation in the formation of ore deposits," *Lithos*, vol. 34, no. 1–3, pp. 1–18, 1995.

[3] S. J. Barnes, A. J. Naldrett, and M. P. Gorton, "The origin of the fractionation of platinum-group elements in terrestrial magmas," *Chemical Geology*, vol. 53, no. 3-4, pp. 303–323, 1985.

[4] J. P. Lorand, "Abundance and distribution of CuFeNi sulfides, sulfur, copper and platinum-group elements in orogenic-type spinel lherzolite massifs of Ariège (Northeastern Pyrenees,

France)," *Earth and Planetary Science Letters*, vol. 93, no. 1, pp. 50–64, 1989.

[5] G. Garuti, C. Gorgoni, and G. P. Sighinolfi, "Sulfide mineralogy and chalcophile and siderophile element abundances in the Ivrea-Verbano mantle peridotites (Western Italian Alps)," *Earth and Planetary Science Letters*, vol. 70, no. 1, pp. 69–87, 1984.

[6] G. Garuti, G. Fershtater, F. Bea, P. Montero, E. V. Pushkarev, and F. Zaccarini, "Platinum-group elements as petrological indicators in mafic-ultramafic complexes of the central and southern Urals: preliminary results," *Tectonophysics*, vol. 276, no. 1–4, pp. 181–194, 1997.

[7] M. Rehkämper, A. N. Halliday, J. G. Fitton, D. C. Lee, M. Wieneke, and N. T. Arndt, "Ir, Ru, Pt, and Pd in basalts and komatiites: new constraints for the geochemical behavior of the platinum-group elements in the mantle," *Geochimica et Cosmochimica Acta*, vol. 63, no. 22, pp. 3915–3934, 1999.

[8] B. M. Jahn, F. Wu, C. H. Lo, and C. H. Tsai, "Crust-mantle interaction induced by deep subduction of the continental crust: geochemical and Sr-Nd isotopic evidence from post-collisional mafic-ultramafic intrusions of the northern Dabie complex, central China," *Chemical Geology*, vol. 157, no. 1-2, pp. 119–146, 1999.

[9] S. S. Sun and W. F. McDonough, "Chemical and isotopic systematics of oceanic basalts: implications for mantle composition and processes," *Magmatism in the Ocean Basins*, pp. 313–345, 1989.

[10] B. R. Hacker, L. Ratschbacher, L. Webb et al., "Exhumation of ultrahigh-pressure continental crust in east central China: late Triassic-Early Jurassic tectonic unroofing," *Journal of Geophysical Research B*, vol. 105, no. 6, pp. 13339–13364, 2000.

[11] J. Zheng, M. Sun, M. F. Zhou, and P. Robinson, "Trace elemental and PGE geochemical constraints of Mesozoic and Cenozoic peridotitic xenoliths on lithospheric evolution of the North China Craton," *Geochimica et Cosmochimica Acta*, vol. 69, no. 13, pp. 3401–3418, 2005.

[12] S. J. Barnes and C. P. Picard, "The behaviour of platinum-group elements during partial melting, crystal fractionation, and sulphide segregation: an example from the Cape Smith Fold Belt, northern Quebec," *Geochimica et Cosmochimica Acta*, vol. 57, no. 1, pp. 79–87, 1993.

[13] B. R. Hacker, L. Ratschbacher, L. Webb, T. Ireland, D. Walker, and D. Shuwen, "U/Pb zircon ages constrain the architecture of the ultrahigh-pressure Qinling-Dabie Orogen, China," *Earth and Planetary Science Letters*, vol. 161, no. 1–4, pp. 215–230, 1998.

[14] X. Yang, "A cold ultramafic intrusive body in Dabieshan area, Anhui Province," *Bulletin of Nanjing Institute of Geology*, vol. 4, pp. 81–95, 1983.

[15] S. Xu, L. Jiang, Y. Liu, and Y. Zhang, "Tectonic framework and evolution of the Dabie Mountains in Anhui, eastern China," *Acta Geologica Sinica*, vol. 5, pp. 221–238, 1992.

[16] C. H. Tsai, J. C. Liou, and W. G. Ernst, "Petrological characterization and tectonic significance of retrogressed garnet peridotites, Raobazhai area, North Dabie Complex, east-central China," *Journal of Metamorphic Geology*, vol. 18, no. 2, pp. 181–192, 2000.

[17] Q. Zhang, B. Ma, R. Liu et al., "A remnant of continental lithospheric mantle above subduction zone: geochemical constraints on ultramafic rock from Raobazhai area, Anhui province," *Science in China Series B*, vol. 38, no. 12, pp. 1522–1529, 1995.

[18] X. C. Zhi, Y. B. Jin, Q. Meng, and T. S. Gao, "Trace element geochemistry of Raobazhai ultramafic complex, North Dabie

Mountain," *Acta Petrologica Sinica*, vol. 20, no. 3, pp. 463–472, 2004.

[19] R. Y. Zhang, J. G. Liou, and B. L. Cong, "Ultrahigh-pressure metamorphosed talc-, magnesite- and Ti-clinohumite-bearing mafic–ultramafic complex, Dabie mountains, east-central China," *Journal of Petrology*, vol. 36, no. 4, pp. 1011–1037, 1995.

[20] V. Chavagnac and B. M. Jahn, "Coesite-bearing eclogites from the Bixiling Complex, Dabie Mountains, China: Sm-Nd ages, geochemical characteristics and tectonic implications," *Chemical Geology*, vol. 133, no. 1–4, pp. 29–51, 1996.

[21] H. Downes, "Formation and modification of the shallow sub-continental lithospheric mantle: a review of geochemical evidence from ultramatic xenolith suites and tectonically emplaced ultramafic massifs of Western and Central Europe," *Journal of Petrology*, vol. 42, no. 1, pp. 233–250, 2001.

[22] K. Gueddari, M. Piboule, and J. Amossé, "Differentiation of platinum-group elements (PGE) and of gold during partial melting of peridotites in the lherzolitic massifs of the Betico-Rifean range (Ronda and Beni Bousera)," *Chemical Geology*, vol. 134, no. 1–3, pp. 181–197, 1996.

[23] J. W. Morgan, "Ultramafic xenoliths: clue to the earth's late accretionary history," *Journal of Geophysical Research*, vol. 91, pp. 12375–12387, 1986.

[24] J. P. Lorand, L. Pattou, and M. Gros, "Fractionation of Platinum-group elements and gold in the upper mantle: a detailed study in Pyrenean orogenic lherzolites," *Journal of Petrology*, vol. 40, no. 6, pp. 957–981, 1999.

[25] W. F. McDonough and F. A. Frey, "Rare earth elements in upper mantle rocks," in *Geochemistry and Mineralogy of the Rare Earth Elements*, B. R. Lipin and G. A. McKay, Eds., vol. 21, pp. 99–145, Mineralogical Society of America, 1989.

[26] R. Y. Zhang, D. Rumble, J. G. Liou, and Q. C. Wang, "Low $\delta^{18}O$, ultrahigh-P garnet-bearing mafic and ultramafic rocks from Dabie Shan, China," *Chemical Geology*, vol. 150, no. 1-2, pp. 161–170, 1998.

[27] P. Kepezhinskas, M. J. Defant, and E. Widom, "Abundance and distribution of PGE and Au in the island-arc mantle: implications for sub-arc metasomatism," *Lithos*, vol. 60, no. 3-4, pp. 113–128, 2002.

[28] G. E. Brügmann, N. T. Arndt, A. W. Hofmann, and H. J. Tobschall, "Noble metal abundances in komatiite suites from Alexo, Ontario and Gorgona Island, Colombia," *Geochimica et Cosmochimica Acta*, vol. 51, no. 8, pp. 2159–2169, 1987.

[29] J. P. Lorand, R. R. Keays, and J. L. Bodinier, "Copper and noble metal enrichments across the lithosphere-asthenosphere boundary of mantle diapirs: evidence from the lanzo lherzolite massif," *Journal of Petrology*, vol. 34, no. 6, pp. 1111–1140, 1993.

[30] J. J. Standish, S. R. Hart, J. Blusztajn, H. J. B. Dick, and K. L. Lee, Abyssal peridotite osmium isotopic composition for Cr-spinel, Geochem. Geophys. Geosyst. 2001GG000161, 2001.

[31] M. Rehkämper, A. N. Halliday, J. Alt, J. G. Fitton, J. Zipfel, and E. Takazawa, "Non-chondritic platinum-group element ratios in oceanic mantle lithosphere: petrogenetic signature of melt percolation?" *Earth and Planetary Science Letters*, vol. 172, no. 1-2, pp. 65–81, 1999.

[32] J. E. Snow and G. Schmidt, "Constraints on Earth accretion deduced from noble metals in the oceanic mantle," *Nature*, vol. 391, no. 6663, pp. 166–169, 1998.

[33] A. Büchl, G. Brügmann, V. G. Batanova, C. Münker, and A. W. Hofmann, "Melt percolation monitored by Os isotopes and HSE abundances: a case study from the mantle section of the Troodos Ophiolite," *Earth and Planetary Science Letters*, vol. 204, no. 3-4, pp. 385–402, 2002.

[34] J. P. Lorand, L. Reisberg, and R. M. Bedini, "Platinum-group elements and melt percolation processes in Sidamo spinel peridotite xenoliths, Ethiopia, East African Rift," *Chemical Geology*, vol. 196, no. 1–4, pp. 57–75, 2003.

[35] J. P. Lorand, G. Delpech, M. Grégoire, B. Moine, S. Y. O'Reilly, and J. Y. Cottin, "Platinum-group elements and the multistage metasomatic history of Kerguelen lithospheric mantle (South Indian Ocean)," *Chemical Geology*, vol. 208, no. 1–4, pp. 195–215, 2004.

[36] J. P. Lorand, A. Luguet, O. Alard, A. Bezos, and T. Meisel, "Abundance and distribution of platinum-group elements in orogenic lherzolites; a case study in a Fontete Rouge lherzolite (French Pyrénées)," *Chemical Geology*, vol. 248, no. 3-4, pp. 174–194, 2008.

[37] L. Reisberg, X. Zhi, J. P. Lorand, C. Wagner, Z. Peng, and C. Zimmermann, "Re-Os and S systematics of spinel peridotite xenoliths from east central China: evidence for contrasting effects of melt percolation," *Earth and Planetary Science Letters*, vol. 239, no. 3-4, pp. 286–308, 2005.

[38] H. Becker, M. F. Horan, R. J. Walker, S. Gao, J. P. Lorand, and R. L. Rudnick, "Highly siderophile element composition of the Earth's primitive upper mantle: constraints from new data on peridotite massifs and xenoliths," *Geochimica et Cosmochimica Acta*, vol. 70, no. 17, pp. 4528–4550, 2006.

[39] L. Ackerman, R. J. Walker, I. S. Puchtel, L. Pitcher, E. Jelínek, and L. Strnad, "Effects of melt percolation on highly siderophile elements and Os isotopes in subcontinental lithospheric mantle: a study of the upper mantle profile beneath Central Europe," *Geochimica et Cosmochimica Acta*, vol. 73, no. 8, pp. 2400–2414, 2009.

[40] Y. Xiao and H.-F. Zhang, "Effects of melt percolation on platinum group elements and Re-Os systematics of peridotites from the Tan-Lu fault zone, eastern North China Craton," *Journal of the Geological Society*, vol. 168, no. 5, pp. 1201–1214, 2011.

[41] M. J. Defant and M. S. Drummond, "Derivation of some modern arc magmas by melting of young subducted lithosphere," *Nature*, vol. 347, no. 6294, pp. 662–665, 1990.

[42] M. S. Drummond, M. J. Defant, and P. K. Kepezhinskas, "Petrogenesis of slab-derived trondhjemite-tonalite-dacite/adakite magmas," *Transactions of the Royal Society of Edinburgh, Earth Sciences*, vol. 87, no. 1-2, pp. 205–215, 1996.

[43] H. Martin, "Adakitic magmas: modern analogues of Archaean granitoids," *Lithos*, vol. 46, no. 3, pp. 411–429, 1999.

[44] L. Zheng, X. Zhi, and L. Reisberg, "Re-Os systematics of the Raobazhai peridotite massifs from the Dabie orogenic zone, eastern China," *Chemical Geology*, vol. 268, no. 1-2, pp. 1–14, 2009.

[45] S.-J. Barnes and W. D. Maier, "The fractionation of Ni, Cu, and the noble metals in silicate and sulphide liquids," in *Dynamic Processes in Magmatic Ore Deposits and Their Application in Mineral Exploration*, R. R. Keays, Ed., Short Course Notes 13, pp. 69–106, Geological Association Canada, 1999.

[46] W. Yi, A. N. Halliday, J. C. Alt et al., "Cadmium, indium, tin, tellurium, and sulfur in oceanic basalts: implications for chalcophile element fractionation in the Earth," *Journal of Geophysical Research B*, vol. 105, no. 8, pp. 18927–18948, 2000.

[47] G. E. Brügmann, N. T. Arndt, A. W. Hofmann, and H. J. Tobschall, "Noble metal abundances in komatiite suites from Alexo, Ontario and Gorgona Island, Colombia," *Geochimica et Cosmochimica Acta*, vol. 51, no. 8, pp. 2159–2169, 1987.

[48] M. F. Zhou, "PGE distribution in 2.7-Ga layered komatiite flows from the Belingwe greenstone belt, Zimbabwe," *Chemical Geology*, vol. 118, no. 1–4, pp. 155–172, 1994.

[49] I. O. Oshin and J. H. Crocket, "Noble metals in Thetford Mines ophiolites, Quebec, Canada—part I: distribution of gold, iridium, platinum, and palladium in the ultramafic and gabbroic rocks," *Economic Geology*, vol. 77, no. 6, pp. 1556–1570, 1982.

[50] C. J. Capobianco, R. L. Hervig, and M. J. Drake, "Experiments on crystal/liquid partitioning of Ru, Rh and Pd for magnetite and hematite solid solutions crystallized from silicate melt," *Chemical Geology*, vol. 113, no. 1-2, pp. 23–43, 1994.

[51] C. Ballhaus, "Is the upper mantle metal-saturated?" *Earth and Planetary Science Letters*, vol. 132, no. 1–4, pp. 75–86, 1995.

[52] R. J. Walker, E. Hanski, J. Vuollo, and J. Liipo, "The Os isotopic composition of Proterozoic upper mantle: evidence for chondritic upper mantle from the Outokumpu ophiolite, Finland," *Earth and Planetary Science Letters*, vol. 141, no. 1–4, pp. 161–173, 1996.

[53] M. Tredoux, N. M. Lindsay, G. Davies, and I. McDonald, "The fractionation of platinum-group elements in magmatic systems, with the suggestion of a novel causal mechanism," *South African Journal of Geology*, vol. 98, no. 2, pp. 157–167, 1995.

[54] C. Ballhaus and P. Sylvester, "Spatial platinum-group-element distribution in magmatic sulfides: implications for the platinum group-element behavior during mantle melting," in *Proceedings of the 7th Annual V.M. Goldschmidt Conference*, pp. 15–16, 1997.

# 16

# 3D Geostatistical Modeling and Uncertainty Analysis in a Carbonate Reservoir, SW Iran

**Mohammad Reza Kamali,**[1] **Azadeh Omidvar,**[2] **and Ezatallah Kazemzadeh**[1]

[1] *Research Institute of Petroleum Industry (RIPI), West Boulevard, Azadi Sport Complex, Tehran 1485733111, Iran*
[2] *Department of Geology, Science and Research Branch, Islamic Azad University, End of Ashrafi Esfahani Highway, Simon Boulevard, P.O. Box 1944964814, Tehran, Iran*

Correspondence should be addressed to Azadeh Omidvar; omidvar.azadeh@yahoo.com

Academic Editor: Agust Gudmundsson

The aim of geostatistical reservoir characterization is to utilize wide variety of data, in different scales and accuracies, to construct reservoir models which are able to represent geological heterogeneities and also quantifying uncertainties by producing numbers of equiprobable models. Since all geostatistical methods used in estimation of reservoir parameters are inaccurate, modeling of "estimation error" in form of uncertainty analysis is very important. In this paper, the definition of Sequential Gaussian Simulation has been reviewed and construction of stochastic models based on it has been discussed. Subsequently ranking and uncertainty quantification of those stochastically populated equiprobable models and sensitivity study of modeled properties have been presented. Consequently, the application of sensitivity analysis on stochastic models of reservoir horizons, petrophysical properties, and stochastic oil-water contacts, also their effect on reserve, clearly shows any alteration in the reservoir geometry has significant effect on the oil in place. The studied reservoir is located at carbonate sequences of Sarvak Formation, Zagros, Iran; it comprises three layers. The first one which is located beneath the cap rock contains the largest portion of the reserve and other layers just hold little oil. Simulations show that average porosity and water saturation of the reservoir is about 20% and 52%, respectively.

## 1. Introduction

The first step in optimizing the use of explored resources is to define the reservoir, which has a determinant role in reservoir management [1]. Definition of a reservoir includes description of empty spaces and size of grains, porosity and permeability of reservoir, identification of facies, sedimentary environment, and description of basin [2].

Three-dimensional models provide the best mechanism for linking all the existing data [3]. Nowadays, efficient three-dimensional simulation is popular in all major oil companies and has become an essential part of normal exploration and production activities. To overcome the inherent two-dimensional limitation of paper, it is necessary to use defined three-dimensional data. Three-dimensional simulation of geological structures enables collection of all the existing data

for a certain project in a united model, by means of which data can be analyzed in software environment [4].

There are several methods for estimation. In a general classification, they can be divided into geostatistical and classical methods. Classical methods are those using classical statistics for estimation, while in geostatistical methods the estimation is made based on spatial structure in the environment [5].

3D geological models play an important role in petroleum engineering. There are different methods for 3D modeling, in each of these methods geological, mathematical, or statistical information is used [6].

Modern specialized software programs can model complicated and nonorderly geological volumes in three dimensions. This is done by using geological maps and construction information for creating a proper model [7].

When calculating reserves using any of the above methods, two calculation procedures may be used: deterministic and/or probabilistic. The deterministic method is by far the most common. The procedure is to select a single value for each parameter to input into an appropriate equation and to obtain a single answer. The probabilistic method, on the other hand, is more rigorous and less commonly used. This method utilizes a distribution curve for each parameter, and through the use of Monte Carlo Simulation, a distribution curve for the answer can be developed. Assuming good data, a lot of qualifying information can be derived from the resulting statistical calculations, such as the minimum and maximum values, the mean (average value), the median (middle value), the mode (most likely value), the standard deviation and the percentiles, see Figures 1 and 2.

The probabilistic methods have several inherent problems. They are affected by all input parameters, including the most likely and maximum values for the parameters. In such methods, one cannot back calculate the input parameters associated with reserves. Only the end result is known but not the exact value of any input parameter. On the other hand, deterministic methods calculate reserve values that are more tangible and explainable. In these methods, all input parameters are exactly known; however, they may sometimes ignore the variability and uncertainty in the input data compared to the probabilistic methods which allow the incorporation of more variance in the data.

In recent years, the quantification, understanding, and management of subsurface uncertainties has become increasingly important for oil and gas companies as they strive to optimize reserve portfolios, make better field development decisions, and improve day-to-day technical operations such as well planning. Stochastic approaches based on the standard volumetric equation are now commonly used for screening and value assessment of hydrocarbon assets. Uncertainty in static volumes and recoverable reserves are quantified by Monte Carlo sampling of probability distributions for the controlling parameters in the volumetric equation. Volumes are calculated by simple multiplication of the sampled values for each of the input distributions. As the Monte Carlo sampling and direct multiplication is very fast, 1000s of Monte Carlo loops can be run to provide reliable output distributions.

Although these approaches are very fast, it is often difficult to estimate the intrinsic dependencies between the input parameters, and they provide no quantification or visualization of the spatial location and variability of the uncertainty. An alternative is to use a 3D model as the basis for the volumetric calculations. This allows the dependencies between the various input parameters to be treated in a realistic manner and provides information on the spatial variability of the uncertainty [8]. The main sources of uncertainty come from the reservoir's geological structure, the variability of petrophysical properties, and the OWC and GOC locations [9, 10]. The probabilistic distribution of Gross Rock Volume (GRV) (MMcu.m) and Stock Tank Oil-in Place (STOIIP) (MMbbls) can then be obtained and used to get unbiased volume estimates and to quantify the risk associated with them [11, 12].

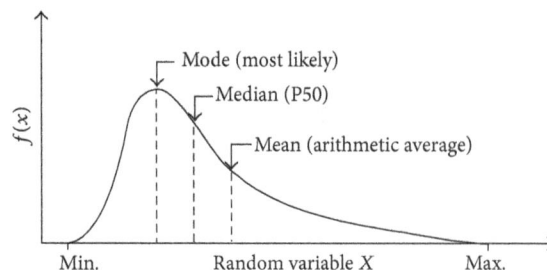

FIGURE 1: Measures of central tendency.

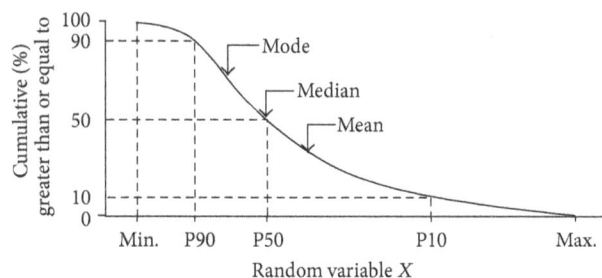

FIGURE 2: Percentiles.

## 2. Geological Setting

The reservoir under study is referred to as Savrak Formation belonging to Zagros petroliferous region. The word "Saravk" is derived from Tange Savrak in the mount Bnatsegna in Khuzestan Province and its type section is located in southwest of the mount Bnagestan.

This Formation is divided into three parts as lower, middle, and upper limestone. Sarvak calcareous Formation in Zagros is usually identified with two shallow and deep facies and spans from Albian to Turonian [13].

## 3. Methodology

This study was carried in two steps. A static model was constructed in the first step and, in order to obtain results with less errors, uncertainty analysis of the resulted model was carried out in the second step.

The existing information for the provision of static model is as follows.

(i) Data obtained from petrophysical logs of 9 wells.

(ii) Well trajectories and underground contour (UGC) map on top of the reservoir.

(iii) Data obtained from core studies and petrographic analyses.

Three-dimensional model generation was performed using the Petrel (Mark of Schlumberger). This software program is able to measure various parameters such as the amount of porosity, fluid permeability, and water saturation in each point by using petrophysical data of wells, such as acoustic, nuclear, and electrical logs and by employing positioning and statistical methods [14].

Modeling is carried out in four main steps as follows.

(1) Data loading and data quality controlling (Data input).

(2) Structural modeling.

(3) Creation of three-dimensional petrophysical models for reservoirs properties (property modeling).

(4) Analysis of uncertainty and volumetric calculations.

Briefly, after loading data in the first step and creating a cohesive model proportionate to the existing map, a blocked three-dimensional network was produced (Figure 3). This network as the main body of the model enables the simultaneous study of construction data and petrophysical properties in order to produce reality-oriented models. Moreover, when producing the petrophysical model, it enables petrophysical parameters defined for each well to be generalized to the entire reservoir.

The candidate reservoir is divided into a blocked network in which the properties of each block such as petrophysical and lithological properties are equal in all volumes and similar properties in the blocks which lack information can be estimated in view of distance between networks and the value of data. Obviously, the higher the amount of this information, the more realistic these estimations will be. The dimensions of blocks of this network in line with $X$ and $Y$ have been set $100*100$ m and are separated with one-meter distance in reservoir zone and with two meters distance in nonreservoir zones after creation of horizons (Figure 4) and in time of layering. Next, in the step of production of petrophysical models of properties, petrophysical logs information was scaled up.

Scale up means attribution of logs recorded values to blocks of three-dimensional network [15]. Scaled up data are processed during variography step in terms of permanency and normality (average of data is zero and their standard deviation is 1) and related variograms are depicted.

There are several methods to assess normal data distribution. The two most common methods are considered below.

The first method is a graphical approach, so that the cumulative frequency curve of data is plotted according to high-level categories on a paper of which one of the axes is arithmetic scale and the other axis a logarithmic scale. Since the cumulative frequency curve of a normal distribution in terms of high-level categories in these coordinates is a straight line, the closer the extension of the sum of points plotted on this paper to straight line, the closer the data distribution to normal. The second method for evaluating normal distribution of data is the qualitative method. For this purpose, it is sufficient to plot the histogram of data [16].

Variogram analysis is an important part of geostatistical modeling. Indeed, variogram shows the mean square difference between the two values as a function of their increment [17]. The value of variogram is calculated by

$$\gamma(h) = \frac{\sum_{i=1}^{n_h}(x_i - x_{i+h})^2}{2n_h}. \tag{1}$$

TABLE 1: Parameters of variograms are delineated for porosity and saturation data.

| Nugget | Sill | Effective range | Variogram | layer | property |
|--------|------|-----------------|-----------|-------|----------|
| 0.126 | 1 | 2605.2 | spherical | Layer 1 | |
| 0.092 | 1 | 2840.1 | spherical | Layer 2 | Porosity |
| 0.277 | 1 | 3800 | spherical | Layer 3 | |
| 0.290 | 1 | 3529.5 | spherical | Layer 1 | |
| 0.095 | 1 | 4047.8 | spherical | Layer 2 | Water saturation |
| 0.032 | 1 | 3843.2 | spherical | Layer 3 | |

In the above equation, $\gamma(h)$ is called semivariogram and $2\gamma(h)$ is called variogram, and $n_h$ is the number of pairs of points with the distance $h$ from each other participating in the variogram. $x_i$ is a grade in point $i$ and $x_{i+h}$ is grade in a point with distance $h$ from the point $i$.

By calculating variogram for different $h$ values, the $\gamma(h)$ diagram in terms of $h$ can be drawn, and the $h$ value is called lag.

For many modeling, the variogram model begins from a nonzero value and increases up to range called effective range (a), eventually reach to the constant value called sill. The effective range is the range in which the data spatial structure is relevant, and outside the range, the data effect is independent from each other [18] (Table 1).

Property modeling is usually used for inherent properties of a parameter. Variogram is a tool for describing these inherent properties [19]. In other words, variogram is a method of analyzing and describing these spatial changes based on the principle that samples close together are more similar than samples far from each other (Figure 5).

Following this three-dimensional model of reservoir properties including porosity and water saturation are modeled by using output variograms of variography step by sequential Gaussian simulation method (one of probability-oriented methods) (Figures 6 and 7).

In sequential simulation methods, the places lacking sample are consecutively visited in a stochastic manner until all places without sample are visited. In each place which has no sample, multiple realizations can be obtained from simulated variable on the basis of estimation as well as local uncertainty and in the order that places without sample are visited and the way in which a value in places without sample are simulated. Two sequential simulation methods which are widely used are sequential index simulation (SIS) and sequential Gaussian simulation (SGS).

Sequential Gaussian simulation is the most widely used geostatistical method in the recent modeling projects. This method is very simple and flexible. Sequential simulation of a variable is carried out in five steps: conversion of main data to a new space; variogram modeling in new space; determination of a stochastic route in order to visit all of the places lacking sample; estimation of places without sample in alternate way and finally reversed conversion of simulated values.

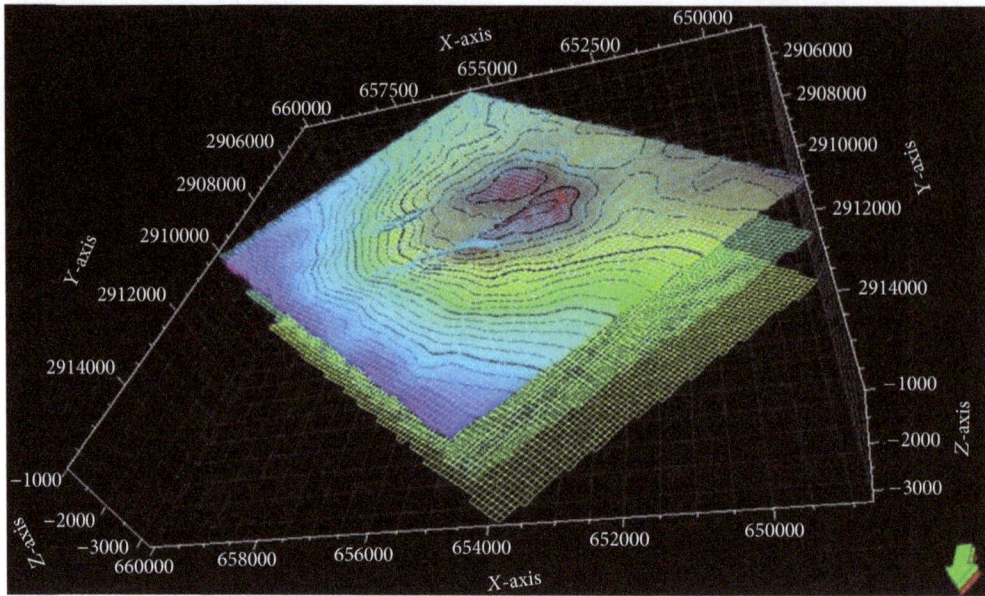

FIGURE 3: Structural grid of the studied reservoir; dimensions of blocks are 100 m * 100 m * 1 m along $X$, $Y$, and $Z$ directions, respectively.

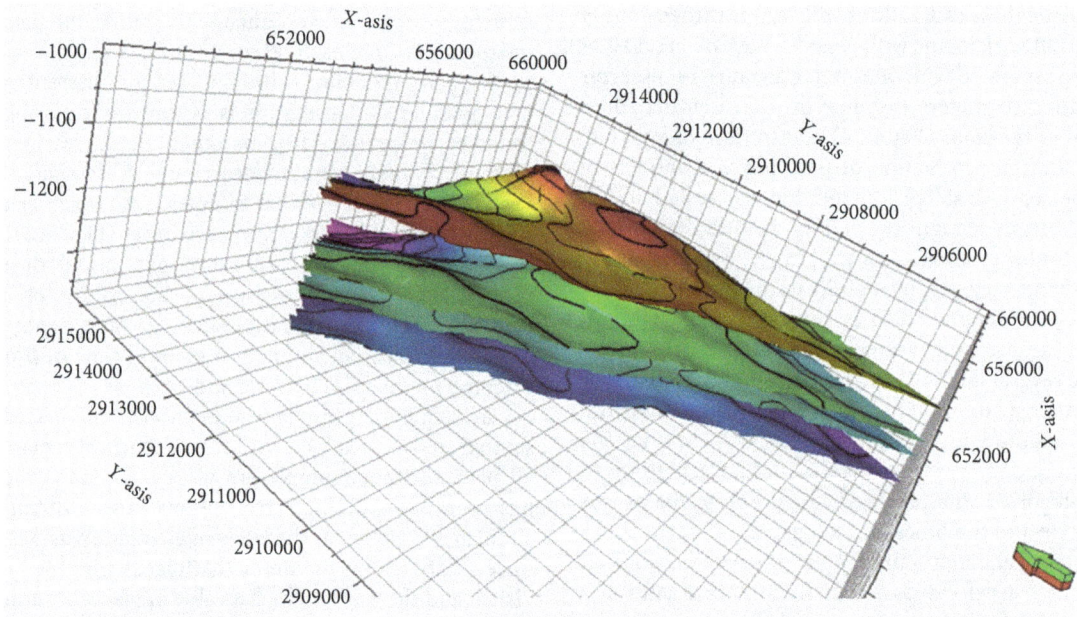

FIGURE 4: Horizons generated in Petrel.

Volumetric calculation produced models are performed on the basis of linear equation. $B_O$ and OWC values have been set 1.2 and −1074 in laboratory and field studies.

For oil reservoirs, the original oil-in place (OOIP) volumetric calculation is [20] as follows.

3.1. Imperial Consider

$$\text{OOIP (STB)} = \text{Rock Volume} * 7,758 * \phi * (1 - Sw) * \frac{1}{B_o},$$

$$(2)$$

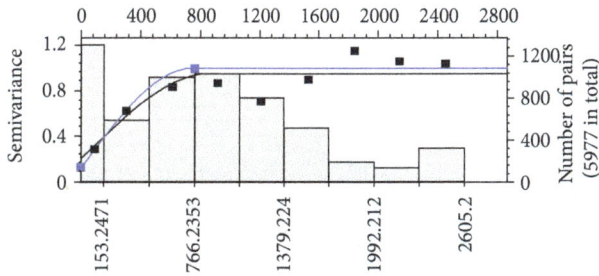

FIGURE 5: Variogram for porosity data in layer number one. Variogram type is spherical, the sill is 1, and the nugget is 0.126.

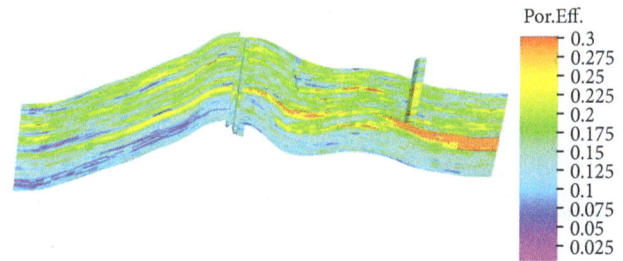

FIGURE 6: Cross-section of porosity distribution which is modeled using SGS algorithm. Note, align along $I$-Direction.

where rock volume (acre feet) is the $A * h$, $A$ is the drainage area, acres, $h$ is the net pay thickness, feet 7,758 is the API Bbl per acre-feet (converts acre-feet to stock tank barrels), $\phi$ is the porosity, fraction of rock volume available to store fluids, $Sw$ is the volume fraction of porosity filled with interstitial water, $Bo$ is the formation volume factor (Reservoir Bbl/STB), $1/B_o$ is the shrinkage (STB/reservoir Bbl), $B_o$ and OWC values have been set 1.2 and $-1074$ in laboratory and field studies.

## 4. Uncertainty Analysis

The available data for oil and gas fields are in general not enough to minimize the uncertainties related to the construction of reservoir models. The understanding of uncertainties involved in reservoir modeling is an essential tool to support decisions in the petroleum industry. The knowledge of uncertainty management related to prediction of hydrocarbon volumes has increased in the last decades, as a result of reliable 3D geological models made available by improvements in computer processing [21]. Lelliott et al. grouped the sources of uncertainties related to geological modeling into: data density (the density of boreholes used to construct the model); data quality (quality of the data used to construct the model, including borehole elevation, sample type, drilling method, and logging quality); geological complexity (geological variability throughout the site); and modeling software.

According to Zabalza-Mezghani et al. [22], the sources of uncertainties, in reservoir engineering, can be classified as anywhere within the reservoir modeling workflow. Such uncertainties are associated with: the static model, up scaling, fluid flow modeling, production data integration, production scheme development, and economic evaluation. These authors classified the different uncertainty behaviors as deterministic, discrete, and stochastic uncertainties [23].

So, the uncertainties inherent to each input data set used to build 3D static reservoir models cannot be expressed in a single deterministic realization. Each of the said calculations has an inherent uncertainty which causes a major uncertainty in reservoir estimation in time of combination. Three-dimensional models also follow this rule and the existence of big elements of uncertainty in modeling steps is undeniable. This amount of error is usually determined by using volumetric equations, but recent software advancements have enabled

the use of three-dimensional models as the foundation of investigating this uncertainty in reservoir.

The use of these three-dimensional models has many advantages over estimation based on direct application of volumetric equations. Among these advantages is that three-dimensional models enable presentation of inherent correlations in a realistic attitude and this results in more accuracy in our estimation of uncertainty and creation of a better foundation for conscious capital management. In a general classification, we can mention analysis of uncertainty of reservoirs impure stone volume, uncertainty of properties of stone and fluid, and uncertainty of simulations of flow in reservoir.

In the present static model, analysis of uncertainty is performed on reservoir stone volume and consequently volumetric calculations (which is affected by the said parameter). To this end, such parameters as reservoir geometric structure (existing horizons and zones), depth of water and oil (OWC), and three-dimensional models produced for porosity and water saturation have been studied. Each of these factors is separately studied and the amount of its effect on volumetric calculation and the amount of oil is determined. For this purpose, cycles are created in which a combination of realizations is created in each investigation by Mont Carlo method and the effect of factor changes on the result of volumetric calculations are measured. Next, the best choice (closest sample to level of 50%) and its seed number are identified from among hundreds of produced samples. The results can be shown in histograms (Figures 8, 9, and 10) in which probability level of 10%, probability level of 50% (reality-oriented model), and probability level of 90% have been identified and CDF curve can be depicted. This way we can study performance of that parameter on the amount of oil.

In analysis of uncertainty, 300 realizations (normally between 50 and 500) for each of parameters under study were produced by Mont Carlo method and the best routes for creation of petrophysical models of reservoir were determined. The realization made by these routes is selected as optimal model and the results of volumetric calculation are reported as estimations with highest probability and lowest error. Finally, after studying the results of three-dimensional simulation of well data, modeling distribution of petrophysical properties and analyzing uncertainty in the present model, the depth of water and oil contact surface

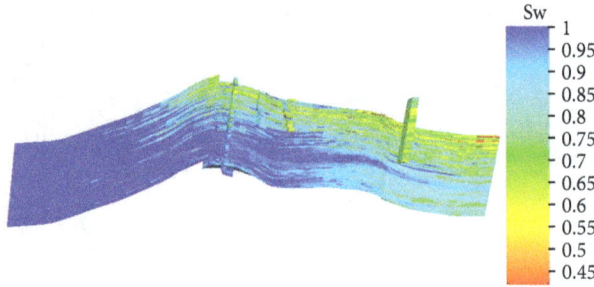

FIGURE 7: Cross-section of water saturation distribution which is modeled using SGS algorithm. Note, align along $I$-Direction.

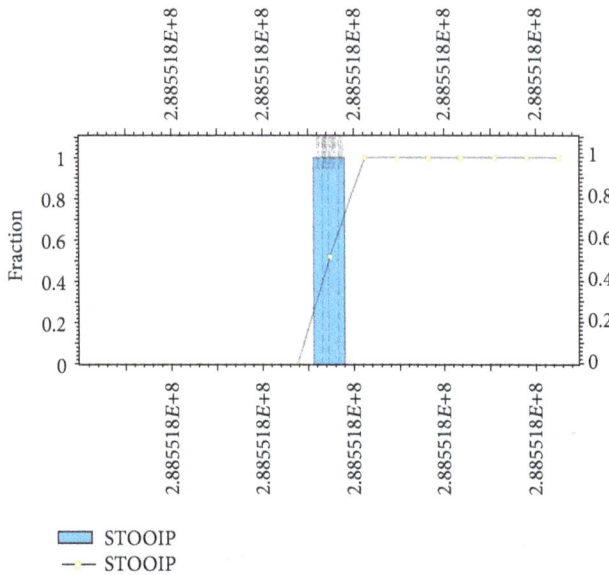

FIGURE 8: Uncertainty analysis, which has performed on reservoir geometry using Monte Carlo method. The figure shows effect of reservoir geometry on calculated STOOIP.

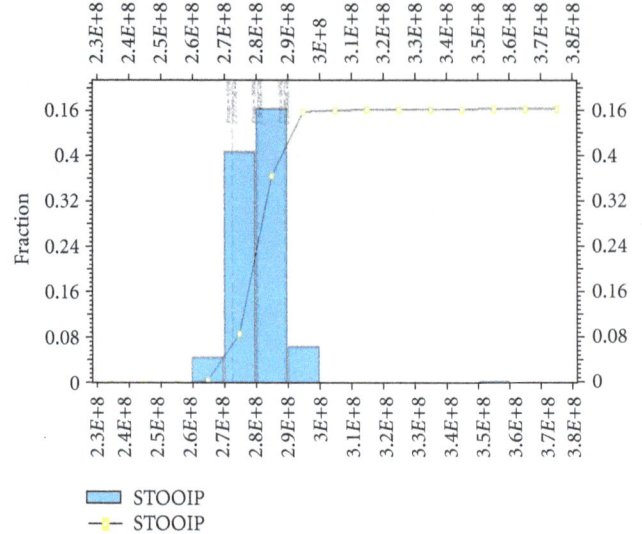

FIGURE 9: Uncertainty analysis, which has performed on reservoir geometry using Monte Carlo method. The figure shows effect of OWC on calculated STOOIP.

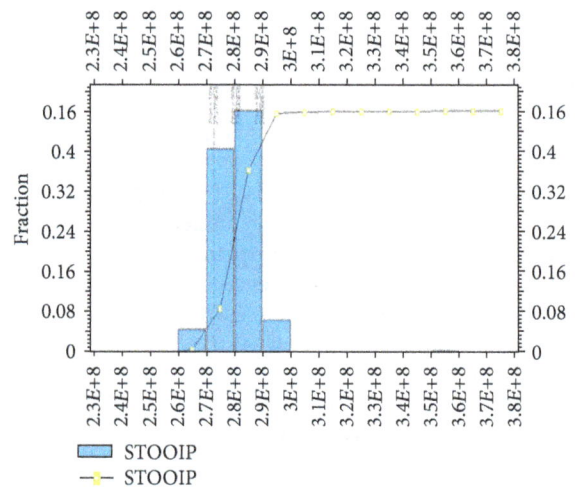

FIGURE 10: Uncertainty analysis, which has performed on reservoir geometry using Monte Carlo method. The figure shows effect of property distribution on calculated STOOIP.

was determined as the most effective factor in creation of uncertainty in the model, meaning that lowest changes in this value have high effect on the amount of estimations because the depth has a direct effect on the volume of reservoir stone (geometrical structure of reservoir) and finally on the amount of estimation of reserve amount. It is evident that in order to minimize this error (uncertainty), the certainty of this value must be increased. To this end, accurate recording of this depth in more wells and more situations in the reservoir can give us a more reliable value (a more frequent value).

## 5. Conclusion

The results of geostatistical simulation and creation of three-dimensional models from petrophysical parameters and analyses of uncertainty in reservoir indicated the following.

(i) Sequential Gaussian simulation is very efficient in determination and investigation of uncertainties of three-dimensional models owing to their high accuracy, absence of softening property and possibility to create many three dimensional realizations in which heterogeneity and scope of changes in variables are shown well.

(ii) According to results of study of uncertainty, the depth of water and oil contact surface was determined as the most effective parameter in calculations. Therefore, it is obvious that special care must be taken in determination and investigation of changes of this surface in reservoir. These changes have direct effect on geometry and total volume of reservoir (bulk volume) and finally on the amount of reserves.

(iii) The results of simulation showed that layer no. 1 holds the highest volume of oil and has constituted oil zone of the reservoir under study.

(iv) Generally in the simulations performed, the average porosity of reservoir was about 20% and water saturation 52%.

(v) In volumetric calculations, amount of reserves in the major layer (layer 1) is estimated 280 million barrels.

# References

[1] M. Nikravesh, "Computational intelligence for geosciences and oil exploration," in *Forging New Frontiers: Fuzzy Pioneers I*, vol. 66, pp. 267–332, California University Press, 2007.

[2] B. Yeten and F. Gümrah, "The use of fractal geostatistics and artificial neural networks for carbonate reservoir characterization," *Transport in Porous Media*, vol. 41, no. 2, pp. 173–195, 2000.

[3] G. Zamora Valcarce, T. Zapata, A. Ansa, and G. Selva, "Three-dimensional structural modeling and its application for development of the El Portón field, Argentina," *AAPG Bulletin*, vol. 90, no. 3, pp. 307–319, 2006.

[4] R. R. Jones, K. J. W. McCaffrey, P. Clegg et al., "Integration of regional to outcrop digital data: 3D visualisation of multi-scale geological models," *Computers and Geosciences*, vol. 35, no. 1, pp. 4–18, 2009.

[5] R. Haining, *Spatial Data Analysis: Theory and Practice*, Cambrige University Press, Cambrige, UK, 2003.

[6] J. W. Jennings Jr., F. J. Lucia, S. C. Ruppel, A. John, and G. Katherine, "3D modeling of startigraphically controlled petrophysical variability in the South Wasson Clear Fork reservoir," in *Proceedings of the SPE Annual Technical Conference and Exhibition*, pp. 2209–2223, San Antonio, Tex, USA, October 2002.

[7] O. Kaufmann and T. Martin, "3D geological modeling from boreholes, cross section and geological maps, application over former natural gas storages in coal mines," *Computers and Geosciences*, vol. 34, pp. 278–290, 2008.

[8] A. MacDonald and J. L. Tollesfrud, "3D reservoir Uncertainty modeling workflows, production and benefits," Roxar, September 2008.

[9] D. P. Hampson, J. S. Schuelke, and J. A. Quirein, "Use of multiattribute transforms to predict log properties from seismic data," *Geophysics*, vol. 66, no. 1, pp. 220–236, 2001.

[10] P. J. Hatchell, "Fault whispers: transmission distortions on prestack seismic reflection data," *Geophysics*, vol. 65, no. 2, pp. 377–389, 2000.

[11] K. Hirsche, J. Porter-Hirsche, L. Mewhort, and R. Davis, "The use and abuse of geostatistics," *Leading Edge*, vol. 16, no. 3, pp. 253–260, 1997.

[12] M. T. Olowokere, "Geostatistical modeling of interval velocity to quantifying hydrocarbon resource in multi-layer reservoir from TMB field, Niger delta," *International Journal of Physical Sciences*, vol. 5, no. 12, pp. 1897–1907, 2010.

[13] H. Motii, "Geology of Iran, Zagros Geology," Geological Survey of Iran, 2009.

[14] Schlumberger, "Petrel Introduction Course," *Seismic-to-Simulation Software petrel introduction course, 2008*.

[15] Schlumberger, "Petrel Introduction Course," *Shlumberger information solutions, 2006*.

[16] M. Abdideh and D. Bargahi, "Designing a 3D model for prediction the top of formation in oil fields using geostatistical methods," *Geocarto International Journal*, vol. 27, pp. 569–579, 2012.

[17] N. Cressie and D. M. Hawkins, "Robust estimation of the variogram," *Journal of the International Association for Mathematical Geology*, vol. 12, no. 2, pp. 115–125, 1980.

[18] R. Corstanje, S. Grunwald, and R. M. Lark, "Inferences from fluctuations in the local variogram about the assumption of stationarity in the variance," *Geoderma*, vol. 143, no. 1-2, pp. 123–132, 2008.

[19] Schlumberger, "Property Modeling Course," *Shlumberger information solutions, 2004*.

[20] L. Dean, "Reservoir engineering for geologists," Part 3—Volumetric Estimation, Reservoir no. 11, pp. 21–23, 2007.

[21] M. R. Lelliott, M. R. Cave, and G. P. Walthall, "A structured approach to the measurement of uncertainty in 3D geological models," *Quarterly Journal of Engineering Geology and Hydrogeology*, vol. 42, no. 1, pp. 95–105, 2009.

[22] I. Zabalza-Mezghani, E. Manceau, M. Feraille, and A. Jourdan, "Uncertainty management: from geological scenarios to production scheme optimization," *Journal of Petroleum Science and Engineering*, vol. 44, no. 1-2, pp. 11–25, 2004.

[23] J. F. Bueno, R. D. Drummond, A. C. Vidal, and S. S. Sancevero, "Constraining uncertainty in volumetric estimation: a case study from Namorado Field, Brazil," *Journal of Petroleum Science and Engineering*, vol. 77, no. 2, pp. 200–208, 2011.

# Permissions

The contributors of this book come from diverse backgrounds, making this book a truly international effort. This book will bring forth new frontiers with its revolutionizing research information and detailed analysis of the nascent developments around the world.

We would like to thank all the contributing authors for lending their expertise to make the book truly unique. They have played a crucial role in the development of this book. Without their invaluable contributions this book wouldn't have been possible. They have made vital efforts to compile up to date information on the varied aspects of this subject to make this book a valuable addition to the collection of many professionals and students.

This book was conceptualized with the vision of imparting up-to-date information and advanced data in this field. To ensure the same, a matchless editorial board was set up. Every individual on the board went through rigorous rounds of assessment to prove their worth. After which they invested a large part of their time researching and compiling the most relevant data for our readers. Conferences and sessions were held from time to time between the editorial board and the contributing authors to present the data in the most comprehensible form. The editorial team has worked tirelessly to provide valuable and valid information to help people across the globe.

Every chapter published in this book has been scrutinized by our experts. Their significance has been extensively debated. The topics covered herein carry significant findings which will fuel the growth of the discipline. They may even be implemented as practical applications or may be referred to as a beginning point for another development. Chapters in this book were first published by Hindawi Publishing Corporation; hereby published with permission under the Creative Commons Attribution License or equivalent.

The editorial board has been involved in producing this book since its inception. They have spent rigorous hours researching and exploring the diverse topics which have resulted in the successful publishing of this book. They have passed on their knowledge of decades through this book. To expedite this challenging task, the publisher supported the team at every step. A small team of assistant editors was also appointed to further simplify the editing procedure and attain best results for the readers.

Our editorial team has been hand-picked from every corner of the world. Their multi-ethnicity adds dynamic inputs to the discussions which result in innovative outcomes. These outcomes are then further discussed with the researchers and contributors who give their valuable feedback and opinion regarding the same. The feedback is then collaborated with the researches and they are edited in a comprehensive manner to aid the understanding of the subject.

Apart from the editorial board, the designing team has also invested a significant amount of their time in understanding the subject and creating the most relevant covers. They scrutinized every image to scout for the most suitable representation of the subject and create an appropriate cover for the book.

The publishing team has been involved in this book since its early stages. They were actively engaged in every process, be it collecting the data, connecting with the contributors or procuring relevant information. The team has been an ardent support to the editorial, designing and production team. Their endless efforts to recruit the best for this project, has resulted in the accomplishment of this book. They are a veteran in the field of academics and their pool of knowledge is as vast as their experience in printing. Their expertise and guidance has proved useful at every step. Their uncompromising quality standards have made this book an exceptional effort. Their encouragement from time to time has been an inspiration for everyone.

The publisher and the editorial board hope that this book will prove to be a valuable piece of knowledge for researchers, students, practitioners and scholars across the globe.

# List of Contributors

**Alexandre Martins Fernandes, Murilo Basso Nolasco and Jefferson Mortatti**
Centro de Energia Nuclear na Agricultura, Universidade de S̄ao Paulo, Avenida Centen´ario, 303, 13416-970 Piracicaba, SP, Brazil

**Christophe Hissler**
Département Environnement et Agrobiotechnologies, Centre de Recherche Public Gabriel Lippmann, 41, rue du Brill, Grand-Duchy of Luxembourg, 4422 Belvaux, Luxembourg

**Cajus Diedrich**
Paleo Logic Private Research Institute, Petra Bezruce 96, 26751 Zdice, Czech Republic

**Hongyuan Zhang and Junlai Liu**
State Key Laboratory of Geological Processes and Mineral Resources, China University of Geosciences, Beijing 100083, China
Faculties of Earth Sciences and Resources, China University of Geosciences, Beijing 100083, China

**Wenbin Wu**
Faculties of Earth Sciences and Resources, China University of Geosciences, Beijing 100083, China

**Quanlin Hou and Qing Liu**
Graduate University of Chinese Academy of Sciences, Beijing 100049, China

**Hongyuan Zhang**
School of Earth Sciences and Resources, China University of Geosciences, Beijing 100083, China

**Xiaohui Zhang and Jun Li**
Institute of Geology and Geophysics, Chinese Academy of Sciences, Beijing 100029, China

**Randall C. Orndorff**
U.S. Geological Survey, 926A National Center, Reston, VA 20192, USA

**Arthur Schmidt Nanni and Luiz Fernando Scheibe**
Universidade Federal de Santa Catarina (UFSC), Campus Universit´ario, Trindade, 88.010-970 Florianópolis, SC, Brazil

**Ari Roisenberg and Antonio Pedro Viero**
Universidade Federal do Rio Grande do Sul (UFRGS), Avenida Bento Gonc̦alves 9500, prédio 43126/103, 91501-970 Porto Alegre, RS, Brazil

**Maria Helena Bezerra Maia de Hollanda**
Centro de Pesquisas Geocronolãgicas (CPGeo), Instituto de Geociências (USP), Universidade de São Paulo (USP), Rua do Lago 562, Cidade Universit´aria, 05508-080 São Paulo, SP, Brazil

**Maria Paula Casagrande Marimon**
Universidade do Estado de Santa Catarina, UDESC-FAED, 88.035-001 Florianópolis, SC, Brazil

**Khalid S. Essa**
Geophysics Department, Faculty of Science, Cairo University, P.O. 12613, Giza, Egypt

**Quanlin Hou and Qing Liu**
Graduate University of the Chinese Academy of Sciences, Beijing 100049, China

**Hongyuan Zhang**
School of Earth Sciences and Resources, China University of Geosciences, Beijing 100083, China

**Jun Li**
Institute of Geology and Geophysics, The Chinese Academy of Sciences, Beijing 100029, China

**Yudong Wu**
MLR Key Laboratory of Metallogeny and Mineral Assessment, Institute of Mineral Resources, CAGS, Beijing 100037, China

**G. A. Dino, M. Fornaro and A. Trentin**
DST, Universit`a degli Studi di Torino, Via Valperga Caluso, 35, 10125 Torino, Italy

**YudongWu**
MLR Key Laboratory of Metallogeny and Mineral Assessment, Institute of Mineral Resources, CAGS, Beijing 100037, China
Key Lab of Computational Geodynamics of Chinese Academy of Sciences and College of Earth Science, Graduate University of the Chinese Academy of Sciences, Beijing 100049, China

**Quanlin Hou, Yiwen Ju and Wei Wei**
Key Lab of Computational Geodynamics of Chinese Academy of Sciences and College of Earth Science, Graduate University of the Chinese Academy of Sciences, Beijing 100049, China

**Daiyong Cao**
Key Laboratory of Coal Resources, China University of Mining and Technology, Beijing 100083, China

**Junjia Fan**
Key laboratory of Basin Structure and Petroleum Accumulation, CNPC and Petro China Research Institute of Petroleum Exploration and Development, Beijing 100083, China

**Piero Comin-Chiaramonti and Angelo De Min**
Mathematics and Geosciences Department, Trieste University, ViaWeiss 8, 34127 Trieste, Italy

**Aldo Cundari**
Geotrack International, 37Melville Road, Brunswick West, Melbourne, VIC 3055, Australia

**Vicente A. V. Girardi, Celso B. Gomes and Claudio Riccomini**
Geosciences Institute, University of São Paulo, Cidade Universit´aria, Rua do Lago 562, 05508-900 São Paulo, SP, Brazil

**Marcia Ernesto**
Astronomical and Geophysical Institute (IAG) of the São Paulo University, Rua do Matão 1226, 05508-090 São Paulo, SP, Brazil

**Xiaoshi Li, Yiwen Ju and Quanlin Hou**
Key Laboratory of Computational Geodynamics, College of Earth Science, Graduate University of Chinese Academy of Sciences, Beijing 100049, China

**Zhuo Li**
State Key Laboratory of Petroleum Resource and Prospecting, China University of Petroleum, Beijing 102249, China

**Junjia Fan**
Key Lab of Basin Structure and Petroleum Accumulation, Petro China Research Institute of Petroleum Exploration and Development, Beijing 100083, China

**CharlesW. Messo, Shukrani Manya and Makenya A. H. Maboko**
Department of Geology, University of Dar es Salaam, P.O. Box 35052, Dar es Salaam, Tanzania

**Qing Liu and Quanlin Hou**
Key Laboratory of Computational Geodynamics, Graduate University of Chinese Academy of Sciences, Chinese Academy of Sciences, 19 Yuquan Road, Beijing 100049, China

**Liewen Xie**
State Key Laboratory of Lithospheric Evolution, Institute of Geology and Geophysics, Chinese Academy of Sciences, P.O. Box 9825, Beijing 10029, China

**Hui Li**
416 Geological Prospecting Party, Bureau of Geology, Mineral Exploration and Development, Hunan Province, Zhuzhou 412007, China

**Shanqin Ni and Yudong Wu**
Institute of Mineral Resources, Chinese Academy of Geological Sciences, 26 Baiwanzhuang Road, Beijing 100037, China

**Mohammad Reza Kamali and Ezatallah Kazemzadeh**
Research Institute of Petroleum Industry (RIPI), West Boulevard, Azadi Sport Complex, Tehran 1485733111, Iran

**Azadeh Omidvar**
Department of Geology, Science and Research Branch, Islamic Azad University, End of Ashrafi Esfahani Highway, Simon Boulevard, P.O. Box 1944964814, Tehran, Iran